WALK BEHIND LAWN MOWER SERVICE MANUAL

(Third Edition)

CONTAINS

- INTRODUCTION & FAMILIARIZATION
- ENGINE & MOWER IDENTIFICATION
- STARTING & OPERATION
- PROBLEMS, REMEDIES & TROUBLESHOOTING
- MAINTENANCE
- ENGINE REPAIR & SERVICE
- SELF-PROPELLED DRIVE SERVICE
- BLADE BRAKE CLUTCH SERVICE
- SAFETY TIPS

Published by

INTERTEC PUBLISHING CORPORATION

P.O. Box 12901, Overland Park Kansas 66212

Library of Congress Catalog Card Number 86-83162

Cover photograph courtesy of
American Honda Motor Co., Inc. Power Equipment Division.

CONTENTS

DUAL DIMENSIONS

This service manual provides specifications in both the U.S. Customary and Metric (SI) systems of measurement. The first specification is given in the measuring system perceived by us to be the preferred system when servicing a particular component, while the second specification (given in parenthesis) is the converted measurement. For instance, a specification of "0.011 inch (0.28 mm)" would indicate that we feel the preferred measurement, in this instance, is the U.S. system of measurement and the metric equivalent of 0.011 inch is 0.28 mm.

INTRODUCTION

Repairing a lawn mower can be an interesting and rewarding experience. Ability will increase as you gain experience. Your selection of tools can limit your ability to perform certain repairs, but only a few are necessary for most service and other tools can be acquired as they are needed. This book explains commonly used procedures and will help you to develop your knowledge of how to fix lawn mowers. Easy to read fundamental theory, basic service procedures and troubleshooting hints are combined with professional service data, specifications and recommendations.

Lawn mowers equipped with gasoline engines have been manufactured for a long time by many different companies. All of the known lawn mower designs and all of the known forms of repair are not included in this book. The procedures that are included are based on sound practices and many can be used when repairing other lawn mowers, engines or even unrelated equipment not specifically included in this book. Many repair procedures should be thought out individually. The cost of repair, the tools required and other factors should be considered before deciding upon a method of repair.

On mowers equipped with safety equipment, MAKE SURE all equipment is connected and in good working condition at all times. The equipment was installed on the mower for a purpose, SAFETY; bypassing or disconnecting safety equipment is not recommended.

TOOLS

The specific tools required to service lawn mowers will depend upon the extent to which you want to service the mower and the origin of the mower manufacturer. Special tools that save time can be a valuable aid but may be too expensive, if used only once.

Standard or metric hand tools are required for most service. Refer to Fig. 2. Wrenches (standard sizes 1/4 to 3/4 in.; metric sizes 8 to 19 mm), screwdrivers, hammer, thickness (feeler) gage, file and ruler will be all that is required for most basic service. An inexpensive puller is recommended for removing some flywheels. The puller may be fabricated, purchased or rented.

Wrench sizes are determined by measuring across the flats of the hex as shown in Fig. 3. Be careful when measuring and be sure the correct wrench is used. The size of the head determines the size of the wrench used and is related to the size of the screw, but is not the screw size. The procedure to determine the correct screw size is outlined in the following section.

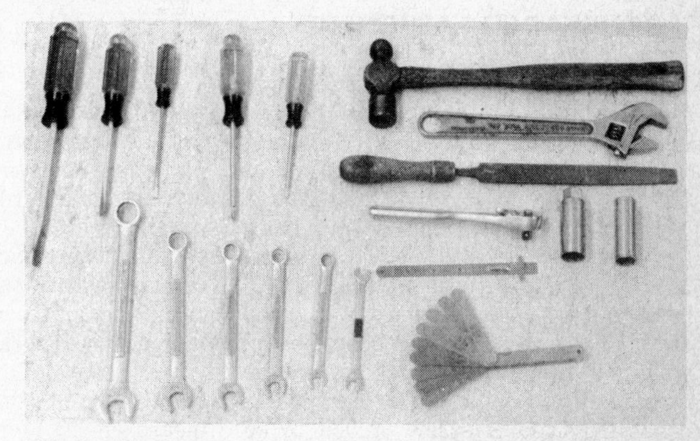

Fig. 2—Most service can be performed using only the tools shown.

Fig. 1—Begin servicing lawn mowers gradually. Don't go in over your head too quickly.

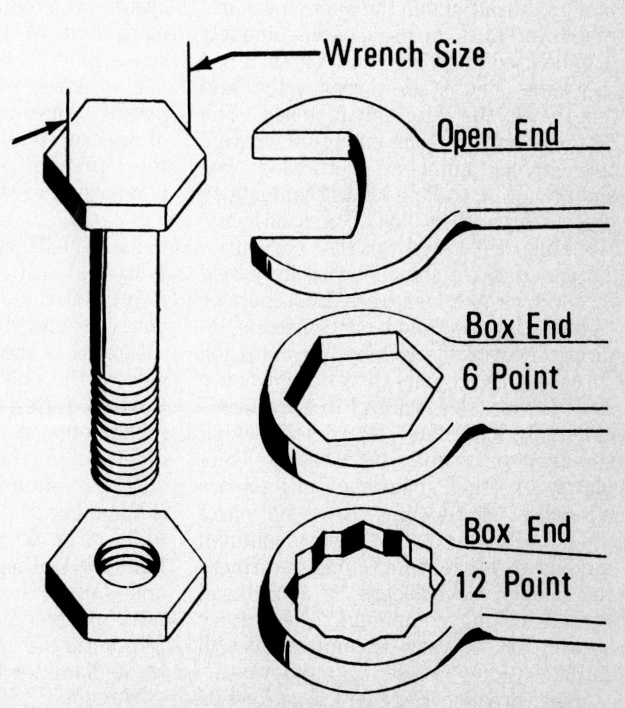

Fig. 3—Wrench size is determined by measuring across the flats of a screw head. This is not the size of the screw.

Part No.	Qty.	Description
418	1	Bracket Assembly - R.H. Handle
	1	Bracket Assembly - L.H. Handle
		Rear Wheel Assembly - W/Tire - 8"
		Front Wheel Assembly - W/Tire - 6"
		Cutter Disc Assembly
		[illegible]ck Assembly
		[illegible]ne Assembly - Tecumseh [illegible] LAV 30
		[illegible]andle Assembly
		[illegible]utter Disc
		[illegible] Mounting

Ref. No.	Part No.	Qty.	Description
24	324007	1	Decal
25	400108	4	Screw - 1/4-20 x 3/4 Hex Hd. Cap
26	400190	3	Screw - 5/16-18 x 1-1/4 Hex Hd. Cap
27	400274	4	Screw - 3/8-16 x 2-1/2 Hex Hd. Cap
28	400440	1	Screw - 1/2-20 x 1-1/4 Hex Hd. Cap
29	409508	2	Screw - No. 2 x 1/4 Type "U" PK Drive
30	441648	4	Screw - 5/16-18 x 1-1/4 Reg. Sq. Carriage
31	443106	3	Nut - 5/16-18 Hex.
32	443110	4	Nut - 3/8-16 Hex.

Fig. 4—A parts catalog for a typical lawn mower. The screw size is sometimes listed as shown at 25, 26, 27, 28, 29 and 30.

THREADED FASTENERS

Screws and bolts are the two common types of threaded fasteners. The primary difference between a screw and a bolt is that a bolt requires a nut to fasten something together. A screw is installed directly into a piece which has a threaded hole to receive the threads. There are many types of screws and bolts but the type which was used originally is the type which should be installed if renewal is necessary. Some manufacturers indicate in their parts books the type of screw or bolt to be used. Refer to Fig. 4.

Be sure to select a fastener which is manufactured under the same measurement system as the piece of equipment it is to be used on. Also make certain it has the same diameter, thread pitch and length as the original fastener. The fastener must be long enough to engage the correct number of threads, but should not be so long that it contacts or obstructs another part. Check all questionable threaded parts carefully. Threaded parts should be easily joined by hand without looseness. Looseness or tight fitting threaded parts usually indicate incorrectly matched or damaged threads. Occasionally threaded parts are tight fitting (self-locking) to prevent accidentally loosening. These self-locking threaded parts may be threaded completely together smoothly using proper wrenches. Self-locking threaded parts are not often used and the most common method to prevent loosening is through the use of lockwashers or a chemical thread locking compound. The service section for the various components will call attention to these locking devices.

Most threaded fasteners described as having "right-hand" threads are installed by turning the fastener in a clockwise direction and removed by turning the fastener in a counterclockwise direction. Fasteners for some special applications are cut with "left-hand" threads and are usually marked with an "L". Install threaded fasteners with "left-hand" threads by turning the fastener counterclockwise and remove by turning the fastener in a clockwise direction.

The size of threaded fasteners manufactured using the "inch" system of measurement is given according to the ISO (International Organization for Standardization) inch system. This system was established in 1964 and is equivalent to the old Unified Thread system.

The size of threaded fasteners manufactured using the "metric" system of measurement is given according to the Optimum Metric Fastener System established in the 1970's.

ISO INCH SYSTEM. The ISO inch system identifies threaded fastener size using three dimensions: length, diameter and thread pitch.

Length is measured from the underside of the bolt head to the end of the threads. Refer to L – Fig. 5.

Diameter is the actual outside diameter of the threads. Refer to D – Fig. 5.

Thread pitch is measured by counting the number of threads in one inch. Refer to TI – Fig. 5. The ISO inch system retains the Unified National Coarse (UNC) and Unified National Fine (UNF) thread pitch designations to provide two different thread pitch possibilities for the same diameter threaded fastener.

Threaded ISO inch system fastener sizes are written as follows: 1/4-20 (UNC) x 1, which indicates a 1/4 inch diameter fastener with 20 threads per inch with a length of 1 inch. The same fastener with UNF threads would be written as follow: 1/4-24 (UNF) x 1, which indicates a 1/4 inch diameter fastener with 24 threads per inch with a length of 1 inch.

Threaded fastener strength is indicated by the number of marks on the fastener head. Refer to Fig. 5. No markings on fastener head indicate a grade 1 or 2 fastener. Two marks indicate a grade 4, three marks indicate a grade 5, five marks indicate a grade 7 and six marks indicate a grade 8 fastener. The fastener strength increases as the number of marks increase. There are many critical fastener applications where fastener strength and hardness have been carefully calculated and under no circumstances should a fastener with a different rating be substituted.

OPTIMUM METRIC FASTENER SYSTEM. The Optimum Metric Fastener System identifies threaded fastener size using three dimensions: length, diameter and thread pitch.

Length and diameter for Optimum Metric Fastener System fasteners are measured in the same manner as an ISO inch system fastener. Refer to L and D – Fig. 5.

Thread pitch for Optimum Metric Fastener System fasteners are measured from one thread peak to another thread peak. Refer to TM – Fig. 5. Most metric fasteners have one standard thread pitch matched to fastener diameter, however, fine or coarse thread pitch may be used in specialized applications.

Threaded fastener strength is indicated by a number stamped on the fastener head. Refer to Fig. 5. The larger the number stamped on fastener head, the greater the strength of the fastener. There are many critical fastener applications where fastener strength and hardness have been carefully calculated and under no circumstances should a fastener with a different rating be substituted.

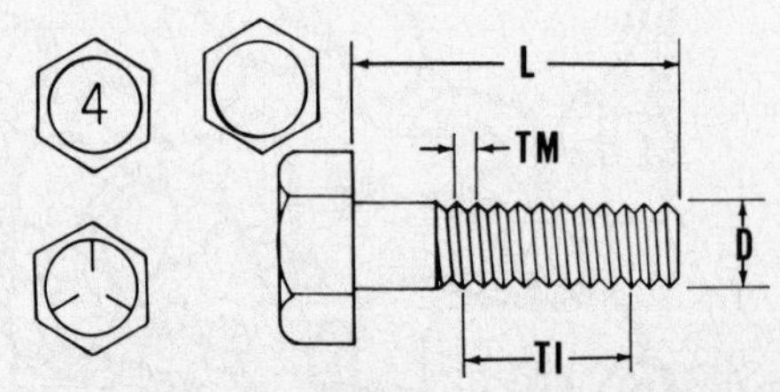

Fig. 5—Screw size is not determined by size of head. Size of fasteners manufactured under the ISO Inch System is determined by length (L), diameter (D) and number of threads per inch (TI). Size of fasteners manufactured under the Optimum Metric Fastener System is determined by length (L), diameter (D) and distance between threads (TM). Refer to text.

ENGINE OPERATING PRINCIPLES

The small gasoline engines used on lawn mowers are technically known as "Internal Combustion Engines." The source of power is heat caused by burning a combustible mixture of petroleum products and air. In a reciprocating engine, this burning takes place in a closed cylinder containing a piston. Expansion of the fuel-air mixture causes the piston to move which in turn causes a shaft (crankshaft) to rotate. A complete series of timed events (or cycle) must take place for the engine to run. These events will all occur once during one complete revolution of the crankshaft for engines called two-stroke engines or during two revolutions of the crankshaft for four-stroke engines. The term strokes refers to the movement of the piston from one extreme to the other (Top to Bottom or Bottom to Top) in the cylinder.

OTTO CYCLE

In a spark ignited engine, a series of five events is required in order for the engine to provide power. This series of events is called the "OTTO CYCLE" (or "Work Cycle") and is repeated as long as work is being done. This series of events which comprise the "Cycle" is as follows:

1. A combustible mixture of fuel and air is pushed into the cylinder.

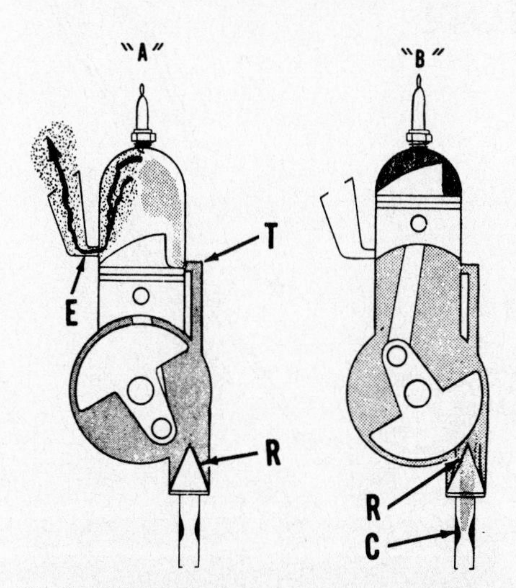

Fig. 6—Schematic diagram of a two-stroke cycle engine operating on the Otto Cycle (spark ignition). View "B" shows piston near top of upward stroke and atmospheric pressure is forcing air through carburetor (C), where fuel is mixed with the air, and the fuel-air mixture enters crankcase through open reed valve (R). In view "A", piston is near bottom of downward stroke and has opened the cylinder exhaust port (E) and transfer (intake) port (T); fuel-air mixture in crankcase has been compressed by downward stroke of engine and flows into cylinder through open port. Incoming mixture helps clean burned exhaust gases from cylinder.

2. The mixture of fuel and air is compressed by the piston moving upward in the cylinder.
3. The compressed fuel-air mixture is ignited by a timed electric spark.
4. The burning air-fuel mixture expands, forcing the piston downward in the cylinder thus converting the chemical energy generated by combustion into mechanical power.
5. The gaseous products formed by the burned fuel-air mixture are exhausted from the cylinder so a new "Cycle" can begin.

The above described five events which comprise the work cycle of an engine are commonly referred to as (1), INTAKE; (2), COMPRESSION; (3), IGNITION; (4), EXPANSION (POWER); and (5), EXHAUST.

The events necessary to complete a cycle must occur at specific times in order to continue repeating these events or "Run."

TWO-STROKE CYCLE

In two-stroke cycle engines, the piston is used as a sliding valve for the cylinder intake and exhaust ports. The intake and exhaust ports are both open when the piston is at the bottom of its downward stroke (bottom dead center or BDC). The exhaust port is open to atmospheric pressure; therefore, the fuel-air mixture must be elevated to a higher than atmospheric pressure in order for the mixture to enter the cylinder. As the crankshaft is turned from BDC and the piston starts on its upward stroke, the intake and exhaust ports are closed and the fuel-air mixture in the cylinder is compressed. When the piston is at or near the top of its upward stroke (top dead center or TDC), an electric spark across the electrode gap of the spark plug ignites the fuel-air mixture. As the crankshaft turns past TDC and the piston starts on its downward stroke, the rapidly burning fuel-air mixture expands and forces the piston downward. As the piston nears bottom of its downward stroke, the cylinder exhaust port is opened and the burned gaseous products from combustion of the fuel-air mixture flows out the open port. Slightly further downward travel of the piston opens the cylinder intake (or transfer) port and a fresh charge of fuel-air mixture is forced into the cylinder. Since the exhaust port remains open, the incoming flow of fuel-air mixture helps clean (scavenge) any remaining burned gaseous products from the cylinder. As the crankshaft turns past BDC and the piston starts on its upward stroke, the cylinder intake and exhaust ports are closed and a new cycle begins.

Since the fuel-air mixture must be elevated to a higher than atmospheric pressure to enter the cylinder of a two-stroke cycle engine, a compressor pump must be used. Coincidentally, downward movement of the piston decreases the volume of the engine crankcase. Thus, a compressor pump is made available by sealing the engine crankcase and connecting the carburetor to a port in the crankcase. When the piston moves upward, volume of the crankcase is increased which lowers pressure within the crankcase to below atmospheric. Air will then be forced through the carburetor, where fuel is mixed with the air, and on into the engine crankcase. In order for downward movement of the piston to compress the fuel-air mixture in the crankcase, a valve must be provided to close the carburetor to crankcase port. In Fig. 6, a reed type inlet valve is shown. Spring steel reeds (R) are forced open by atmospheric pressure as shown in view "B" when the piston is on its upward stroke and pressure in the crankcase is below atmospheric. When the piston reaches TDC, the reeds close as shown in view "A" and fuel-air mixture is trapped in the crankcase to be compressed by downward movement of the piston. The cylinder intake (or transfer) port (T) connects the crankcase compression chamber to the cylinder; the transfer port is the cylinder intake port through which the compressed fuel-air mixture in the crankcase is transferred to the cylinder when the piston is at bottom of stroke as shown in view "A."

Due to rapid movement of the fuel-air mixture through the crankcase, the crankcase cannot be used as a lubricating oil sump because the oil would be carried into the cylinder. Lubrication is accomplished by mixing a small amount of oil with fuel; thus, lubricating oil for the engine moving parts is carried into the crankcase with the fuel-air mixture. Normal lubricating oil to fuel mixture ratios vary from one part of oil mixed with 16 to 50 parts of fuel by volume. In all instances, manufacturer's recommendations for oil type and fuel-oil mixture ratio should be observed.

Two-stroke engines are characterized by their light weight. Some Ariens, Clinton, Jacobsen, Lawn Boy and Sensation mowers are equipped with two-stroke engines and some other mower companies use Tecumseh two-stroke engines.

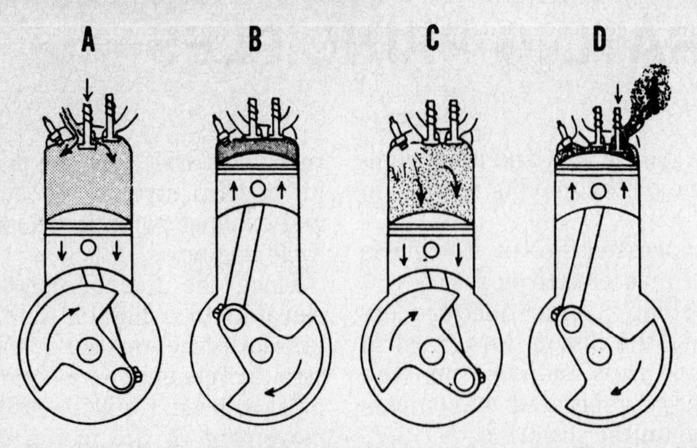

Fig. 7—Schematic diagram of four-stroke cylcle engine operating on the Otto (spark ignition) cycle. In view "A", piston is on first downward (intake) stroke and atmospheric pressure is forcing fuel-air mixture from carburetor into cylinder through open intake valve. In view "B", both valves are closed and piston is on its first upward stroke compressing the fuel-air mixture in cylinder. In view "C", spark across electrodes of spark plug has ignited fuel-air mixture and heat of combustion rapidly expands the burning gaseous mixture forcing the piston on its second downward (expansion or power) stroke. In view "D", exhaust valve is open and piston on its second upward (exhaust) stroke forces the burned mixture from cylinder. A new cycle then starts as in view "A."

FOUR-STROKE CYCLE

The five events of the Otto cycle take place in four strokes of the piston (two revolutions of the crankshaft) in four-stroke engines. Therefore, a power stroke occurs on alternate downward strokes of the piston.

In view "A" of Fig. 7, the piston is on the first downward stroke of the cycle. The mechanically operated intake valve has opened the intake port and, as the downward movement of the piston has reduced the air pressure in the cylinder to below atmospheric pressure, air is forced through the carburetor, where fuel is mixed with the air, and into the cylinder through the open intake port. The intake valve remains open and the fuel-air mixture continues to flow into the cylinder until the piston reaches bottom of its downward stroke. As the piston starts on its first upward stroke, the mechanically operated intake valve closes and, since the exhaust valve is closed, the fuel-air mixture is compressed as in view "B."

Just before the piston reaches the top of its first upward stroke, a spark at the spark plug electrodes ignites the compressed fuel-air mixture. As the engine crankshaft turns past top center, the burning fuel-air mixture expands rapidly and forces the piston downward on its power stroke as shown in view "C." As the piston reaches the bottom of the power stroke, the mechanically operated exhaust valve starts to open and as the pressure of the burned fuel-air mixture is higher than atmospheric pressure, it starts to flow out the open exhaust port. As the engine crankshaft turns past bottom center, the exhaust valve is almost completely open and remains open during the upward stroke of the piston as shown in view "D." Upward movement of the piston pushes the remaining burned fuel-air mixture out of the exhaust port. Just before the piston reaches the top of its second upward or exhaust stroke, the intake valve opens and the exhaust valve closes. The cycle is completed as the crankshaft turns past top center and a new cycle begins as the piston starts downward as shown in view "A."

MOWER FAMILIARIZATION

Learn the correct names for common parts. Notice that different manufacturers may call similar parts or parts that serve the same function by different names. Refer to Fig. 9.

The name of the manufacturer of the mower and engine is sometimes displayed on a decal or plate, but sometimes identification must be accomplished by noticing small details of construction peculiar to one manufacturer. To order parts and for some service, the complete model and serial numbers must be known.

ROTARY MOWERS

GENERAL. The engine of a rotary mower is used to turn a cutting blade. The blade may be constructed in several different ways, but most are solid, one piece steel. The blade may be straight or curved to conform to deck configuration. Refer to view of blade in Fig. 11. Blades are not all constructed the same and will have mounting holes to fit a particular engine. Be sure the correct blade is being used on the mower.

Fig. 9—Drawing of typical mower showing most of the typical parts.

1. Handlebar
2. Handlebar attaching brackets
3. Engine
4. Mower deck
5. Wheel support plates
6. Key
7. Blade drive clutch
8. Friction washers
9. Blade
10. Clamp washer
11. Lockwasher
12. Cap screw
13. Wheel
14. Bushing
15. Washer

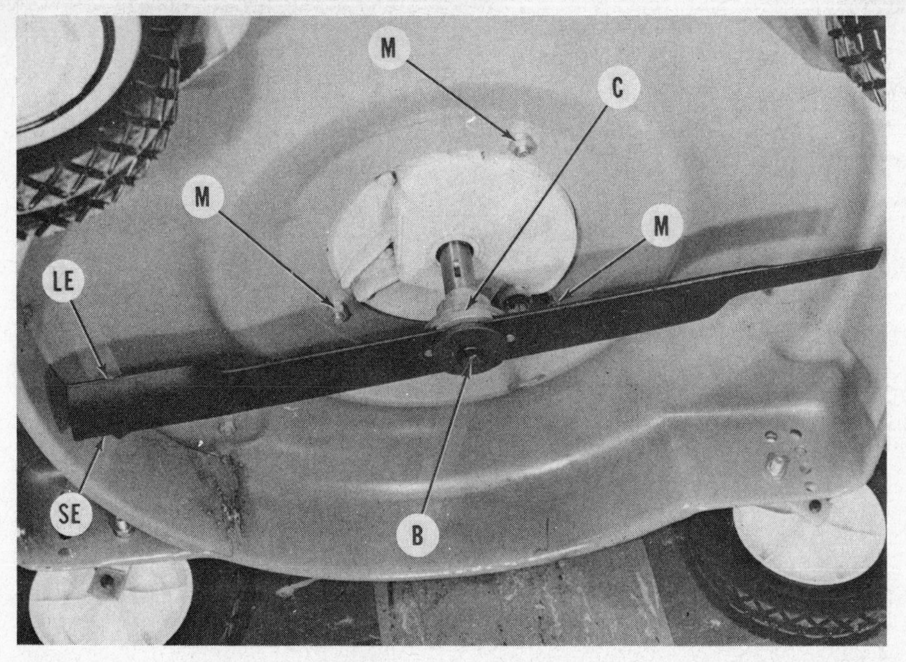

Fig. 11—The cutting blade must be installed securely and correctly for the engine to operate properly.

B. Blade attaching screw
C. Blade clutch
LE. Lift edge of blade (up)
M. Engine mounting bolts
SE. Sharp cutting edge (down)

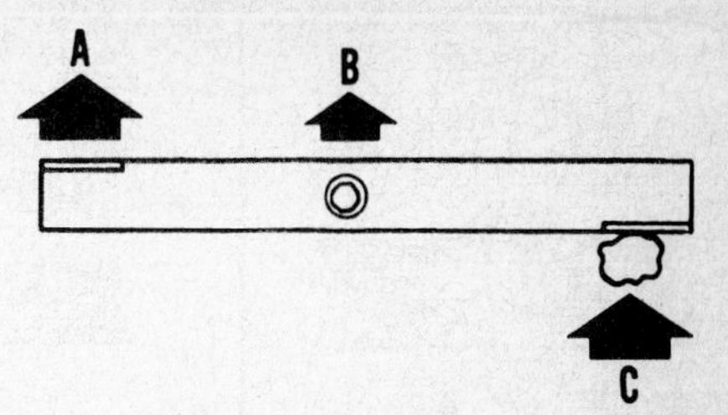

Fig. 14—The inertia of blade (A) when other end hits a solid object (C) results in a side force (B) against the crankshaft.

Most rotary mower blades have a bent edge on the rear of the blade. This is the "lift" portion of the blade and it creates a suction as the blade is moving to lift the grass for cutting and to discharge cut grass. Blades with a lot of lift are designed for cutting thick grasses such as Bermuda and Zoysia and when using a grass catcher. Low lift blades are for use with the thinner grasses such as Blue grass and Fescue.

Several methods may be used to drive a rotary mower blade. The simplest method is to attach the blade to the end of the engine's crankshaft. When the rotary blade is mounted remotely, a V-belt is generally used to transfer power from the engine to the blade.

Regardless of the type of blade drive used, the blade must be mounted correctly to cut properly. Lift edge on blades so equipped must be pointing up. On all blades the blade must be installed with the sloping portion of the cutting edge up and the flat side of the cutting edge down. Refer to Fig. 12.

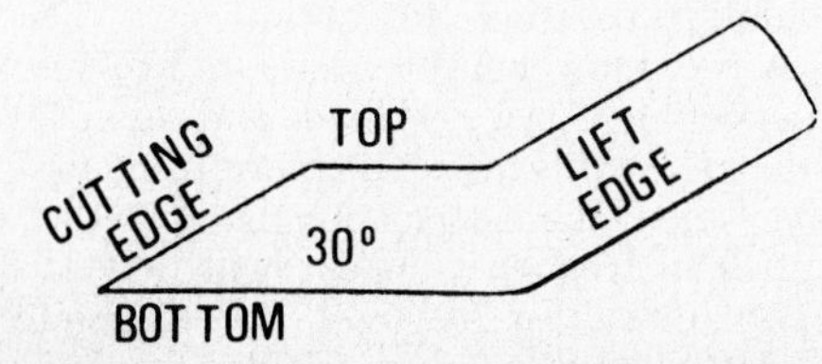

Fig. 12—Drawing of typical blade cross-section identifying parts of blade. The cutting edge should form approximately a 30° angle with bottom.

On models so equipped, slip clutches are designed to allow the blade to turn when sufficient force is exerted against the blade.

Two types of clutches are generally used. Clutch shown at (A and C–Fig. 13) have two lugs (1) which engage holes in the blade. The lugs will break off when the blade is forced to turn in a direction different than the clutch body. Clutch lugs should be inspected periodically as vibration and wear will cause lugs to fail during normal use of mower. Clutch shown at (B) consists of a clutch body and washers (2) which fit on both sides of blade. Blade can slip between washers when force is exerted on blade. Washers may be made of rubber, fiber or steel. If washers become loose, blade may move during normal operation and create enough heat to deform washers. Be sure blade is secured tightly.

Slip clutches will allow the blade to turn around the crankshaft end but they may not prevent the crankshaft from bending if the end of the blade strikes an immovable object. The inertia of the free end of the blade keeps the free end of the blade moving until the crankshaft stops it. Note in Fig. 14 that the force generated by the inertia of the free end of the blade is against the crankshaft and not around it. The slip clutch protects the crankshaft from damaging forces rotating around it but not against it.

If a slip clutch is not used to attach the blade to the crankshaft, the blade may be mounted rigidly to the crankshaft. A single screw may hold blade between ridges of the adapter (E–Fig. 15) or screws and nuts may be used to secure blade to the adapter as shown at (D). The blade is mounted rigidly and is not permitted to slip at all.

Some late model mowers may be equipped with a blade clutch, refer to appropriate repair section for servicing information.

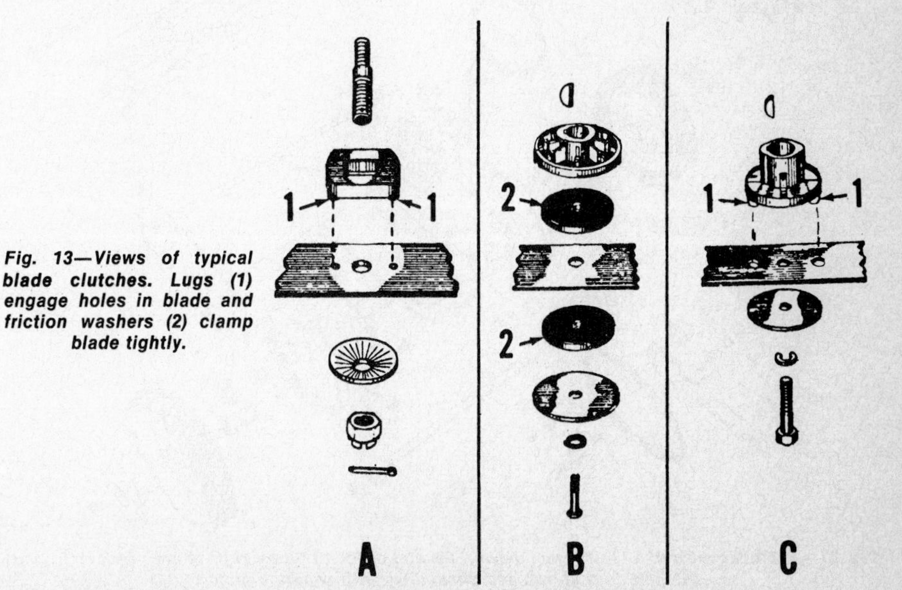

Fig. 13—Views of typical blade clutches. Lugs (1) engage holes in blade and friction washers (2) clamp blade tightly.

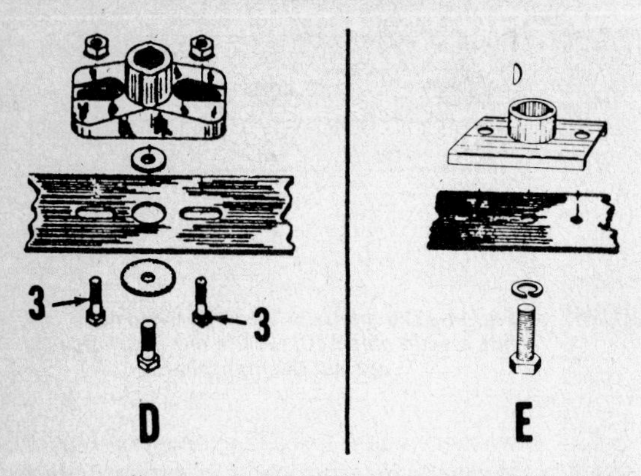

Fig. 15—Two types of rigid blade mounts. Type "D" attaches the blade with three screws. Type "E" attaches blade with one screw but ridges on the mount keep blade from turning.

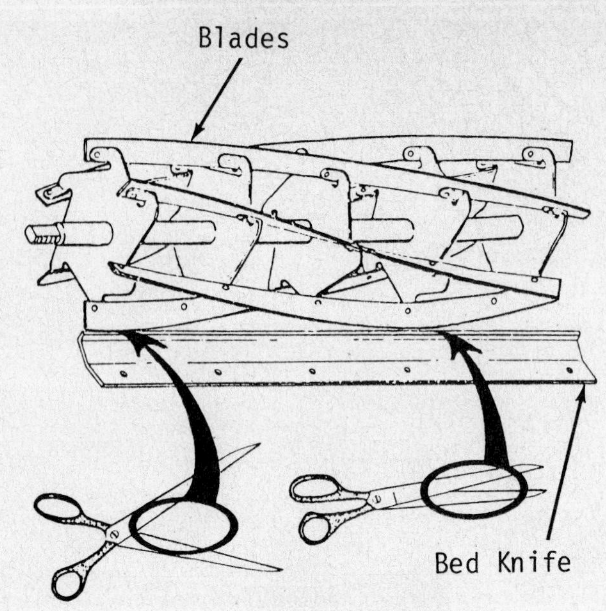

Fig. 18—The cutting action of the blade and bed knife of a reel mower corresponds to the shearing action of scissors.

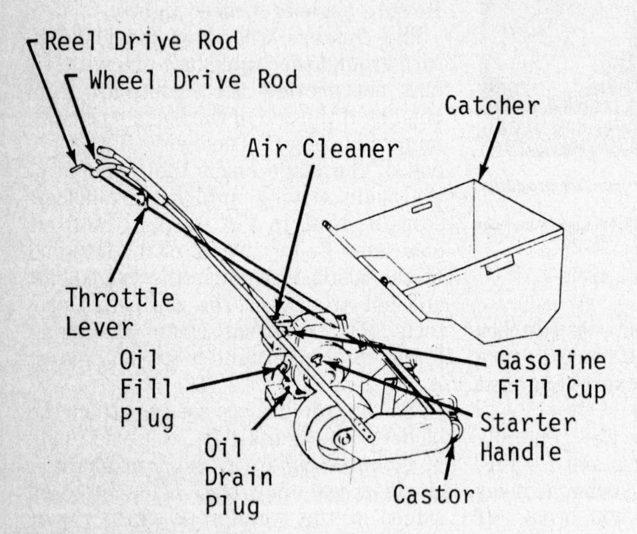

Fig. 16—Drawing of a King O'Lawn reel type mower showing some parts.

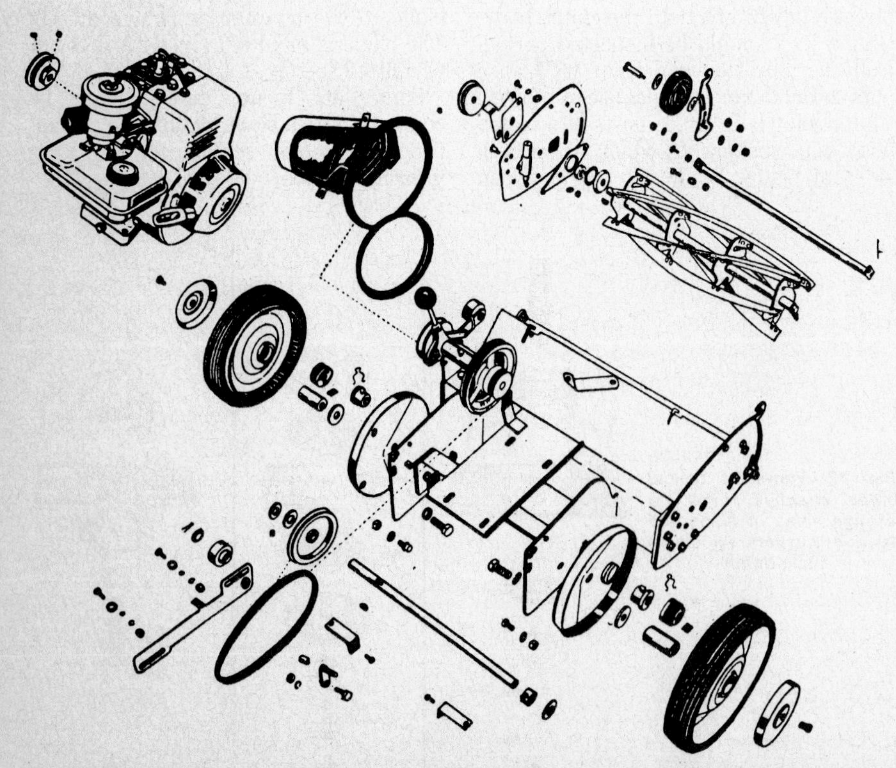

Fig. 17—Exploded view of a typical reel mower. Several different types of drive are used, including the belt drive shown and chain drive with separate clutch.

REEL MOWERS

Sharp, properly adjusted reel type mowers (Fig. 16 through 18) in good operating condition offer several advantages over conventional rotary mowers. The quality of cut possible with a reel type mower is unequaled, especially for semi-formal/formal turf areas and with certain grass varieties such as fine leaved Bermuda, Bent Grass and Zoysia which require short height. Grooming for best appearance is easily accomplished with a reel type mower without scalping. Reel mowers also require less power to operate than a rotary mower with same width of cut and propulsion type.

The chief disadvantages of reel mowers are: Reel mowers require fairly smooth ground that is free from rocks, sticks, wire and other debris which could cause damage to the mower. The height of the grass to be cut is limited by the diameter of the reel. If grass is too high, the mower will knock the grass down and run over it without cutting. Reel mowers generally require more care while operating and often involve more maintenance than rotary mowers. These disadvantages are usually not very critical when grooming a regularly maintained formal turf area.

A reel mower uses a scissors-like action between each reel blade and the cutter bar to cut grass. The more frequent the "snips" the better and smoother the cut. The frequency of cut will depend upon drive speed, reel speed, reel diameter and number of blades. The clips should be often enough to mow without leaving a riffled or corrugated effect.

ENGINE AND MOWER IDENTIFICATION

The following paragraphs provide information to help determine the manufacturer of the engine used on mowers manufactured by a particular company. To correctly identify an engine, first determine what company manufactured the engine and then refer to the correct engine section for engine model and serial number locations.

If the manufacturer of the mower being serviced is not listed, identify engine manufacturer and refer to the correct engine section for engine model and serial number locations.

ACE

Ace walk-behind lawn mowers are equipped with four-stroke Briggs & Stratton engines. Mower model and serial number plate is attached to rear portion of mower deck.

Refer to appropriate engine service section for engine model and serial number location.

AIRCAP

Aircap walk-behind lawn mowers are equipped with four-stroke Briggs & Stratton engines. Mower model and serial number plate is attached to rear portion of mower deck. Refer to Fig. AP.

Refer to appropriate engine service section for engine model and serial number location.

Fig. AP—View showing location of mower model and serial numbers (N) on Aircap walking lawn mowers.

ALLIS-CHALMERS

Allis-Chalmers, walk-behind lawn mowers are equipped with four-stroke Briggs & Stratton engines. Mower model and serial number plate is attached to rear portion of mower deck.

Refer to appropriate engine service section for engine model and serial number location.

ARIENS

Ariens walk-behind lawn mowers are equipped with four-stroke engines manufactured by either Briggs & Stratton or Tecumseh, or a two-stroke engine manufactured by Sachs. Mower model and serial numbers are located on a label on mower deck similar to Fig. AR.

Refer to appropriate engine service section for engine model and serial number location.

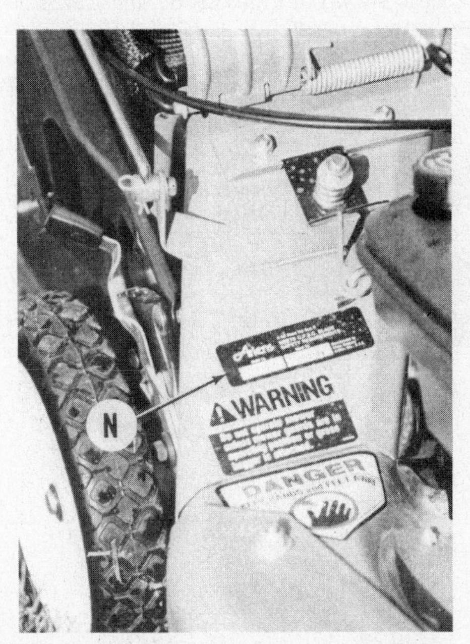

Fig. AR—View showing location of mower model and serial numbers (N) on Ariens walking lawn mowers.

ATLAS

Atlas walk-behind lawn mowers are equipped with four-stroke Briggs & Stratton engines. Mower model and serial number plate is attached to rear portion of mower deck.

Refer to appropriate engine service section for engine model and serial number location.

BOLENS

Bolens walk-behind lawn mowers are equipped with four-stroke Briggs & Stratton or Tecumseh engines. Mower model and serial number plate is attached to rear portion of mower deck.

Refer to appropriate engine service section for engine model and serial number location.

BUNTON/GOODALL

Bunton/Goodall walk-behind lawn mowers are equipped with four-stroke Briggs & Stratton or Tecumseh engines. Mower model and serial number plate is attached to rear portion of mower deck.

Refer to appropriate engine service section for engine model and serial number location.

CRAFTSMAN

Craftsman walk-behind lawn mowers are equipped with four-stroke Briggs & Stratton engines or four-stroke Craftsman engines manufactured by Tecumseh. Mower model and serial number plate is attached to rear portion of mower deck.

Refer to the Craftsman engine service section for correct engine model identification and cross-reference to appropriate Tecumseh engine model or the appropriate engine service section for Briggs & Stratton engine.

JOHN DEERE

John Deere walk-behind lawn mowers are equipped with four-stroke Briggs & Stratton engine or a two-stroke Lawn Boy engine. Mower model and serial number plate is attached to rear of mower deck.

Refer to appropriate engine service section for engine model and serial number location.

DYNAMARK

Dynamark walk-behind lawn mowers are equipped with four-stroke Briggs & Stratton or Tecumseh engines. Mower model and serial number plate is attached to rear portion of mower deck.

Refer to appropriate engine service section for engine model and serial number location.

FORD

Ford walk-behind lawn mowers are equipped with four-stroke Briggs & Stratton engines. Mower model and serial number plate is attached to rear portion of mower deck.

Refer to appropriate engine service section for engine model and serial number location.

GILSON

Gilson walk-behind lawn mowers are equipped with four-stroke Briggs and

Stratton or Tecumseh engines. Mower model and serial number plate is attached to rear portion of mower deck.

Refer to appropriate engine service section for engine model and serial number location.

HONDA

Honda walk-behind lawn mowers are equipped with four-stroke engines manufactured by Honda Motor Company to metric specifications. Mower model and serial numbers are located as shown in Fig. HN.

Refer to appropriate Honda engine service sections for engine identification and engine model and serial number location.

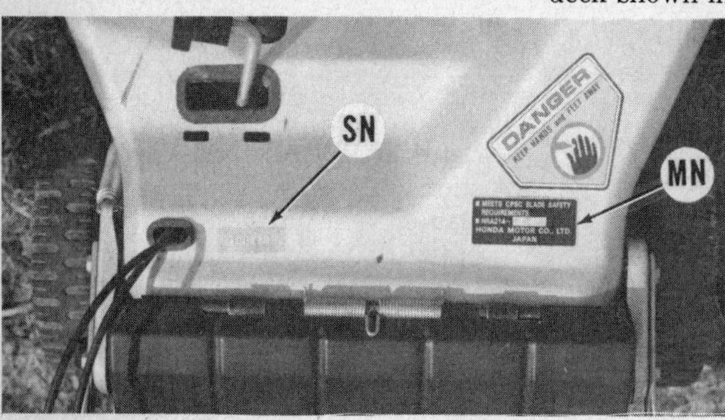

Fig. HN—View showing location of mower serial number (SN) and model number (MN) on Honda walking lawn mowers.

INTERNATIONAL HARVESTER

International Harvester walk-behind lawn mowers are equipped with four-stroke engines manufactured by Briggs & Stratton or two-stroke engines manufactured by Lawn Boy.

The two-stroke engine used on 1974 and earlier mowers is similar to equivalent Lawn Boy models. Mower model and serial numbers are located on a plate attached to mower deck or battery box of electric start models. Engine model number is located on plate attached to fin of cylinder head.

The four-stroke engine used on 1975 and later mowers is manufactured by Briggs & Stratton. The engine serial number and model number are stamped on the cooling shroud (blower housing).

Refer to appropriate Briggs & Stratton or Lawn Boy engine service section.

JACOBSEN

Jacobsen walk-behind lawn mowers are equipped with four-stroke engines manufactured by Briggs & Stratton or Tecumseh, or a two-stroke engine manufactured by Jacobsen. Mower model and serial numbers are located on a label attached to mower deck similar to Fig. JA.

Refer to appropriate engine service section for engine model and serial number location.

KING O'LAWN

King O'Lawn walk-behind lawn mowers are equipped with four-stroke Briggs & Stratton engine. Mower model and serial number plate is attached to rear of mower deck.

Refer to appropriate engine service section for engine model and serial number location.

LAWN BOY

Lawn Boy walk-behind lawn mowers are equipped with two-stroke engines manufactured by the Lawn Boy Division of the Outboard Marine Corporation. Mower model and serial number is located on plate attached to the mower deck shown in Figs. LMB1 and LMB2.

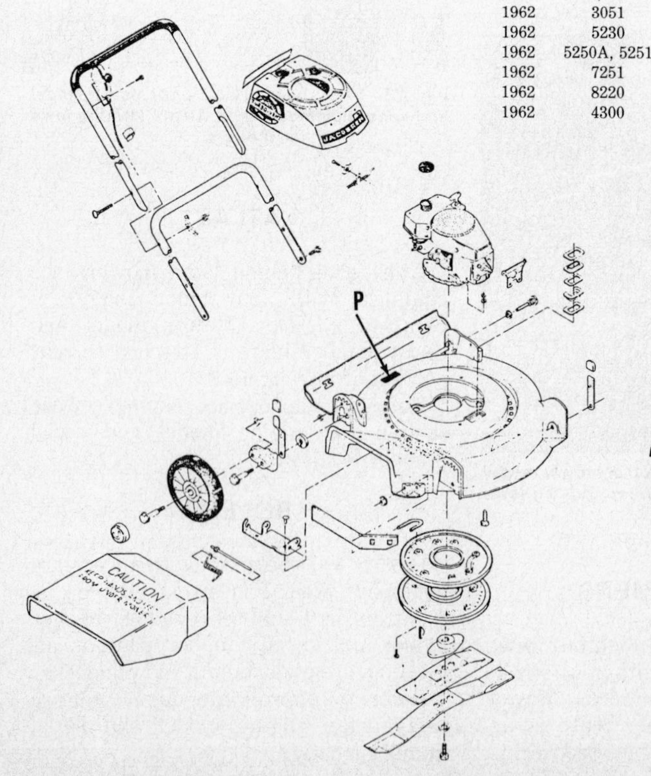

Fig. JA—View of Jacobsen rotary mower showing location of nameplate (P) for identifying mower and engine.

The following is a partial listing of mower models, year produced and original engine model installed.

YEAR	MOWER MODEL	SIZE OF CUT	ENGINE MODEL
1955	8F20	18 in.	C-10
1955	1F20	21 in.	C-10
1956	5000	18 in.	C-12
1956	7000	21 in.	C-12
1957	3100	18 in.	C-20
1957	5100	18 in.	C-12
1957	7100	21 in.	C-12
1957	6100	18 in.	C-40
1957	8100	21 in.	C-40
1958	3200, 3200A	18 in.	C-21
1958	5100X	18 in.	C-12
1958	7100X	21 in.	C-12
1958	5200	18 in.	C-13
1958	7200	21 in.	C-13
1958	6200	18 in.	C-41
1958	8200	21 in.	C-41
1959	3210	18 in.	C-22
1959	3050	18 in.	C-70
1959	7050	21 in.	C-70
1959	5210	18 in.	C-14
1959	7210	21 in.	C-14
1959	6210	18 in.	C-41
1959	8210	21 in.	C-41
1959	5250	19 in.	C-60
1959	19H	21 in.	C-75
1959	89H	18 in.	C-75
1960	3050	18 in.	C-70
1960	5210	18 in.	C-14
1960	7210	21 in.	C-14
1960	5250	19 in.	C-60
1960	7250	21 in.	C-15
1960	8210	21 in.	C-41
1960	19H	21 in.	C-75
1960	89H	18 in.	C-75
1960	18PL	18 in.	C-14M
1960	21PL	21 in.	C-14M
1960	21SPL	21 in.	C-41M
1961	3050	18 in.	C-70
1961	5210, 5210A	18 in.	C-14, C-17
1961	7210, 7210A	21 in.	C-14, C-17
1961	5250	19 in.	C-60
1961	7250	21 in.	C-15, 16
1961	8210, 8210A	21 in.	C-41, 43
1962	3051	18 in.	C-73
1962	5230	19 in.	C-17
1962	5250A, 5251	19 in.	C-61
1962	7251	21 in.	C-18
1962	8220	21 in.	C-43
1962	4300	24 in.	C-18

YEAR	MOWER MODEL	SIZE OF CUT	ENGINE MODEL
1963	3052	18 in.	C-76
1963	5231	19 in.	D-400
1963	7252	21 in.	D-400
1963	8221	21 in.	D-440
1963	4301	24 in.	D-400
1964	3052	18 in.	C-76
1964	5232	19 in.	D-401
1964	7214	21 in.	D-401
1964	7253	21 in.	D-401
1964	8222	21 in.	D-441
1964	4301A	24 in.	D-401
1964	6250	21 in.	C-19
1964	9274	21 in.	D-451
1964	9244	24 in.	D-451
1965	3052	18 in.	C-76
1965	5233	19 in.	D-402
1965	7215	21 in.	D-402
1965	7254, 7254-WB	21 in.	D-402
1965	8223	21 in.	D-442
1965	4302	24 in.	D-402
1965	6251	21 in.	C-19
1965	9245	21 in.	D-452, D-402
1965	9275	24 in.	D-452, D-402
1966	3053	18 in.	D-430
1966	5234	19 in.	D-403
1966	7255, 7216	21 in.	D-403
1966	7001	21 in.	C-18
1966	8224	21 in.	D-443
1966	8001	21 in.	C-43
1966	6252	21 in.	C-19
1967	3054	18 in.	D-430
1967	5235	19 in.	D-404
1967	7217, 7256	21 in.	D-404
1967	8225, 8226	21 in.	D-444
1967	6252	21 in.	C-19
1967	3001	18 in.	C-77
1967	3002	18 in.	C-78
1967	5001	19 in.	C-18
1967	7001	21 in.	C-18
1967	8001	21 in.	C-43
1967	8001, 8002	21 in.	C-43
1968	8226	21 in.	D-445
1968	8226E	21 in.	D-445E
1968	8227	21 in.	D-446
1968	8227E	21 in.	D-446E
1968	7218, 7257	21 in.	D-405
1968	7257E	21 in.	D-405E
1968	7219	21 in.	D-406
1968	7258E	21 in.	D-406E
1968	5236	19 in.	D-405
1968	5237	19 in.	D-406
1968	7002	21 in.	C-18
1968	8003	21 in.	C-43
1968	5002	19 in.	C-18
1968	7010	21 in.	D-430
1969	8228	21 in.	D-447
1969	8228E	21 in.	D-447E
1969	7220, 7259	21 in.	D-407
1969	7259E	21 in.	D-407E
1969	5238	19 in.	D-407
1969	7003	21 in.	C-18
1969	7011	21 in.	D-431
1969	8004	21 in.	C-44
1969	5003	19 in.	C-18
1969	3055	18 in.	D-430
1969	6275	21 in.	D-475
1970/71	8229, 8229S	21 in.	D-448
1970/71	8229E	21 in.	D-448E
1970/71	7221, 7260, 7260S	21 in.	D-408
1970/71	7260E	21 in.	D-408E
1970/71	5239, 5269, 5269S	19 in.	D-408
1970/71	7004	21 in.	C-18B
1970/71	8005	21 in.	C-45
1970/71	5004	19 in.	C-18B
1970/71	3056	18 in.	D-432
1970/71	3003	18 in.	C-79
1970/71	6275	21 in.	D-476
1970/71	6252	21 in.	C-19B
1970/71	7012	21 in.	D-432
1971	5020	19 in.	D-432
1971	7020, 7080	21 in.	D-432
1971	8020	21 in.	D-420
1972	8230E	21 in.	D-640E
1972	8230	21 in.	D-640
1972	8229B	21 in.	D-448
1972	7261E	21 in.	D-600E
1972	7261, 7222	21 in.	D-600

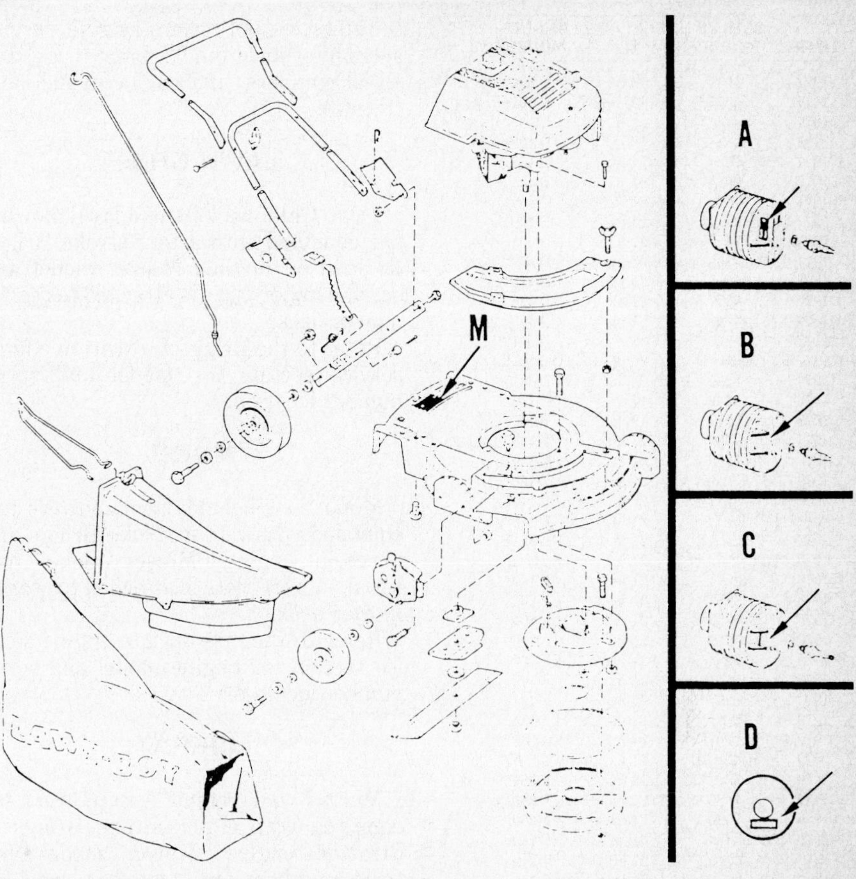

Fig. LMB1—View of Lawn Boy rotary mower showing location of plate (M) identifying mower model. Engine identification numbers are located on identification tag (A), ribs (B and C) or starter cup housing (D).

YEAR	MOWER MODEL	SIZE OF CUT	ENGINE MODEL
1972	5270, 5240	19 in.	D-600
1972	7260A	21 in.	D-408
1972	5269A, 5239A	19 in.	D-408
1972	7081, 7021	21 in.	D-433
1972	5080, 5021	19 in.	D-433
1972	8021	21 in.	D-420
1972	6253, 6275	21 in.	D-476
1972	3057	18 in.	D-433
1973	8231E	21 in.	D-640E
1973	8231	21 in.	D-640
1973	8229C	21 in.	D-448
1973	7262E	21 in.	D-600E
1973	7262, 7223	21 in.	D-600
1973	5271, 5241	19 in.	D-600
1973	7260B	21 in.	D-408
1973	5269B, 5239B	19 in.	D-408
1973	7082, 7022	21 in.	D-433
1973	5081, 5022	19 in.	D-433
1973	8250	21 in.	D-480
1973	6254, 6276	21 in.	D-476
1973	3058	18 in.	D-433
1974	8232E	21 in.	D-640E
1974	8232	21 in.	D-640
1974	8229E	21 in.	D-448
1974	7263E	21 in.	D-660E
1974	7263, 7224	21 in.	D-600
1974	5272, 5242	19 in.	D-600
1974	7260C	21 in.	D-408
1974	5269C, 5239C	19 in.	D-408
1974	7083, 7023	21 in.	D-433
1974	5063, 5023	19 in.	D-433
1974	8251	21 in.	D-480
1974	6254, 6276	21 in.	D-476
1975	8233	21 in.	D-601
1975	8229E	21 in.	D-449
1975	7264, 7225	21 in.	D-601
1975	5273, 5243	19 in.	D-601
1975	7260D	21 in.	D-409
1975	5269D, 5239D	19 in.	D-409

YEAR	MOWER MODEL	SIZE OF CUT	ENGINE MODEL
1975	7084, 7024	21 in.	D-433
1975	5064, 5024	19 in.	D-433
1975	8252	21 in.	D-481
1975	6255, 6277	21 in.	D-476
1976	8234	21 in.	D-641
1976	8229G	21 in.	D-449
1976	7265	21 in.	D-601
1976	7226	21 in.	D-601
1976	5274	19 in.	D-601
1976	5244	19 in.	D-601
1976	7260F	21 in.	D-409
1976	5269F	19 in.	D-409
1976	5239F	19 in.	D-409
1976	7085	21 in.	D-433
1976	7025	21 in.	D-433
1976	5065	19 in.	D-433
1976	5025	19 in.	D-433
1976	8253	21 in.	D-481
1976	6255	21 in.	D-476
1976	6277	21 in.	D-476
1977	5024	19 in.	D-433
1977	5064	19 in.	D-433

Fig. LMB2—View showing plate located at rear of mower deck identifying mower model and serial number.

YEAR	MOWER MODEL	SIZE OF CUT	ENGINE MODEL
1977	5245	19 in.	D-601
1977	5275	19 in.	D-601
1977	5239G	19 in.	D-409
1977	5269G	19 in.	D-409
1977	7024	21 in.	D-433
1977	7084	21 in.	D-433
1977	7221G	21 in.	D-409
1977	7260G	21 in.	D-409
1977	7227	21 in.	D-601
1977	7266	21 in.	D-601
1977	6255	21 in.	D-476
1977	6277	21 in.	D-476
1977	8229H	21 in.	D-449
1977	8235	21 in.	D-641
1977	8235AE	21 in.	D-641AE
1977	8255	21 in.	D-481
1978	5247	19 in.	F-100
1978	5277	19 in.	F-100
1978	7229	21 in.	F-100
1978	7268	21 in.	F-100
1978	R7268	21 in.	F-100
1978	8237	21 in.	F-140
1978	R8237	21 in.	F-140
1978	8237AE	21 in.	F-140AE
1978	R8237AE	21 in.	F-140AE
1978	8270	21 in.	F-140
1978	6257	21 in.	F-200
1978	6279	21 in.	F-200
1979	7229	21 in.	F-100
1979	7268	21 in.	F-100
1979	R7268	21 in.	F-100
1979	8237	21 in.	F-141
1979	R8237	21 in.	F-141
1979	8238AE	21 in.	F-141AE
1979	R8238AE	21 in.	F-141AE
1979	5026	19 in.	F-200
1979	7086	21 in.	F-200
1979	7025	21 in.	F-200
1979	8350	21 in.	F-240
1979	8310	21 in.	F-240
1979	6290	21 in.	F-200
1979	6279	21 in.	F-200
1979	4550	21 in.	F-100
1979	4570	21 in.	F-100
1979	8650	21 in.	F-140
1979	8670	21 in.	F-140
1979	4500	20 in.	D-410
1979	8600	20 in.	D-410
1980	5247	19 in.	F-101
1980	7229	21 in.	F-101
1980	7268	21 in.	F-100
1980	R7268	21 in.	F-100
1980	8237	21 in.	F-140
1980	R8237	21 in.	F-140AE
1980	8238AE	21 in.	F-140AE
1980	4551	21 in.	F-101
1980	4571	21 in.	F-101
1980	8651	21 in.	F-140
1980	8671	21 in.	F-141
1980	6259	21 in.	F-200
1980	6300	21 in.	F-201
1980	8400	21 in.	F-241
1980	4501	20 in.	D-411
1980	8601	20 in.	D-411
1981	5247	19 in.	F-100
1981	7229	21 in.	F-100
1981	7268	21 in.	F-100
1981	8237	21 in.	F-140
1981	8238AE	21 in.	F-140AE
1981	4571	21 in.	F-101
1981	8671	21 in.	F-141
1981	8671AE	21 in.	F-142AE
1981	6259	21 in.	F-200
1981	6300	21 in.	F-201
1981	8401	21 in.	F-241
1981	4502	20 in.	F-340
1981	8602	20 in.	F-340
1982	5247	19 in.	F-100
1982	7229	21 in.	F-100
1982	7268	21 in.	F-100
1982	8237	21 in.	F-140
1982	8238AE	21 in.	F-140AE
1982	4571	21 in.	F-101
1982	8671	21 in.	F-141
1982	8671AE	21 in.	F-142AE
1982	6259	21 in.	F-200
1982	8401	21 in.	F-241
1982	4502	20 in.	F-340
1982	8602	20 in.	F-340

Refer to appropriate engine service section for location of engine model and serial numbers and engine model identification.

LAWN CHIEF

Lawn Chief walk-behind lawn mowers are equipped with a four-stroke Briggs & Stratton engine. Mower model and serial number plate is attached to rear of mower deck.

Refer to the Briggs & Stratton engine service section for model and serial number location.

MONO

Mono walk-behind lawn mowers are equipped with a four-stroke Briggs and Stratton engine. Mower model and serial number plate is attached to rear of mower deck.

Refer to the appropriate engine service section for engine model and serial number location.

MURRAY

Murray walk-behind lawn mowers are equipped with a four-stroke Briggs & Stratton engine. Mower model and serial numbers are located on label on mower deck as shown in Fig. M.

Refer to the appropriate engine service section for engine model and serial number location.

Fig. M—View showing location of model and serial numbers (N) on Murray walking lawn mowers.

MTD LAWNFLITE

MTD Lawnflite walk-behind lawn mowers are equipped with a four-stroke Briggs & Stratton engine. Mower model and serial number plate is attached to rear of mower deck.

Refer to appropriate engine service section for engine model and serial number location.

J.C. PENNEY

J.C. Penney walk-behind lawn mowers are equipped with a four-stroke Briggs & Stratton engine or a two-stroke Lawn Boy engine. Mower model and serial number plate is attached to rear of mower deck.

Refer to appropriate engine service section for engine model and serial number location.

ROPER

Roper walk-behind lawn mowers are equipped with a four-stroke Briggs & Stratton engine. Mower model and serial number plate is attached to rear of mower deck.

Refer to appropriate engine service section for engine model and serial number location.

SENSATION

Sensation walk-behind mowers are equipped with four-stroke engines manufactured by Briggs & Stratton, Clinton or Honda, or a two-stroke engine manufactured by Kawasaki. Mower model and serial numbers are located on plate attached to rear of mower deck.

Refer to appropriate engine service section for engine model and serial number location.

SIMPLICITY

Simplicity walk-behind lawn mowers are equipped with four-stroke engines manufactured by Briggs & Stratton or Tecumseh. Mower model and serial number plate is attached to rear of mower deck.

SNAPPER

Snapper walk-behind lawn mowers are equipped with four-stroke engines manufactured by Briggs & Stratton or

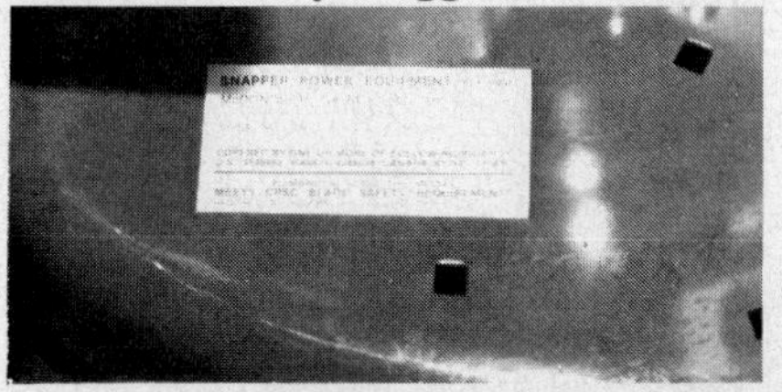

Fig. S1—View showing plate located at rear of mower deck identifying Snapper mower model and serial number.

Tecumseh. Mower model and serial numbers are located on plate attached to rear of mower deck as shown in Fig. S1.

Refer to appropriate engine service section for engine model and serial number location.

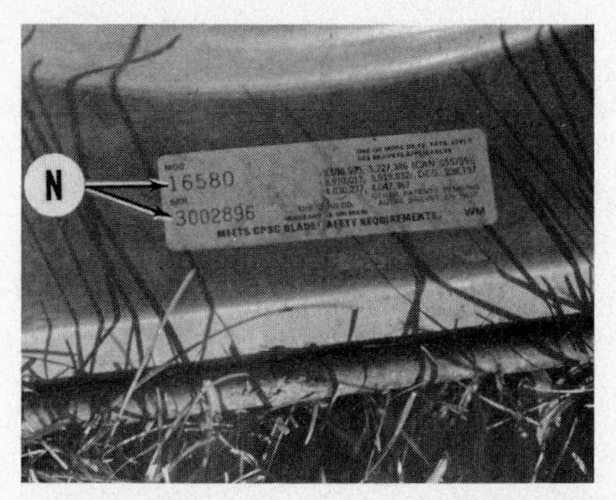

Fig. TO—View showing location of model and serial numbers (N) on Toro walking lawn mowers.

TORO

Toro walk-behind lawn mowers are equipped with four-stroke Briggs & Stratton, Tecumseh or Toro (manufactured by Tecumseh) engines or two-stroke Suzuki or Toro (manufactured by Suzuki) engines. Mower model and serial numbers are located on plate attached to mower deck as shown in Fig. TO.

Refer to appropriate engine service section for engine model and serial number location.

WARDS

Wards walk-behind lawn mowers are equipped with four-stroke Briggs & Stratton or Tecumseh engines. Mower model and serial number plate is attached to rear of mower deck.

Refer to appropriate engine service section for engine model and serial number location.

WHITE

White walk-behind lawn mowers are equipped with a four-stroke Briggs & Stratton engine. Mower model and serial number plate is attached to rear of mower deck.

Refer to appropriate engine service section for engine model and serial number location.

YARD MAN

Yard Man walk-behind lawn mowers are equipped with four-stroke Briggs & Stratton or Tecumseh engines. Mower model and serial number plate is attached to rear of mower deck.

Refer to appropriate engine service section for engine model and serial number location.

STARTING AND OPERATION

Some standard procedures should be carefully followed. Most of the suggested procedures are in the interest of safety; however, some can prevent unnecessary damage to the mower or other property.

1. Never operate the engine in an enclosed area. Exhaust gases contain carbon monoxide, an odorless and deadly poison. Operate engine only where there is adequate ventilation.

2. Do not operate mower on top of loose rocks, sand or other material which may be thrown by the blade. Small objects may be hurled at high velocity by the blade and cause personal injury or property damage. Refer to Fig. 20.

3. Never fill the gas tank while engine is running. Avoid spilling gasoline on hot engine. Rinse all spilled fuel off with water before attempting to start engine. Refer to Fig. 21.

4. Always keep nuts, bolts and screws tight. Loose or missing parts can change or interfere with the operation of the engine or mower. Damage or injury can sometimes be easily avoided by making sure that all parts are securely in place.

5. Always disconnect and ground the wire from spark plug before inspecting, or repairing any part of mower or engine. The engine may accidentally start causing serious damage or injury if the spark plug wire remains attached. Refer to Fig. 22.

6. Always check the oil level before starting engine. On mowers with four-stroke engine, be sure that crankcase is

Fig. 20—Never start the mower on top of rocks, sand, gravel, etc. These small objects will be hurled by the blade.

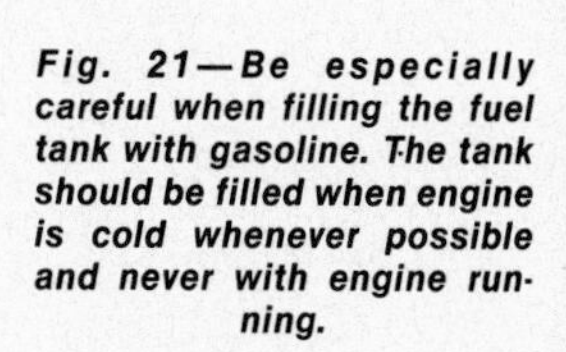

Fig. 21—Be especially careful when filling the fuel tank with gasoline. The tank should be filled when engine is cold whenever possible and never with engine running.

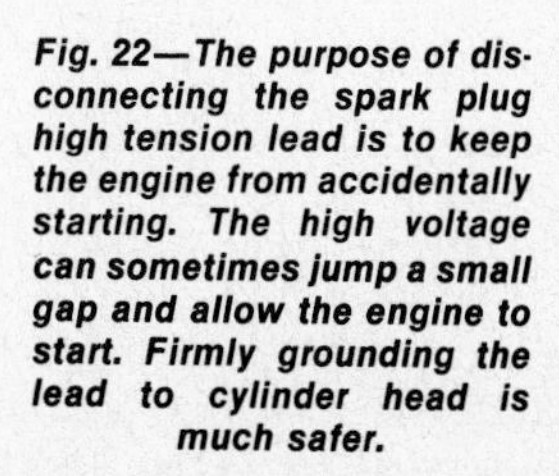

Fig. 22—The purpose of disconnecting the spark plug high tension lead is to keep the engine from accidentally starting. The high voltage can sometimes jump a small gap and allow the engine to start. Firmly grounding the lead to cylinder head is much safer.

filled and that mower is level when checking. Total capacity is only about 1¼ pints (0.6L) and the loss of only a small quantity of oil can destroy the engine. On mowers with a two-stroke engine, the oil for engine lubrication is mixed with the fuel. Operating the engine for even a short time without correct amount of oil mixed with the gasoline may extensively damage the engine.

7. Keep the engine clean. The engine is cooled by the circulation of air. Damage can result from overheating if the flow of air is restricted.

Fig. 25—Views of two oil containers with typical marking indicating the "SAE" viscosity and the "API SERVICE" rating.

Oil and Fuel (4-Stroke Engines). Position mower so engine is level. Remove the oil level plug (P–Fig. 23) and observe the level of the oil in the threaded hole. The hole should be maintained at top of the hole to the point of overflowing. Total capacity of oil is only 1¼ pints (0.6L) and it is important that it is full. Do not attempt to run engine even for a short time with less than the recommended amount of oil.

Some engines are equipped with a dipstick attached to a fill plug (Fig. 24). Procedure for checking is similar, except that oil should be up to the full mark on dipstick.

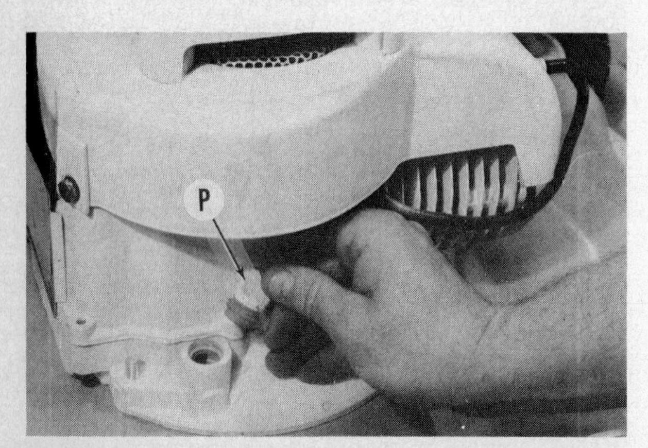

Fig. 23—Always check oil level before starting engine.

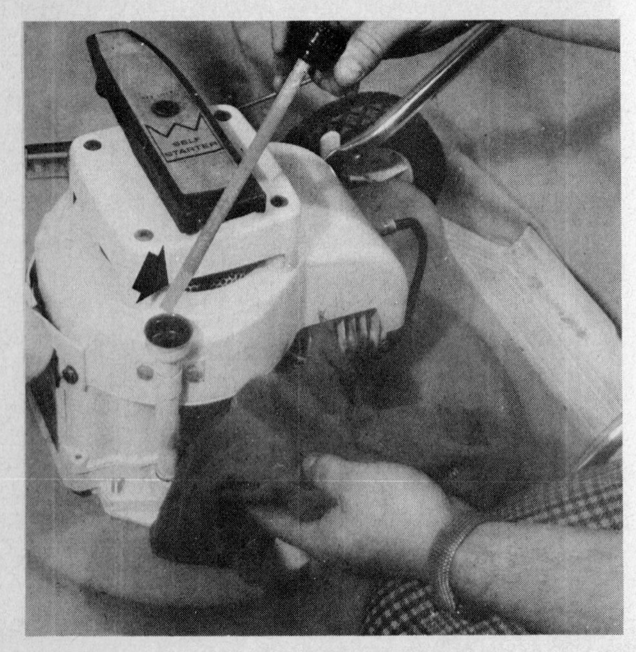

Fig. 24—A dipstick is attached to the fill plug shown for checking level of oil.

On all four-stroke engines, pour oil in slowly. The size of the fill openings and the vent passages are small and some time is required for oil to flow into the engine.

Manufacturers recommend single-viscosity oil with an API service classification MS, SC, SD or SE (Fig. 25). SAE 30 oil should be used in summer or when temperature is above 32°F (0°C). Use SAE 10W oil in temperatures of 32°F (0°C) and below. Most manufacturers will accept 10W-30 multi-viscosity oil as a substitute, however, previously used oil, poor quality oil or 10W-40 oil should not be used.

Manufacturers recommend regular or unleaded gasoline with an octane rating of at least 87. Fuel which contains alcohol in any percentage should not be used as alcohol damages some fuel system components.

Fuel should be stored in clean, clearly marked containers. Fuel should not be kept for long periods of time as water condensation and fuel deterioration will occur. Octane rating will also be lower in old fuel.

Oil and Fuel (2-Stroke Engines). Internal parts of two-stroke engines are lubricated by oil mixed with the gasoline. Failure to mix the correct amount or type of oil with the gasoline will result in damage. It is important to follow the engine manufacturers recommendation as to oil type and mixing ratio.

Excessive oil or the wrong type oil will cause low power, plug fouling and excessive carbon build-up. Insufficient amount of oil will result in inadequate lubrication and will result in internal damage similar to any engine operating without oil–Rapid, Severe and Extensive. The recommended ratio of oil to be mixed with the gasoline is listed in the individual engine section of this manual.

Oil should be mixed with gasoline in a separate container before it is poured into the fuel tank. The following table may be useful in mixing correct quantities of oil and gasoline to provide a specific ratio.

Ratio	Gasoline	Oil
10:1	.63 Gallon	½ Pint (237mL)
14:1	.88 Gallon	½ Pint (237mL)
16:1	1.00 Gallon	½ Pint (237mL)
20:1	1.25 Gallons	½ Pint (237mL)
30:1	1.88 Gallons	½ Pint (237mL)
32:1	2.00 Gallons	½ Pint (237mL)
50:1	3.13 Gallons	½ Pint (237mL)
10:1	1 Gallon	.79 Pint (379mL)
14:1	1 Gallon	.57 Pint (270mL)
16:1	1 Gallon	.50 Pint (237mL)
20:1	1 Gallon	.40 Pint (189mL)
30:1	1 Gallon	.27 Pint (126mL)
32:1	1 Gallon	.25 Pint (118mL)
50:1	1 Gallon	.16 Pint (76mL)

Often the engine manufacturer recommends a specific type or brand of oil. Other types may perform satisfactorily; however, the type recommended was tested and was proven to be acceptable.

Manufacturer may recommend acceptable oils according to BIA TC (Boating Industry Association Two-Cycle) standards. BIA ratings are listed on the oil container (Fig. 25A) and most containers are marked on the side to provide a means of properly mixing the correct amount of oil with gasoline to provide the proper fuel:oil mixture ratio.

Manufacturers recommend regular or unleaded gasoline with an octane rating of at least 87. Fuel which contains alcohol in any percentage should not be used as alcohol damages some fuel system components.

Fuel should be stored in clean, clearly marked containers. Fuel should not be kept for long periods of time as water condensation and fuel deterioration will occur. Octane rating will also be lower in old fuel.

Fig. 26—The chances of catching fire are greater when refueling a hot engine than with a cold engine.

Fig. 27—Tall grass can cause enough resistance to prevent the engine from starting.

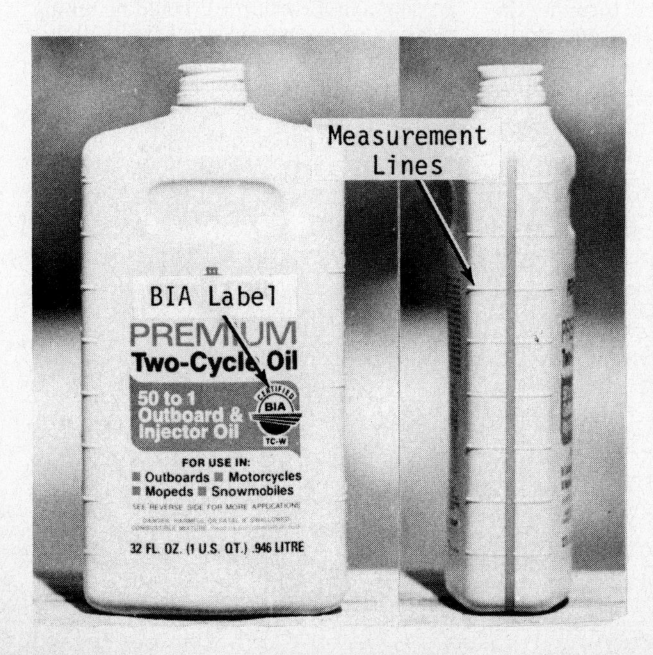

Fig. 25A—View showing oil container marked with BIA TC (Boating Industry Association Two-Cycle) ratings. Oil containers are also marked for US customary and metric measurements to aid proper mixing ratio.

Fill The Fuel Tank. It is much safer to fill the fuel tank when the engine is cold. Refer to Fig. 26. BE SURE to use manufacturers recommended grade of gasoline, at an octane rating minimum or above. On two-stroke engines, be sure that the proper amount of oil has been mixed with the gasoline. Use only clean, fresh gasoline. Always store fuel safely in a properly marked steel container. Prolonged storage of gasoline may produce deposits in the storage tank (can) or the engine fuel tank which can prevent an engine from running. Gasoline may be stored for short lengths of time, but never use gasoline that has been stored over the winter. Even if the engine runs, the quality of the gasoline is poor and may cause engine damage. Old stale gasoline has a different odor than fresh gasoline, and may be an early clue to the cause of poor performance.

Locate the Mower in a Safe, Practical Area. Mowers are difficult and sometimes impossible to start when standing in tall, thick grasses. Refer to Fig. 27. The resistance to turn may be enough to prevent the starter from operating fast enough to cause the engine to run. Do not compensate for this condition by starting the mower in a rock roadway. The rocks can be thrown

Fig. 27A—Move the mower to area where blade is not obstructed, when starting the engine.

great distances with great force by the rotating blade. Select the place to start the engine carefully with much consideration to safety.

Set The Controls. Make sure all control levers, cables and safety equipment are in good and free working condition. A lever or cable that is damaged or binding may not provide correct engine component control. A faulty safety switch may prevent engine from starting when properly engaged or allow engine to start when safety equipment is not correctly positioned.

Control is required for three basic engine functions: Ignition; Engine Speed (throttle); Starting Enrichment (choke). The ignition must be turned on or, on models so equipped, the safety equipment must be positioned in starting location. On models without controlled carburetor functions, the throttle should be opened slightly and the choke should be actuated unless the engine is already warm. Often a single control (E–Fig. 28) interconnects these three operations and may be operated from the handlebar by a remote control lever.

NOTE: The remote control levers and cable may not move the engine controls correctly, always check for proper movement at engine if engine fails to start.

Operate Starter. The starters used may be as simple as a rope wrapped around a pulley or may be an electric starter operated by turning a key or pressing a switch. Upon operation of the starter the engine should begin to run. Move manual choke controls, if so equipped, to open the choke soon after engine starts. If control was set at an indicated "CHOKE" or "START" position, move the control to "RUN" position.

Stopping The Mower. On some models equipped with safety control handle, releasing handle will cause an ignition kill switch to be engaged and engine will die. On other models, ignition control must be moved to "STOP" or "OFF" position for engine to die.

Cleaning Mower Deck. Some mowers are provide with a hose connection (P–Fig. 29) to ease cleaning the mower. Attach hose and operate engine with water running into the clean out connection for a short period before stopping engine. On all mowers, clean the mower before storing.

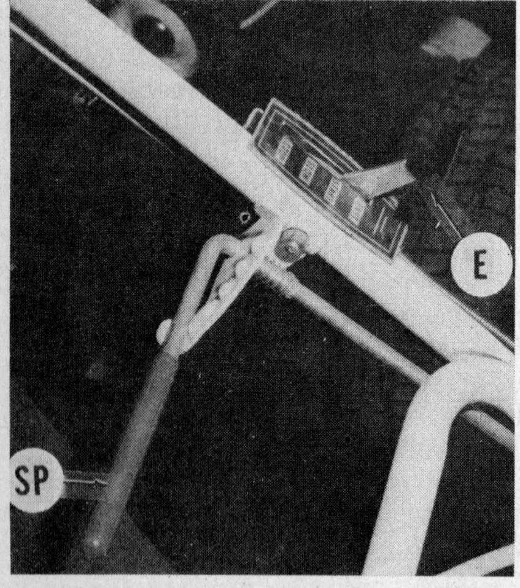

Fig. 28—Be sure controls are properly set before attempting to start.

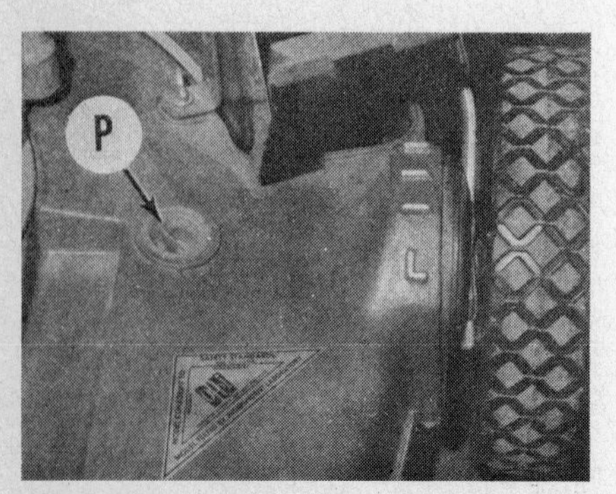

Fig. 29—Many mowers are equipped with a clean out port (P) to help clean the mower.

PROBLEMS AND REMEDIES

Repair is divided into three areas: Recognition that a problem exists; Locating the source of the problem; and Correcting the trouble.

While troubles of a serious nature will occasionally develop, many defects are minor and can be easily remedied. Often correcting a minor problem can prevent or eliminate major failure at a later date. Develop the habit of following a planned sequence in the diagnosis and elimination of trouble. Try to determine the cause of trouble and the remedy by logically thinking out the known facts. The following may be helpful in locating some problems.

1. DETERMINE THE SPECIFIC PROBLEM to be corrected, then outline in your mind (or on paper) the probable cause(s).
2. ELIMINATE ALTERNATIVE POSSIBILITIES by checking out probable causes in a logical order.
3. CORRECT THE PROBLEM, then go one step further by checking out and correcting possible contributing causes or side effects. As an example, plugged passages in a carburetor could be an indication of a contaminated fuel supply. The trouble may reoccur frequently until the fuel tank (and possibly the storage container) is cleaned.

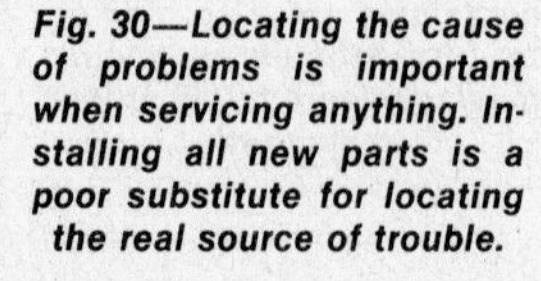

Fig. 30—Locating the cause of problems is important when servicing anything. Installing all new parts is a poor substitute for locating the real source of trouble.

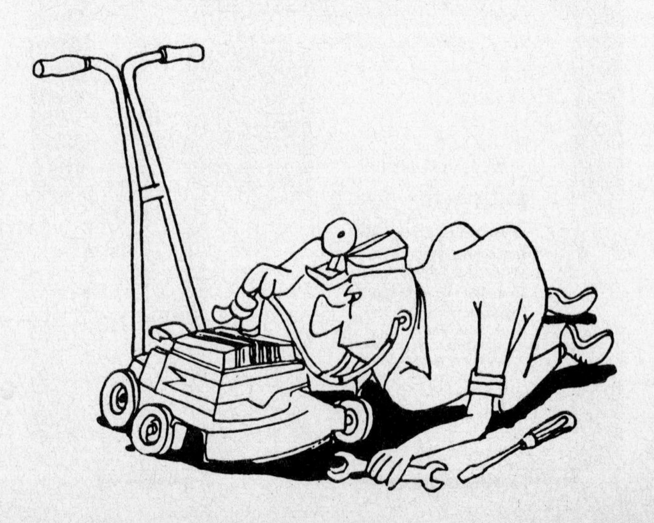

Many troubles are obvious (loose, missing or broken parts, etc.) for which the correction is also evident. The following paragraphs list some common malfunctions, together with procedural recommendations for pinpointing the cause.

NOTE: It should be pointed out that some trouble may result from a primary and one or more secondary causes of a cumulative nature, and that correcting the primary cause may get the engine back in service without correcting the underlying cause and an early second failure could result. It should also be stated that the list contains only the more common malfunctions and causes, and that the troubleshooting procedure can not be limited to only those described.

NOISES

Noise is one of the most significant of mechanical danger signals. Listen for unusual noise and for changes in sounds any piece of equipment makes, then try to determine the cause of the sound.

Knocking or rattling noises can be caused by loose mounting bolts or other loose bolts. Sharp knocking noises which appear to come from within the engine usually indicate serious trouble. Stop the engine and carefully examine the unit to determine the cause.

A bent engine crankshaft, a bent blade or a loose blade may cause extremely loud noises. A bent blade or crankshaft is caused by striking a solid or nearly solid object. The noise is usually accompanied by severe vibration. The noise may be a rapid knocking or a scraping sound. Diagnose trouble by turning blade slowly by hand after removing spark plug and grounding the spark plug high tension wire.

CAUTION: Spark plug wire MUST be disconnected for safety to prevent engine from accidentally starting while fingers are in contact with the blade. Removing the plug also permits crankshaft to turn more easily without the interference of the compression stroke. The spark plug wire should be grounded (to eliminate the possibility of coil damage due to an incomplete secondary circuit and to prevent accidental starting while turning the blade).

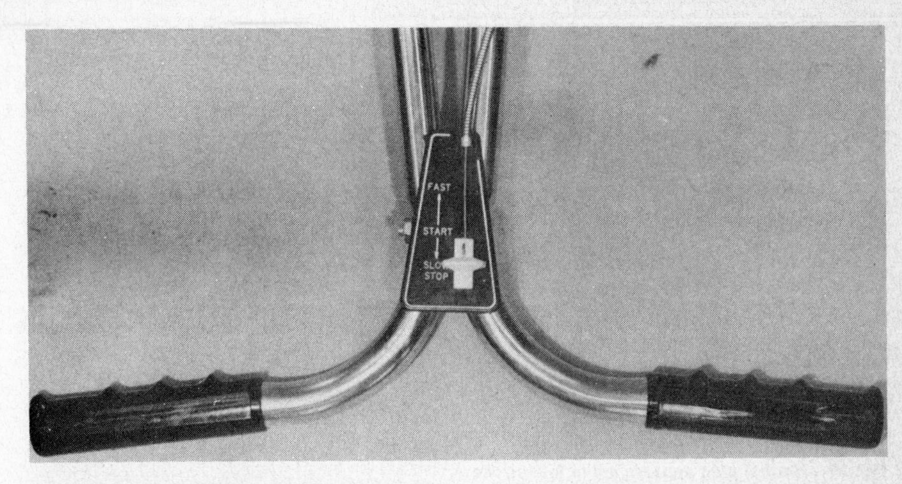

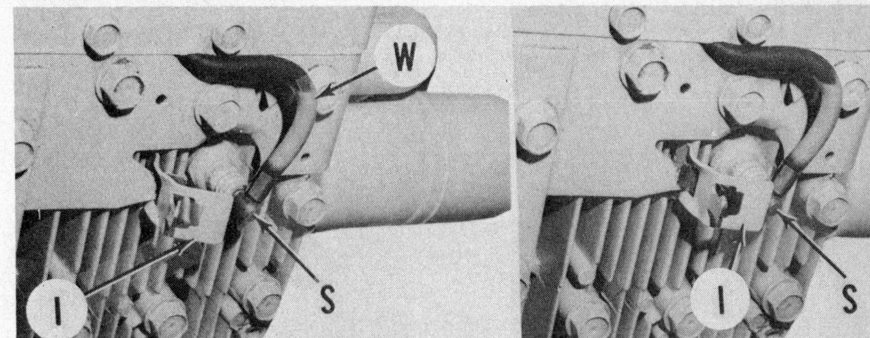

Fig. 32—Stop switch (I) can be against spark plug lead or high tension lead (W) can be shorted out. Either condition will cause engine not to start.

ENGINE WILL NOT START

Failure to start is usually caused by fuel not reaching the cylinder or spark not occurring at the spark plug. Do not neglect to check the obvious causes (fuel tank empty, stop switch actuated, safety switch malfunctioning, spark plug fouled, etc.).

Test for Fuel and Spark. Two essentials necessary for the engine to run are: FUEL and a timed SPARK. Remove the spark plug and check condition of the spark plug. The spark plug should have odor of gasoline and can be slightly damp appearing. Spark plug should not

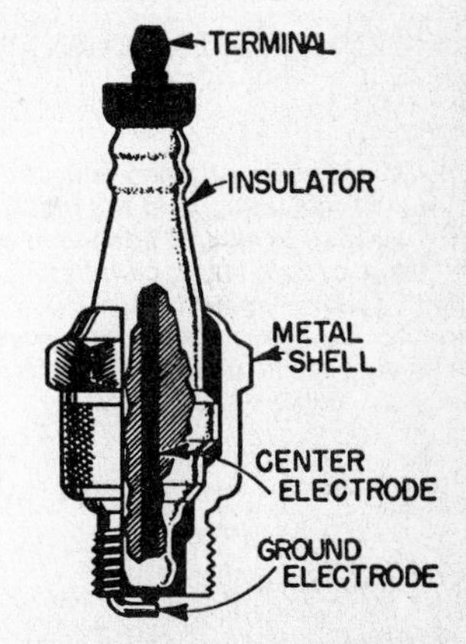

Fig. 33—Cross-sectional drawing of spark plug showing construction and nomenclature. The gap between ground electrode and center electrode is where spark should occur.

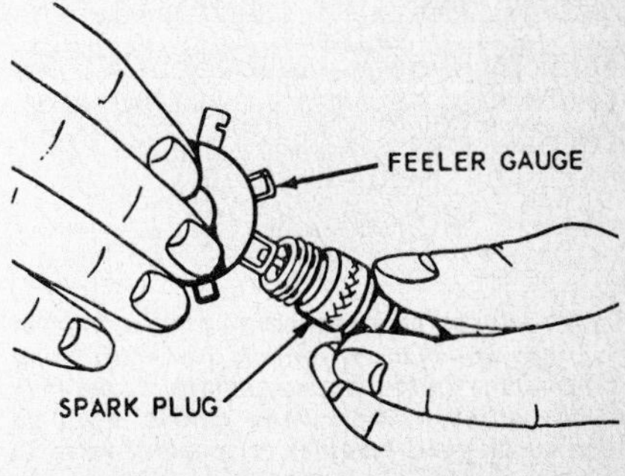

Fig. 34—Be sure to check spark plug electrode gap with correct thickness gage and adjust gap if necessary. Special spark plug gap gages are available from most tool sources.

Fig. 31—The purpose of disconnecting the spark plug high tension lead is to keep the engine from accidentally starting. The high voltage can sometimes jump a small gap and allow the engine to start. Firmly grounding the lead to cylinder head is much safer.

Fig. 35—Normal plug appearance in four-stroke cycle engine. Insulator is light tan to gray in color and electrodes are not burned. Renew plug at regular intervals as recommended by engine manufacturer.

Fig. 36—Appearance of spark plug from four-stroke engine indicating cold fouling. Cause of cold fouling may be use of a too-cold plug, excessive idling or light loads, carburetor adjusted too "rich" or low engine compression. Similar appearance of spark plug from two-stroke engine would be considered normal after several hours of operation.

Fig. 37—Appearance of spark plug indicating wet fouling; a wet, black oily film is over entire firing end of plug. On four-stroke engines, cause may be oil getting by worn valve guides, worn oil rings or plugged breather or breather valve in tappet chamber. On two-stroke engine, cause may be incorrect type or amount of oil mixed with the gasoline or may be caused by wrong (too cold) spark plug installed.

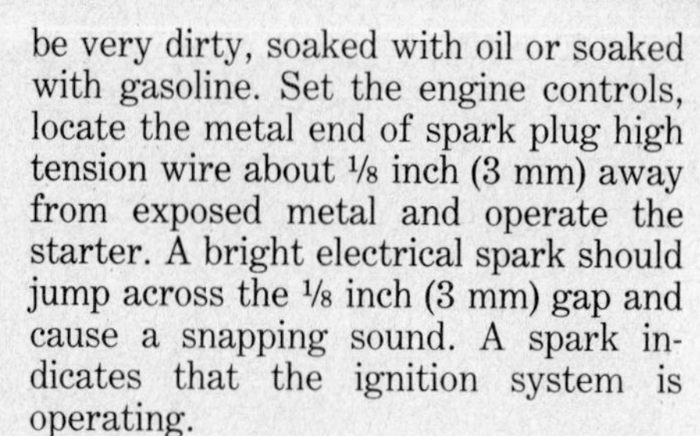

be very dirty, soaked with oil or soaked with gasoline. Set the engine controls, locate the metal end of spark plug high tension wire about 1/8 inch (3 mm) away from exposed metal and operate the starter. A bright electrical spark should jump across the 1/8 inch (3 mm) gap and cause a snapping sound. A spark indicates that the ignition system is operating.

NOTE: A variety of test plugs are available to test ignition system. Refer to Fig. 39 and Fig. 39A.

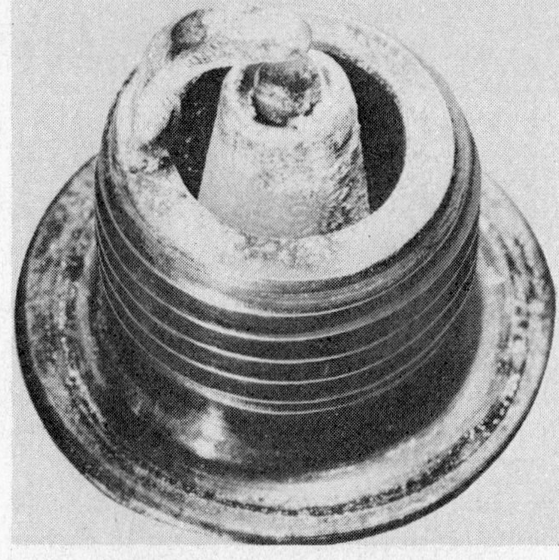

Fig. 38—Appearance of spark plug indicating overheating. Check for plugged cooling fins, or bent or damaged blower housing, engine being operated without all shields in place or other causes of engine overheating. Also can be caused by too lean a fuel-air mixture or spark plug not tightened properly.

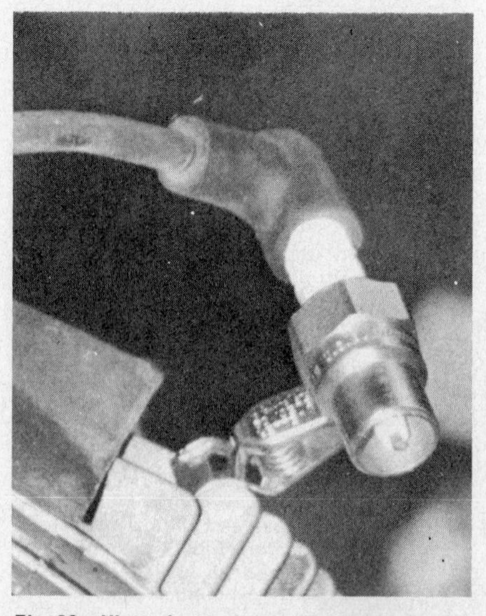

Fig. 39—View of test plug available to test ignition system.

If the ignition will not provide a spark, check the ignition electrical system further to locate trouble. Especially check condition of the engine stop switch wiring for shorts, correct operation of safety switches and condition of breaker points and condenser (P–Fig. 40 or Fig. 41) on models so equipped.

If the test indicates a good spark, but the removed spark plug is dry with no evidence of fuel, be sure that the engine is getting fuel. The carburetor choke valve should be closed when attempting to start. Check the carburetor for proper operation.

NOTE: Internal damage can also cause the engine not to start and the plug to remain dry, but the carburetor will more often be at fault.

If the test indicates a good spark and the removed spark plug is wet, clean or install a new spark plug and attempt to start.

If in doubt about spark plug condition, install a new plug.

NOTE: Before installing plug, be sure to check electrode gap with proper gage and, if necessary, adjust to value given in engine repair section of this manual. DO NOT guess or check gap with a "thin dime"; a few thousandths variation from correct spark plug electrode gap can make an engine run unsatisfactorily, or under some conditions, not start at all.

Fig. 39A—View of test plug which allows compression to be pumped up against spark plug. Scale on tester helps determine which component of ignition system is malfunctioning by firing test plug at different compression levels.

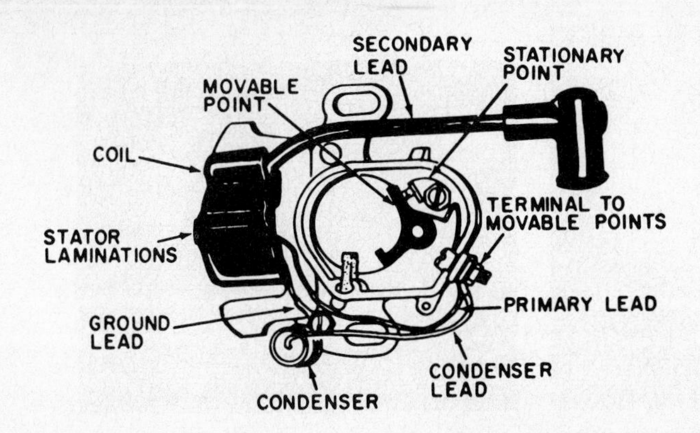

Fig. 40—View of the coil, stator, condenser and ignition breaker points (P) with flywheel removed.

Fig. 41—View of the soft metal flywheel key (K) and the ignition breaker points (P) with flywheel removed.

The choke may be closed too long, flooding the engine with gasoline. Flooding can be caused by a malfunctioning automatic choke or manually controlled choke used improperly.

Fuel Tank Not Filled. Check the fuel tank to be sure that it is filled with gasoline. Water or other fluid in the tank can prevent the engine from running. Be sure that tank is filled with only clean fresh gasoline or fuel mix. Water does not mix with gasoline and will appear as different colored areas floating in the gasoline (W–Fig. 42). Clean water in gasoline may appear lighter in color, but water can be colored by rust and be darker than the gasoline. A large amount of water in the fuel tank can cover the bottom of the tank uniformly and may be more difficult to identify.

Controls Incorrectly Set. On models so equipped, check position of lever (E–Fig. 43) on engine as well as the remote control handle. Be sure the engine controls operate to both extremes of travel when remote control handle is moved. Carburetor functions on some models are automatically controlled; only the ignition system circuit is controlled by the operator.

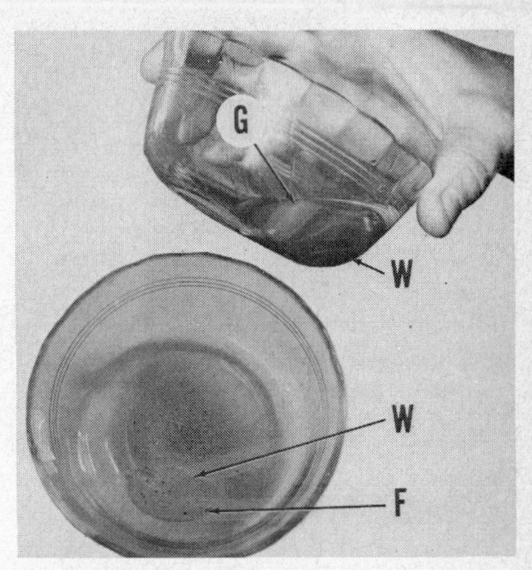

Fig. 42—Water (W) in the fuel (G) can often be noticed visually. Foreign particles (F) including dirt and metal may also be found visually.

On models equipped with a choke, carburetor butterfly should be closed when attempting to start a cold engine. Remove air cleaner and visually check position of choke if questioned.

NOTE: Be sure air cleaner is installed before operating the mower.

Be sure spark high tension wire (W–Fig. 44) is tightly connected on spark plug terminal and spark plug shorting stop switch (I) on models so equipped is not in contact with spark plug terminal.

Blade Loose. On some models the cutting blade is used as part of the flywheel weight. If the engine tries to start but backfires or just does not quite start, check for a loose blade (Fig. 49). Check blade mounting and tighten if necessary to prevent the blade from slipping easily. If the blade is tight, remove the flywheel, then check the condition of the condenser and the gap of the contact breaker points on models so equipped. Inspect for a sheared flywheel key (K–Fig. 41) or excessive wear in crankshaft or flywheel groove.

On later models equipped with blade brake clutch, a loose blade will not cause this condition as the blade is inoperative until the control handle is engaged.

NOTE: Checking blade tightness should be part of routine maintenance. Be sure spark plug wire is disconnected and sufficiently grounded to prevent engine from accidentally starting.

Fig. 43—Be sure engine controls (E) operate properly.

Faulty Starter or Bent Crankshaft. The starter has to turn the engine fast enough to start. A damaged starter, a bent crankshaft or other problems that can slow down the starter may prevent the engine from starting. Check for wire, rope or string wrapped around engine crankshaft and blade. Do not try to start engine with mower in tall grass. Remove the spark plug and see if the engine turns freely with the plug removed.

ENGINE HARD TO START

The same conditions that cause an engine not to start can cause and engine to be hard to start.

Be sure the engine controls are set correctly and all safety switches are operating properly on models so equipped. On models with manual controls, it can be very difficult to start the engine with the controls set on STOP or IDLE. Some mower controls manually operate the choke and the engine may be hard to start if the choke is not closed on a cold engine.

The fuel tank should be filled with clean, fresh gasoline. Dirt, water or anything else in the fuel tank can make the engine hard to start. On four-stroke engines, make sure the fuel is not a mixture of gasoline and oil that should be used in a two-stroke engine.

On some models the cutting blade is used as part of the flywheel weight. If the engine tries to start but backfires or just does not quite start, check for a loose blade (Fig. 49). Check blade mounting and tighten if necessary to prevent the blade from slipping easily. Some engines will not start if the blade is removed or is loose.

Inspect for a sheared flywheel key (K – Fig. 48) or excessive wear in crankshaft or flywheel groove.

The starter must be able to turn the engine fast enough to start. Any damage to engine or starter that causes slow turning may result in hard starting.

ENGINE PERFORMANCE IS ROUGH AND/OR LOW ON POWER

Lack of power without apparent miss may be caused by incorrect carburetor adjustment, incorrect governor adjustment, improper ignition timing, plugged air cleaner, plugged muffler or malfunctioning carburetor or governor.

An intermittent miss may be caused by malfunctioning ignition system or fuel system, but the presence of a miss rules out the governor or ignition timing as a primary factor.

The most common cause of lack of power is actually not a loss of power at all; that is, a dull blade. More power is required to beat the grass with a dull blade than is required to cut the grass with a sharp blade.

Fuel System

Fuel Tank Almost Empty. The engine will run poorly, just before the tank is out of gas. Be sure that the fuel tank is full. Other problems can cause the engine to be affected the same as an empty gas tank. Fuel passages or orifices in the carburetor may be clogged, the carburetor may be loose, the adjustment needle may be turned in too far, etc.

Fig. 45—Sometimes it is best to remove the air cleaner and check to be sure the choke valve is closed during starting and open when engine is running.

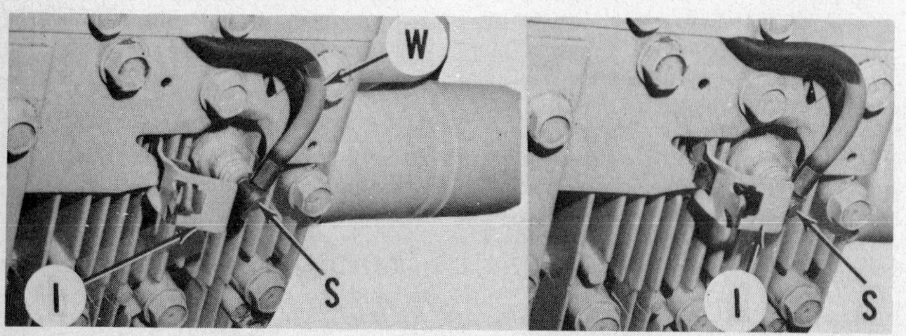

Fig. 44—Stop switch (I) can be against spark plug lead or high tension lead (W) can be shorted out. Either condition can cause engine not to start.

Fig. 46—The air cleaner screw (1) must be installed on some models before running engine. Never leave air cleaner off except during temporary check.

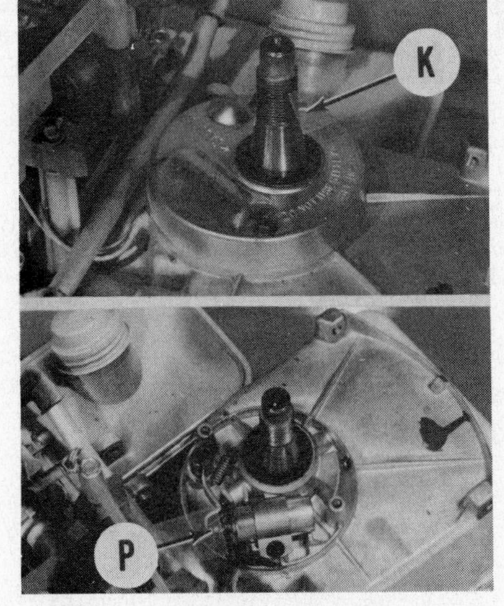

Fig. 48—View of flywheel key (K) and ignition breaker points (P) with flywheel removed.

Fig. 50—Grass and dirt (B) packed between cooling fins will cause engine to overheat.

Choke May Not Be Open Completely. The choke may be stuck, binding or incorrectly adjusted. This over rich (too much gasoline) condition is usually accompanied by gasoline smell and black exhaust smoke.

Air Cleaner Plugged. A clogged air cleaner will also cause the engine to draw in too much gasoline. The air cleaner can be checked by temporarily removing the unit, however, unit **must** be reinstalled or renewed before actual use of the mower. **An engine operating without an air cleaner can be worn out in a very short time under dusty conditions.**

Contaminated Fuel. A frequent cause of intermittent miss or loss of power due to fuel is water or dirt in the system or a plugged fuel strainer. Disconnect fuel line and check for a full flow of clear fuel when shutoff valve is opened. If water or dirt is found (Fig. 42), discard the fuel and remove and clean tank and lines as well as carburetor.

Ignition System

Spark Plug Dirty or Loose. Check the condition of the spark plug, clean old plug or install new plug. The quickest way to check out the ignition system is to use one of the test plugs as shown in Fig. 39 or Fig. 39A.

The presence of a strong, steady spark is not conclusive proof of satisfactory ignition performance but it is reasonable proof. An intermittent miss caused by ignition problems is usually accompanied by a weak spark.

Cutting Blade

Loose Blade, Dull Blade or Blade Installed Upside Down. The bottom leading edge (SE–Fig. 49) of the outer end of blade must be sharp and the blade must be firmly secured to the engine crankshaft. A dull blade requires more effort (Power) to cut than a sharp blade. An engine in good condition may seem to be low on power if the blade is not kept sharp. A loose blade, especially if the blade is dull, may slip more easily than it cuts.

Other Miscellaneous Causes of Low Power

Engine Overheating. Cooling fins, packed with grass (B–Fig. 50) is the usual cause of overheating. A carburetor that is damaged or has the fuel mixture incorrectly adjusted may not be delivering enough fuel to the engine. This lean fuel condition will cause high combustion temperatures which will in turn cause the engine to overheat.

Low Compression. Loose screws (S–Fig. 51) attaching cylinder head to cylinder or a damaged (blown) cylinder head gasket will permit compression to leak. Broken or worn piston rings or damaged valves may also cause leakage.

Fig. 49—View of typical blade installation. Be sure blade is tight and correctly installed.
B. Blade retaining screw
C. Clutch
LE. Lift edge of blade (up)
M. Engine mount bolts
SE. Sharp edge (down)

Fig. 51—Cylinder head screws (S) should be tightened to specified torque in proper sequence. Refer to appropriate engine service section.

TROUBLESHOOTING CHART

SYMPTOM	CAUSE	CORRECTION
Fails to Start		
No Fuel To Carburetor	No Fuel In Tank	Fill Tank
	Shut-off Valve Closed	Open Valve
	Intake Screen Plugged	Clean Screen
	Fuel Line Plugged	Clean Line
No Fuel To Cylinder	No Fuel To Carburetor	See Above
	Float Stuck	Clean Carburetor
	Inlet Needle Stuck	Clean Carburetor
	Incorrect Float Level	Adjust Float
	Choke Inoperative	Overhaul Carburetor
	Fuel Passages Clogged	Overhaul Carburetor
	No Compression (4-Stroke)	Overhaul Engine
	Crankcase Leak (2-Stroke)	Install New Seals/Gaskets
	Reed Valve Leaking (2-Stroke)	Install New Valve
	Exhaust Plugged (2-Stroke)	Clean Exhaust
Engine Flooded	Float Stuck	Clean Carburetor
	Incorrect Float Level	Adjust Float
	Overchoked	Purge Engine
	No Spark At Plug	See Below
No Spark At Plug	Plug Fouled	Clean Plug
	Incorrect Plug Gap	Adjust Gap
	No Spark At Plug Wire	See Below
No Spark At Plug Wire	Points Coated	Clean Points
	Flywheel Key Sheared	Install New Key
	Points Not Opening	Adjust Points
	Points Not Closing	Adjust Points
	Points Burned	Install New Points
	Kill Switch Shorted	Repair Switch
	Condenser Shorted	Install New Condenser
	Coil Shorted	Install New Coil
	Flywheel Magnets Coated	Clean Magnets
	Incorrect Air Gap	Adjust Gap
	CDI Module Bad	Install New Module
	Safety Switch Malfunction	Repair/Renew Switch
	Cranking Speed Slow	Repair As Needed
No Compression	Valves Stuck (4-Stroke)	Free Valves
	Valves Burned (4-Stroke)	Install New Valves
	Exhaust Plugged (2-Stroke)	Clean Exhaust
	Piston Damaged	Overhaul Engine
	Cylinder Damaged	Overhaul Engine
	Connecting Rod Damaged	Overhaul Engine
Lacks Power		
Engine Smokes	Rich Fuel Mixture	Adjust Carburetor/Controls
	Air Filter Plugged	Clean Air Filter
	Worn Piston Rings (4-Stroke)	Overhaul Engine
	Too Much Oil In Fuel Mix (2-Stroke)	Use Correct Mix Ratio
Lean Fuel Mixture	Improper Carburetor Adjustment	Adjust Carburetor
	Manifold Gasket Leaking (4-Stroke)	Install New Gasket
	Tank Vent Plugged	Clean Vent
	Tank Screen Plugged	Clean Screen
	Crankcase Leaking (2-Stroke)	Install New Seals/Gaskets
Partial Miss	Weak Spark	Overhaul Magneto
	Lean Fuel Mixture	See Above
Engine Overheats		
	Engine Dirty	Clean Engine
	Low Oil Level (4-Stroke)	Add Oil
	Wrong Oil/Gasoline Mix (2-Stroke)	Use Correct Mix
	Engine Overloaded	Reduce Load/Sharpen Blade
	Cooling Fins Missing/Broken	Install New Parts
	Shrouds Missing	Install Shrouds
	Lean Fuel Mixture	See Above

Engine Vibrates Excessively. The engine crankshaft is usually bent, such as shown in Fig. 53, but vibration can also be caused by blade out of balance, bent or loose. Loose engine mount bolts (M–Fig. 49) can cause mower to vibrate more than normal.

Camshaft Incorrectly Installed. Refer to the appropriate engine service section for correct installation and timing of camshaft and crankshaft. Timing will not change unless gears are damaged or block is disassembled.

ENGINE SURGES

If engine surges (rythmically speeds up and slows down) while running, the trouble is probably caused by incorrect carburetor adjustment, incorrect governor adjustment or malfunctioning carburetor or governor. Refer to the appropriate engine service section in this manual for specific adjustment procedure.

ENGINE STARTS THEN STOPS

If engine starts, runs a minute or two, then dies, check the fuel tank cap. An improper (unvented) cap may have been accidentally installed on the engine fuel tank. The fuel tank cap is equipped

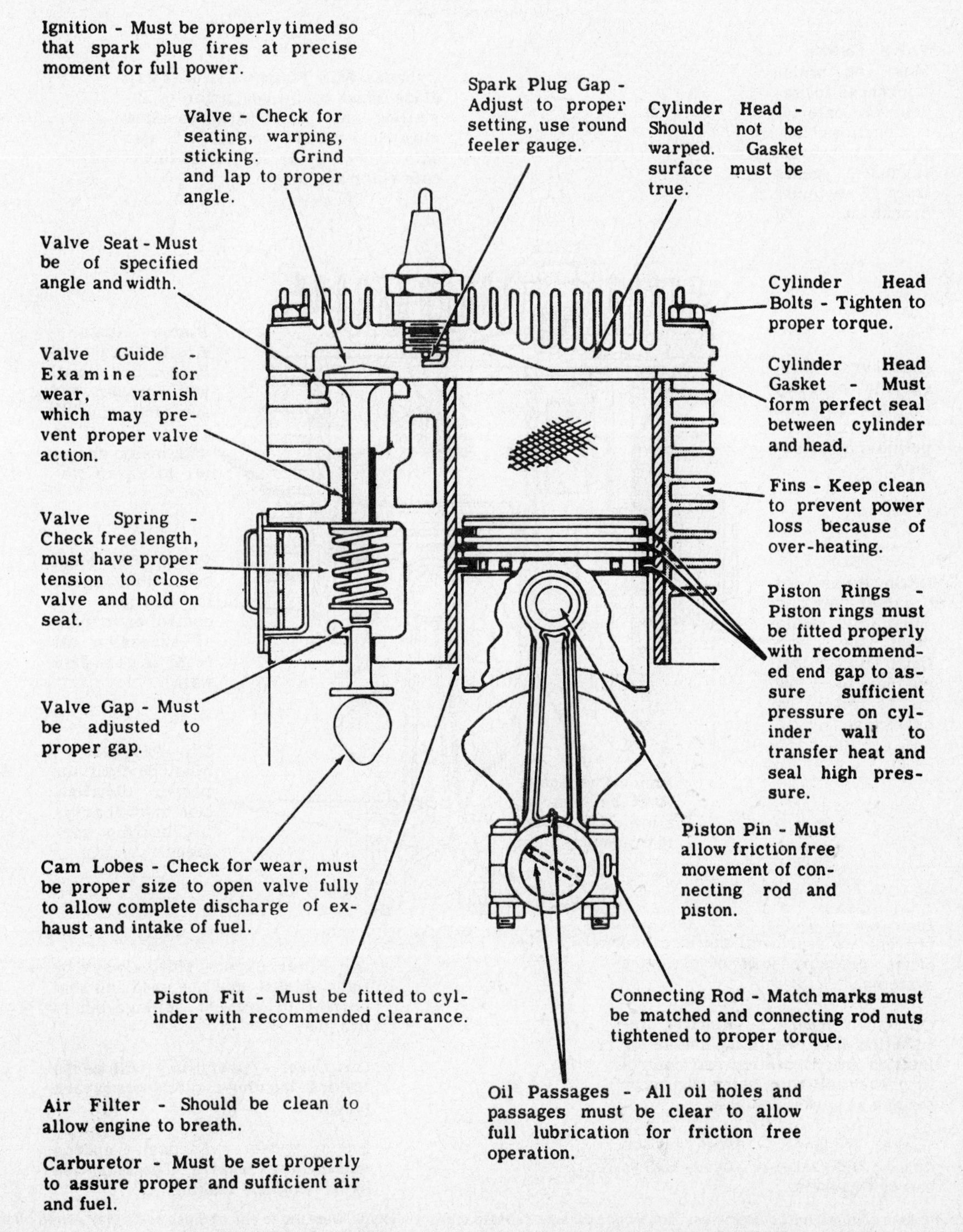

Fig. 52—Some causes of low power from four-stroke engine are indicated above.

with a baffled vent hole (V–Fig. 54) which prevents fuel from slopping out but permits air to enter as the fuel is used. If vent is not present (or is plugged), fuel will cease to flow when sufficient vacuum builds in tank.

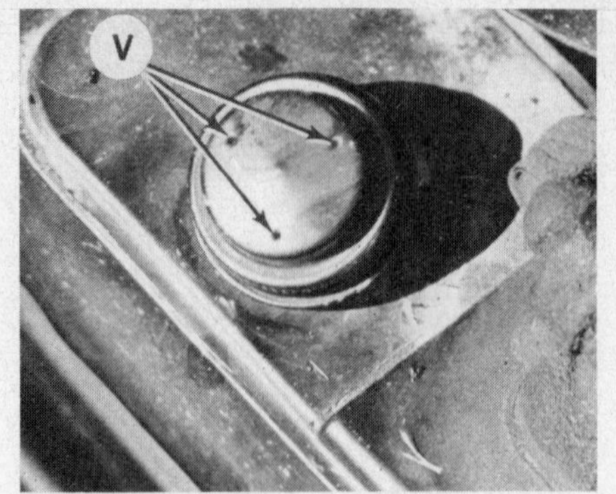

Fig. 54—Gas cap should have vent holes (V) to allow entrance of air.

ENGINE USES TOO MUCH OIL

Excessive oil consumption in a four-stroke engine can easily destroy an engine. The total amount of oil is small and loss of only a small amount will greatly increase the operating temperature of the engine. The high temperature is especially critical at some of the bearings. High temperatures may cause an engine power loss and excessively high temperatures will cause engine damage. Refer to the causes of oil consumption suggested in Fig. 53

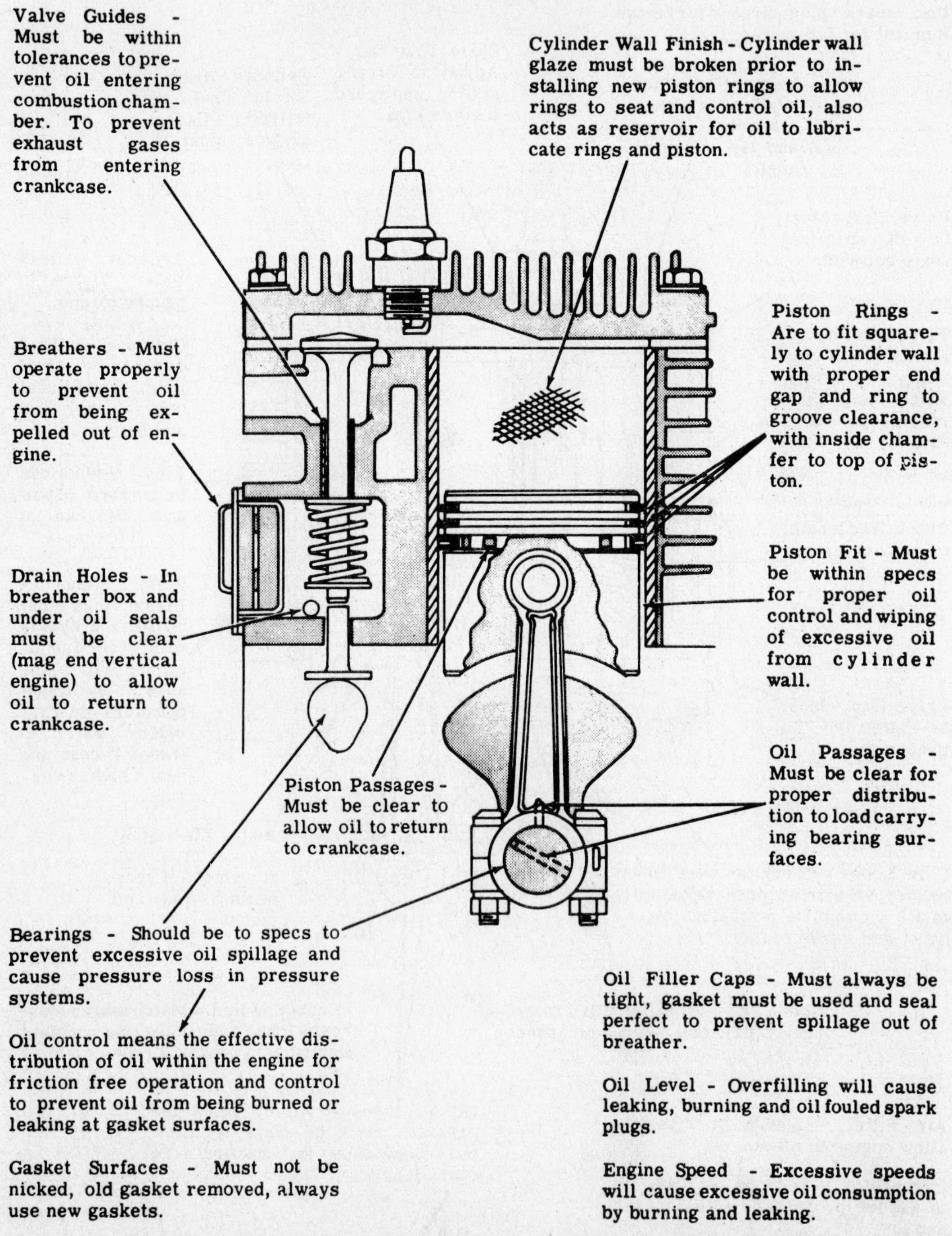

Fig. 53—In four-stroke engines, the compartments containing oil are sealed to prevent leakage. Damage to any of these seals may permit oil to enter an area of the engine where it does not belong or may permit the oil to leak out of the engine.

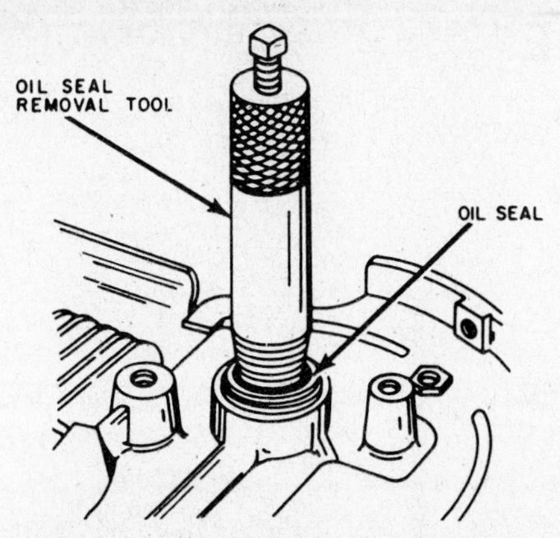

Fig. 55—View of commercially available special seal removal tool to remove oil seals. The only practical repair for a damaged oil seal is installing a new one.

Fig. 57—The purpose of disconnecting the spark plug high tension lead is to keep the engine from accidentally starting. The high voltage can sometimes jump a small gap and allow the engine to start. Firmly grounding the lead to cylinder head is much safer.

SAFETY

Personal safety for yourself, those around you or those who will operate equipment in the future should be the number one concern when using or servicing power equipment of any type. Carefully think out every step before performing service work or using power equipment.

There is no substitute for common sense, caution and forethought when working on or using power equipment. The following list of safety suggestions is provided as an aid to develop your own personal safety code to follow when working on or using power equipment.

1. ALWAYS read operator's manual and be thoroughly familiar with all controls and safety features.

2. NEVER remove, alter or disconnect any safety equipment, switches, warning or instructional decals. Never operate, or allow another person to operate, equipment which has had safety features removed, altered or disconnected.

3. NEVER leave mower unattended while engine is running.

4. NEVER alter engine governor or speed control to increase engine rpm above manufacturer's specifications.

5. ALWAYS wait until all engine or blade motion has stopped before refueling, removing or installing grasscatchers, guards or optional equipment.

6. ALWAYS properly ground spark plug wire before performing any maintenance on mower.

7. NEVER operate engine in an enclosed room or where ventilation is not adequate.

8. ALWAYS wear appropriate clothing, shoes and eye protection when servicing or operating mower.

9. ALWAYS store fuel in approved containers which are clearly marked as to contents. Always use caution when handling fuel.

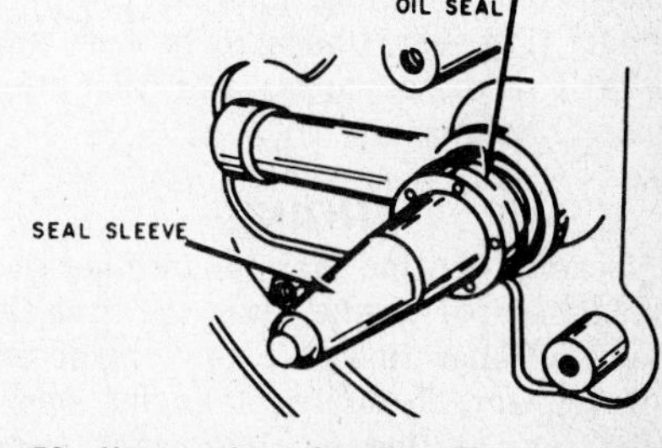

Fig. 58—Use caution when installing new seal. Sharp edges of crankshaft can cut a new seal while installing. Use a sleeve as shown or tape to protect seal from sharp edges of shaft while installing.

10. NEVER allow irresponsible persons or people not instructed in proper safety and equipment use to operate equipment.

The list of safety suggestions is limitless. Remember, YOU as a service person or as the end user of power equipment are responsible for personal safety and the safety of those around you.

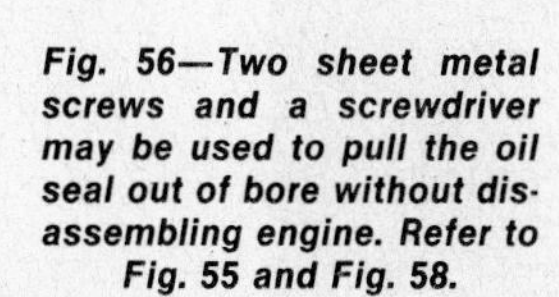

Fig. 56—Two sheet metal screws and a screwdriver may be used to pull the oil seal out of bore without disassembling engine. Refer to Fig. 55 and Fig. 58.

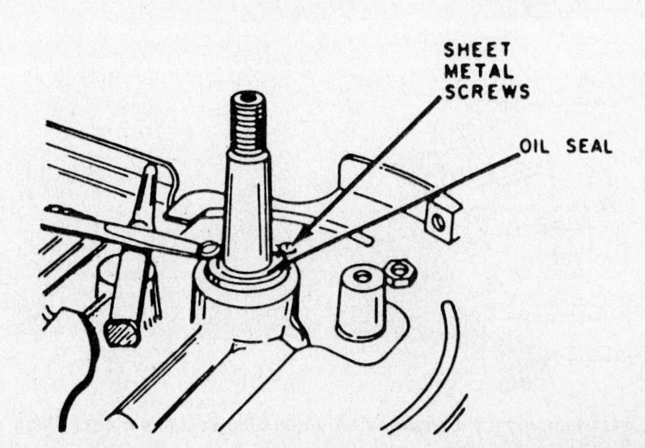

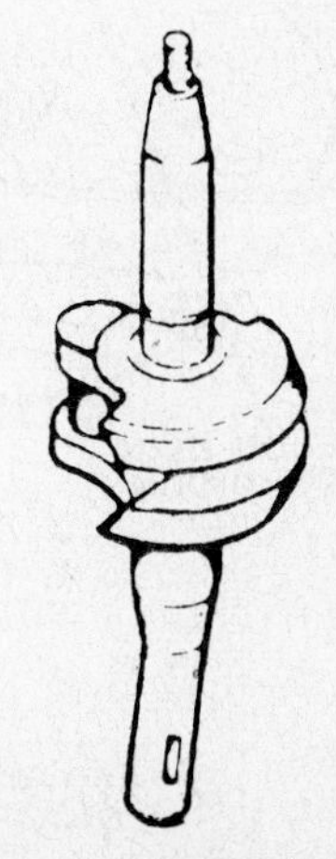

Fig. 59—A bent crankshaft must be renewed to ensure correct and safe operation of engine.

ROUTINE MAINTENANCE

Proper operation and care can prevent damage that would need to be repaired. Lawn mowers were designed to cut grass; not steel, iron, wire, rocks, concrete, rope, etc. Improper use can cause damage, even if accidentally done. Check the mower regularly (but especially following any misuse) and keep the mower clean.

Clean and inspect the complete mower for general condition. It is nearly impossible to inspect a dirty mower (Fig. 67). Lubricate the axles at the center of the wheels. Some wheels are permanently lubricated and some have grease fittings, but many are lubricated by oiling axle at center of wheel (Fig. 68). Check to be sure that wheels are not loose and that they track straight.

ENGINE

Several routine maintenance services should be performed after the first two hours of operation of a new engine and after every 25 hours of engine operation or at least once each year thereafter. This service interval should be considered minimal. Never attempt to extend this time. Small lawns which require fifteen minutes to mow may not accumulate 25 hours on the mower but service should be accomplished at least once every year. Mowers that are operated in dusty or dirty conditions should be serviced more often than every 25 hours of running. This routine service at 25 hour intervals should also be supplemented by normal checks

Fig. 62—View of two oil containers with typical markings indicating the "SAE" viscosity and "API SERVICE" rating.

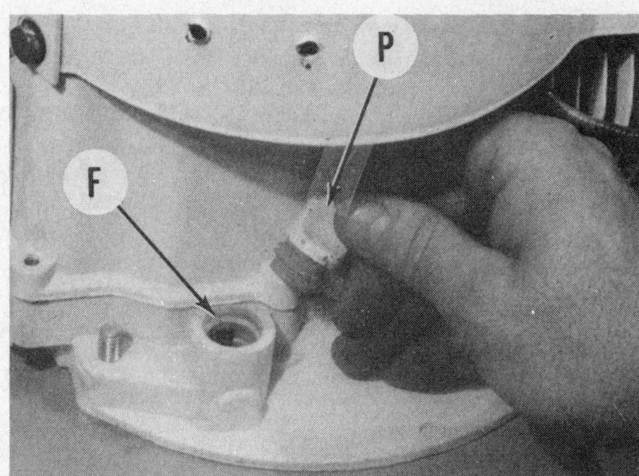

Fig. 63—View of typical filler opening (F) with plug (P) removed.

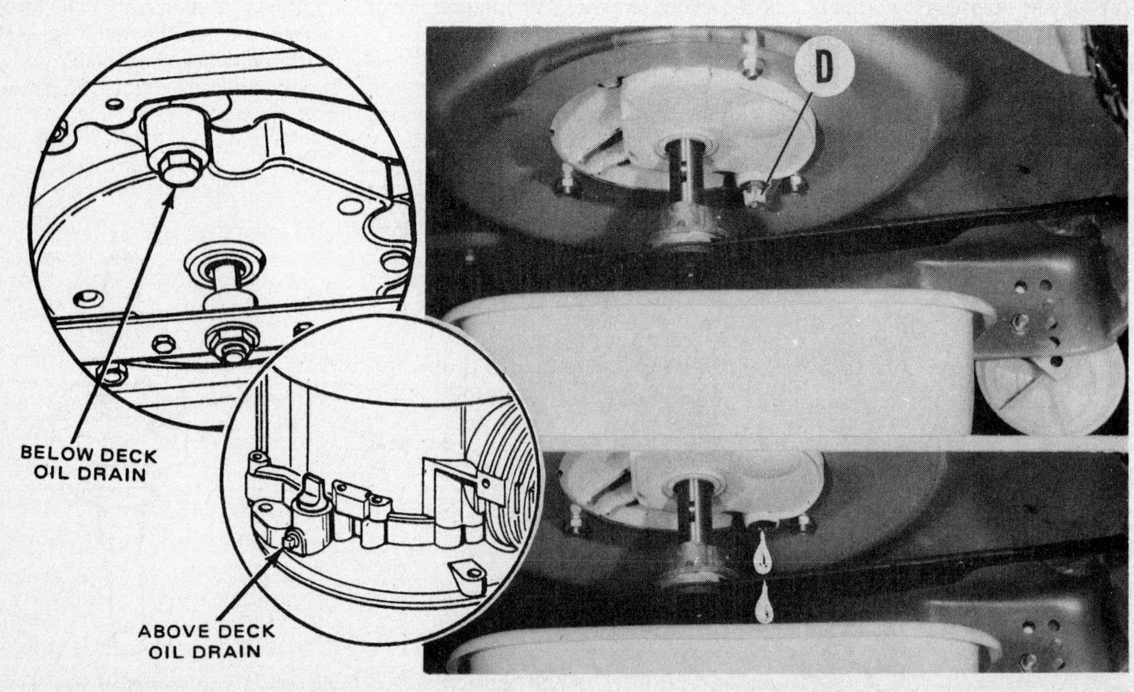

Fig. 61—Oil is drained by removing plug (D). Be sure to retighten plug securely before filling with new oil.

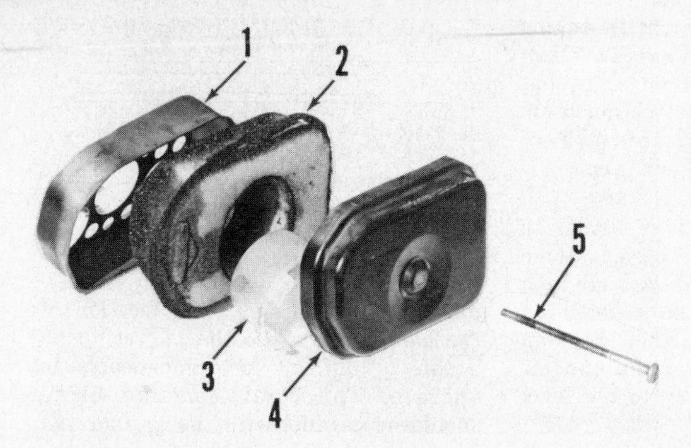

Fig. 64—View showing a typical foam type air cleaner. Clean all parts after 25 hours of normal operation. Clean more often in dusty conditions.

1. Housing
2. Foam element
3. Plastic insert
4. Cover
5. Screw

before starting and while cleaning before storing.

At least every 25 hours of operation, change the engine oil and clean the air filter.

Drain oil after running engine and while the oil is still warm. Disconnect and ground the wire from the spark plug to prevent accidental starting. Locate mower on level surface above a container suitable for containing old oil, then remove plug.

NOTE: The oil drain plug is above the deck on some models and below deck on other models.

Oil is necessary for both lubrication and cooling. Running the engine with not enough oil will quickly cause overheating. The heat will, of course, result in damage to internal engine parts. Reinstall and tighten the drain plug sufficiently to prevent it from working loose. Fill engine crankcase through filler opening with a good quality oil marked SAE 30, API SERVICE MS, SC, SD or SE. Capacity of many engines is only 1¼ pints (0.6L) which is slightly less than three-quarters of one quart. Check oil level after filling.

Fig. 67—Dirty mowers may be impossible to inspect. Long accumulations are difficult to remove.

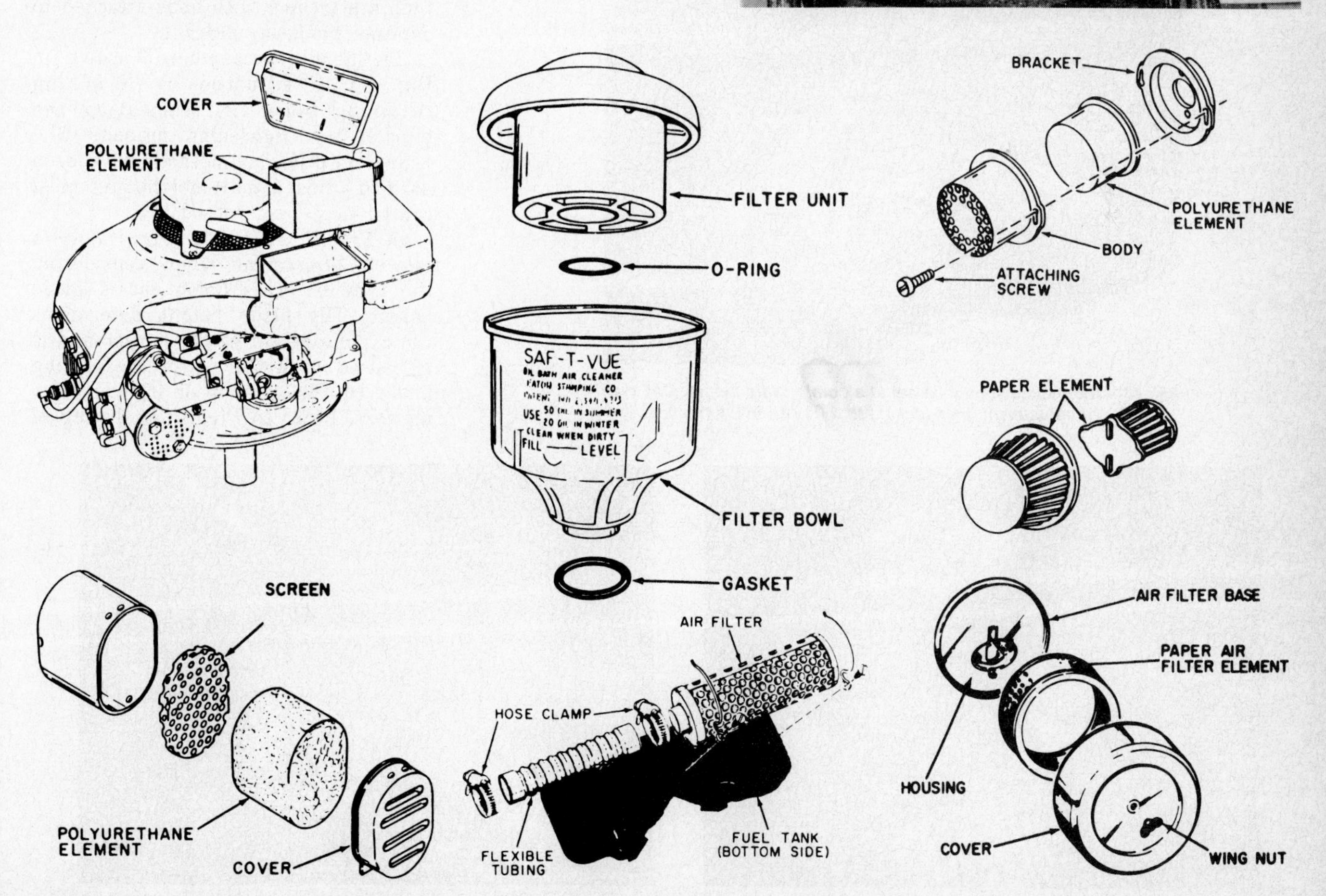

Fig. 65—View showing typical air cleaners. Clean all parts every 25 hours of operation. Paper filter elements can be blown out with compressed air from the inside, but should not be washed. Polyurethane elements can be washed in detergent or kerosene, then dried.

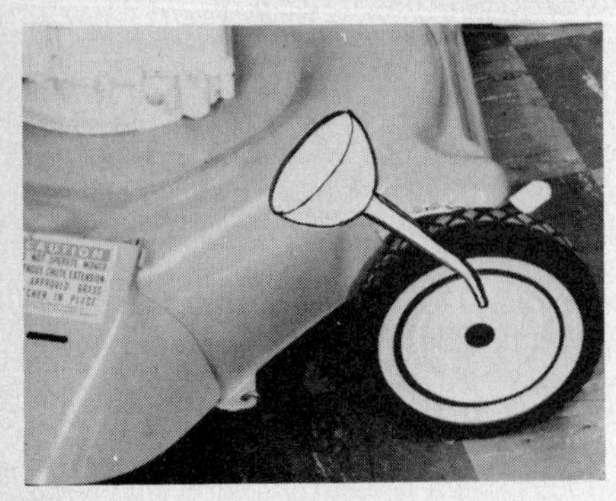

Fig. 68—The wheels of most mowers can be kept lubricated with oil. Some are equipped wtih grease fitting.

Do not overfill, but be sure that oil is to "FULL" mark on dipstick or to top of fill opening or low filler opening. Install fill plug. Reconnect high tension wire to spark plug after all other service is completed.

Remove any covers necessary, then remove the air cleaner assembly from the engine. Clean the container and element thoroughly. Wash all parts except paper element in detergent or kerosene and dry completely. Saturate foam (polyurethane) element in engine oil, then squeeze out **all** excess. Foam should have light tint of oil color, but dark appearance indicates too much oil. The foam contains a labyrinth of passages through which air passes. If these passages become clogged with dirt or with oil, the result is similar to operating with the choke closed. Some mowers may be equipped with oil bath air cleaners which can be cleaned by removing, disassembling, then cleaning with a suitable solvent. Refill the air cleaner cup with engine oil to the level indicated on the side.

Run engine until the fuel tank is completely out of gas before storing the mower for the winter. Any gasoline left in the tank during the off season will begin to evaporate and leave a residue. Some chemical products are available which can be added to the gasoline before storage, but these will not be necessary if the tank is empty.

ROTARY MOWERS

Some adjustments must be correctly set in order to assure proper performance of the mower. The location of the handle can cause the mower to be unsafe or difficult for some people to operate. The rotary cutting blade should be parallel with the ground and the blade should cut the grass at the correct height.

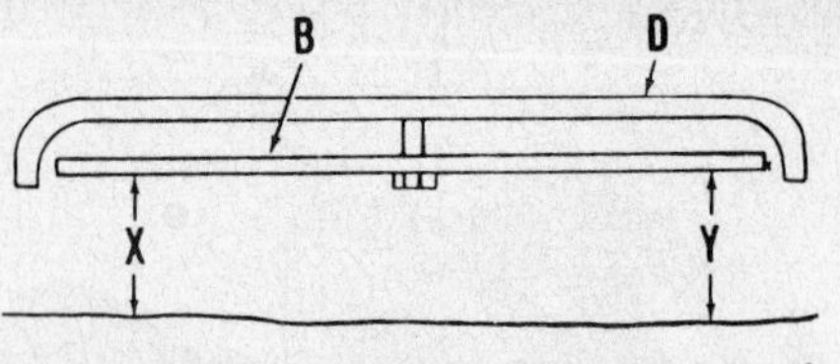

Fig. 72—Blade height (X) should be set correctly for type of grass being cut. Blade is at (B) and mower deck at (D).

HANDLE POSITION. Handle position is adjustable on most lawn mowers for operator comfort and safety. The mower handle is attached to a bracket on the mower deck. Notice in Fig. 70 that a retaining lug (L) secures the handle in the bracket. The handle height is changed by locating lug (L) in different hole (H).

Height of the handle of some mowers is adjusted with offset holes drilled at attaching ends of handle or lugs are installed off center. Handle height can be changed only by removing handle, then reattaching with ends attached to opposite mounting brackets.

The handle arrangement shown in Fig. 71 is adjusted by loosening retaining bolts (B), relocating the bracket, then tightening the bolts (B).

Several different methods have been used to adjust handle height, but most can be easily understood.

CUTTING HEIGHT. A critical lawn mower adjustment when considering the type of grass being cut is blade height. The blade height determines the cutting height of the mower and is measured from the tip of the cutting blade to the ground on rotary lawn mowers. Refer to Fig. 72. A listing of

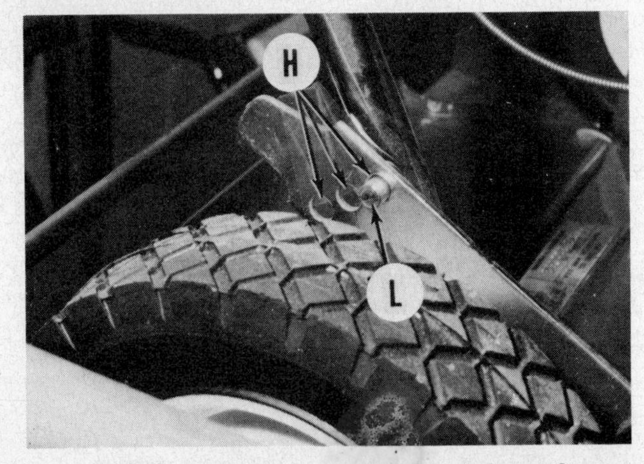

Fig. 70—The handle should be set at a comfortable height. One type of adjustment uses lugs (L) that fit into holes (H).

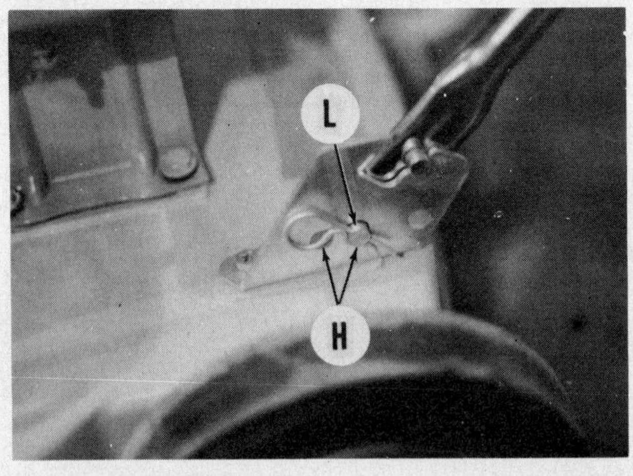

Fig. 71—Handlebar mounting brackets are sometimes slotted to permit adjustment of handlebar height.

Fig. 73—Quick blade height adjustment is often accomplished by locating wheel adjustment levers (L) in notches (N).

Fig. 75—The purpose of disconnecting the spark plug high tension lead is to keep the engine from accidentally starting. The high voltage can sometimes jump a small gap shown at right allowing the engine to start. Firmly grounding the lead to cylinder head is much safer.

recommended cuts for some popular grasses is as follows:

Cutting Height	Grass Type
½ to 1 inch	Bermuda, Zoysia
1 to 2 inches	Red fescue, Meadow fescue, Kentucky bluegrass Perennial ryegrass
1½ to 3 inches	Bahia grass, Tall fescue, St. Augustine

Note that the preceding list of grasses is only a general guide to some of the commonly grown grasses in home lawns. Special strains of these grasses may require a different cutting height than shown.

Adjustment is usually accompanied by moving the wheels. Some mowers have wheels that are quickly adjusted by moving the lever (L—Fig. 73) from one notch (N) to another. Others provide several holes (H—Fig. 74) for mounting wheels at different heights. The blade will usually be parallel with the ground if all of the wheels are adjusted alike.

BLADE INSPECTION. Blade inspection should first included operation of mower. Mower vibration may be caused by a bent or unbalanced blade. Unbalanced blades can be balanced, but a bent blade must be replaced. Notice if mower cuts poorly or irregularly which indicates blade damage. Check for poor grass discharge which would indicate lift on blade is not functioning properly.

CAUTION: Ignition wire to spark plug should be disconnected and grounded before performing any inspection or servicing of the blade or underside of mower.

Raise the mower and check the blade for firm mounting. Several different methods for mounting the blade have been used, but none should permit the blade to slip when checking by hand.

Remove blade from mower and clean thoroughly with a wire brush. Inspect cutting and lift areas of blade. Damaged cutting edges can usually be sharpened following procedure in SHARPENING section. Lift areas must not be mis-shaped or broken. Damaged lift areas affect cutting and blade's ability to discharge grass. Check blade straightness. Inspect blade mounting hole or holes. Holes should not be deformed as blade can be installed off-center and cause vibration. Inspect

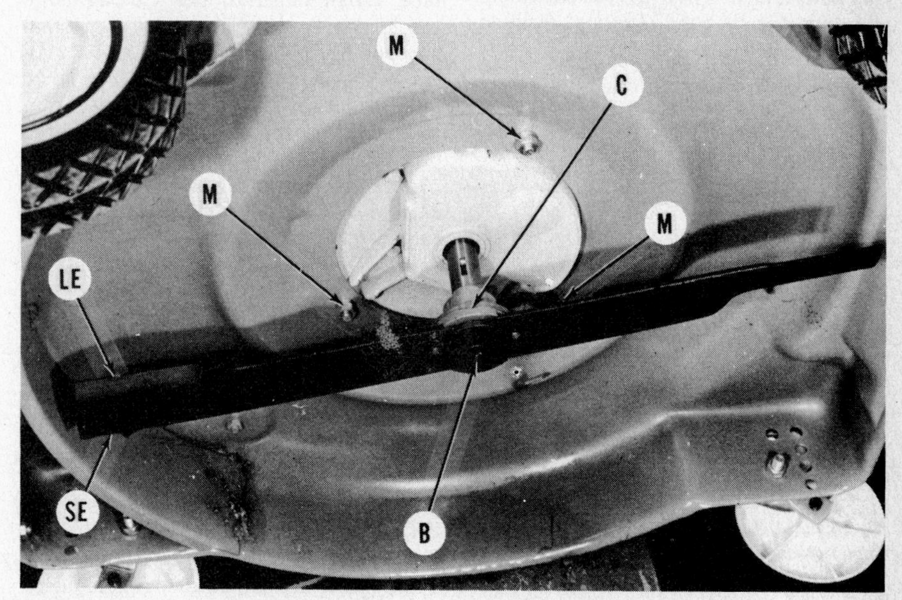

Fig. 76—The cutting blade must be installed securely and correctly.

B. Blade attaching screw
C. Blade clutch
LE. Lift edge of blade (up)
M. Engine mounting bolts
SE. Sharp cutting edge (down)

Fig. 74—View showing deck with several wheel attachment holes (H) for adjusting blade height.

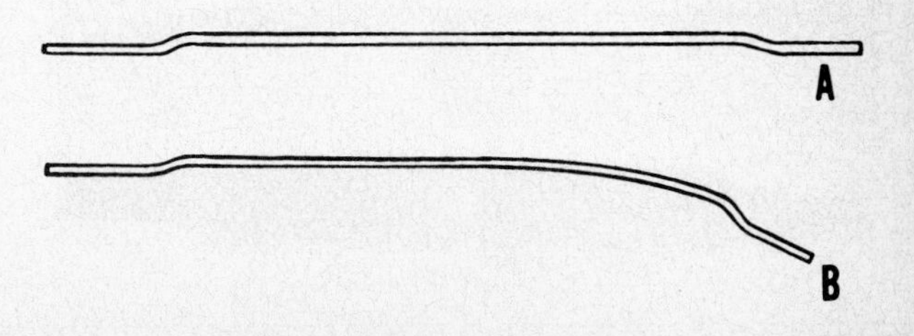

Fig. 77—Be sure blade is straight and not bent. The blade may be shaped as shown in "A" but obviously bent blades "B" should be discarded.

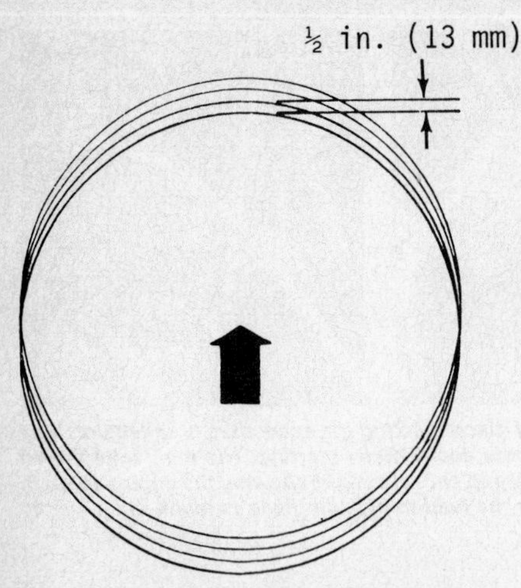

Fig. 78—At normal walking speeds only the outside edge of a sharp blade is used to cut the grass. A dull blade may never actually cut the grass but only beat it throughout the length of the blade.

retaining bolts or pins. Worn bolts or pins must be replaced with new bolts or pins. Inspect mounting holes in blade and install new blade if holes are worn or damaged.

Visually inspect blade for cracks around mounting hole (or holes) and at the ends. Hang blade by mounting hole and strike blade with another metal object. A solid blade free of cracks will ring like a bell while a cracked blade will emit a dull sound. Discard cracked blades.

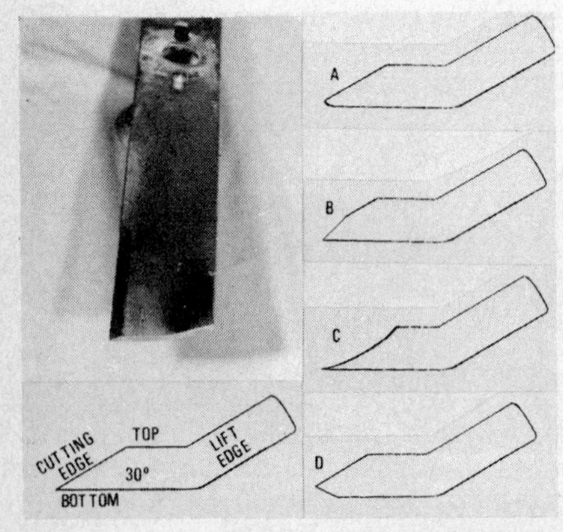

Fig. 79—Views of typical blade. View "A" shows typical cross section of dull blade. View "B" is incorrect because of sharpening at too much angle. View "C" is sharpened at less than 30° angle. View "D" shows incorrect method of sharpening caused by grinding lower edge of blade.

SHARPENING. Blade sharpening can be accomplished using a file or grinder. Satisfactory results are obtained using a file, but much more time is required as opposed to using a power tool, especially if the blade is nicked severely. Blade should be removed from mower to properly sharpen blade.

A blade needs to be sharp only at the outer end. A 21-inch blade rotating at 3000 rpm will cut grass with only the outer ½-inch of each blade end when the mower is traveling at normal walking speed. Refer to Fig. 78. A mower with a 21-inch blade rotating at 3000 rpm would have to travel at 18 miles-per-hour if the cutting edges are not to overlap. It is not necessary, therefore, to sharpen more than the outer three inches of the cutting edge.

Rotary mower blades are usually sharpened to an angle of 30 degrees as shown in Fig. 79. This is an ideal situation which cannot always be attained on a well used or nicked blade without a great deal of sharpening. Blades wear on the underside which forms a curve into the cutting edge as shown at (A). If the blade is not worn excessively, the desired 30° edge angle and flat underside can be obtained without excessive grinding. If the blade is well worn, be sure to face off the underside as well as sharpening the top face. Small nicks to the cutting edge can be removed in the course of normal sharpening, but deep indentations may cause sharpening problems. Most large gouges do not affect cutting operation as they are located towards the center of the blade and not in the outer ½-inch of the blade which does the cutting. A great deal of grinding is required to remove the gouge as well as grind down the opposite cutting edge to make the two edges symmetrical. Satisfactory cutting performance can usually be obtained by smoothing and removing burrs from the gouge and then sharpening remainder of cutting edge. The blade will cut with no discernible difference and the size of the gouge will be reduced after each sharpening until it is removed.

Cutting edges should be sharpened to be as symmetrical as possible to aid blade balancing. Cutting edge should be

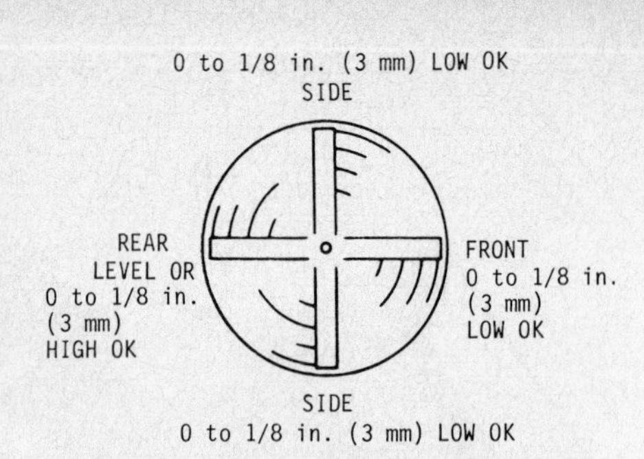

Fig. 81—The blade should be nearly level with the ground. A blade that is 1/8 inch (3 mm) higher in rear is permissible, but blade should never be lower in rear than in front.

sharpened to retain original shape. After sharpening, blade must be balanced as outlined in BALANCING section.

BALANCING. An unbalanced blade can cause severe vibration problems. Blades which are out of balance by only an ounce at one end will create objectionable vibration in a mower.

The most popular blade balancer is the cone type shown in Fig. 80. Before balancing blade, be sure blade is clean and properly sharpened. Place blade on cone with cone located in the center mounting hole of the blade. Blade should be balanced in all directions; end to end and front to back. Heavy parts of blade will be lower than remainder of blade and the heavy part must have metal removed to balance blade. Resharpen cutting edges if necessary for balance. Do not grind away lift edges.

After completing balancing operation, mount blade on mower and tighten mounting bolts. Measure distance from end of mower blade to mower deck, then rotate blade and measure distance from same point on mower deck to other end of blade. For correct dynamic blade balance, distances should be as close as possible.

BLADE TRACKING AND HEIGHT. The rotary blade should cut on a plane parallel to the level mower deck. If the blade or engine crankshaft is bent, the blade will cut unevenly. To check blade track, measure the distance from the end of the blade to the bottom of the mower deck. Rotate the blade and measure the distance from the same point on the mower deck to the other end of the blade. Difference in distances should not vary more than 1/16-inch on mowers with straight crankshafts. Check blade for straightness and replace if bent. Blade run-out on a mower with an engine crankshaft which was bent and straightened may exceed 1/16-inch and cutting perfor-

Fig. 80—View of a commonly used cone type balancer. Alignment of the two parts of cone indicate balanced blade. The unit shown is available from Frederick Mfg. Co., Inc., 1400 Agnes St., Kansas City, Mo. 64127.

mance is the only criteria which may be used.

Blade height refers to the position of the blade in the mower deck. The rotary mower blade must always operate within the mower deck and never extend below the deck. Blade height should be the same within 1/8-inch (Fig. 81) at all points around the mower deck. This can be measured from the mower deck to the blade end. Blade height is adjusted by placing washers of required thickness between the engine mounting bosses and the mower deck. Engine mount bolts must be removed to install or remove washers.

REEL MOWER

The cutting blades (1—Fig. 85) must be kept sharp and the bed knife (2), sometimes called cutter bar, should be properly adjusted. When properly adjusted, the reel should rotate easily with light finger pressure and the reel blades should just touch the bed knife, throughout their entire length. On some models, the reel is moved to adjust the cutting surface, on other mowers the bed knife can be moved to just contact the reel blades. Accurate adjustments is impossible if the mower frame is loose. Be sure frame is firmly assembled and all screws are tight before attempting to adjust reel to bed knife. If adjusted too tight, the cutting surfaces of blades and bed knife will be quickly worn and damaged.

Accurate sharpening requires special equipment to assure that each blade on the reel is equidistant from reel axis throughout the entire length of the blade and that the cutting edge is slightly higher than the trailing edge. The cutting edge of the bed knife will be rippled if operated with reel adjusted too tight.

Check the cutting surface of each blade and the bed knife. The reel should turn easily by hand with the blades lightly contacting the bed knife. Small nicks can be removed by carefully using a fine file or stone, but special equipment is necessary for most accurate sharpening. Be sure that drive chains, belts and clutches are correctly adjusted. Check all assembly and mounting screws to be sure that none are missing or loose. Correct alignment and blade to bed knife adjustment necessitate rigid assembly of the mower frame.

Nicks and other damage to the cutting surfaces of the blades and bed knife can be caused by improper adjustment or some other cause, but are most often the result of trying to mow wire, rocks, sticks, etc. Much damage can be prevented by thoroughly checking the lawn area before mowing and removing debris such as rocks, wire, etc. before beginning to mow.

Lubricate mower with grease injected into fittings approximately every 16 hours of operation. Chains, wheels and other bushings, not lubricated by grease should be lubricated by coating with SAE 20 oil every eight hours of operation. All suggested lubrication intervals should be reduced if operated in dusty or dirty conditions and should never be left unattended longer than one year even if operated less than recommended times. Dirt accelerates wear and unit should be kept clean at all times. Be especially careful not to allow any oil on "V" belt drives or clutches.

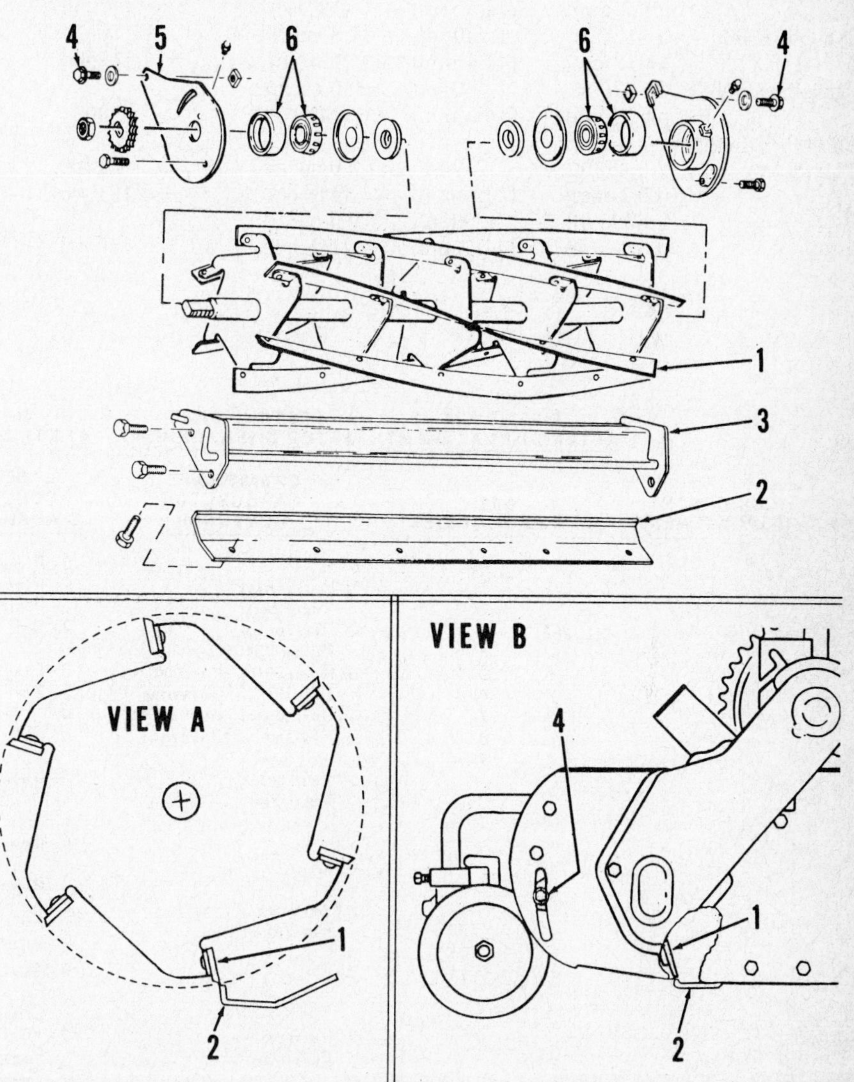

Fig. 85—Views of grass cutting parts of typical reel mower. Adjustment of the mower shown is accomplished by moving the reel after loosening screws (4) so that blades (1) contact bed knife (2) throughout entire length. View "A" shows angle of bed knife and blade cutting surfaces. Partially cut away view "B" shows adjustment of a typical King O'Lawn reel type mower.

1. Reel blade
2. Bed knife
3. Bed bar
4. Adjustment screw
5. Reel bearing holder
6. Reel bearings

BRIGGS & STRATTON

Model	Bore	Stroke	Displacement
6B, Early 60000	2.3125 in. (58.7 mm)	1.500 in. (38.1 mm)	6.3 cu. in. (103 cc)
Late 60000	2.3750 in. (60.3 mm)	1.500 in. (38.1 mm)	6.7 cu. in. (109 cc)
8B, 80000	2.3750 in. (60.3 mm)	1.750 in. (44.5 mm)	7.8 cu. in. (127 cc)
81000, 82000	2.3125 in. (60.3 mm)	1.750 in. (44.5 mm)	7.8 cu. in. (127 cc)
92000, 93500	2.5625 in. (65.1 mm)	1.750 in. (44.5 mm)	9.0 cu. in. (148 cc)
94000, 94500, 94900	2.5625 in. (65.1 mm)	1.750 in. (44.5 mm)	9.0 cu. in. (148 cc)
95500	2.5625 in. (65.1 mm)	1.750 in. (44.5 mm)	9.0 cu. in. (148 cc)

Model	Bore	Stroke	Displacement
100000	2.5000 in. (63.1 mm)	2.125 in. (54.0 mm)	10.4 cu. in. (169 cc)
110900, 111200, 111900	2.7813 in. (70.6 mm)	1.875 in. (47.6 mm)	11.4 cu. in. (186 cc)
112200	2.7813 in. (70.6 mm)	1.875 in. (47.6 mm)	11.4 cu. in. (186 cc)
113900	2.7813 in. (70.6 mm)	1.875 in. (47.6 mm)	11.4 cu. in. (186 cc)
13000	2.5625 in. (65.1 mm)	2.438 in. (60.9 mm)	12.6 cu .in. (203 cc)
131400, 131900	2.5625 in. (65.1 mm)	2.438 in. (60.9 mm)	12.6 cu. in. (203 cc)
140000	2.7500 in. (69.9 mm)	2.375 in. (60.3 mm)	14.1 cu. in. (231 cc)

CUBIC INCH DISPLACEMENT

6, 8, 9, 10, 11, 13, 14, 17, 19, 20, 23, 24, 25, 30, 32

FIRST DIGIT AFTER DISPLACEMENT BASIC DESIGN SERIES	SECOND DIGIT AFTER DISPLACEMENT CRANKSHAFT, CARBURETOR GOVERNOR	THIRD DIGIT AFTER DISPLACEMENT BEARINGS, REDUCTION GEARS & AUXILIARY DRIVES	FOURTH DIGIT AFTER DISPLACEMENT TYPE OF STARTER
0	0 -	0 - Plain Bearing.	0 - Without Starter
1	1 - Horizontal Vacu-Jet	1 - Flange Mounting Plain Bearing	1 - Rope Starter
2	2 - Horizontal Pulsa-Jet	2 - Ball Bearing	2 - Rewind Starter
3	3 - Horizontal Flo-Jet (Pneumatic Governor)	3 - Flange Mounting Ball Bearing	3 - Electric - 110 Volt, Gear Drive
4	4 - Horizontal Flo-Jet (Mechanical Governor)	4 -	4 - Elec. Starter-Generator - 12 Volt, Belt Drive
5	5 - Vertical Vacu-Jet	5 - Gear Reduction (6 to 1)	5 - Electric Starter Only - 12 Volt, Gear Drive
6	6 -	6 - Gear Reduction (6 to 1) Reverse Rotation	6 - Alternator Only *
7	7 - Vertical Flo-Jet	7 -	7 - Electric Starter, 12 Volt Gear Drive, with Alternator
8	8 -	8 - Auxiliary Drive Perpendicular to Crankshaft	8 - Vertical-pull Starter
9	9 - Vertical Pulsa-Jet	9 - Auxiliary Drive Parallel to Crankshaft	

* Digit 6 formerly used for "Wind-Up" Starter on 60000, 80000 and 92000 Series

EXAMPLES

To identify Model 100202:

10	0	2	0	2
10 Cubic Inch	Design Series 0	Horizontal Shaft - Pulsa-Jet Carburetor	Plain Bearing	Rewind Starter

Similarly, a Model 92998 is described as follows:

9	2	9	9	8
9 Cubic Inch	Design Series 2	Vertical Shaft - Pulsa-Jet Carburetor	Auxiliary Drive Parallel to Crankshaft	Vertical Pull Starter

Fig. B59—Explanation of numerical code used by Briggs & Stratton to identify engine and optional equipment from model number.

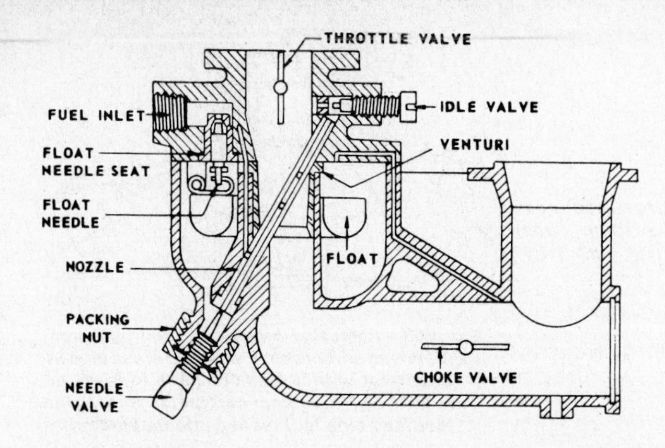

Fig. B60—Cross-sectional view of typical B&S "two-piece" carburetor. Before separating upper and lower body sections, loosen packing nut and unscrew nut and needle valve as a unit. Then, using special screwdriver, remove nozzle. Refer to text.

ENGINE IDENTIFICATION

Engines covered in this section have aluminum cylinder blocks with either plain aluminum cylinder bore or with a cast iron sleeve integrally cast into the block.

Early production of the 60000 model engine were of the same bore and stroke as the 6B model engine. The bore on 60000 engine was changed from 2.3125 inches (58.7 mm) to 2.3750 inches (60.3 mm) at serial number 5810060 on engines with plain aluminum bore, and at serial number 5810030 on engines with a cast iron sleeve.

Refer to Fig. B59 for chart explaining engine numerical code to identify engine model. Always furnish correct engine model and model number when ordering parts or service information.

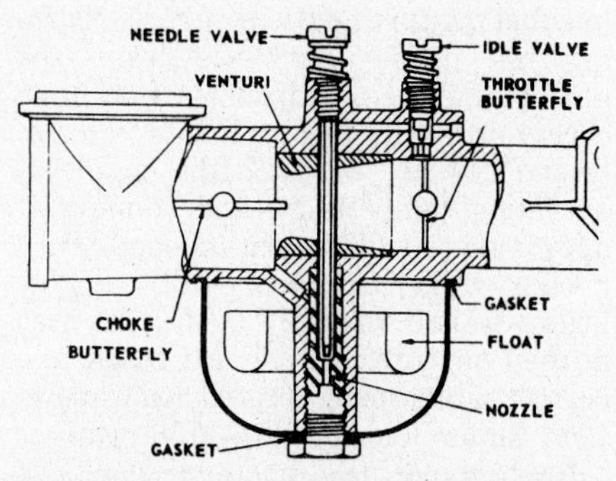

Fig. B64—Cross-sectional view of typical B&S small "one-piece" float type carburetor. Refer to Fig. B65 for disassembly views.

Fig. B61—Checking upper body of "two-piece" carburetor for warpage. Refer to text.

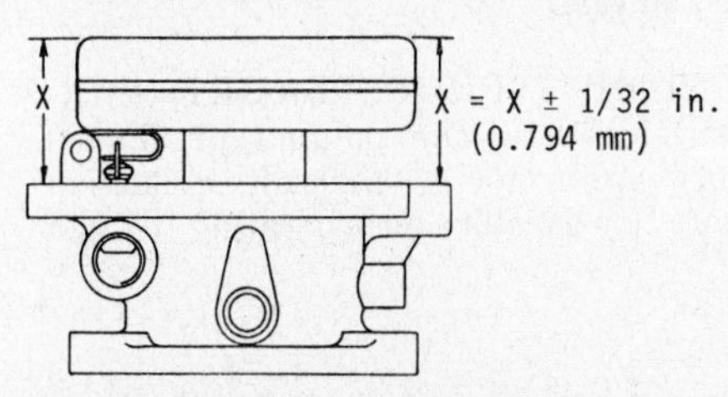

Fig. B62—Carburetor float setting should be within specifications shown. To adjust float setting, bend tang with needlenose pliers as shown in Fig. B63.

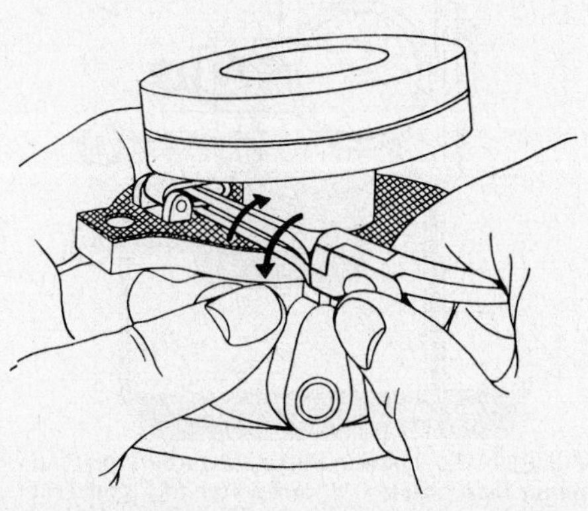

Fig. B63—Bending tang with needlenose pliers to adjust float setting. Refer to Fig. B62 for method of checking float setting.

MAINTENANCE

SPARK PLUG. Recommended spark plug for all models is a Champion J8, or equivalent. If resistor type plugs are necessary to decrease radio interference, use Champion RJ8 or equivalent. Electrode gap for all models is 0.030 inch (0.76 mm).

CAUTION: Briggs & Stratton does not recommend using abrasive blasting method to clean spark plugs as this may introduce some abrasive material into the engine which could cause extensive damage.

FLOAT TYPE (FLO-JET) CARBURETORS. Three different float type carburetors are used. They are called a "two-piece" (Fig. B60), a small "one-piece" (Fig. B64) or a large "one-piece" (Fig. B66) carburetor depending upon the type of construction.

Float type carburetors are equipped with adjusting needles for both idle and power fuel mixtures. Counterclockwise rotation of the adjusting needles richens the mixture. For initial starting adjustment, open the main needle valve (power fuel mixture) 1½ turns on the two-piece carburetor and 2½ turns on the small one-piece carburetor. Open the idle needle ½ to ¾ turn on the two-piece carburetor and 1½ turns on the small one-piece carburetor. On the large one-piece carburetor, open both needle valves 1-1/8 turns.

Make final adjustments with engine at operating temperature and running. Set the speed control for desired operating speed, turn main needle clockwise until engine misses, and then turn it counterclockwise just past the smooth operating point until the engine begins to run unevenly. Return the speed control to idle position and adjust the idle speed

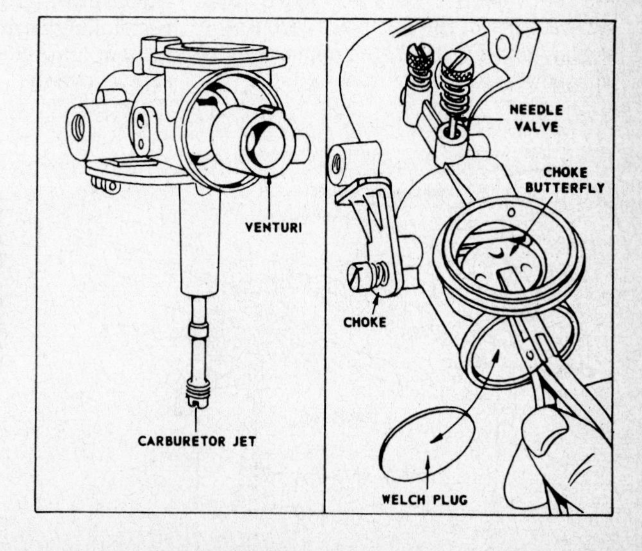

Fig. B65—Disassembling the small "one-piece" float type carburetor. Pry out welch plug, remove choke butterfly (disc), remove choke shaft and needle valve; venturi can then be removed as shown in left view.

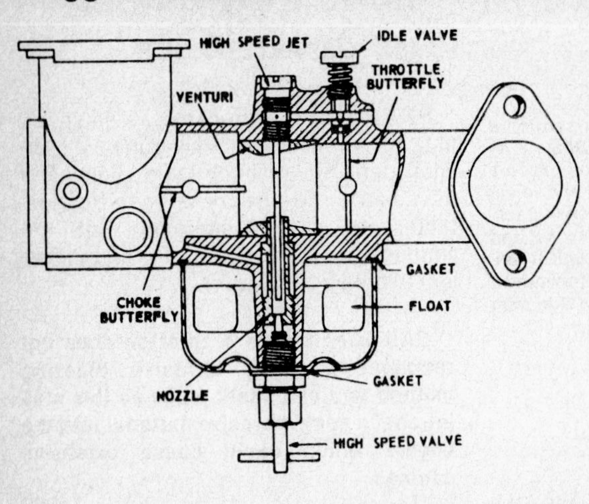

Fig. B66—Cross-sectional view of B&S large "one-piece" float type carburetor.

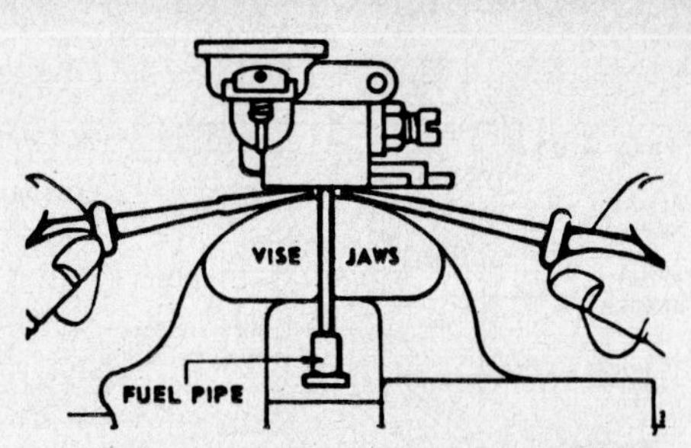

Fig. B68—Removing brass fuel feed pipe from suction type carburetor. Press new brass pipe into carburetor until it projects 2-9/32 to 2-5/16 inches (57.9-58.7 mm) from carburetor face. Nylon fuel feed pipe is threaded into carburetor.

stop screw until the engine idles at 1750 rpm. Adjust the idle needle valve until the engine runs smoothly. Reset the idle speed stop screw if necessary. The engine should then accelerate without hesitation. If engine does not accelerate properly, turn the main needle valve counterclockwise slightly to provide a richer fuel mixture.

The float setting on all float type carburetors should be within dimensions shown in Fig. B62. If not, bend the tang on float as shown in Fig. B63 to adjust float setting. If any wear is visible on the inlet valve or the inlet valve seat, install a new valve and seat assembly. On large one-piece carburetors, the renewable inlet valve seat is pressed into the carburetor body until flush with the body.

NOTE: The upper and lower bodies of the two-piece float type carburetor are locked together by the main nozzle. Refer to cross-sectional view of carburetor in Fig. B60. Before attempting to separate the upper body from the lower body, loosen packing nut and unscrew nut and needle valve. Then, using special screwdriver (B&S tool 19061 or 19062), remove nozzle.

If a 0.002 inch (0.05 mm) feeler gage can be inserted between upper and lower bodies of the two-piece carburetor as shown in Fig. B61, the upper body is warped and should be renewed.

Check the throttle shaft for wear on all float type carburetors. If 0.010 inch (0.25 mm) or more free play (shaft-to-bushing clearance) is noted, install new throttle shaft and/or throttle shaft bushings. To remove worn bushings, turn a 1/4 inch × 20 tap into bushing and pull bushing from body casting with the tap. Press new bushings into casting by using a vise and if necessary, ream bushings with a 7/32 inch drill bit.

SUCTION TYPE (VACU-JET) CARBURETORS. A typical suction type (Vacu-Jet) carburetor is shown in Fig. B67. This type carburetor has only one fuel mixture adjusting needle. Turning the needle clockwise leans the air:fuel mixture. Adjust suction type carburetors with fuel tank approximately one-half full and with the engine at operating temperature and running at approximately 3000 rpm, no-load. Turn needle valve clockwise until engine begins to run unevenly from a too rich air:fuel mixture. This should result in a correct adjustment for full load operation. Adjust idle speed to 1750 rpm.

To remove the suction type carburetor, first remove carburetor and fuel tank as an assembly, then remove carburetor from fuel tank. When reinstalling carburetor on fuel tank, use a new gasket and tighten retaining screws evenly.

The suction type carburetor has a fuel feed pipe extending into fuel tank. The pipe has a check valve to allow fuel to feed up into the carburetor but prevents fuel from flowing back into the tank. If check valve is inoperative and cleaning in alcohol or acetone will not free the check valve, renew the fuel feed pipe. If feed pipe is made of brass, remove as shown in Fig. B68. Using a vise, press new pipe into carburetor so it extends from 2-9/32 to 2-5/16 inches (57.9-58.7 mm) from carburetor body. If pipe is made of nylon (plastic), screw pipe out of carburetor body with wrench. When installing new nylon feed pipe, be careful not to overtighten.

NOTE: If soaking carburetor in cleaner for more than one-half hour, be sure to remove all nylon parts and "O" ring, if used, before placing the carburetor in cleaning solvent.

PUMP TYPE (PULSA-JET) CARBURETORS. The pump type (Pulsa-Jet) carburetor is basically a suction type carburetor incorporating a fuel

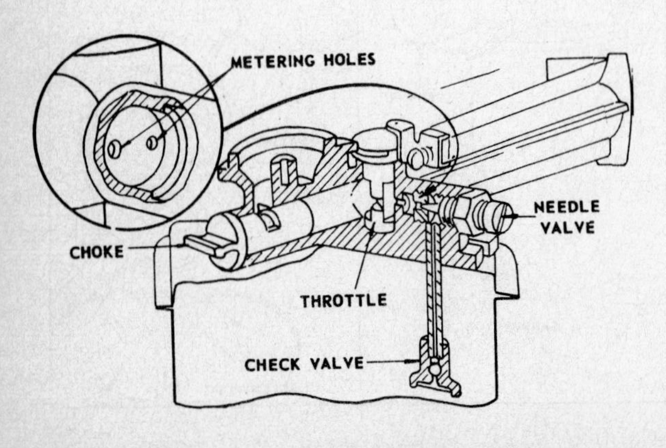

Fig. B67—Cutaway view of typical suction type (Vacu-Jet) carburetor. Inset shows fuel metering holes which are accessible for cleaning after removing needle valve. Be careful not to enlarge the holes when cleaning them.

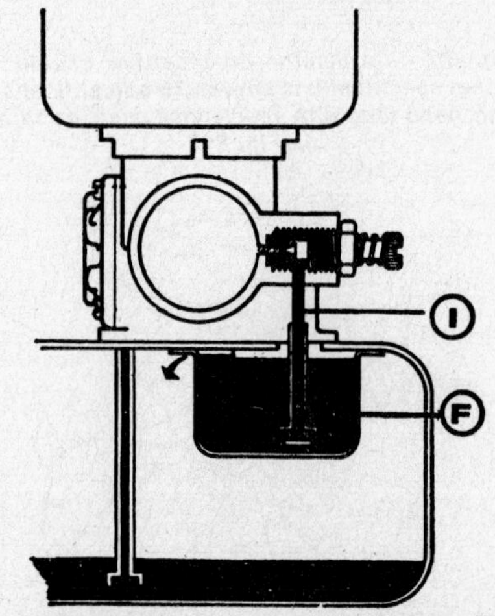

Fig. B69—Fuel flow in Pulsa-Jet carburetor. Fuel pump incorporated in carburetor fills constant level sump (F) below carburetor and excess fuel flows back into tank. Fuel is drawn from sump through inlet (I) past fuel mixture adjusting needle by vacuum in carburetor.

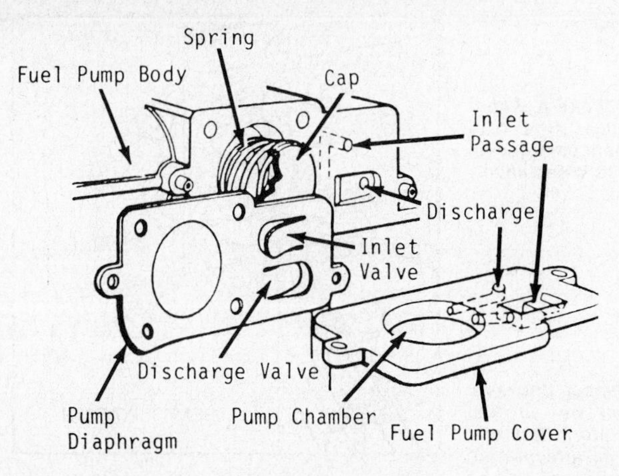

Fig. B70—Exploded view of fuel pump that is incorporated in Pulsa-Jet carburetor except those used on 82900, 92900, 94900, 110900, 111900, 112200 and 113900 models; refer to Fig. B71.

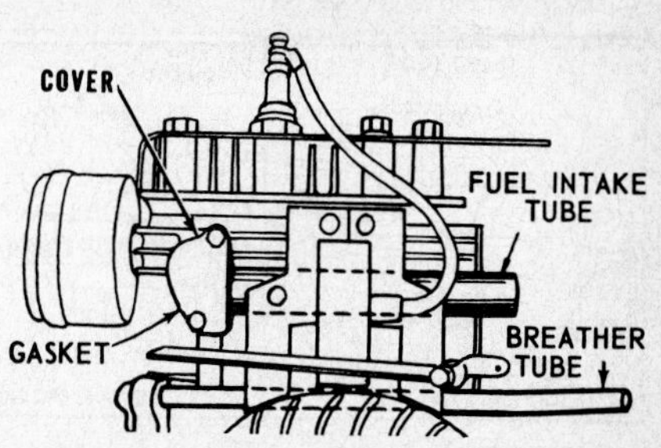

Fig. B74—On 82000 models, intake tube is threaded into intake port of engine; a gasket is placed between intake port cover and intake port.

pump to fill a constant level fuel sump in top of fuel tank. Refer to schematic view in Fig. B69. This makes a constant air:fuel mixture available to engine regardless of fuel level in tank. Adjustment of the pump type carburetor fuel mixture needle valve is the same as outlined for suction type carburetors in previous paragraph, except that fuel level in tank is not important.

To remove the pump type carburetor, first remove the carburetor and fuel tank as an assembly; then, remove carburetor from fuel tank. When reinstalling carburetor on fuel tank, use a new gasket or pump diaphragm as required and tighten retaining screws evenly.

Fig. B70 shows an exploded view of the pump unit used on all carburetors except those for Models 82900, 92900, 94900, 110900, 111900, 112200 and 113900 the pump diaphragm is placed between the carburetor and fuel tank as shown in Fig. B71.

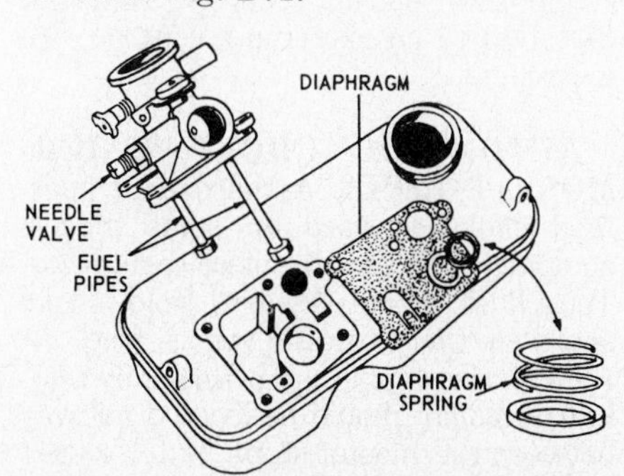

Fig. B71—On Models 82900, 92900, 94900, 110900, 111900, 112200 and 113900, pump type (Pulsa-Jet) carburetor diaphragm is installed between carburetor and fuel tank.

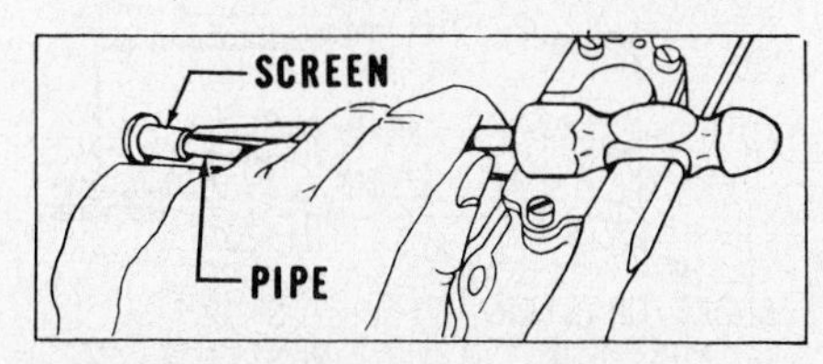

Fig. B72—To renew screen housing on pump type carburetor with brass feed pipes, drive old screen housing from pipe as shown. To hold pipe, clamp lightly in a vise.

The pump type carburetor has two fuel feed pipes. The long pipe feeds fuel into the pump portion of the carburetor from which fuel then flows to the constant level fuel sump. The short pipe extends into the constant level sump and feeds fuel into the carburetor venturi via fuel mixture needle valve.

As check valves are incorporated in the pump diaphragm, fuel feed pipes on pump type carburetors do not have a check valve. However, if the fuel screen in lower end of pipe is broken or clogged and cannot be cleaned, the pipe or screen housing can be renewed. If pipe is made of nylon, pipe snaps into place and considerable force is required to remove or install pipe. Be careful to not damage new pipe. If pipe is made of brass, clamp pipe lightly in a vise and drive old screen housing from pipe with a screwdriver or small chisel as shown in Fig. B72. Drive a new screen housing onto pipe with a soft faced hammer.

NOTE: If soaking carburetor in cleaner for more than one-half hour, be sure to remove all nylon parts and "O" ring, if used, before placing carburetor in cleaning solvent.

NOTE: On engine Models 82900, 92900, 94900, 95500, 110900, 111900 and 113900, be sure air cleaner retaining screw is in place if engine is being operated (during tests) without air cleaner installed. If screw is not in place, fuel will lift up through the screw hole and enter carburetor throat as the screw hole leads directly into the constant level fuel sump.

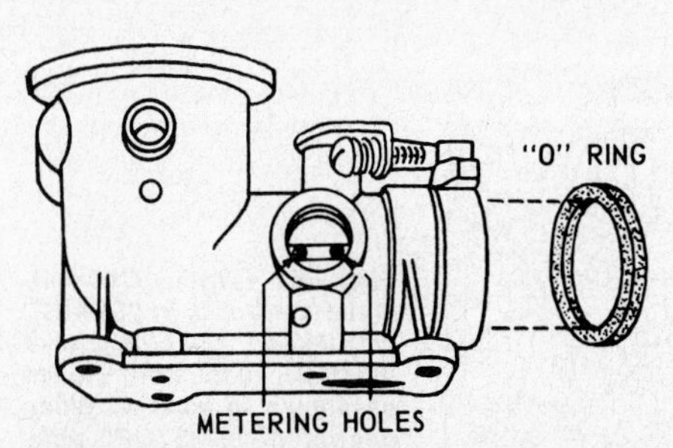

Fig. B73—Metering holes in pump type carburetors are accessible for cleaning after removing fuel mixture needle valve. On models with intake pipe, carburetor is sealed to pipe with "O" ring.

INTAKE TUBE. Models 82000, 92000, 93500, 94500, 94900, 95500, 100900, 110900, 111900, 113900, 130900 and 131900 have an intake tube between carburetor and engine intake port. Carburetor is sealed to intake tube with an "O" ring as shown in Fig. B73.

On Model 82000 engines, the intake tube is threaded into the engine intake port. A gasket is used between the engine intake port cover and engine casting. Refer to Fig. B74.

On Models 92000, 93500, 94500, 94900, 95500, 110900, 111900 and 113900 engines, the intake tube is bolted to the engine intake port and gasket is used between the intake tube and engine casting. Refer to Fig. B75. On Models 100900, 130900 and 131900 intake tubes are attached to engine in similar manner.

CHOKE-A-MATIC CARBURETOR CONTROLS. Engines equipped with float, suction or pump type carburetors may be equipped with a control unit with which the carburetor choke, throttle and magneto grounding switch are operated from a single lever (Choke-A-Matic carburetors). Refer to Figs. B76 through B82 for views showing the different

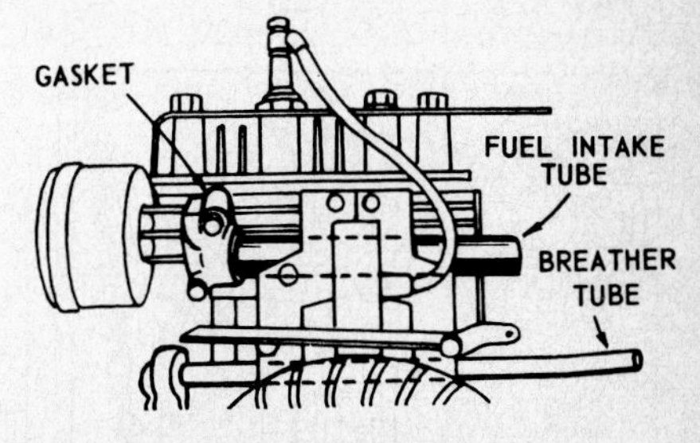

Fig. B75—On 92000, 93500, 94500, 94900, 95500, 110900, 111900 and 113900 models, fuel intake tube is bolted to engine intake port and a gasket is placed between tube and engine. On 100900, 130900 and 131900 vertical crankshaft models, intake tube and gasket are similar.

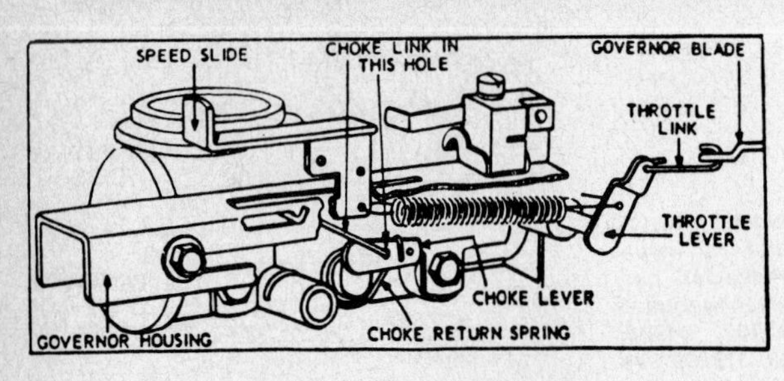

Fig. B76—Choke-A-Matic control on float type carburetor. Remote control can be attached to speed slide.

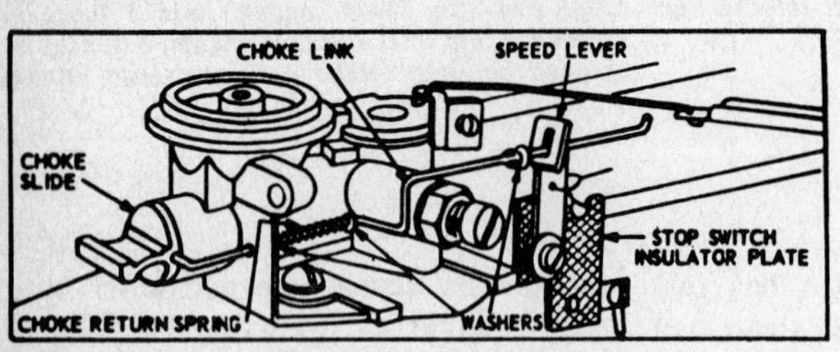

Fig. B77—Typical Choke-A-Matic control on suction type carburetor. Remote control can be attached to speed lever.

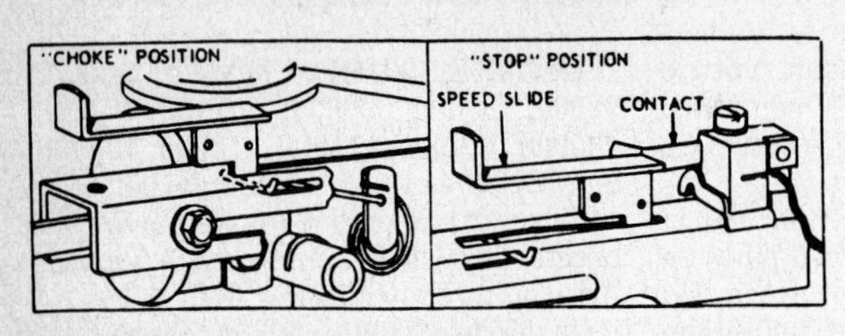

Fig. B78—Choke-A-Matic control in choke and stop positions on float carburetor.

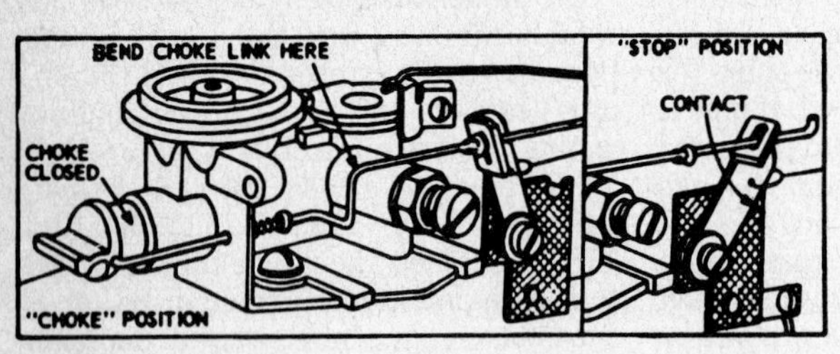

Fig. B79—Choke-A-Matic controls in choke and stop positions on suction carburetor. Bend choke link if necessary to adjust control.

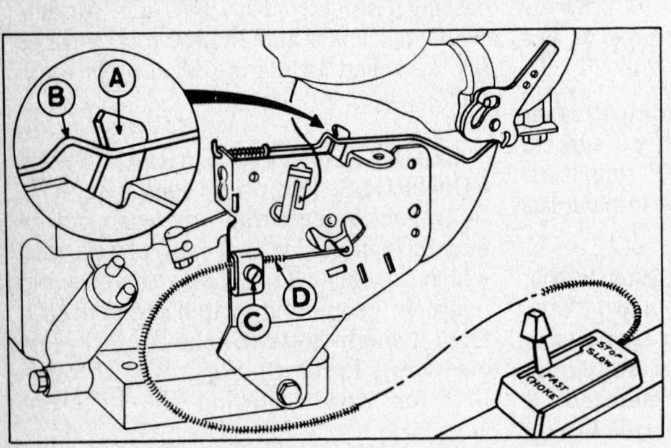

Fig. B80—On Choke-A-Matic control shown, choke actuating lever (A) should just contact choke link or shaft (B) when control is at "FAST" position. If not, loosen screw (C) and move control wire housing (D) as required.

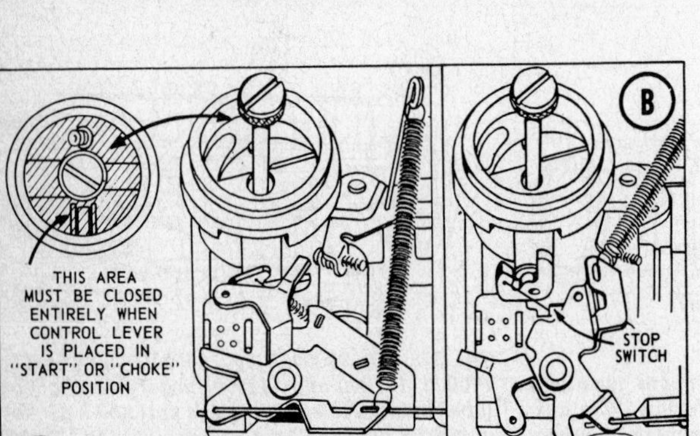

Fig. B81—When Choke-A-Matic control is in "START" or "CHOKE" position, choke must be completely closed as shown in view A. When control is in "STOP" position, arm should contact stop switch (view B).

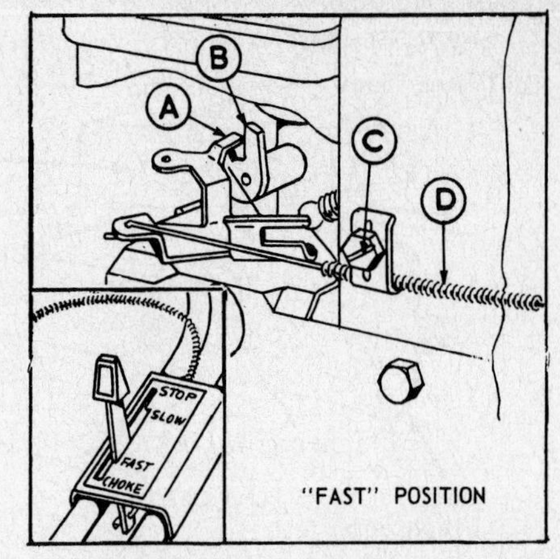

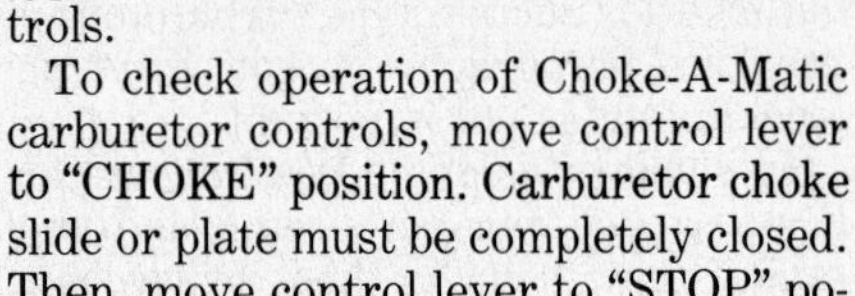

Fig. B82—On Choke-A-Matic controls shown, lever (A) should just contact choke shaft arm (B) when control is in "FAST" position. If not, loosen screw (C) and move control wire housing (D) as required, then tighten screw.

types of Choke-A-Matic carburetor controls.

To check operation of Choke-A-Matic carburetor controls, move control lever to "CHOKE" position. Carburetor choke slide or plate must be completely closed. Then, move control lever to "STOP" position. Magneto grounding switch should be making contact. With the control lever in "RUN", "FAST" or "SLOW" position, carburetor choke should be completely open. On units with remote controls, synchronize movement of remote lever to carburetor control lever by loosening screw (C – Fig. B80 or Fig. B82) and moving control wire housing (D) as required; then, tighten screw to clamp the housing securely. Refer to Fig. B83 to check remote control wire movement.

AUTOMATIC CHOKE (THERMOSTAT TYPE). A thermostat operated choke is used on some models equipped with the two-piece carburetor. To adjust choke linkage, hold choke shaft so thermostat lever is free. At room temperature, stop screw in thermostat collar should be located midway between thermostat stops. If not, loosen stop screw, adjust the collar and tighten stop screw. Loosen set screw (S – Fig.

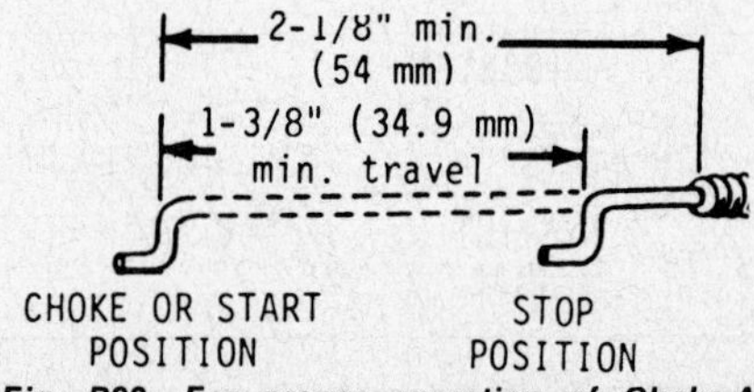

Fig. B83—For proper operation of Choke-A-Matic controls, remote control wire must extend to dimension shown and have a minimum travel of 1-3/8 inches (34.9 mm).

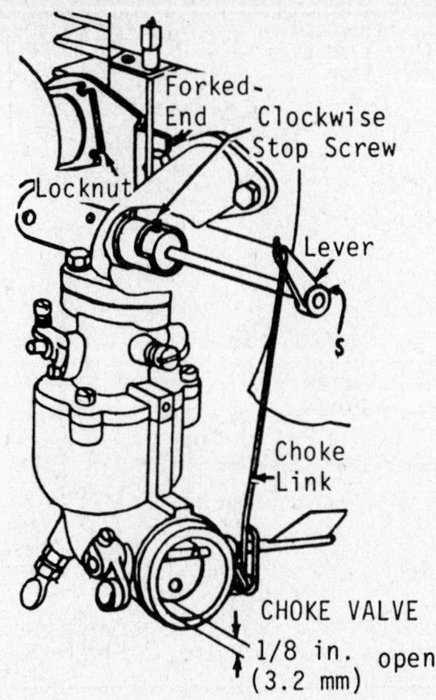

Fig. B84—Automatic choke used on some models equipped with "two-piece" Flo-Jet carburetor showing unit in "HOT" position.

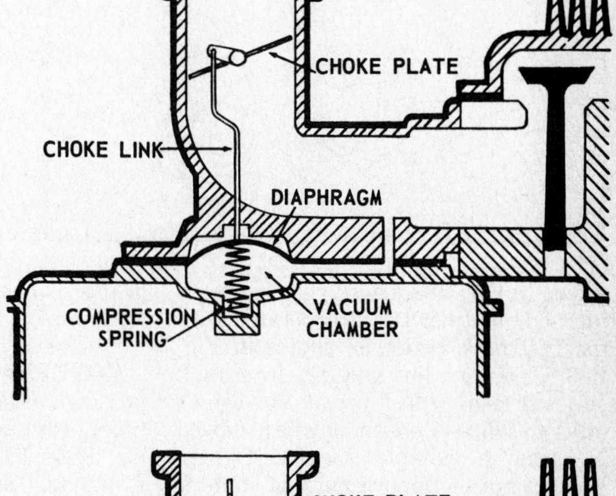

Fig. B86—Diagram showing vacuum operated automatic choke used on some 92000, 93500, 94500, 94900, 95500, 110900, 111900 and 113900 model vertical crankshaft engines in closed (engine not running) position.

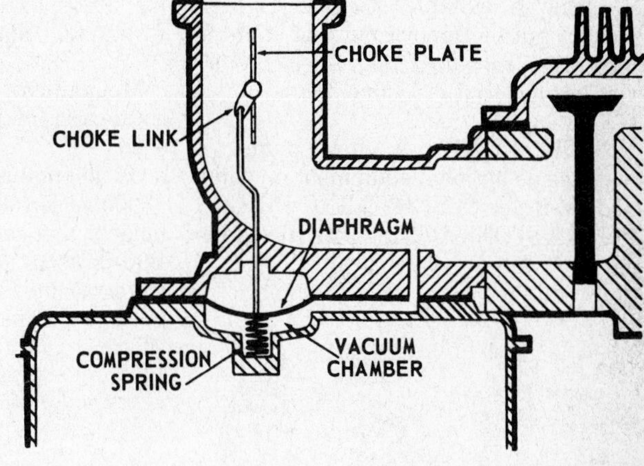

Fig. B87—Diagram showing vacuum operated automatic choke in open (engine running) position.

B84) on thermostat lever. Then, slide lever on shaft to ensure free movement of choke unit. Turn thermostat shaft clockwise until stop screw contacts thermostat stop. While holding shaft in this position, move shaft lever until choke is open exactly 1/8 inch (3.17 mm) and tighten lever set screw. Turn thermostat shaft counterclockwise until stop screw contacts thermostat stop as shown in Fig. B85. Manually open choke valve until it stops against top of choke link opening. At this time, choke valve should be open at least 3/32 inch (2.38 mm), but not more than 5/32 inch (3.97 mm). Hold choke valve in wide open position and check position of counterweight lever. Lever should be in a horizontal position with free end towards right.

AUTOMATIC CHOKE (VACUUM TYPE). A spring and vacuum operated automatic choke is used on some 92000, 93500, 94500, 94900, 95500, 110900, 111900 and 113900 vertical crankshaft engines. A diaphragm under carburetor is connected to the choke shaft by a link. The compression spring works against the diaphragm, holding choke in closed position when engine is not running. See Fig. B86. As engine starts, increased vacuum works against the spring and pulls the diaphragm and choke link down, holding choke in open (running) position shown in Fig. B87.

During operation, if a sudden load is applied to engine or a lugging condition develops, a drop in intake vacuum occurs, permitting choke to close partially. This provides a richer fuel mixture to meet the condition and keeps the engine running smoothly. When the load condition has been met, increased vacuum returns choke valve to normal running (fully open) position.

FUEL TANK OUTLET. Small models with float type carburetors are equipped with a fuel tank outlet as

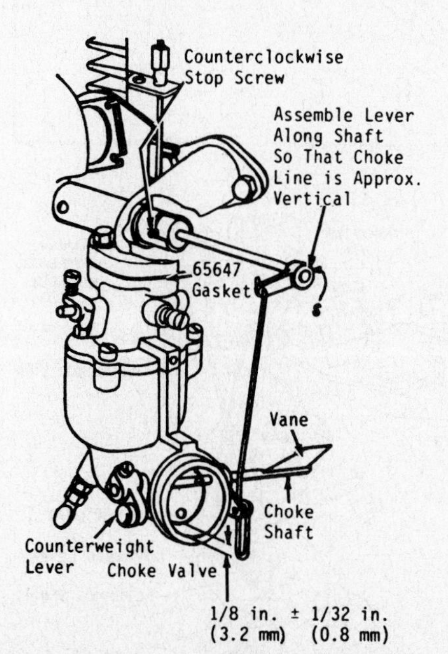

Fig. B85—Automatic choke on "two-piece" Flo-Jet carburetor in "COLD" position.

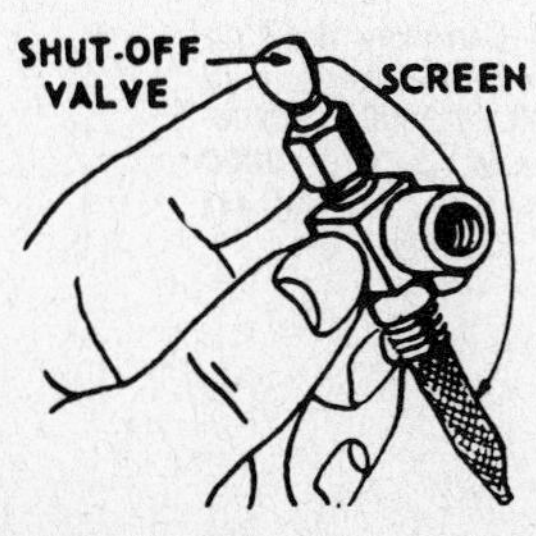

Fig. B88—Fuel tank outlet used on smaller engines with float type carburetor.

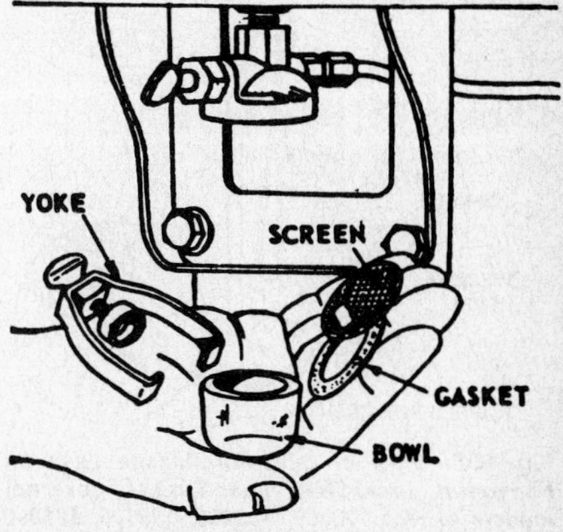

Fig. B89—Fuel tank outlet used on larger B&S engines.

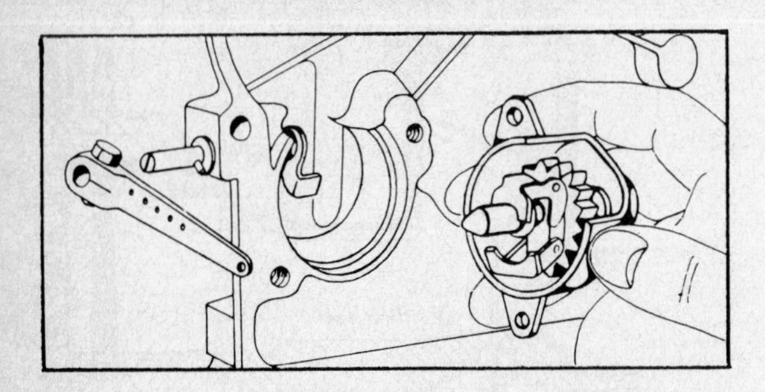

Fig. B90—Removing governor unit (except on 100000, 112200, 130000, 131000 and late 140000 models) from inside crankcase cover on horizontal crankshaft models. Refer to Fig. B91 for exploded view of governor.

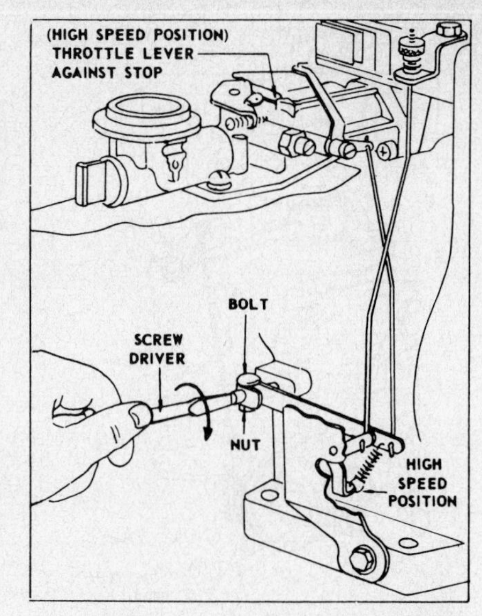

Fig. B95—Linkage adjustment on 100000, 112200, 130000, 131000 and late 140000 horizontal crankshaft mechanical governor models; refer to text for procedure.

shown in Fig. B88. On larger engines, a fuel sediment bowl is incorporated with the fuel tank outlet as shown in Fig. B89. Clean any lint and dirt from tank outlet screens with a brush. Varnish or other gasoline deposits may be removed by using a suitable solvent. Tighten packing nut or remove nut and shut-off valve, then renew packing if leakage occurs around shut-off valve stem.

FUEL PUMP. A fuel pump is available as optional equipment on some models. Refer to SERVICING BRIGGS & STRATTON ACCESSORIES section in this manual for fuel pump service information.

GOVERNOR. All models are equipped with either a mechanical (flyweight type) or an air vane (pneumatic) governor. Refer to the appropriate paragraph for model being serviced.

Mechanical Governor. Three different designs of mechanical governors are used.

On all engines except 100000, 112200, 130000, 131000 and all late 140000 models, a governor unit as shown in Fig. B90 is used. An exploded view of this governor unit is shown in Fig. B91. The governor housing is attached to inner side of crankcase cover and the governor gear is driven from the engine camshaft gear. Use Figs. B90 and B91 as a

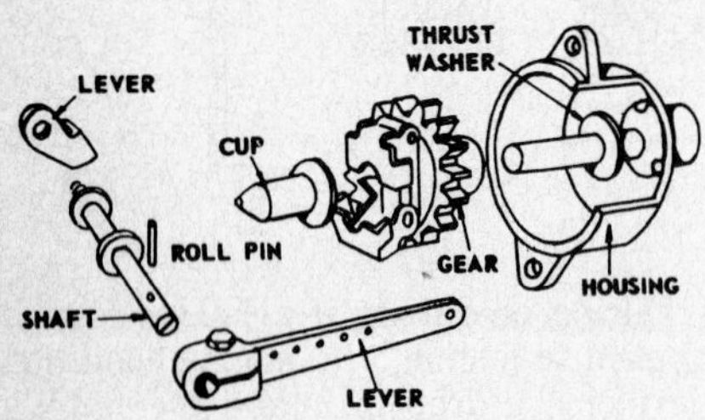

Fig. B91—Exploded view of govenor unit used on all horizontal crankshaft models (except 100000, 112200, 130000, 131000 and late 140000 models) with mechanical governor. Refer also to Figs. B90 and B92.

Fig. B93—Installing crankcase cover on 100000, 112200, 130000, 131000 and late 140000 models with mechanical governor. Governor crank (C) must be in position shown. A thrust washer (W) is placed between governor (G) and crankcase cover.

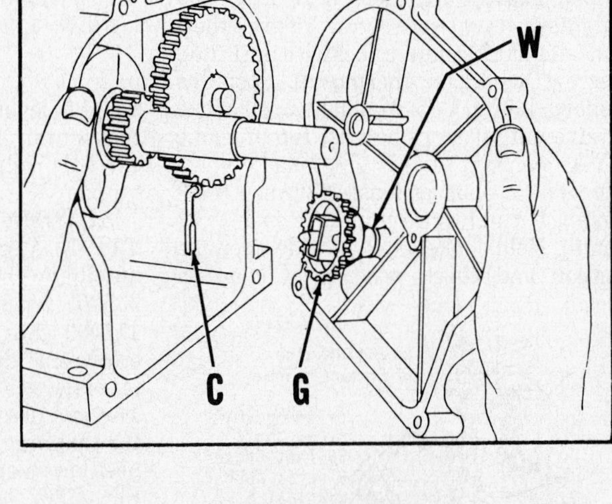

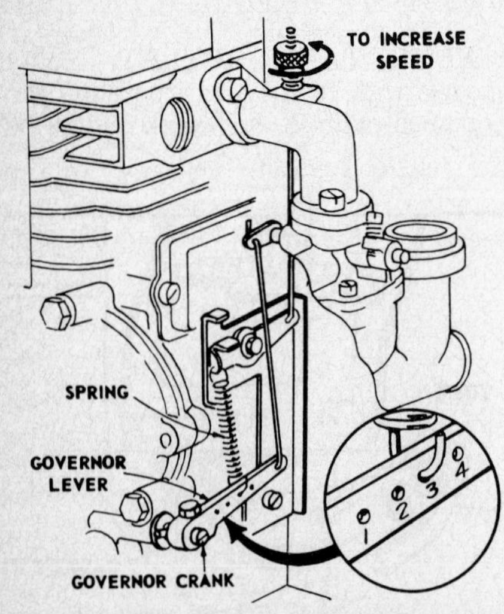

Fig. B92—View of governor linkage used on horizontal crankshaft mechanical governor models except 100000, 112200, 130000, 131000 and late 140000 models. Governor spring should be hooked in governor lever as shown in inset.

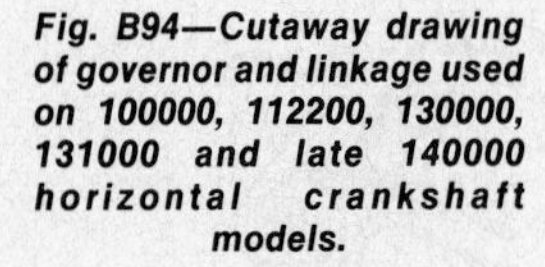

Fig. B94—Cutaway drawing of governor and linkage used on 100000, 112200, 130000, 131000 and late 140000 horizontal crankshaft models.

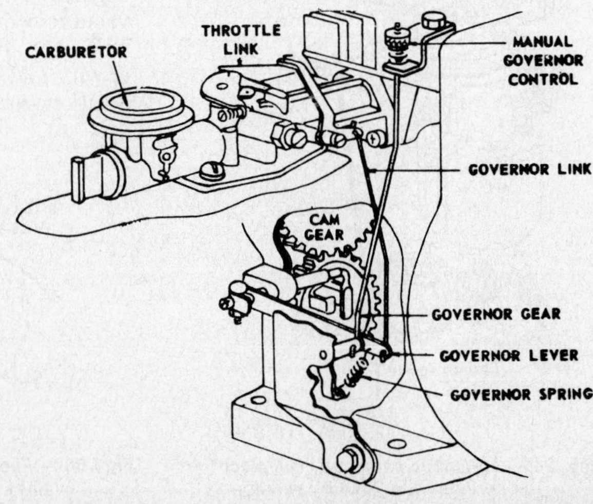

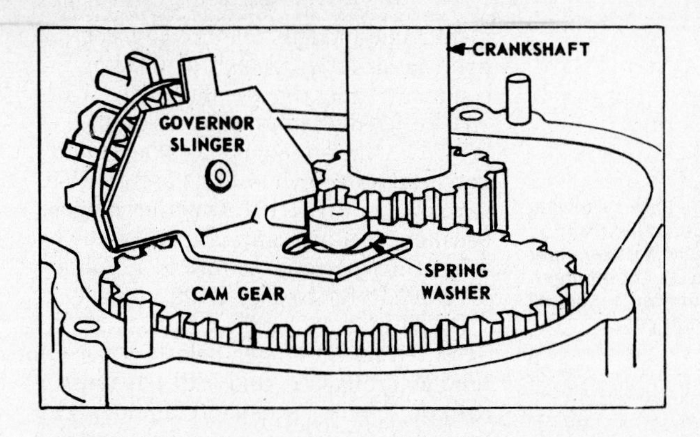

Fig. B96—View showing 95500, 100000, 113900, 130000, 131000 and 140000 vertical crankshaft model mechanical governor unit. Drawing is of lower side of engine with oil sump (engine base) removed; note spring washer location used on 100000, 130000 and 131000 models only.

disassembly and assembly guide. Renew any parts that bind or show excessive wear. After governor is assembled, refer to Fig. B92 and adjust linkage by loosening screw clamping governor lever to governor crank. Turn governor lever counterclockwise so carburetor throttle is in wide open position. Hold lever and turn governor crank as far counterclockwise as possible, then tighten screw clamping lever to crank. Governor crank can be turned with screwdriver. Check linkage to be sure it is free and that the carburetor throttle will move from idle to wide open position.

On 100000, 112200, 130000, 131000 and late 140000 horizontal crankshaft models, the governor gear and weight unit (G–Fig. B93) is supported on a pin in engine crankcase cover and the governor crank is installed in a bore in the engine crankcase. A thrust washer (W) is placed between governor gear and crankcase cover. When assembling crankcase cover to crankcase, be sure governor crank (C) is in position shown in Fig. B93. After governor unit and crankcase cover is installed, refer to Fig. B94 for installation of linkage. Before attempting to start engine, refer to Fig. B95 and adjust linkage by loosening bolt clamping governor lever to governor crank. Set control lever in high speed position. Using a screwdriver, turn governor crank as far clockwise as possible and tighten governor lever clamp bolt. Check to be sure carburetor throttle can be moved from idle to wide open position and that linkage is free.

On vertical crankshaft 95500, 100000, 113900, 130000, 131000 and 140000 with mechanical governor, the governor weight unit is integral with the lubricating oil slinger and is mounted on lower end of camshaft gear as shown in Fig. B96. With engine upside down, place governor and slinger unit on camshaft gear as shown, place spring washer (100000, 130900 and 131900 models only) on camshaft gear and install engine base on crankcase. Assemble linkage as shown in Fig. B97; then, refer to Fig. B98 and adjust linkage by loosening governor lever to governor crank clamp bolt, place control lever in high speed position, turn governor crank with screwdriver as far as possible and tighten governor lever clamping bolt.

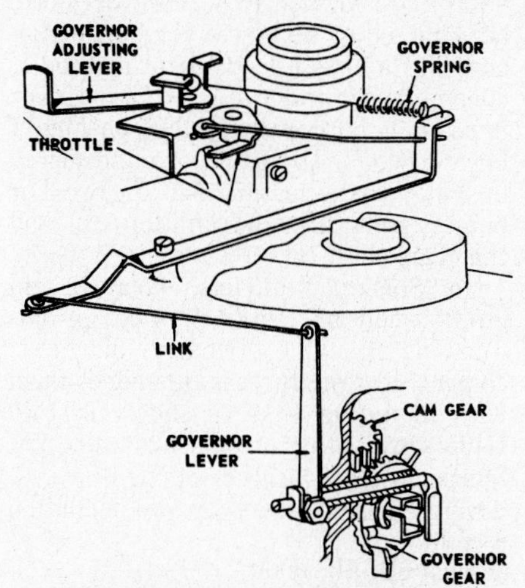

Fig. B97—Schematic drawing of 95500, 100000, 113900, 130000, 131000 and 140000 mechanical governor and linkage used on vertical crankshaft mechanical governed models.

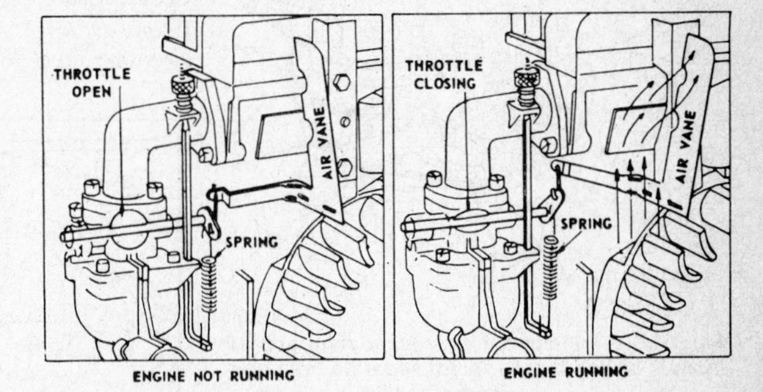

Fig. B99—Views showing operating principle of air vane (penumatic) governor. Air from flywheel fan acts against air vane to overcome tension of governor spring; speed is adjusted by changing spring tension.

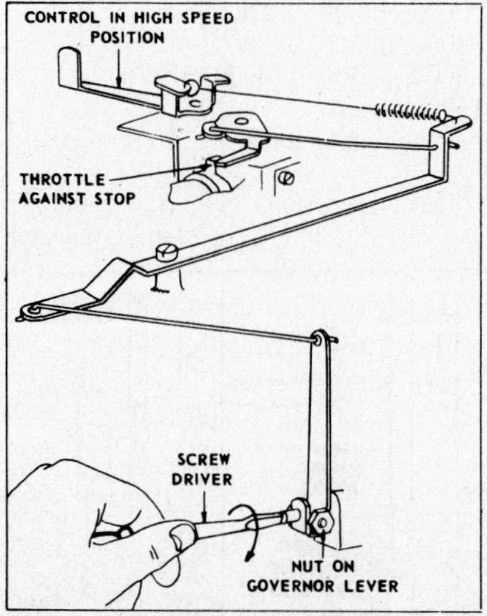

Fig. B98—View showing adjustment of 95500, 100000, 113900, 130000, 131000 and 140000 vertical crankshaft mechanical governor; refer to text for procedure.

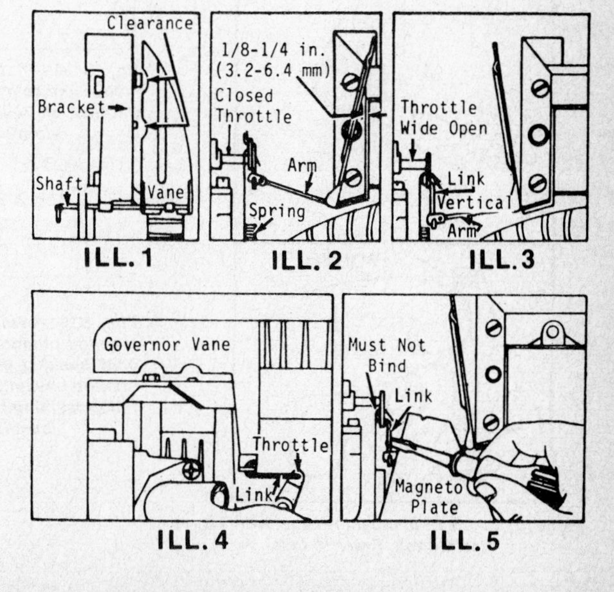

Fig. B100—Air vane governors and linkage. ILL. 1; the governor vane should be checked for clearance in all positions. ILL. 2; the vane should top 1/8 to 1/4 inch (3.175-6.35 mm) from the magneto coil. ILL. 3; with wide open throttle, the link connecting vane arm to throttle lever should be in a vertical position on vertical cylinder engines and in a horizontal position (ILL. 4) on horizontal cylinder engines. Bend link slightly (ILL. 5) to remove any binding condition in linkage.

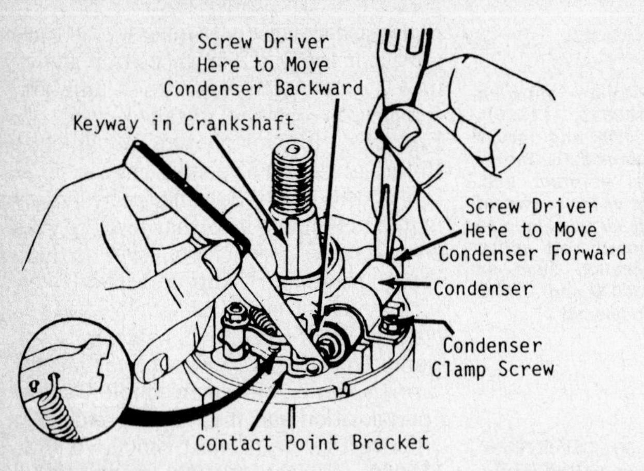

Fig. B101—View showing breaker point adjustment on models having breaker point integral with condenser. Move condenser to adjust point gap.

Air Vane (Pneumatic) Governor. Refer to Fig. B99 for views showing typical air vane governor operation.

The vane should stop 1/8 to ¼ inch (3.17-6.35 mm) from the magneto coil (Fig. B100, illustration 2) when the linkage is assembled and attached to carburetor throttle shaft. If necessary to adjust, spring the vane while holding the shaft. With wide open throttle, the link from the air vane arm to the carburetor throttle should be in a vertical position on horizontal crankshaft models and in a horizontal position on vertical crankshaft models. (Fig. B100, illustration 3 and 4.) Check linkage for binding. If binding condition exists, bend links slightly to correct. Refer to Fig. B100, illustration 5.

NOTE: Some engines are equipped with a nylon governor vane which does not require adjustment.

MAGNETO. Breaker contact gap on all models with magneto type ignition is 0.020 inch (0.51 mm). On all models except "Sonoduct" (vertical crankshaft with flywheel below engine), breaker points and condenser are accessible after removing engine flywheel and breaker cover. On "Sonoduct" models, breaker points are located below breaker cover on top side of engine.

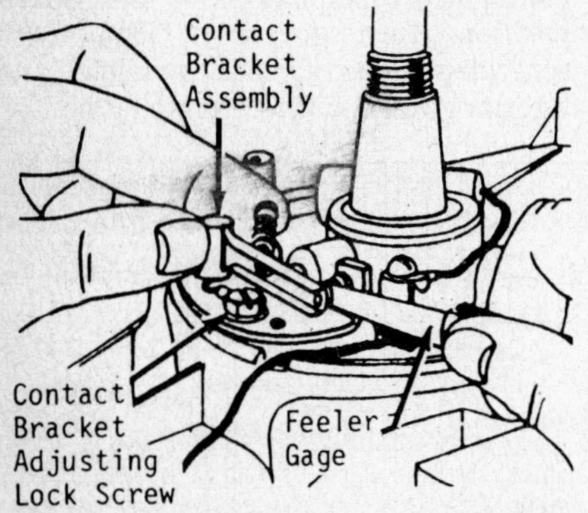

Fig. B102—Adjustment of breaker point gap on models having breaker point separate from condenser.

Fig. B102A—Insert plunger into bore with groove toward top. Refer to text.

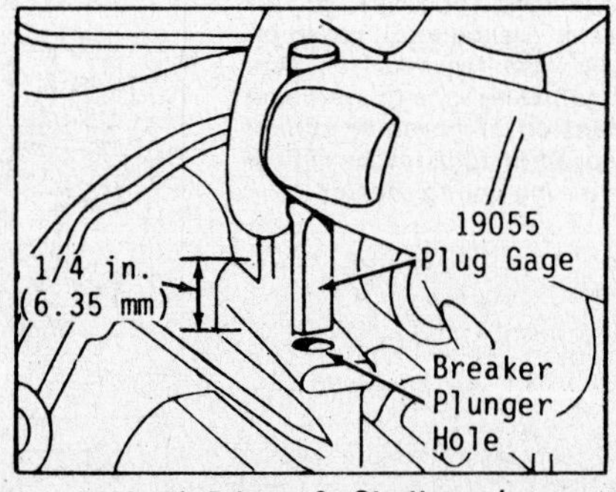

Fig. B103—If Briggs & Stratton plug gage #19055 can be inserted in breaker plunger bore a distance of ¼ inch (6.35 mm) or more, bore is worn and must be rebushed.

On some models, one breaker contact point is an integral part of the ignition condenser and the breaker arm pivots on a knife edge retained by a slot in pivot post. On these models, breaker contact gap is adjusted by moving the condenser as shown in Fig. B101. On other models, breaker contact gap is adjusted by relocating position of breaker contact bracket. Refer to Fig. B102.

On all models, breaker contact arm is actuated by a plunger held in a bore in engine crankcase and rides against a cam on engine crankshaft. Plunger can be removed after removing breaker points. Renew plunger if worn to a length of 0.870 inch (22.1 mm) or less. Plunger must be installed with grooved end to the top to prevent oil seepage (Fig. B102A). Check breaker point plunger bore wear in crankcase with B&S plug gage 19055. If plug gage will enter bore ¼ inch (6.35 mm) or more, bore should be reamed and a bushing installed. Refer to Fig. B103 for method of checking bore and to Fig. B104 for steps in reaming bore and installing bushing if bore is worn. To ream bore and install bushing, it is necessary that the breaker points, armature and ignition coil and the crankshaft be removed.

On "Sonoduct" models, armature and ignition coil are inside flywheel on bottom side of engine. Armature air gap is correct if armature is installed flush with mounting boss as shown in Fig. B105. On all other models, armature and ignition coil are located outside flywheel. Armature-to-flywheel air gap should be as follows:

Models 6B, 8B, 6000–
Two Leg Armature . . . 0.006-0.010 in. (0.15-0.25 mm)
Three Leg Armature . . 0.012-0.016 in. (0.30-0.41 mm)

Models 80000, 82000, 92000, 93000, 93500, 94000, 94500, 94900, 95000, 95500, 110000, 110900–
Two Leg Armature . . . 0.006-0.010 in. (0.15-0.25 mm)
Three Leg Armature . . 0.012-0.016 in. (0.30-0.41 mm)

Fig. B104—Views showing reaming plunger bore to accept bushing (left view), installing bushing (center) and finish reaming bore (right) of bushing.

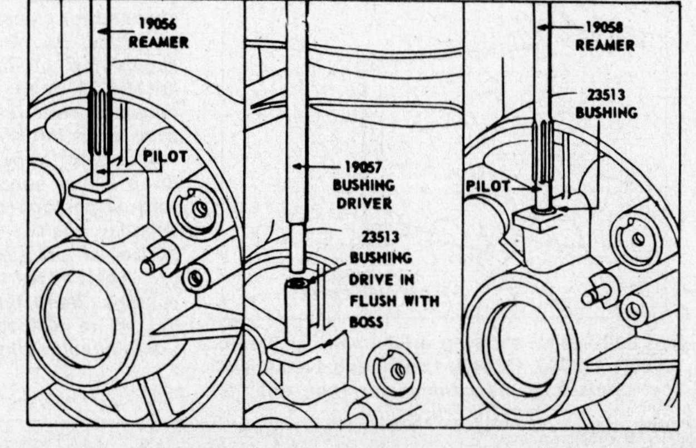

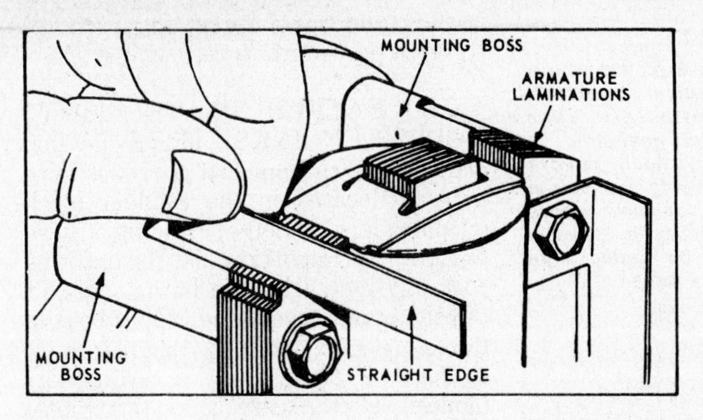

Fig. B105—On "Sonoduct" models, align armature core with mounting boss for proper magneto air gap.

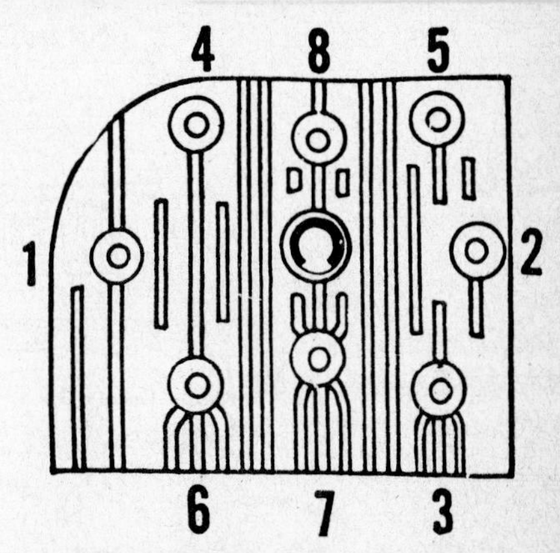

Fig. B106—Cylinder head screw tightening sequence. Long screws are used in positions 2, 3 and 7.

Models 111200, 111900,
112200, 113900 . . . . 0.010-0.016 in. (0.25-0.41 mm)
Models 100000,
130000, 131000–
Two Leg Armature . . . 0.010-0.014 in. (0.25-0.36 mm)
Three Leg Armature . . 0.012-0.016 in. ((0.30-0.41 mm)
Model 140000–
Two Leg Armature . . . 0.010-0.014 in. (0.25-0.36 mm)
Three Leg Armature . . 0.016-0.019 in. (0.41-0.48 mm)

MAGNETRON IGNITION. Magnetron ignition is a self-contained breakerless ignition system. Flywheel does not need to be removed except to check or service keyways or crankshaft key.

To check spark, remove spark plug. Connect spark plug cable to B&S tester, part 19051, and ground remaining tester lead to cylinder head. Spin engine at 350 rpm or more. If spark jumps the 0.166 inch (4.2 mm) tester gap, system is functioning properly.

To remove armature and Magnetron module, remove flywheel shroud and armature retaining screws. Use a 3/16 inch (4.76 mm) diameter pin punch to release stop switch wire from module. To remove module, unsolder wires, push module retainer away from laminations and remove module. See Fig. B105A.

Resolder wires for reinstallation and use Permatex or equivalent to hold ground wires in position.

Adjust armature air gap to 0.010-0.014 inch (0.25-0.36 mm).

LUBRICATION. Vertical crankshaft engines are lubricated by an oil slinger wheel driven by the cam gear. On early 6B and 8B models, the oil slinger wheel was mounted on a bracket attached to the crankcase. Renew the bracket if pin on which gear rotates is worn to 0.490 inch (12.45 mm). Renew steel bushing in hub or gear if worn. On later model vertical crankshaft engines, the oil slinger wheel, pin and bracket are an integral unit with the bracket being retained by the lower end of the engine camshaft.

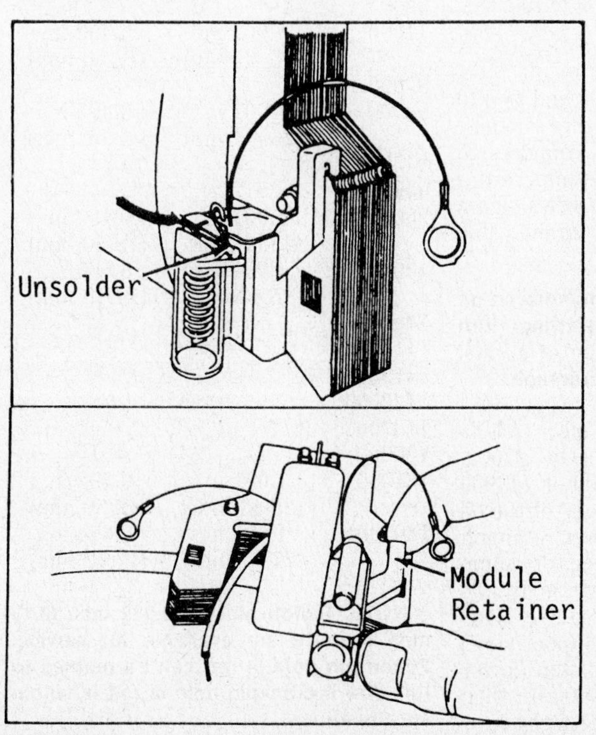

Fig. B105A—Wires must be unsoldered to remove Magnetron ignition module. Refer to text.

On 100000, 130000 and 131000 models, a spring washer is placed on lower end of camshaft between bracket and oil sump boss. Renew the oil slinger assembly if teeth are worn on slinger gear or gear is loose on bracket. On horizontal crankshaft engines, a splash system (oil dipper on connecting rod) is used for engine lubrication.

Check oil level at five hour intervals and maintain at bottom edge of filler plug or to FULL mark on dipstick.

Recommended oil change interval for all models is every 25 hours of normal operation.

Manufacturer recommends using oil with an API service classification of SC, SD, SE, or SF. Use SAE 10W-40 oil for temperatures above 20°F (−7°C) and SAE 5W-30 oil for temperatures below 20°F (−7°C).

Crankcase capacity for all aluminum cylinder engines with displacement of 9 cubic inches (147 cc) or below, is 1¼ pint (0.6 L).

Crankcase capacity of vertical crankshaft aluminum cylinder engines with displacement of 11 cubic inches (186 cc) is 1¼ pint (0.6 L).

Crankcase capacity for all 10 and 13 cubic inch (169 and 203 cc) vertical crankshaft, aluminum cylinder engines is 1¾ pint (0.8 L).

Crankcase capacity for all 10 and 13 cubic inch (169 and 203 cc) horizontal crankshaft, aluminum cylinder engines is 1¼ pint (0.6 L).

Crankcase capacity for 14 cubic inch (231 cc) vertical crankshaft, aluminum cylinder engines is 2¼ pint (1.1 L).

Crankcase capacity for 14 cubic inch (231 cc) horizontal crankshaft, aluminum cylinder engines if 2¾ pint (1.3 L).

Crankcase capacity for engines with cast iron cylinder liner is 3 pints (1.4 L).

CRANKCASE BREATHER. The crankcase breather is built into the

Fig. B107—Exploded view of typical vertical crankshaft model with air vane (pneumatic) governor. To remove flywheel, remove blower housing and starter unit; then, unscrew starter clutch housing (5). Flywheel can then be pulled from crankshaft.

1. Snap ring
2. Washer
3. Ratchet
4. Steel balls
5. Starter clutch
6. Washer
7. Flywheel
8. Breaker cover
9. Breaker point spring
10. Breaker arm & pivot
11. Breaker plunger
12. Condenser clamp
13. Coil spring (primary wire retainer)
14. Condenser
15. Governor air vane & bracket assy.
16. Spark plug wire
17. Armature & coil assy.
18. Air baffle
19. Spark plug grounding switch
20. Cylinder head
21. Cylinder head gasket
22. Cylinder block
23. Crankshaft oil seal
24. Cylinder shield
25. Flywheel key
26. Gasket
27. Breather & tappet chamber cover
28. Breather tube assy.
29. Coil spring
30. Crankshaft
31. Cam gear & shaft
32. Piston rings
33. Piston
34. Connecting rod
35. Rod bolt lock
36. Piston pin retaining rings
37. Piston pin
38. Intake valve
39. Valve springs
40. Valve spring keepers
41. Tappets
42. Exhaust valve
43. Gasket
44. Oil slinger assy.
45. Oil sump (engine base)
46. Crankshaft oil seal

engine valve cover. The mounting holes are offset so the breather can only be installed one way. Rinse breather in solvent and allow to drain. A vent tube connects the breather to the carburetor air horn on certain model engines for extra protection against dusty conditions.

REPAIRS

CYLINDER HEAD. When removing cylinder head, be sure to note the position from which each of the different length screws was removed. If screws are not reinstalled in the same holes when installing the head, it will result in screws bottoming in some holes and not enough thread contact in others. Lubricate the cylinder head screws with graphite grease before installation. Do not use sealer on head gasket. When installing cylinder head, tighten all screws lightly and then retighten them in sequence shown in Fig. B106 to 165 in.-lbs. (19 N·m) on 140000 models and to 140 in.-lbs. (16 N·m) on all other models. Run the engine for 2 to 5 minutes to allow it to reach operating temperature and retighten the head screws again following the sequence and torque value specified.

NOTE: When checking compression on models with "Easy-Spin" starting, turn engine opposite the direction of normal rotation. See CAMSHAFT paragraph.

OIL SUMP REMOVAL, AUXILARY PTO MODELS. On 92000, 93500, 94900, 95500, 110900 or 113900 models with auxiliary pto, one of the oil sump (engine base) to cylinder retaining screws is installed in the recess in sump for the pto auxiliary drive gear. To remove the oil sump, refer to Fig. B108 and remove the cover plate (upper view) and then remove the shaft stop (lower left view). The gear and shaft can then be moved as shown in lower right view to allow removal of the retaining screws. Reverse procedure to reassemble.

OIL BAFFLE PLATE. 140000 MODEL ENGINES. Model 140000 engines with mechanical governor have a baffle located in the cylinder block (crankcase). When servicing these engines, it is important that the baffle be correctly installed. The baffle must fit tightly against the valve tappet boss in the crankcase. Check for this before installing oil sump (vertical crankshaft models) or crankcase cover (horizontal crankshaft models).

CONNECTING ROD. The connecting rod and piston are removed from cylinder head end of block as an assembly. The aluminum alloy connecting rod rides directly on the induction hardened crankpin. The rod should be rejected if the crankpin hole is scored or out-of-round more than 0.0007 inch (0.018 mm) or if the piston pin hole is scored or out-of-round more than 0.0005 inch (0.013 mm). Wear limit sizes are given in the following chart. Reject the connecting rod if either the crankpin or piston pin hole is worn to, or larger than the sizes given in the following table.

REJECT SIZES FOR CONNECTING ROD

Model	Bearing Bore	Pin Bore
6B, 60000 . . .	0.876 in. (22.25 mm)	0.492 in. (12.50 mm)
8B, 80000 . . .	1.001 in. (25.43 mm)	0.492 in. (12.50 mm)
82000, 92000, 11000 . . .	1.001 in. (25.43 mm)	0.492 in. (12.50 mm)
92000, 93500 . . .	1.001 in. (25.43 mm)	0.492 in. (12.50 mm)
94500, 94900, 95500 . . .	1.001 in. (25.43 mm)	0.492 in. (12.50 mm)
100000 . .	1.001 in. (25.43 mm)	0.555 in. (14.10 mm)
110900, 111200, 111900, 112200, 113900, 130000, 131000 . .	1.001 in. (25.43 mm)	0.492 in. (12.50 mm)
140000 . .	1.095 in. (27.81 mm)	0.674 in. (17.12 mm)

NOTE: Piston pins of 0.005 inch (0.13 mm) oversize are available for service. Piston pin hole in rod can be reamed to this size if crankpin hole in rod is within specifications.

Connecting rod must be reassembled so match marks on rod and cap are aligned. Tighten connecting rod bolts to 165 in.-lbs. (19 N•m) for 140000 models and to 100 in.-lbs. (11 N•m) for all other models.

PISTON, PIN AND RINGS. Pistons for use in engines having aluminum bore ("Kool Bore") are not interchangeable with those for use in cylinders having cast-iron sleeve. Pistons may be identified as follows: Those for use in cast-iron sleeve cylinders have a plain, dull aluminum finish, have an "L" stamped on top and use an oil ring expander. Those for use in aluminum bore cylinders are chrome plated (shiny finish), do not have an identifying letter and do not use an oil ring expander.

Reject pistons showing visible signs of wear, scoring and scuffing. If, after cleaning carbon from top ring groove, a new top ring has a side clearance of 0.007 inch (0.18 mm) or more, reject the piston. Reject piston or hone piston pin hole to 0.005 inch (0.13 mm) oversize if pin hole is 0.0005 inch (0.013 mm) or more out-of-round, or is worn to a diameter of 0.554 inch (14.07 mm) or more on 100000 engines, 0.673 inch (17.09 mm) or more on 140000 engines, or 0.491 inch (12.47 mm) or more on all other models.

If the piston pin is 0.0005 inch (0.013 mm) or more out-of-round, or is worn to a diameter of 0.552 inch (14.02 mm) or smaller on 100000 engines, 0.671 inch (17.04 mm) or smaller on 140000 engines, or 0.489 inch (12.42 mm) or smaller on all other models, reject pin.

The piston ring gap for new rings should be 0.010-0.025 inch (0.25-0.64 mm) for models with aluminum cylinder bore and 0.010-0.018 inch (0.25-0.46 mm) for models with cast-iron cylinder bore. On aluminum bore engines, reject compression rings having an end gap of 0.035 inch (0.80 mm) or more and reject oil rings having an end gap of 0.045 inch (1.14 mm) or more. On cast-iron bore engines, reject compression rings having an end gap of 0.030 inch (0.75 mm) or more and reject oil rings having an end gap of 0.035 inch (0.90 mm) or more.

Pistons and rings are available in several oversizes as well as standard.

A chrome ring set is available for slightly worn standard bore cylinders. Refer to note in CYLINDER section.

CYLINDER. If cylinder bore wear is 0.003 inch (0.76 mm) or more or is 0.0025 inch (0.06 mm) or more out-of-round, cylinder must be rebored to next larger oversize.

The standard cylinder bore sizes for each model are given in the following table.

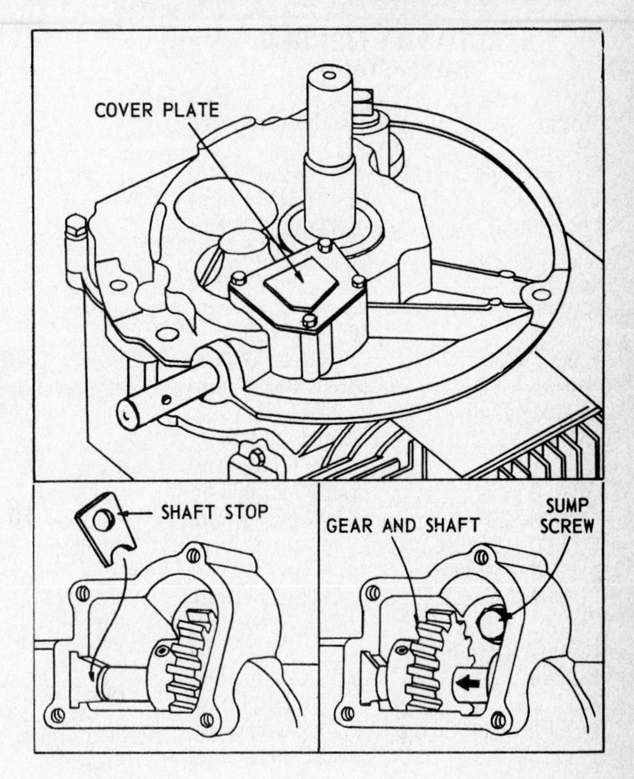

Fig. B108—To remove oil sump (engine base) on 92000, 93500, 94900, 95500, 110900 or 113900 models with auxiliary pto, remove cover plate, shaft stop and retaining screw as shown.

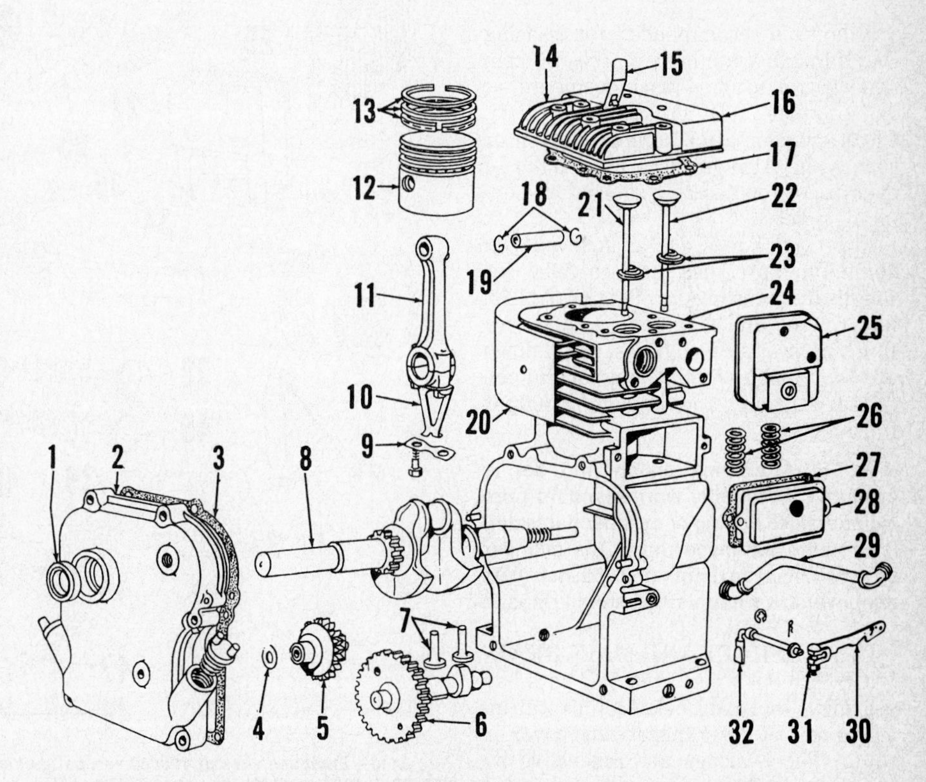

Fig. B109—Exploded view of 100000 horizontal crankshaft engine assembly. Except for 112200, 130000, and late 140000 models, other horizontal crankshaft models with mechanical governor will have governor unit as shown in Fig. B90; otherwise, construction of all other horizontal crankshaft models is similar.

1. Crankshaft oil seal
2. Crankcase cover
3. Gasket
4. Thrust washer
5. Governor assy.
6. Cam gear & shaft
7. Tappets
8. Crankshaft
9. Rod bolt lock
10. Oil dipper
11. Connecting rod
12. Piston
13. Piston rings
14. Cylinder head
15. Spark plug ground switch
16. Air baffle
17. Cylinder head gasket
18. Piston pin retaining rings
19. Piston pin
20. Air baffle
21. Exhaust valve
22. Intake valve
23. Valve spring retainers
24. Cylinder block
25. Muffler
26. Valve springs
27. Gaskets
28. Breather & tappet chamber cover
29. Breather pipe
30. Governor lever
31. Clamping bolt
32. Governor crank

STANDARD CYLINDER BORE SIZES

Model	Cylinder diameter
6B, Early 60000	2.3115-2.3125 in. (58.71-58.74 mm)
Late 60000	2.3740-2.3750 in. (60.30-60.33 mm)
8B, 80000, 81000, 82000	2.3740-2.3750 in. (60.30-60.33 mm)
92000, 93500	2.5615-2.5625 in. (65.06-65.09 mm)
94500, 94900, 95500	2.5615-2.5625 in. (65.06-65.09 mm)
100000	2.4990-2.5000 in. (63.47-63.50 mm)
110900, 111200, 111900, 112200, 113900	2.7802-2.7812 in. (70.62-70.64 mm)
130000, 131400, 131900	2.5615-2.5625 in. (65.06-65.09 mm)
140000	2.7490-2.7500 in. (69.82-69.85 mm)

A hone is recommended for resizing cylinders. Operate hone at 300-700 rpm and with an up and down movement that will produce a 45° crosshatch pattern. Clean cylinder after honing with oil or soap suds. Always check availability of oversize piston and ring sets before honing cylinder.

Approved hones are as follows: For aluminum bore, use Ammco 3956 for rough and finishing or Sunnen AN200 for rough and Sunnen AN500 for finishing. For sleeved bores, use Ammco 4324 for rough and finishing, or Sunnen AN100 for rough and Sunnen AN300 for finishing.

NOTE: A chrome piston ring set is available for slightly worn standard bore cylinders. No honing or cylinder deglazing is required for these rings. The cylinder bore can be a maximum of 0.005 inch (0.01 mm) oversize when using chrome rings.

CRANKSHAFT AND MAIN BEARINGS. Except where equipped with ball bearings, the main bearings are an integral part of the crankcase and cover or sump. The bearings are renewable by reaming out the crankcase and cover or sump bearing bores and installing service bushings. The tools for reaming the crankcase and cover or sump, and for installing the service bushings are available from Briggs & Stratton. If the bearings are scored, out-of-round 0.0007 inch (0.018 mm) or more, or are worn to or larger than the reject sizes in the following table, ream the bearings and install service bushings.

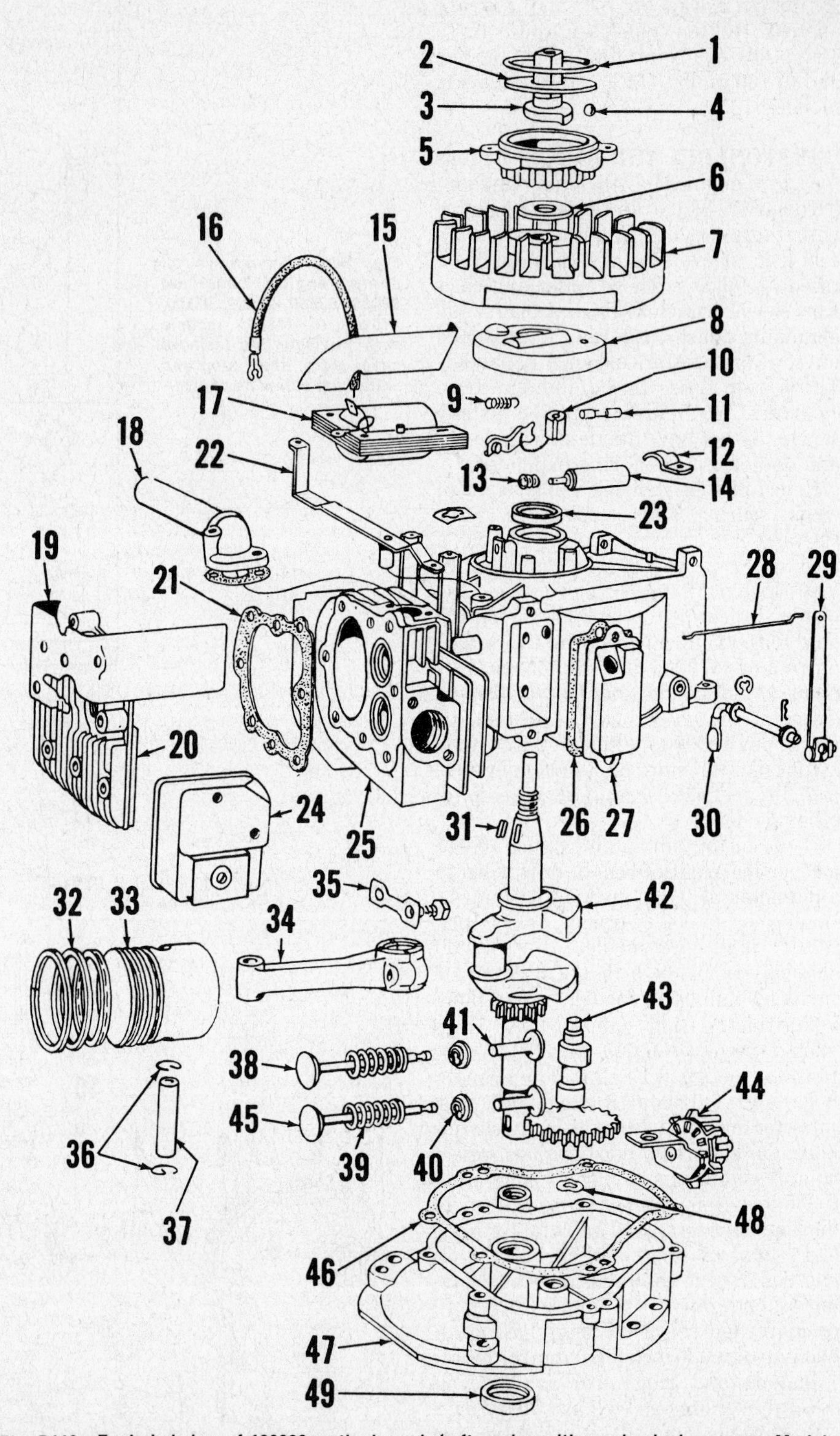

Fig. B110—Exploded view of 100000 vertical crankshaft engine with mechanical governor. Models 130000, 131000 and 140000 vertical crankshaft models with mechanical governor are similar. Refer to Fig. B107 for typical vertical crankshaft model with air vane governor.

1. Snap ring
2. Washer
3. Starter ratchet
4. Steel balls
5. Starter clutch
6. Washer
7. Flywheel
8. Breaker cover
9. Breaker arm spring
10. Breaker arm & pivot
11. Breaker plunger
12. Condenser clamp
13. Primary wire retainer spring
14. Condenser
15. Air baffle
16. Spark plug wire
17. Armature & coil assy.
18. Intake pipe
19. Air baffle
20. Cylinder head
21. Cylinder head gasket
22. Linkage lever
23. Crankshaft oil seal
24. Muffler
25. Cylinder block
26. Gasket
27. Breather & tappet chamber cover
28. Governor link
29. Governor crank
30. Governor crank
31. Flywheel key
32. Piston rings
33. Piston
34. Connecting rod
35. Rod bolt lock
36. Piston pin retaining rings
37. Piston pin
38. Intake valve
39. Valve springs
40. Valve spring retainers
41. Tappets
42. Crankshaft
43. Cam gear
44. Governor & oil slinger assy.
45. Exhaust valve
46. Gasket
47. Oil sump (engine base)
48. Thrust washer
49. Crankshaft oil seal

MAIN BEARING REJECT SIZES

Model	Magneto Bearing	PTO Bearing
6B, 60000 ...	0.878 in. (22.30 mm)	0.878 in.* (22.30 mm)*
8B, 80000 ...	0.878 in. (22.30 mm)	0.878 in.* (22.30 mm)*
81000, 82000 ...	0.878 in. (22.30 mm)	0.878 in.* (22.30 mm)*
92000, 93500, 94000, 94500, 94900, 95500 ...	0.878 in. (22.30 mm)	0.878 in.* (22.30 mm)*
100000 ..	0.878 in. (22.30 mm)	1.003 in. (25.48 mm)
110900, 111200, 111900, 112200, 113900 ..	0.878 in. (22.30 mm)	0.878 in.* (22.30 mm)*
130000, 131000 ..	0.878 in. (22.30 mm)	1.003 in. (25.48 mm)
140000 ..	1.004 in. (25.50 mm)	1.185 in. (30.10 mm)

*All models equipped with auxiliary drive unit have a main bearing rejection size for main bearing at pto side of 1.003 inch (25.48 mm)

Bushings are not available for all models; therefore, on some models the sump must be renewed if necessary to renew the bearing.

Main bearing journal diameter for flywheel and pto side main bearing journals on all standard models with plain bearings, except 130000, 131000 and 140000 models, is 0.873 inch (22.17 mm). On models equipped with auxiliary drive unit, pto side main bearing journal diameter is 0.998 inch (25.35 mm). Main bearing journal diameter at flywheel side of 130000 and 131000 models is 0.873 inch (22.17 mm) and journal diameter at pto side is 0.998 inch (25.35 mm). Main bearing journal diameter at flywheel side of 140000 model is 0.997 inch (25.32 mm) and journal diameter at pto side is 1.179 inch (29.95 mm).

Crankpin journal rejection size for 6B and 60000 models is 0.870 inch (22.10 mm), crankpin journal rejection size for 140000 model is 1.090 inch (27.69 mm) and crankpin journal rejection size for all other models if 0.996 inch (25.30 mm).

Ball bearing mains are a press fit on the crankshaft and must be removed by pressing the crankshaft out of the bearing. Reject ball bearing if worn or rough.

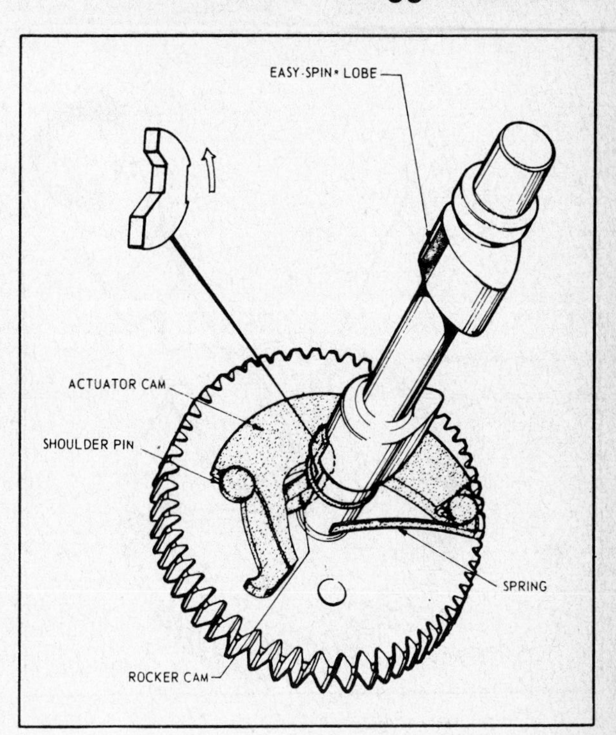

Fig. B111 – Compression release camshaft used on 111200, 111900 and 112200 engines. At cranking speed, spring holds actuator cam inward against the rocker cam and rocker cam is forced above exhaust cam surface.

Expand new bearing by heating it in oil and install it on crankshaft with seal side towards crankpin journal.

Crankshaft end play on all models is 0.002-0.008 inch (0.05-0.20 mm). At least one 0.015 inch cover or sump gasket must be in place. Additional gaskets in several sizes are available to aid in end play adjustment. If end play is over 0.008 inch (0.20 mm), metal shims are available for use on crankshaft between crankshaft gear and cylinder.

Refer to VALVE TIMING section for proper timing procedure when installing crankshaft.

CAMSHAFT. The camshaft and camshaft gear are an integral casting on all models. The camshaft and gear should be inspected for wear on the journals, cam lobes and gear teeth. On all standard models except 110900, 111200, 111900, 112200 and 113900 models, if camshaft journal diameter is 0.498 inch (12.65 mm) or less, reject camshaft.

On 110900, 111200, 111900, 112200 and 113900 models, camshaft journal rejection size for journal on flywheel side is 0.436 inch (11.07 mm) and for pto side journal rejection size is 0.498 inch (12.65 mm)

On models with "Easy-Spin" starting, the intake cam lobe is designed to hold the intake valve slightly open on part of the compression stroke. Therefore, to check compression, the engine must be turned backwards.

"Easy-Spin" camshafts (cam gears) can be identified by two holes drilled in the web of the gear. Where part number of an older cam gear and an "Easy-Spin" cam gear are the same (except for an "E" following the "Easy-Spin" part number), the gears are interchangeable.

The camshaft used on 11200, 11900 and 112200 models is equipped with the "Easy-Spin" intake lobe and a mechanically operated compression release on the exhaust lobe. With engine stopped or at cranking speed, the spring holds the actuator cam weight inward against the rocker cam. See Fig. B111. The rocker cam is held slightly above the exhaust cam surface which in turn holds the exhaust valve open slightly during compression stroke. This compression release greatly reduces the power needed for cranking.

When engine starts and rpm is increased, the actuator cam weight moves outward overcoming the spring pressure. See Fig. B111A. The rocker cam is rotated below the cam surface to provide normal exhaust valve operation.

Refer to VALVE TIMING section for proper timing procedure when installing camshaft.

VALVE SYSTEM. Valve tappet clearance (cold) for 94500, 94900 and 95500 models is 0.004-0.006 inch (0.10-0.15 mm) for intake valve and 0.007-0.009 inch (0.18-0.23 mm) for exhaust valve.

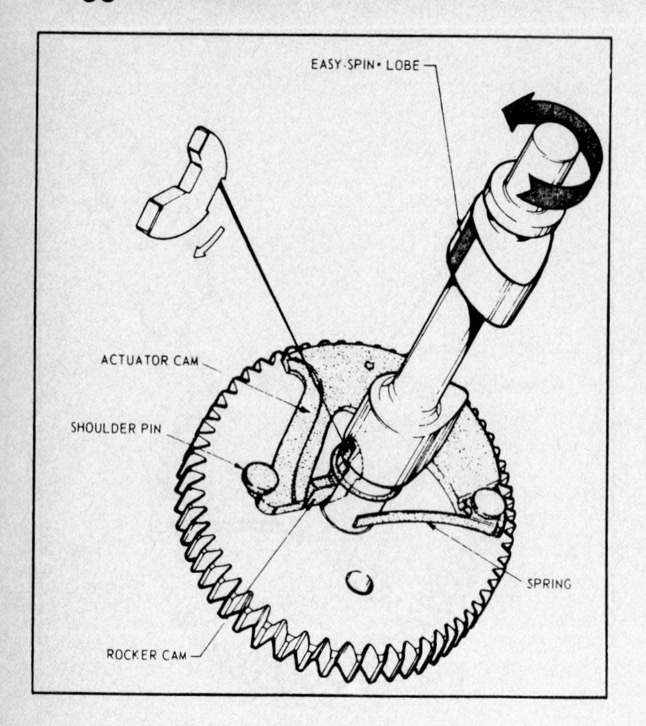

Fig. B111A—When engine starts and rpm is increased, actuator cam weight moves outward allowing rocker cam to rotate below exhaust cam surface.

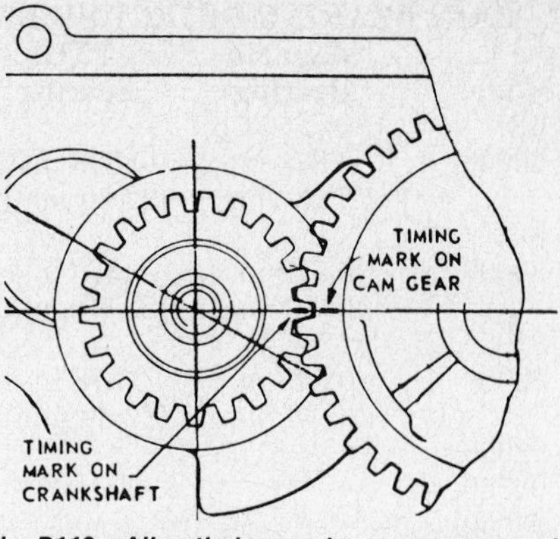

Fig. B113—Align timing marks on cam gear and crankshaft gear on plain bearing models.

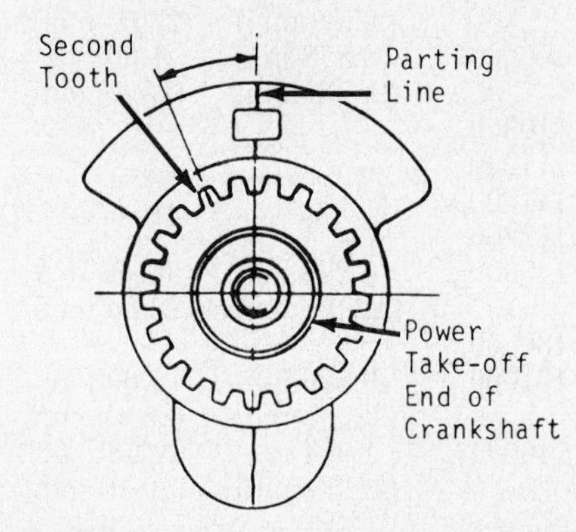

Fig. B114—Location of tooth to align with timing mark on cam gear if mark is not visible on crankshaft gear.

Valve tappet clearance (cold) for all other models is 0.005-0.007 inch (0.13-0.18 mm) for intake valve and 0.009-0.011 inch (0.23-0.28 mm) for exhaust valve.

Valve tappet clearance is adjusted on all models by carefully grinding off end of valve stem to increase clearance or by grinding valve seats deeper and/or renewing valve or lifter to decrease clearance.

Valve face and seat angle should be ground at 45°. Renew valve if margin is 1/16 inch (1.588 mm) or less. Seat width should be 3/64 to 1/16 inch (1.119-1.588 mm).

The valve guides on all engines with aluminum blocks are an integral part of the cylinder block. To renew the valve guides, they must be reamed out and a bushing installed.

On all models except 140000 model, ream guide out first with reamer (B&S part 19064) only about 1/16 inch (1.588 mm) deeper than length of bushing (B&S part 63709). Press in bushing with driver (B&S part 19065) until it is flush with top of guide bore. Finish ream the bushing with a finish reamer (B&S part 19066).

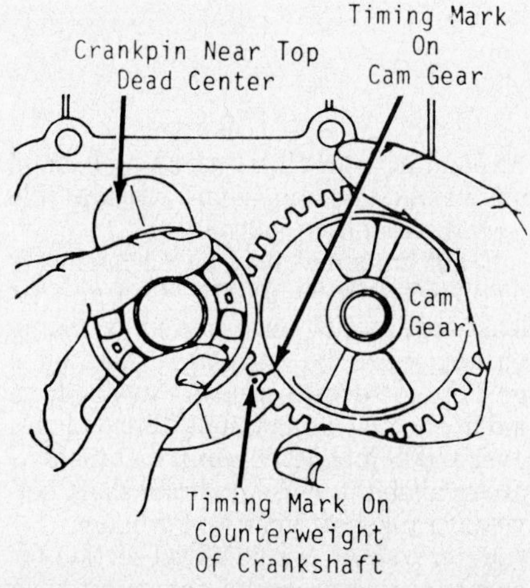

Fig. B112—Align timing marks on cam gear with mark on crankshaft counterweight on ball bearing equipped models.

On 140000 model, ream guide out with reamer (B&S part 19183) only about 1/16 inch (1.588 mm) deeper than top end of flutes on reamer. Press in bushings with soft driver as bushing is finish reamed at the factory.

VALVE TIMING. On engines equipped with ball bearing mains, align the timing mark on the cam gear with the timing mark on the crankshaft counterweight as shown in Fig. B112. On engines with plain bearings, align the timing marks on the camshaft gear with the timing mark on the crankshaft gear as shown in Fig. B113. If the timing mark is not visible on the crankshaft gear, align the timing mark on the camshaft gear with the second tooth to the left of the crankshaft counterweight parting line as shown in Fig. B114.

SERVICING BRIGGS & STRATTON ACCESSORIES

WINDUP STARTER

STARTER OPERATION (EARLY UNITS). Refer to Figs. B115 and B116. Turning the knob (17) to "CRANK" position holds the engine flywheel stationary. Turning the crank (1) winds up the power spring within the spring and cup assembly (8). Spring tension is held by the clutch ratchet (15) and the crank ratchet pawl (2). On some starters, an additional ratchet pawl (2A – Fig. B116) allows the crank to be reversed when there is not enough room for a full turn of crank. Approximately 5½ turns of crank are required to fully wind the power spring. Turning the holding knob (17) to "START" position releases the engine flywheel allowing the power spring to turn the engine.

NOTE: Do not turn knob to "CRANK" position when engine is running.

If engine does not start readily, check to be sure choke is in fully closed position and that ignition system is delivering a spark.

STARTER OPERATION (LATE UNITS). Late production windup starter assemblies have a control lever (Fig. B117) instead of a flywheel holder (17 – Fig. B115 and B116). Moving the control lever to "CRANK" position locks the hub of the starter spring assembly (see lower view, Fig. B117) allowing starter spring to be wound up by turning crank. Moving the control lever to "START" position releases the spring assembly hub. The spring unwinds, engaging the starter ratchet, and cranks up engine.

CAUTION: Never work on engine or driven machinery if starter is wound up. If attached machinery such as mower blade is jammed up so starter will not turn engine, remove blower housing and starter assembly from engine to allow starter to unwind.

OVERHAUL (EARLY UNIT). Remove starter and blower housing from engine. Check condition of flywheel and flywheel holder assembly. Check action of clutch ratchet (15 – Fig. B115 or B116). Remove snap ring (11) to disassemble clutch unit. Bend tangs of blower housing back to remove retainer (9). Remove screw (3) retaining crank and ratchet assembly to power spring cup (8).

Flywheel holder assembly may be renewed as a unit using new rivets and washers (19) after cutting old assembly from blower housing. Knob (17) and snap ring (18) are available separately.

On early models, power spring and cup (8A) are available separately. On later units, spring and cup (8) are available only as an assembly.

Renew parts as necessary and reassemble as follows: Place ratchet (15)

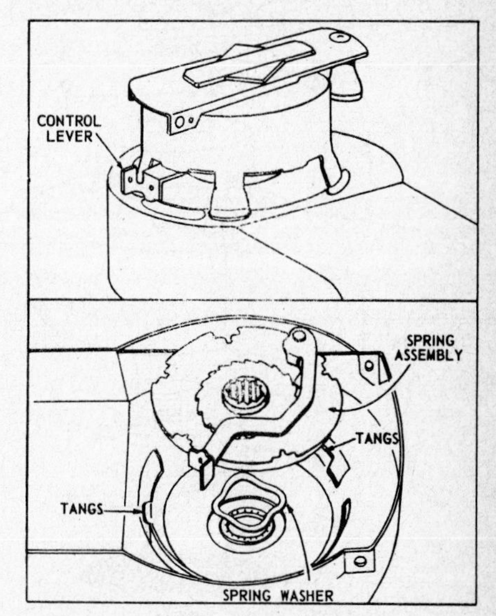

Fig. B117 – On late production units, control lever (top view) locks hub of spring assembly (bottom) to allow spring to be wound up by turning crank. To remove spring assembly, first remove screw (3 – Fig. B118), washer (4) and crank (1). Bend tangs of blower housing out as shown in bottom view and lift cup, spring and control lever assembly from housing. Note spring washer placed between cup and washer.

in cup (13), drop the balls (12) in beside ratchet and install washer (16) and snap ring (11). Reassemble blower housing and starter unit in reverse of disassembly procedure.

OVERHAUL (LATE UNIT). Refer to Fig. B118 and move control lever (L) to "START" position. If starter is wound up and will not turn engine, hold crank (1) securely and loosen screw (3). Remove blower housing and starter unit from engine and, if not already done, remove screw (3) and crank (1). Turn blower housing over and bend tangs outward as shown in bottom view of Fig. B117; then lift cup, spring and control lever assembly from the blower housing.

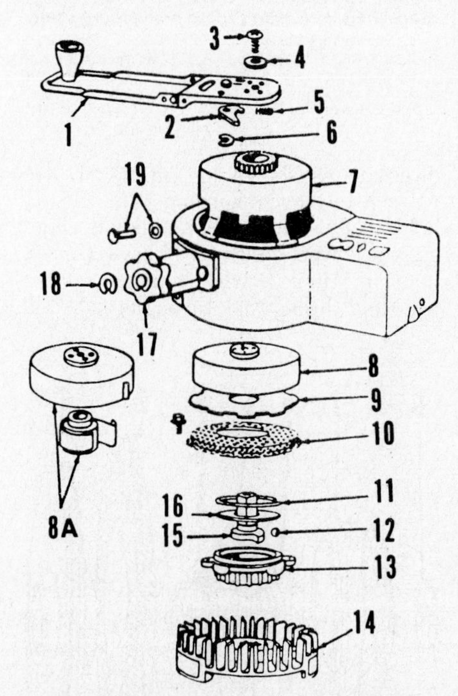

Fig. B115 – Exploded view of typical early production windup starter. Refer to Fig. B116 for exploded view of starter using a ratchet type crank and to Fig. B117 for late production starter.

1. Crank
2. Pawl
3. Screw
4. Washer
5. Spring
6. Snap ring
7. Blower housing
8. Power spring & cup assy.

8A. Power spring & cup assy.

9. Retainer plate
10. Screen
11. Snap ring
12. Clutch balls
13. Clutch housing
14. Flywheel
15. Starter ratchet
16. Ratchet cover
17. Flywheel holder
18. Snap ring
19. Rivet & burr

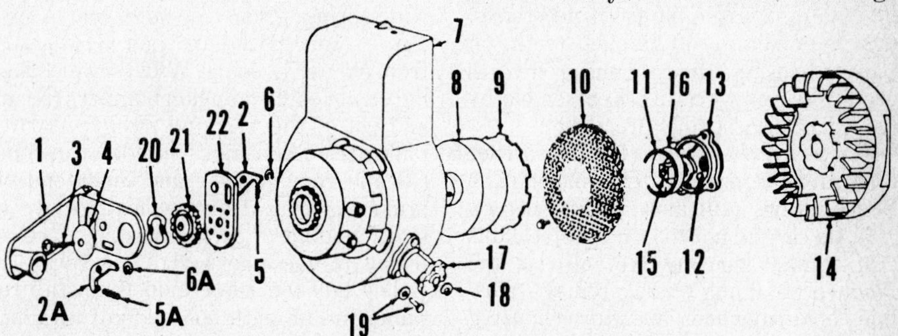

Fig. B116 – Exploded view of typical Briggs & Stratton windup starter equipped with ratchet type windup crank. Starter is used where space does not permit full turn of crank.

1. Crank
2. Pawl

2A. Pawl

3. Screw
4. Washer
5. Spring

5A. Spring

6. Snap ring

6A. Snap ring

7. Blower housing
8. Power spring & cup assy.
9. Retainer plate
10. Screen
11. Snap ring
12. Clutch balls
13. Clutch housing
14. Flywheel
15. Clutch ratchet
16. Ratchet cover
17. Flywheel holder
18. Snap ring
19. Rivet & burr
20. Wave washer
21. Ratchet
22. Pawl housing

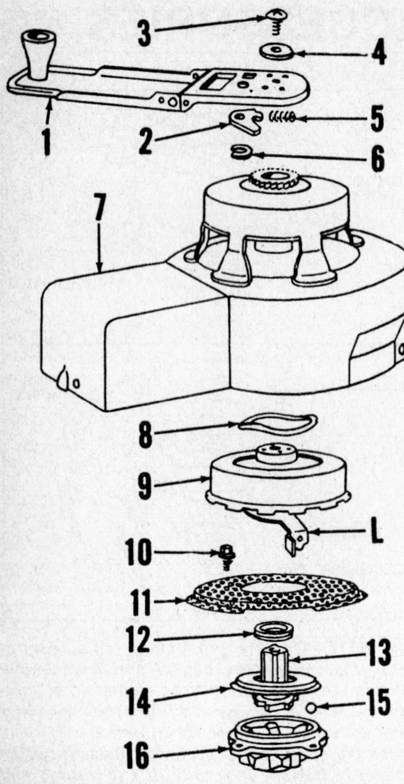

Fig. B118—Exploded view of late production windup starter assembly. Control lever (L) locks ratchet drive hub to allow starter to be wound up and releases hub to crank engine.

L. Control lever
1. Crank
2. Pawl
3. Screw
4. Washer
5. Spring
6. Snap ring
7. Blower housing
8. Spring washer
9. Cup, spring & release assy.
10. Sems screws
11. Rotating screen
12. Seal
13. Starter ratchet
14. Ratchet cover
15. Steel balls
16. Clutch housing

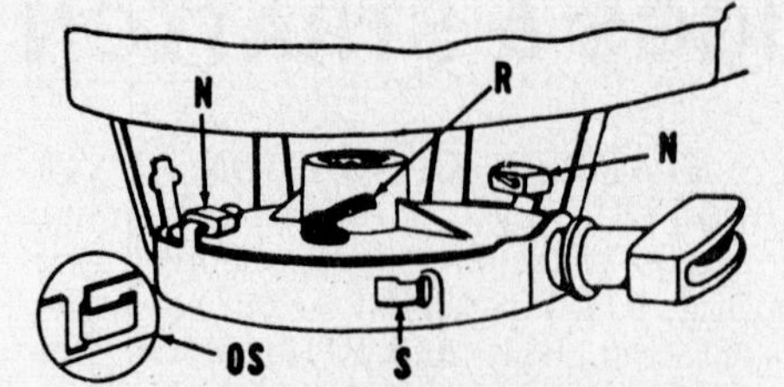

Fig. B120—View of Briggs & Stratton rewind starter assembly.

N. Nylon bumpers
OS. Old style spring
R. Starter rope
S. Rewind spring

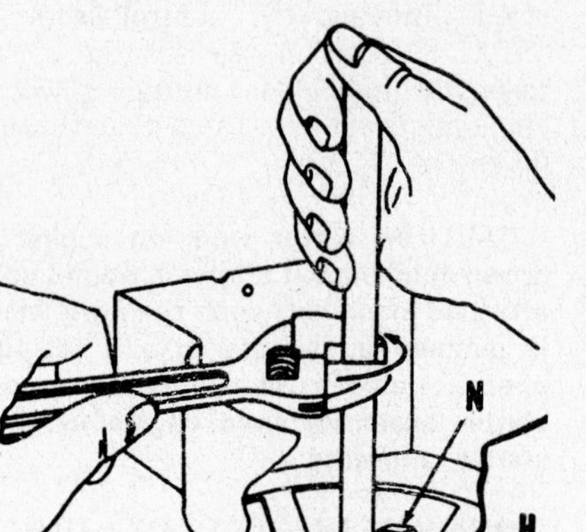

Fig. B121—Using square shaft and wrench to windup the rewind starter spring. Refer to text.

H. Hole in starter pulley
N. Nylon bumpers

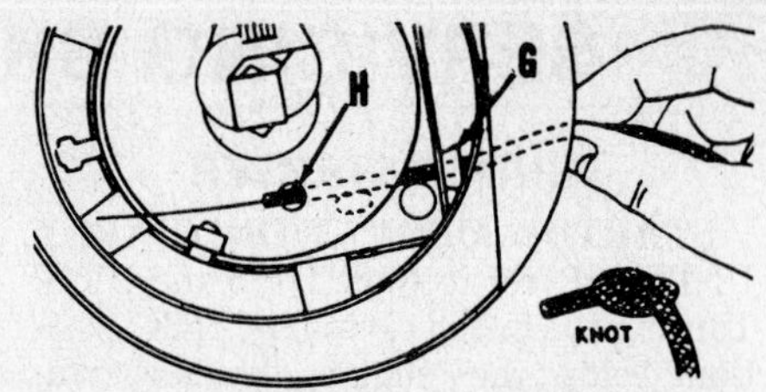

Fig. B122—Threading starter rope through guide (G) in blower housing and hole (H) in starter pulley with wire hooked in end of rope. Tie knot in end of rope as shown.

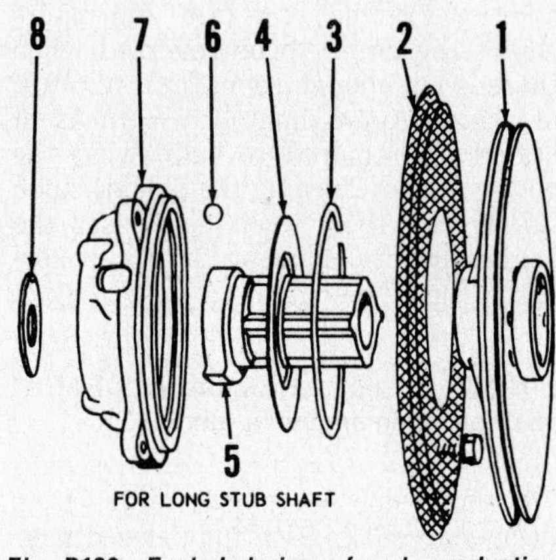

Fig. B123—Exploded view of early production starter clutch unit. Refer to Fig. B126 for view of "long stub shaft". A late type unit (Fig. B125) should be installed when renewing "long" crankshaft with "short" (late production) shaft.

1. Starter rope pulley
2. Rotating screen
3. Snap ring
4. Ratchet cover
5. Starter ratchet
6. Steel balls
7. Clutch housing (flywheel nut)
8. Spring washer

CAUTION: Do not attempt to disassemble the cup and spring assembly; unit is serviced as a complete assembly only.

When reassembling unit, renew the spring washer (8–Fig. B118) if damaged in any way. Grease the mating surfaces of spring cup and blower housing and stick spring washer in place with grease. Renew pawl (2) or pawl spring (5) on crank assembly (1) if necessary. Check condition of ratchet teeth on blower housing. Renew housing if teeth are broken or worn off. Reassemble by reversing disassembly procedure.

Starter ratchet cover (14) and ratchet (13) can be removed after removing the Sems screws (10) and rotating screen (11). Be careful not to lose the steel balls (15). Clutch housing (16) also is the flywheel retaining nut. To remove housing, hold flywheel and turn housing counterclockwise. A special wrench (B&S 19114) is available for removing and installing housing.

To reassemble, first be sure spring washer is in place on crankshaft with cup (hollow) side towards the flywheel. Install starter clutch housing and tighten securely. Place starter ratchet on crankshaft and drop the steel balls in place in housing. Reinstall ratchet cover with new seal (12) if required. Reinstall the rotating screen and install blower housing and starter assembly.

REWIND STARTERS

OVERHAUL. To renew broken rewind spring, grasp free outer end of spring (S–Fig. B120) and pull broken end from starter housing. With blower housing removed, bend nylon bumpers (N) up and out of the way and remove starter pulley from housing. Untie knot in rope (R) and remove rope and inner end of broken spring from pulley. Apply a small amount of grease on inner face of pulley, thread inner end of new spring in pulley hub (on older models, install retainer in hub with split side of retainer away from spring hook) and place pulley in housing. Renew nylon bumpers if necessary and bend bumpers down to within 1/8 inch (3.175 mm) of the pulley. Insert a 3/4-inch (19 mm) square bar in pulley hub and turn pulley approximately 13½ turns in a counterclockwise direction as shown in Fig. B121. Tie wrench to blower housing with wire to hold pulley so hole (H) in pulley is aligned with rope guide (G–Fig. B122) in housing. Hook a wire in inner end of rope and thread rope through guide and

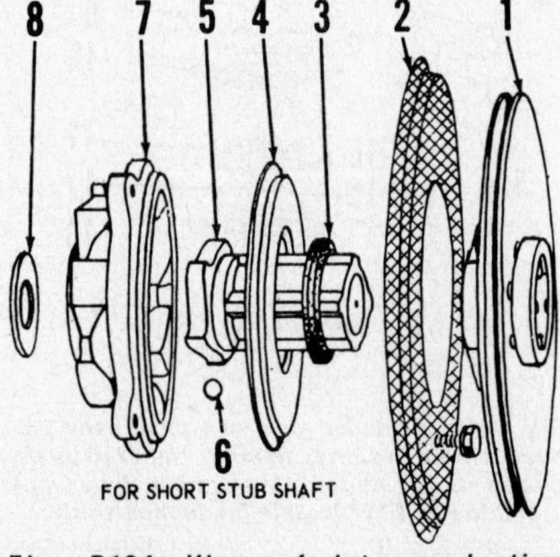

Fig. B124—View of late production sealed starter clutch unit. Late unit can be used with "short stub shaft" only. Refer to Fig. B126. Refer to Fig. B125 for cutaway view of ratchet (5).

1. Starter rope pulley
2. Rotating screen
3. Rubber seal
4. Ratchet cover
5. Starter ratchet
6. Steel balls
7. Clutch housing (flywheel nut)
8. Spring washer

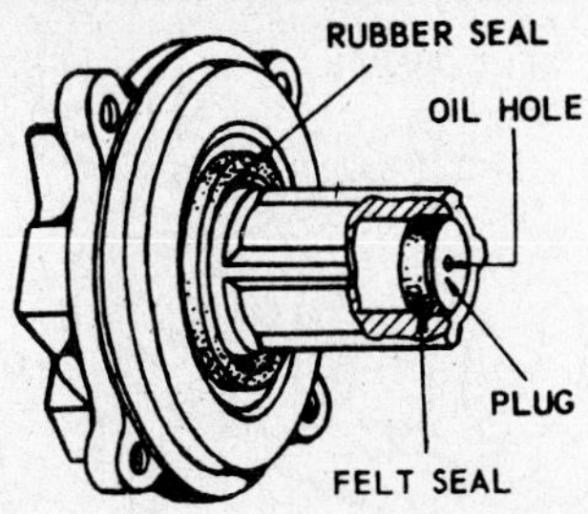

Fig. B125 – Cutaway view showing felt seal and plug in end of late production starter ratchet (5 – Fig. B124).

hole in pulley. Tie a knot in rope and release the pulley allowing the spring to wind the rope into the pulley groove.

To renew starter rope only, it is not generally necessary to remove starter pulley and spring. Wind up the spring and install new rope as outlined in preceding paragraph.

Two different types of starter clutches have been used. Refer to exploded view of early production unit in Fig. B123 and exploded view of late production unit in Fig. B124. The outer end of the late production ratchet (refer to cutaway view in Fig. B125) is sealed with a felt and a retaining plug and a rubber ring is used to seal ratchet to ratchet cover.

To disassemble early type starter clutch unit, refer to Fig. B123. Remove snap ring (3). Lift ratchet (5) and cover (4) from starter housing (7) and crankshaft. Be careful not to lose the steel balls (6). Starter housing (7) is also flywheel retaining nut. To remove housing, first remove screen (2) and using B&S flywheel wrench 19114, unscrew housing from crankshaft in counterclockwise direction. When reinstalling housing, be sure spring washer (8) is placed on crankshaft with cup (hollow) side towards flywheel. Install starter housing and tighten securely. Reinstall rotating screen. Place ratchet on crankshaft and into housing, then insert the steel balls. Reinstall cover and retaining snap ring.

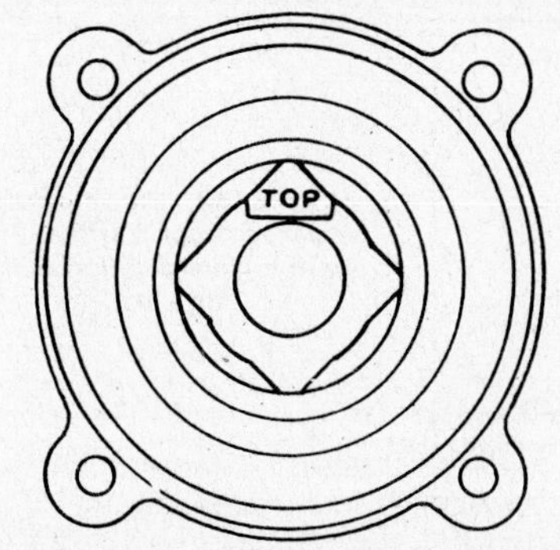

Fig. B128 – When installing blower housing and starter assembly, turn starter ratchet so word "TOP" stamped on outer end of ratchet is towards engine cylinder head.

To disassemble late starter clutch unit, refer to Fig. B124. Remove rotating screen (2) and starter ratchet cover (4). Lift ratchet (5) from housing and crankshaft and extract the steel balls (6). If necessary to remove housing (7), hold flywheel and unscrew housing in counterclockwise direction using B&S flywheel wrench 19114. When installing housing, be sure spring washer (8) is in place on crankshaft with cup (hollow) side towards flywheel. Tighten housing securely. Inspect felt seal and plug in outer end of ratchet. Renew ratchet if seal or plug if damaged as these parts are not serviced separately. Lubricate the felt with oil and place ratchet on crankshaft. Insert the steel balls and install ratchet cover, rubber seal and rotating screen.

NOTE: Crankshafts used with early and late starter clutches differ. Refer to Fig. B126. If renewing early (long) crankshaft with late (short) shaft, also install late type starter clutch unit. If renewing early starter clutch with late type unit, the crankshaft must be shortened to the dimension shown for short shaft in Fig. B126. Also, hub of starter rope pulley must be shortened to ½-inch (12.7 mm) dimension shown in Fig. B127. Bevel end of crankshaft after removing the approximate ⅜ inch (9.525 mm) from shaft.

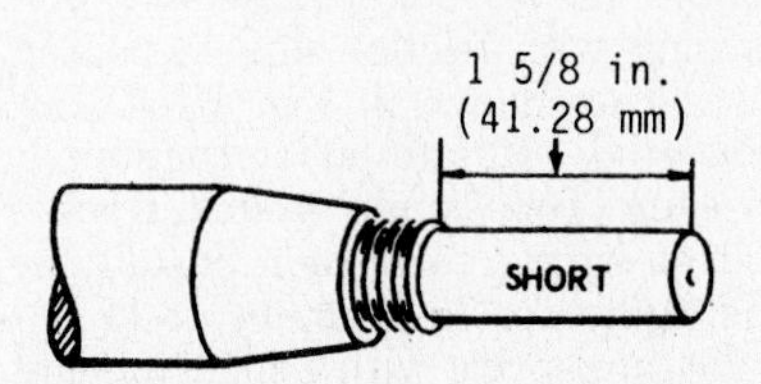

Fig. B126 – Crankshaft with short stub (top view) must be used with late production starter clutch assembly. Early crankshaft (bottom view) can be modified by cutting off stub end to the dimension shown in top view and beveling end of shaft to allow installation of late type clutch unit.

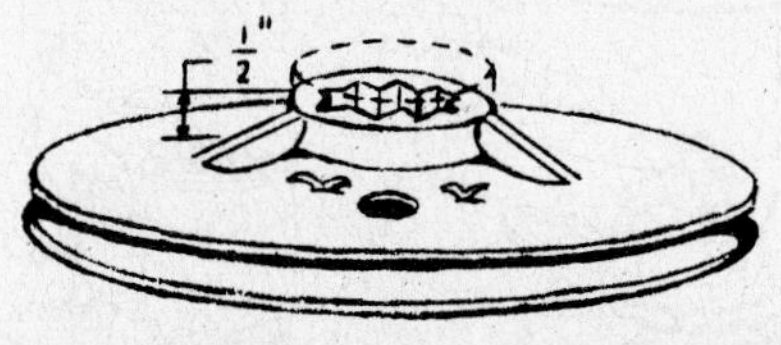

Fig. B127 – When installing a late type starter clutch unit as replacement for early type, either install new starter rope pulley or cut hub of old pulley to dimension shown.

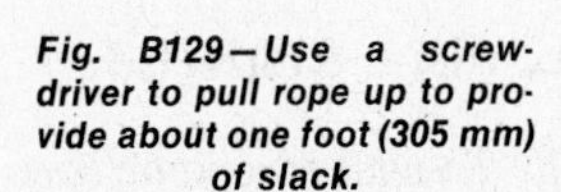

Fig. B129 – Use a screwdriver to pull rope up to provide about one foot (305 mm) of slack.

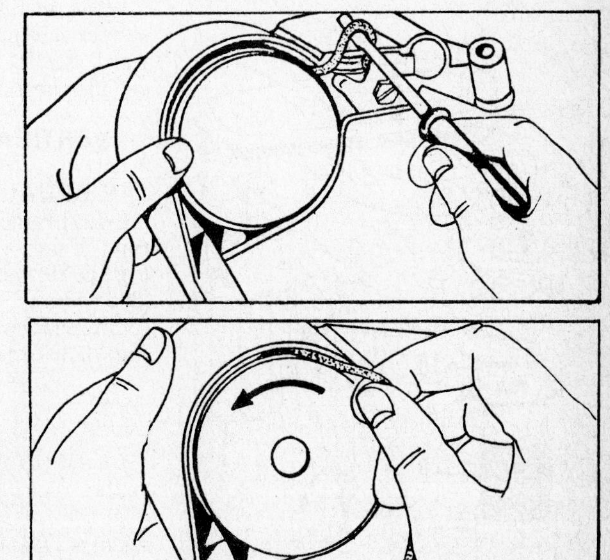

Fib. B130 – Rotate pulley and rope 3 turns counterclockwise to remove spring tension from rope.

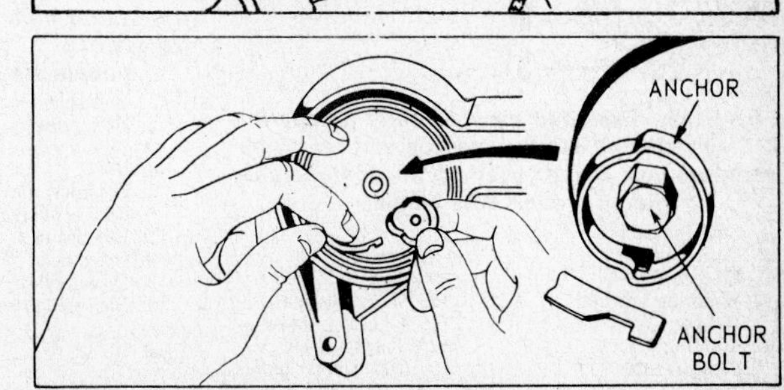

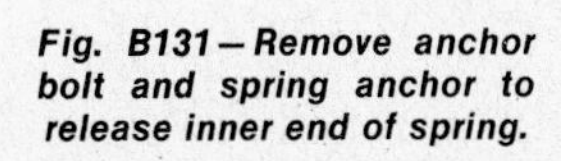

Fig. B131 – Remove anchor bolt and spring anchor to release inner end of spring.

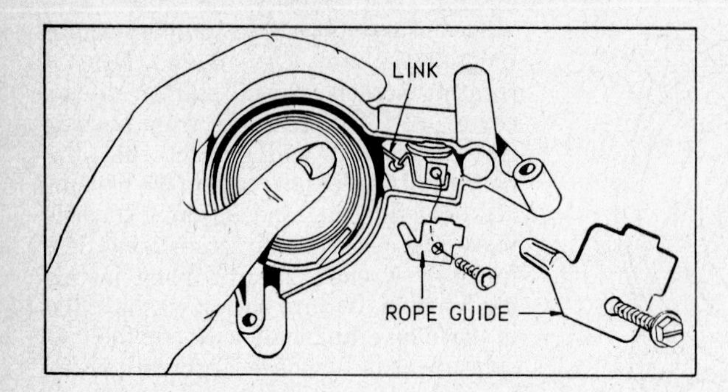

Fig. B132 — Rope guide removed from Vertical Pull starter. Note position of friction link.

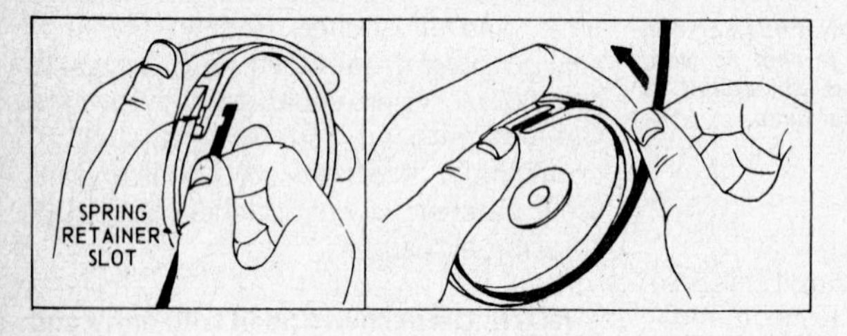

Fig. B133 — Hook outer end of spring in retainer slot, then coil the spring counterclockwise in the housing.

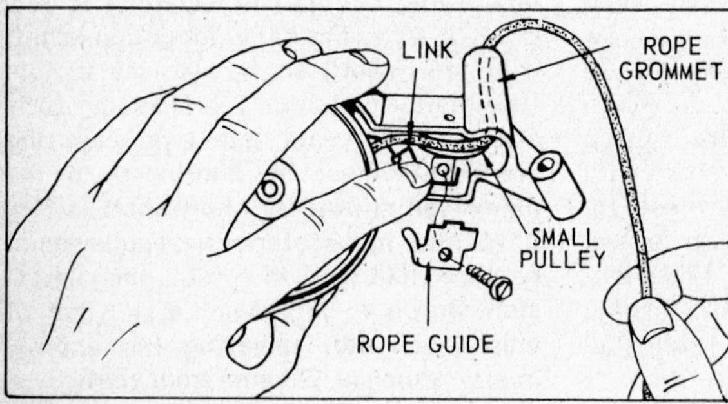

Fig. B134 — When installing pulley assembly in housing, make sure friction link is positioned as shown, then install rope guide.

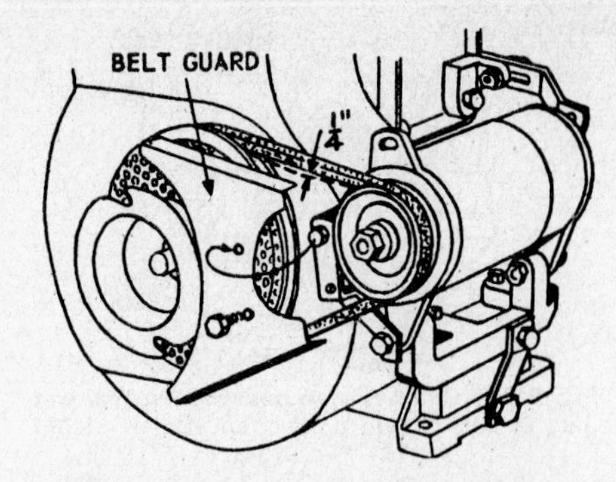

Fig. B137 — View showing starter-generator belt adjustment on models so equipped. Refer to text.

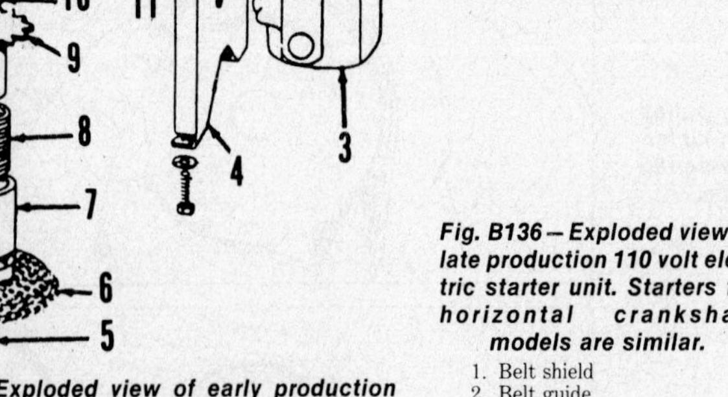

Fig. B135 — Exploded view of early production 110 volt electric starter used on vertical crankshaft models. Starter used on horizontal crankshaft models was similar.

1. Chain shield
2. Drive chain
3. Electric motor
4. Mounting bracket
5. Washer
6. Rotating screen
7. Shaft adapter
8. Spring
9. Sprocket & hub
10. Thrust washer
11. Adjusting screw
12. Guard washer
13. Rope starter pulley

When installing blower housing and starter assembly, turn starter ratchet so word "TOP" on ratchet is towards cylinder head.

VERTICAL PULL STARTER

OVERHAUL. To renew rope or spring, first remove all spring tension from rope. Using a screwdriver as shown in Fig. B129, pull rope up about one foot (305 mm). Wind rope and pulley counterclockwise 3 turns as shown in Fig. B130 to remove all tension. Carefully pry off the plastic spring cover. Refer to Fig. B131 and remove anchor bolt and spring anchor. Carefully remove spring from housing. Unbolt and remove rope guide. Note position of the friction link in Fig. B132. Remove old rope from pulley.

It is not necessary to remove the gear retainer unless pulley or gear is damaged and renewal is necessary. Clean and inspect all parts. The friction link should move the gear to both extremes of its travel. If not, renew the linkage assembly.

Install new spring by hooking end of retainer slot and winding until spring is coiled in housing. See Fig. B133. Insert new rope through the housing and into the pulley. Tie a small knot, heat seal the knot and pull it tight into the recess in pulley. Install pulley assembly in the housing with friction link in pocket of casting as shown in Fig. B134. Install rope guide. Rotate pulley counterclockwise until rope is fully wound. Hook free end of spring to anchor, install screw and tighten to 75-90 in.-lbs. (8-11 N·m). Lubricate spring with a small amount of engine oil. Snap the plastic spring cover in place. Preload spring by pulling rope up about one foot (305 mm), then winding rope and pulley 2 or 3 turns clockwise.

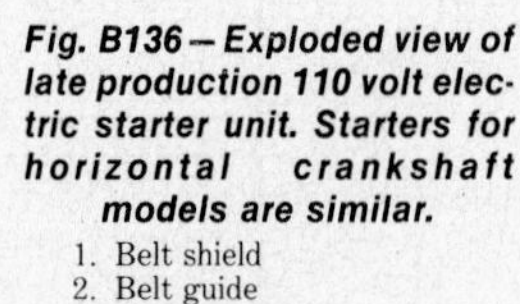

Fig. B136 — Exploded view of late production 110 volt electric starter unit. Starters for horizontal crankshaft models are similar.

1. Belt shield
2. Belt guide
3. Clutch assy.
4. Cover
5. Commutator brush assy.
6. Connector plug
7. Electric motor
8. Drive belt
9. Drive pulley
10. Blower housing

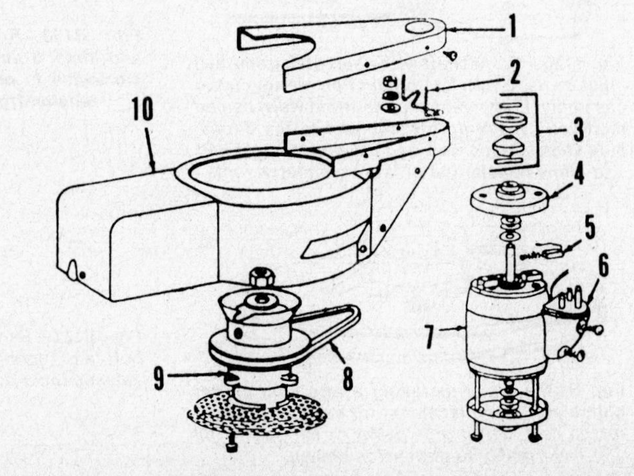

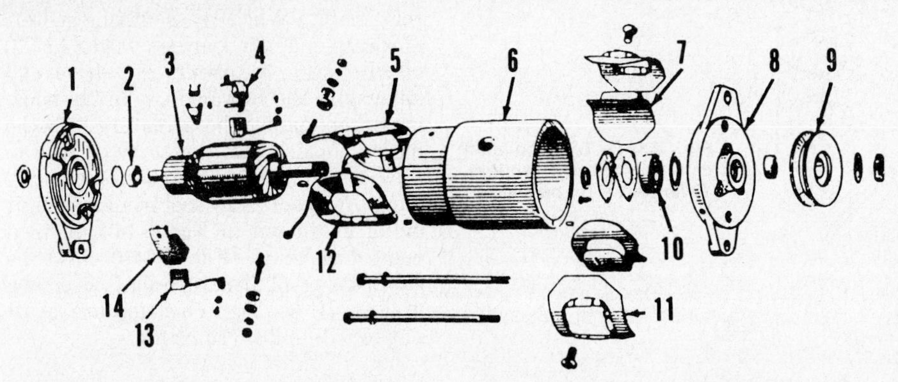

Fig. B138—Exploded view of Delco-Remy starter-generator unit used on some B&S engines.

1. Commutator end frame
2. Bearing
3. Armature
4. Ground brush holder
5. Field coil
6. Frame
7. Pole shoe
8. Drive end frame
9. Pulley
10. Bearing
11. Field coil insulator
12. Field coil
13. Brush
14. Insulated brush holder

110-VOLT ELECTRIC STARTERS (CHAIN AND BELT DRIVE)

Early production 110-volt electric starter with chain drive is shown in Fig. B135. Later type with belt drive is shown in Fig. B136. Starters for horizontal crankshaft engines are similar to units shown for vertical shaft engines.

Chain (2–Fig. B135) on early models is adjusted by changing position of nuts on adjusting stud (11) so chain deflection is approximately 1/4 inch (6.35 mm) between sprockets. There should be about 1/32-inch (0.8 mm) clearance between spring (8) and sprocket (9) when unit is assembled. Clutch unit (7, 8 and 9) should disengage when engine starts.

Belt (8–Fig. B135) tension is adjusted by shifting starter motor in slotted mounting holes so clutch unit (3) on starter motor will engage belt and turn engine, but so belt will not turn starter when clutch is disengaged.

CAUTION: Always connect starter cord to starter before plugging cord into 110-volt outlet and disconnect cord from outlet before removing cord from starter connector. Do not run the electric starter for more than 60 seconds at a time.

12-VOLT STARTER-GENERATOR UNIT

The combination starter-generator functions as a cranking motor when the starting switch is closed. When engine is operating and with starting switch open, the unit operates as a generator. Generator output and circuit voltage for the battery and various operating requirements are controlled by a current-voltage regulator. On units where voltage regulator is mounted separately from generator unit, do not mount regulator with cover down as regulator will not function in this position. To adjust belt tension, apply approximately 30 pounds (14 kg) pull on generator adjusting flange and tighten mounting bolts. Belt tension is correct when a pressure of 10 pounds (5 kg) applied midway between pulleys will deflect belt 1/4 inch (6.35 mm). See Fig. B137. On units equipped with two drive belts, always renew belts in pairs. A 50-ampere hour capacity battery is recommended. Starter-generator units are intended for use in temperatures above 0° F (−18° C). Refer to Fig. B138 for exploded view of starter-generator. Parts and service on the starter-generator are available at authorized Delco-Remy service centers.

12-VOLT ELECTRIC STARTER (BELT DRIVE)

Refer to Fig. B139. Adjust position of starter (10) so clutch will engage belt and turn engine when starter is operated, but so belt will not turn starter when engine is running. The 12-volt electric starter is intended for use in temperatures above 15° F (−9° C). Driven equipment should be disengaged before using starter. Parts and service on starter motor are available at Autolite service centers.

CAUTION: Do not use jumper (booster) battery as this may result in damage to starter motor.

GEAR DRIVE STARTERS

Two types of gear drive starters may be used, a 110 volt AC starter or a 12 volt DC starter. Refer to Fig. B140 for an exploded view of starter motor. A properly grounded receptacle should be used with power cord connected to 110 volt AC starter motor. A 32 ampere hour capacity battery is recommended for use with 12 volt DC starter motor.

To renew a worn or damaged flywheel ring gear, drill out retaining rivets using a 3/16-inch drill. Attach new ring gear

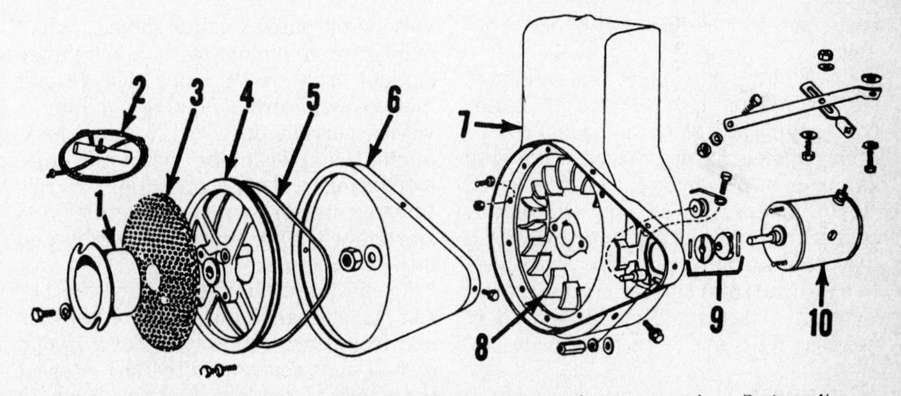

Fig. B139—Exploded view of Briggs & Stratton 12 volt starter used on some engines. Parts and/or service on starter motor are available through an authorized Autolite service center.

1. Rope pulley
2. Starter rope
3. Rotating screen
4. Starter driven pulley
5. Drive belt
6. Belt guard
7. Blower housing
8. Flywheel
9. Starter clutch
10. Starter motor.

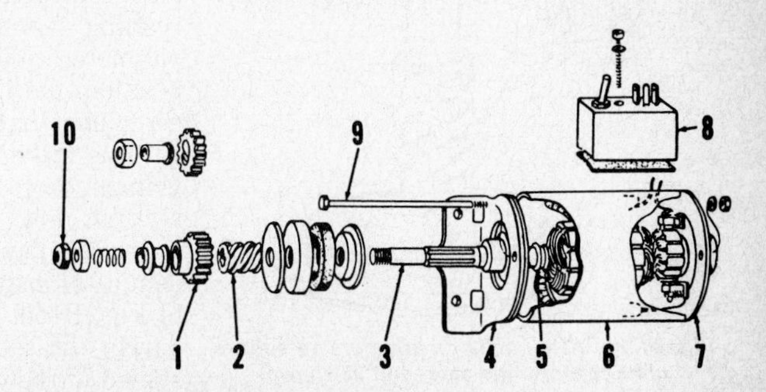

Fig. B140—View of 110 volt AC starter motor. 12 volt DC starter motor is similar. Rectifier and switch unit (8) is used on 110 volt motor only.

1. Pinion gear
2. Helix
3. Armature shaft
4. Drive cap
5. Thrust washer
6. Housing
7. End cap
8. Rectifier & switch
9. Bolt
10. Nut

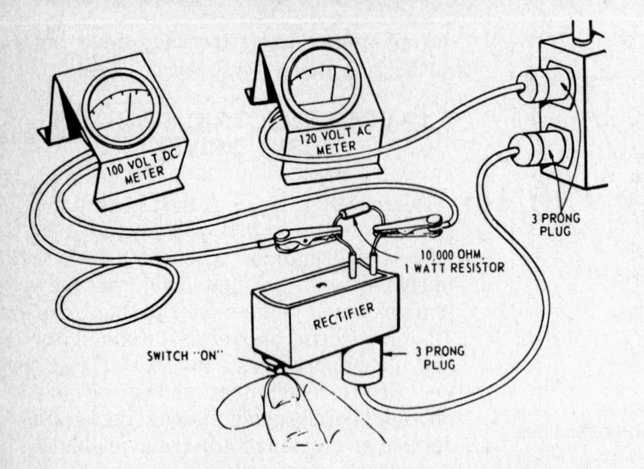

Fig. B141—Test connections for 110 volt rectifier. Refer to text for procedure.

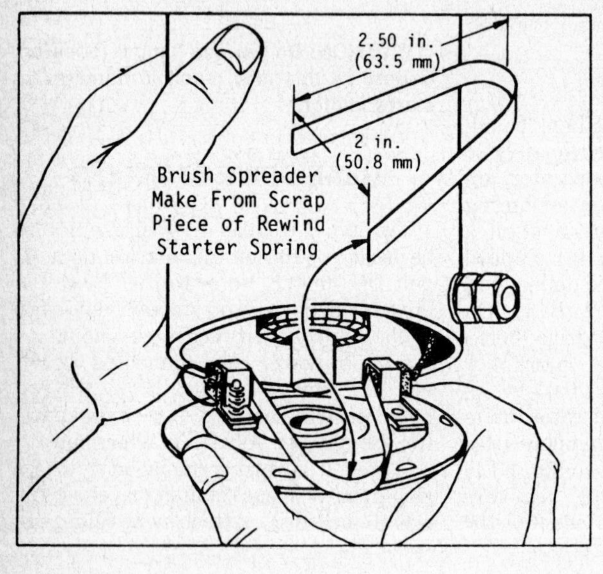

Fig. B142—Tool shown may be fabricated to hold brushes when installing motor and cap.

using screws provided with new ring gear.

To check for correct operation of starter motor, remove starter motor from engine and place motor in a vise or other holding fixture. Install a 0-5 amp ammeter in power cord to 110 volt AC starter motor. On 12 volt DC motor, connect a 6 volt battery to motor with a 0-50 amp ammeter in series with positive (+) line from battery starter motor. Connect a tachometer to drive end of starter. With starter activated on 110 volt motor, starter motor should turn at 5200 rpm minimum with a maximum current draw of 3½ amps. The 12 volt motor should turn at 5000 rpm minimum with a current draw of 25 amps maximum. If starter motor does not operate satisfactorily, check operation of rectifier or starter switch. If rectifier and starter switch are good, disassemble and inspect starter motor.

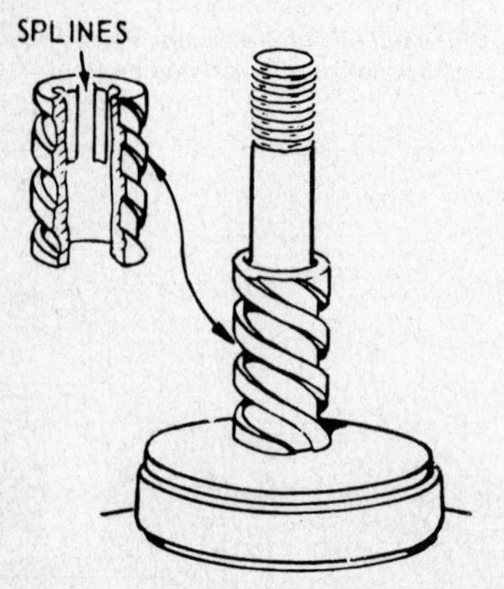

Fig. B143—Install helix on armature so splines of helix are to the outer end as shown.

To check the rectifier used on the 110 volt AC starter motor, remove rectifier unit from starter motor. Solder a 10,000 ohm, 1 watt resistor to the DC internal terminals of the rectifier as shown in Fig. B141. Connect a 0-100 range DC voltmeter to resistor leads. Measure the voltage of the AC outlet to be used. With starter switch in "OFF" position, a zero reading should be shown on DC voltmeter. With starter switch in "ON" position, the DC voltmeter should show a reading that is 0-14 volts lower than AC line voltage measured previously. If voltage drop exceeds 14 volts, renew rectifier unit.

Disassembly of starter motor is evident after inspection of unit and referral to Fig. B140. Note position of bolts (9) during disassembly so they can be installed in their original positions during reassembly. When reassembling motor, lubricate end cap bearings with SAE 20 oil. Be sure to match the drive cap keyway to the stamped key in the housing when sliding the armature into the motor housing. Brushes may be held in their holders during installation by making a brush spreader tool from a piece of metal as shown in Fig. B142. Splined end of helix (2–Fig. B140) must be towards end of armature shaft as shown in Fig. B143. Tighten armature shaft nut to 170 in-lbs. (19 N·m).

SAFETY EQUIPMENT

BAND BRAKE. Flywheel band brake is designed to stop the blade within three seconds after the operator releases the safety control handle. (Fig. B144).

DISASSEMBLY. Refer to Fig. B145 for identification of parts. Disconnect spark plug wire and securely ground wire end. Remove brake control bracket cover, then loosen cable clamp mounting screw and withdraw cable from control lever. On models equipped with electric start, disconnect battery wires and remove battery mounting screws, then withdraw battery. Remove starter motor housing cover to allow access to blower housing mounting screws. On all models remove blower housing and rotating screen mounting screws, then withdraw components from engine assembly. Bend control lever tang (Fig.

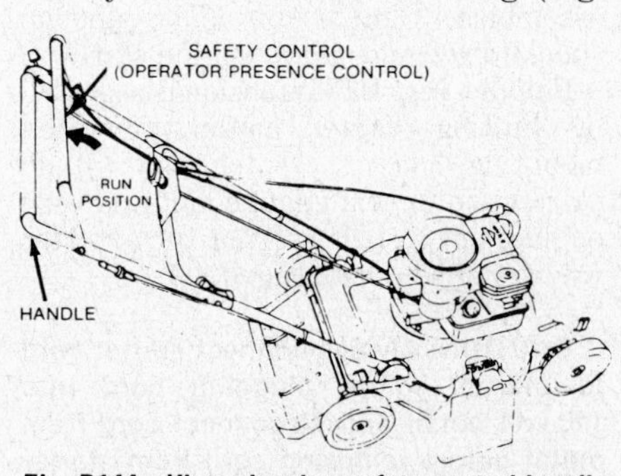

Fig. B144—View showing safety control handle and extended rope start typical of the type used on most models.

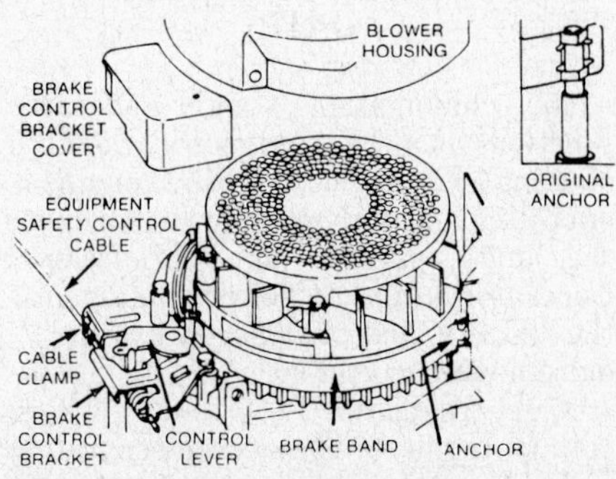

Fig. B145—View showing typical brake band components.

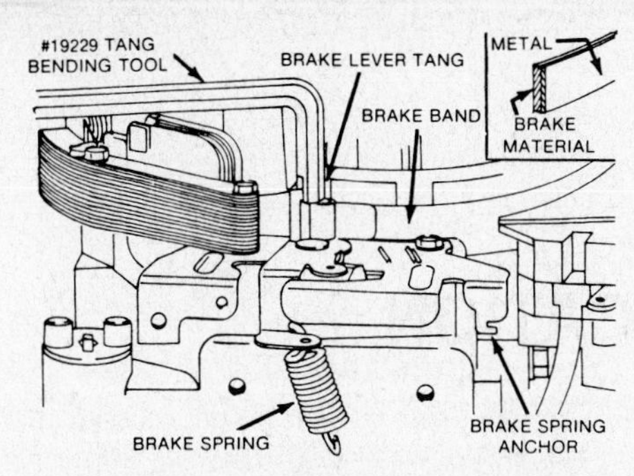

Fig. B146—Remove brake band using Briggs & Stratton tool No. 19229 as outlined in text.

B146) using Briggs & Stratton tool 19229 to clear band brake loop. Release brake spring tension on lever, then remove brake band. Brake band should be renewed if damaged or brake material is worn to less than 0.030 inch (0.8 mm) thickness.

Renew all components as needed, then reassemble in reverse order of disassembly. Refer to the following ADJUSTMENT section for brake band adjustment.

ADJUSTMENT. Loosen control bracket mounting screws (B – Fig. B147) and on electric start models alternator mounting screw (A). Unhook brake spring from control bracket spring anchor. Install special gage link tool (Briggs & Stratton number 19256) in control lever and cable clamp screw hole as shown in Fig. B147, then reinstall brake spring. Move control bracket in the direction as shown by the arrow (Fig. B147) until tension on gage link tool is **JUST** released, then tighten screw "A" and screws "B". Remove gage link tool.

Using a 7/8 inch socket and a torque wrench, measure the amount of force required to rotate the flywheel with brake applied. Flywheel should not rotate when **less than** 45 in.-lbs. (4 N·m) is applied. If force required is less than specified, check the following for damage, misalignment or incorrect adjustment: brake band lining contour or wear, brake band anchors, control bracket, brake spring and anchor. Repair as needed, then repeat brake band application test. When brake band is released, engine should turn freely without any restricted movement.

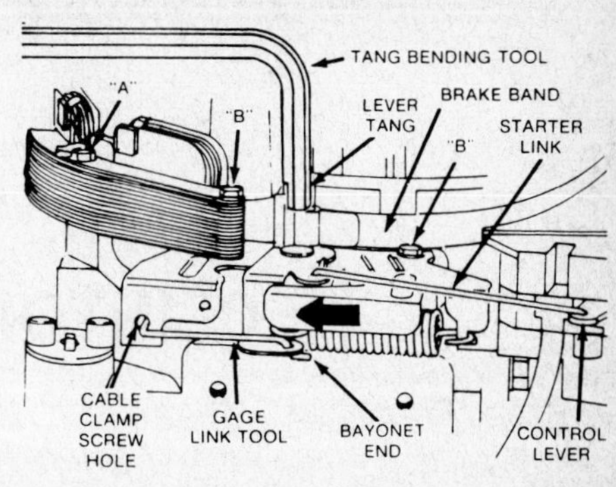

Fig. B147—Adjust brake band using procedure outlined in text.

NOTE: Be sure on models so equipped, engine stop switch is functioning correctly before placing mower back into service.

CLINTON

CLINTON ENGINES CORPORATION
Maquoketa, Iowa

CLINTON ENGINE IDENTIFICATION INFORMATION

To obtain the correct service replacement parts when overhauling Clinton engines, it is important that the engine be properly identified as to:

1. Model number
2. Variation number
3. Type letter

A typical nameplate from the model number series engines prior to 1961 is shown in Fig. CL1. In this example, the following information is noted from the nameplate:

1. Model number – **B-760**
2. Variation number – **AOB**
3. Type letter – **B**

In some cases, the model number may be shown as in the following example:

D-790-2124

Thus, "D-790" would be the model number of a D-700-2000 series engine in which the digits "2124" would be the variation number.

In late 1961, the identification system for Clinton engines was changed to be acceptable for use with IBM inventory record systems. A typical nameplate from a late production engine is shown in Fig. CL2.

In this example, the following information is noted from the nameplate:

1. Model number – 405-0000-000
2. Variation number – 070
3. Type letter – D

In addition, the following information may be obtained from the model number on the nameplate:

First digit – identifies type of engine, i.e., 4 means four-stroke engine and 5 means two-stroke engine.

Second and third digits - completes basic identification of engine. Odd numbers will be used for vertical shaft engines and even numbers for horizontal shaft engines; i.e., 405 would indicate a four-stroke vertical shaft engine and 500 would indicate a two-stroke horizontal shaft engine.

Fourth digit – indicates type of starter as follows:

0 – Rewind starter
1 – Rope starter
2 – Impluse starter
3 – Crank starter
4 – 12 Volt electric starter
5 – 12 Volt starter-generator
6 – 110 Volt electric starter
7 – 12 Volt generator
8 – unassigned to date
9 – Short block assembly.

Fifth digit – indicates bearing type, etc., as follows:

0 – Standard bearing
1 – Aluminum or bronze sleeve bearing with flange mounting surface and pilot diameter on engine mounting face for mounting equipment concentric to crankshaft center line
2 – Ball or roller bearing
3 – Ball or roller bearing with flange mounting surface and pilot diameter on engine mounting face for mounting equipment concentric to crankshaft center line
4 – Numbers 4 through 9 are unassigned to date

Sixth digit – indicates auxiliary power takeoff and speed reducers as follows:

0 – Without auxiliary power pto or speed reducer
1 – Auxiliary power takeoff
2 – 2:1 speed reducer
3 – not assigned to date
4 – 4:1 speed reducer
5 – not assigned to date
6 – 6:1 speed reducer
7 – numbers 7 through 9 are unassigned to date

Seventh digit – if other than "0" will indicate a major design change.

Eighth, ninth & tenth digits – identfies model variations.

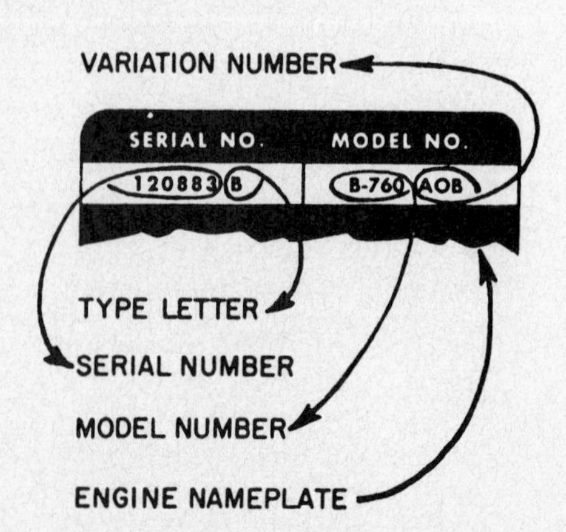

Fig. CL1—Typical nameplate from Clinton engine manufactured prior to late 1961.

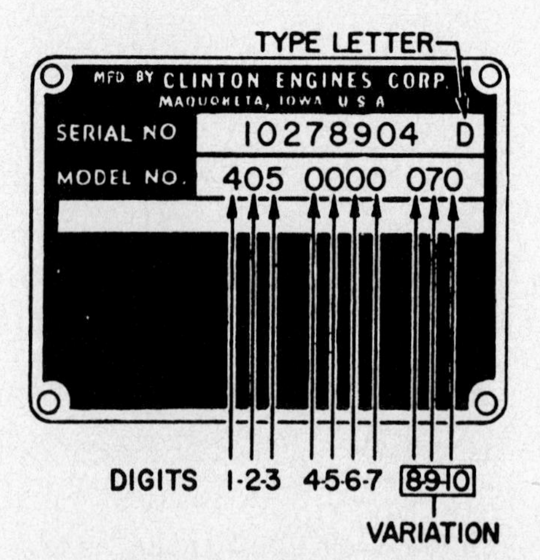

Fig. CL2—Typical nameplate from Clinton engine after model identification system was changed in late 1961. First seven digits indicate basic features of engine. Engines with "Mylar" (plastic) nameplate have engine model and serial numbers stamped on the cylinder air deflector next to the nameplate.

CLINTON

Model	Bore	Stroke	Displacement
E-65 CW, E-65 CCW	2.125 in. (53.2 mm)	1.625 in. (41.3 mm)	5.8 cu. in. (92 cc)
200, A-200, AVS-200-1000, VS-200, VA-200-1000, VS-200-2000, VS-200-3000	1.875 in. (47.6 mm)	1.625 in. (41.3 mm)	4.5 cu. in. (74 cc)
VS-200-4000	2.125 in. (53.2 mm)	1.625 in. (41.3 mm)	5.8 cu. in. (92 cc)
290	1.875 in. (47.6 mm)	1.625 in. (41.3 mm)	4.5 cu. in. (74 cc)
A-400	2.125 in. (53.2 mm)	1.625 in. (41.3 mm)	5.8 cu. in. (92 cc)
A-400-1000, AVS-400, CVS-400-1000, VS-400, VS-400-1000, VS-400-2000, VS-400-3000, VS-400-4000	2.125 in. (53.2 mm)	1.625 in. (41.3 mm)	5.8 cu. in. (92 cc)
A-460*	2.125 in. (53.2 mm)	1.625 in. (41.3 mm)	5.8 cu. in. (92 cc)
490, A-490	2.125 in. (53.2 mm)	1.625 in. (41.3 mm)	5.8 cu. in. (92 cc)
500-0100-000, 501-0000-000, 501-0001-000, GK-590	2.125 in. (53.2 mm)	1.625 in. (41.3 mm)	5.8 cu. in. (92 cc)
502-0308-000, 503-0308-000	2.375 in. (60.3 mm)	1.750 in. (44.5 mm)	7.8 cu. in. (127 cc)
502-0309-000, 503-0309-000	2.500 in. (63.5 mm)	1.750 in. (44.5 mm)	8.6 cu. in. (141 cc)

*Reduction gear

ENGINE IDENTIFICATION

Clinton two-stroke engines are of aluminum alloy die-cast construction with an integral cast-in iron cylinder sleeve. Reed type inlet valves are used on all models. Refer to CLINTON ENGINE IDENTIFICATION INFORMATION section preceding this section.

MAINTENANCE

SPARK PLUG. Champion H10J or equivalent spark plug is recommended for use in E-65 CW and E-65 CCW models.

Champion H11 or H11J or equivalent spark plug is recommended for use in 200, A-200, AVS-200, AVS-200-1000, VS-200, VS-200-1000, VS-200-2000, VS-200-3000 (types A and B), 290, A-400, A-400-1000 (type A), AVS-400, AVS-400-1000, BVS-400, VS-400-1000, VS-400-2000, VS-400-3000, VS-400-4000 (types A and B), 490 and A-490-1000 (type A) models.

Champion J12J or equivalent plug is recommended for VS-200-3000 (type C), VS-200-4000, A-400-1000 (types B, C, D, E, F), CVS-400-1000, VS-400-4000 (types C, D, E, F), A-490-1000 (types B, C, D, E, F), 500-0100-000, 501-000-000 and 501-001-000 models.

Autolite A7NX or equivalent spark plug is recommended for use in 502-0308-000, 502-0309-000, 503-0308-000 and 503-0309-000 models.

On all models, set electrode gap at 0.030 inch (0.76 mm). Use graphite on threads when installing spark plug and tighten spark plug to 275-300 in.-lbs. (31-34 N·m).

CARBURETOR. Several different carburetors have been used on Clinton two-stroke engines. Refer to the appropriate following paragraphs.

Clinton (Walbro) LM Float Type Carburetor. Clinton (Walbro designed) LMB and LMG float type carburetors are used. Refer to Fig. CL5 for identification and exploded view of typical model.

Initial adjustment of idle (9) and main (22) fuel needles from a lightly seated position is 1¼ turns open for each nee-

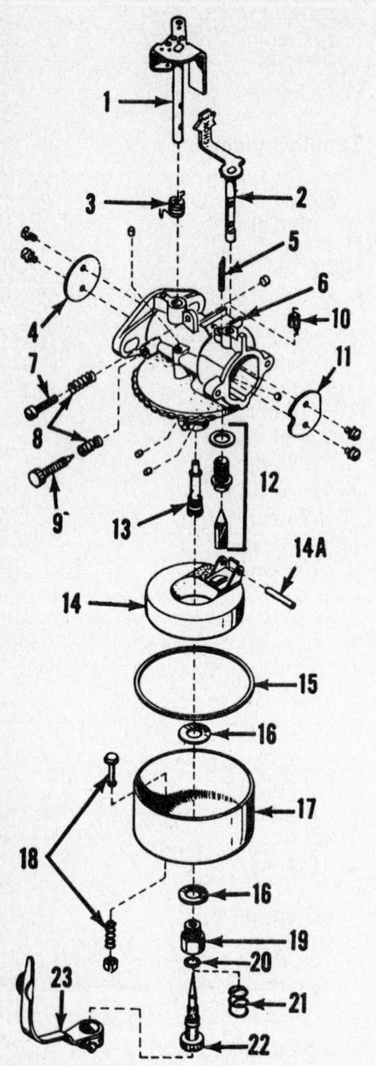

Fig. CL5—Exploded view of typical LMG series carburetor. LMV and LMB series are similar.

1. Throttle shaft
2. Choke shaft
3. Spring
4. Throttle plate
5. Spring
6. Carburetor body
7. Idle stop screw
8. Springs
9. Idle fuel needle
10. Spring
11. Choke plate
12. Inlet needle & seat
13. Main nozzle
14. Float
14A. Float pin
15. Gasket
16. Gaskets
17. Float bowl
18. Drain valve
19. Retainer
20. Seal
21. Spring
22. Main fuel needle
23. Lever (optional)

dle. Final adjustments are made with engine at operating temperature and running. Operate engine at rated speed under load and adjust main fuel mixture needle for smoothest engine operation. Operate engine at idle speed and adjust idle fuel mixture needle for smoothest

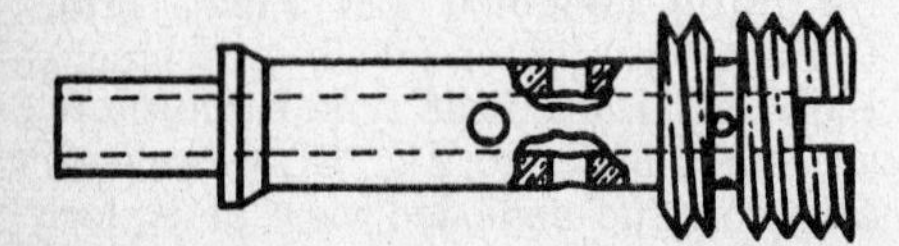

Fig. CL6—When original main fuel nozzle is removed from a LM series carburetor, it must be discarded and service type nozzle shown installed.

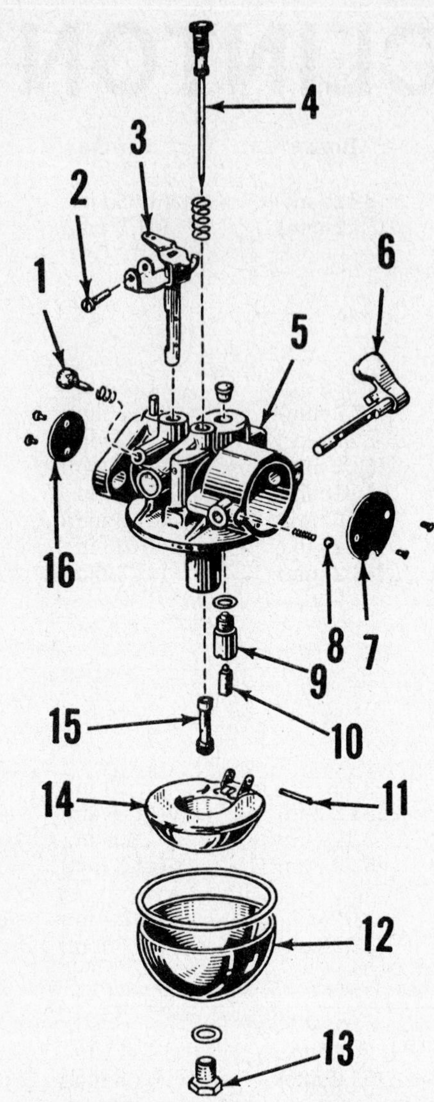

Fig. CL7—Exploded view of a typical Carter Model N carburetor. Design of float and float bowl may vary from that shown.

1. Idle fuel needle
2. Idle stop screw
3. Throttle shaft
4. Main fuel needle
5. Carburetor body
6. Choke shaft
7. Choke plate
8. Detent ball
9. Inlet valve seat
10. Inlet needle
11. Float pin
12. Float bowl
13. Retainer
14. Float
15. Main nozzle
16. Throttle plate

engine idle. Check acceleration from slow to fast idle and open idle needle approximately 1/8 turn (counterclockwise) if necessary for proper acceleration.

When overhauling LMB or LMG carburetors, discard main fuel nozzle (13) if removed and install a service type nozzle (see Fig. CL6). Do not install old nozzle.

To check float setting, invert carburetor throttle body and float assembly. Measure distance between top of free side of float and float bowl mating surface of carburetor. Measurement should be 5/32 inch (3.97 mm). Carefully bend float lever tang which contacts fuel inlet needle if necessary to obtain correct float level.

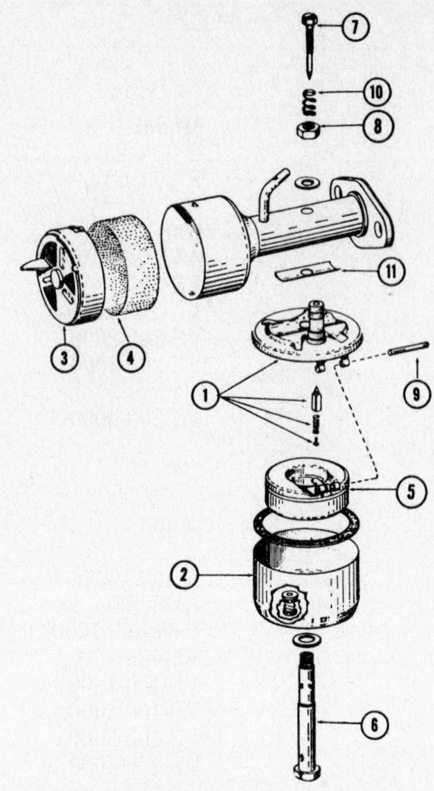

Fig. CL8—Exploded view of float type carburetor used on 501-0000-000 model series engine.

1. Inlet valve & seat
2. Float bowl
3. Choke assy.
4. Air cleaner element
5. Float
6. Main nozzle
7. Fuel needle
8. Nut
9. Float pin
10. Spring
11. Gasket

Float drop should be 3/16 inch (4.76 mm). Carefully bend tabs on float hinge to adjust float drop.

Carter Model N Float Type Carburetor. Refer to Fig. CL7 for identification and exploded view of Carter Model N float type carburetor. The following models have been used:

N-2003-S
N-2029-S
N-2087-S
N-2171-S
N-2457-S

Initial adjustment of fuel mixture needles on Models N-2003-S and N-2029-S from a lightly seated position is ¾ turn open for idle fuel needle (1) and 1 turn open for main fuel mixture needle (4).

Initial adjustment of fuel mixture needles on Models N-2087-S, N-2171-S and N-2457-S is 1½ turns open for idle mixture needle (1) and ½ turn open for main mixture needle (4).

Final adjustments on all models are made with engine at operating temperature and running. Operate engine at rated speed under load and adjust main fuel mixture needle for smoothest engine operation. Operate

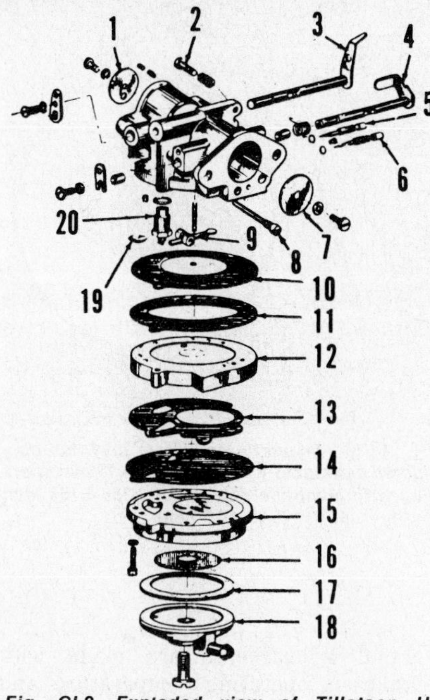

Fig. CL9—Exploded view of Tillotson HL diaphragm type carburetor.

1. Choke plate
2. Idle stop screw
3. Choke shaft
4. Throttle shaft
5. Main fuel needle
6. Idle fuel needle
7. Throttle plate
8. Inlet lever pivot
9. Inlet lever
10. Diaphragm
11. Gasket
12. Cover
13. Gasket
14. Pump diaphragm
15. Pump body
16. Screen
17. Gasket
18. Cover
19. Welch plug
20. Inlet valve & seat

engine at idle speed and adjust idle needle for smooth engine idle then open needle slightly to run a slightly over-rich mixture.

Float setting on Models N-2003-S and N-2457-S is 13/64-inch (5.16 mm) clearance between outer edge of casting and free end of flat. Float setting on Models N-2029-S, N-2087-S and N-2171-S is 11/64-inch (4.37 mm) clearance between outer edge of casting and free end of float. On all models, carefully bend lip on float hinge if necessary to obtain correct float level.

On Model N-2457-S, low speed jet and high speed fuel nozzle are permanently installed. Do not attempt to remove these parts.

Carter Model NS Float Type Carburetor. Refer to Fig. CL8 for identification and exploded view of Carter Model NS float type carburetor. This carburetor is designed for engine operation between 3000 and 3800 rpm and is not equipped with an idle fuel mixture adjusting needle.

Initial adjustment of main fuel mixture needle (7) from a lightly seated position is 1 turn open. Final adjustments are made with engine at operating temperature. Operate engine at rated speed and adjust main fuel mixture needle to obtain smoothest engine operation and acceleration.

Fig. CL10—Exploded view of Brown CP carburetor.

1. Choke plate
2. Idle stop screw
3. Filter plug
4. Throttle shaft
5. Gasket
6. Carburetor body
7. Idle jet
8. Throttle plate
9. Inlet lever spring
10. Gasket
11. Diaphragm
12. Cover
13. Gasket
14. Diaphragm
15. Pump cover
16. Gasket
17. Inlet fitting
18. Inlet lever
19. Inlet valve
20. Idle fuel needle
21. Choke shaft
22. Main fuel needle
23. Inlet lever pivot

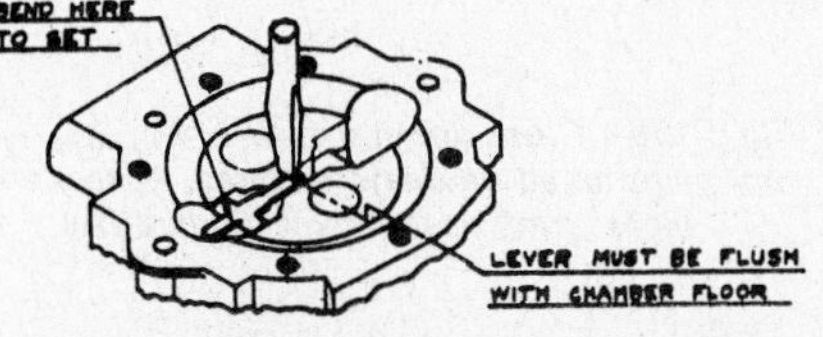

Fig. CL11—Method of checking and setting fuel inlet lever on Brown diaphragm type carburetors.

To check float level, invert carburetor throttle body. Measure between outer edge of bowl cover and top surface of free end of float. Measurement should be 13/64 inch (5.16 mm). Fuel inlet needle, seat and spring are integral parts of bowl cover and available as an assembly only.

Tillotson HL Diaphragm Type Carburetor. Refer to Fig. CL9 for identification and exploded view of Tillotson HL diaphragm type carburetor. Models HL-13A and Hl-44A are used. Clockwise rotation of mixture needles on all models leans the fuel mixture.

Initial adjustment of idle (6) and main (5) fuel mixture needles from a lightly seated position is ¾ turn open for idle mixture needle and 1 to 1¼ turns open for main fuel mixture needle.

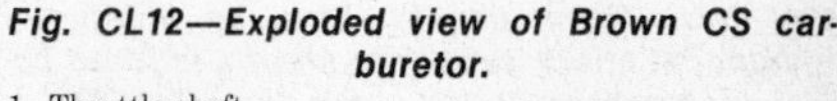

Fig. CL12—Exploded view of Brown CS carburetor.

1. Throttle shaft
2. Throttle plate
3. Inlet lever pivot
4. Carburetor body
5. Choke plate
6. Idle stop screw
7. Inlet needle valve & seat
8. Inlet lever
9. Inlet lever spring
10. Diaphragm
11. Gasket
12. Cover
13. Gasket
14. Pump diaphragm
15. Pump cover
16. Screen
17. Gasket
18. Inlet fitting
19. Connection
20. Choke shaft
21. Main fuel nozzle
22. Idle fuel needle

Final adjustments are made with engine at operating temperature and running. Operate engine at rated speed and adjust main fuel mixture needle for smoothest engine operation, then open needle slightly to provide an over-rich fuel mixture. Operate engine at idle speed and adjust idle mixture screw for smoothest engine idle. If engine fails to accelerate smoothly, open idle mixture needle slightly as necessary.

Brown CP Diaphragm Type Carburetor. Refer to Fig. CL10 for identification and exploded view of Brown Model 3-CP diaphragm type carburetor. Idle mixture needle has a slotted head and main fuel mixture needle has a "T" head. Note fuel must pass through a felt filter and screen, located under plug (3), to enter the diaphragm chamber.

Initial adjustment of idle (20) and main (22) fuel mixture needles from a lightly seated position is ½ to ¾ turn open for each needle.

Final adjustments are made with engine at operating temperature and running. Operate engine at rated speed

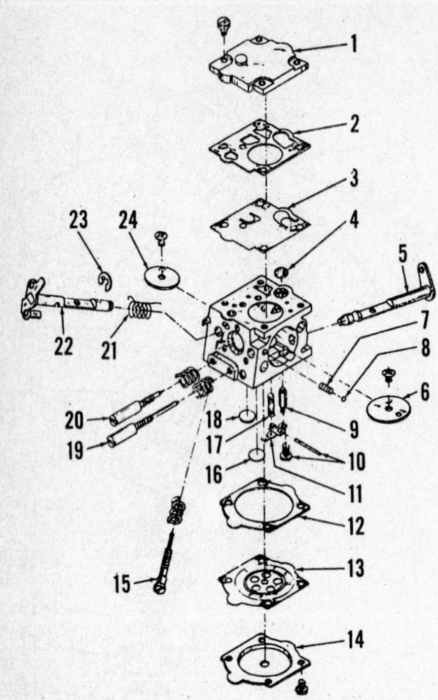

Fig. CL13—Exploded view of typical Walbro SDC, diaphragm type carburetor. Model SDC-35 used on 502-0308-000 engines has choke (5) and throttle (22) levers of slightly different configuration from those shown. Also, an additional diaphragm check valve (not shown) is fitted between gasket (2) and pump diaphragm (3).

1. Fuel pump cover
2. Gasket
3. Pump diaphragm
4. Fuel inlet screen
5. Choke shaft
6. Choke plate
7. Choke detent spring
8. Choke detent ball
9. Inlet valve needle
10. Lever pin retainer
11. Metering lever
12. Metering diaphragm gasket
13. Metering diaphragm
14. Diaphragm cover
15. Idle speed screw & spring
16. Welch plug
17. Metering lever spring
18. Welch plug
19. Main fuel needle & spring
20. Idle mixture needle & spring
21. Throttle return spring
22. Throttle shaft
23. Shaft retainer
24. Throttle plate

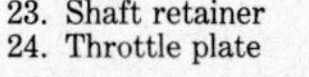

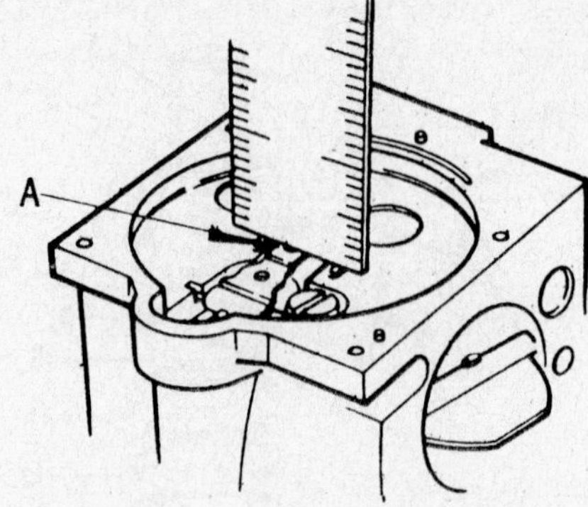

Fig. CL14—View of metering lever adjustment procedure on SDC-34 carburetors. Check engagement of inlet valve needle and diaphragm hook to lever. Ensure spring under lever is correctly seated. Metering lever should just touch straightedge (A).

Fig. CL16—Magneto condition may be considered satisfactory if it will fire an 18 mm spark plug with electrode gap set at 0.156-0.187 inch (3.97-4.76 mm).

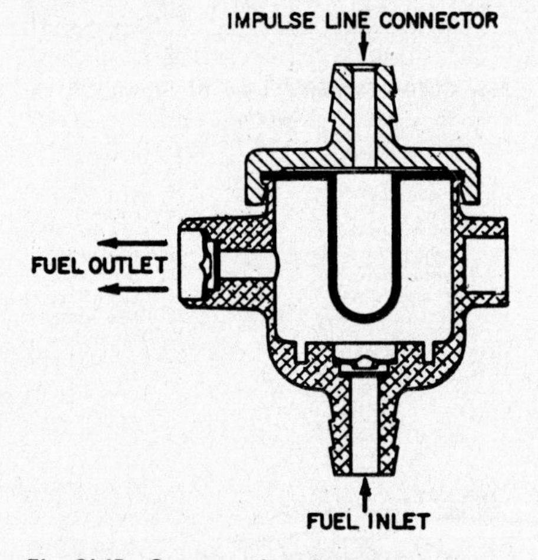

Fig. CL15—Cross section view of impulse type fuel pump used on some engines. Renew complete pump assembly if inoperative.

and adjust main fuel mixture needle for smoothest engine operation, then open mixture needle slightly to provide an over-rich fuel mixture. Operate engine at idle speed and adjust idle mixture needle for smooth engine idle. Check acceleration and open main fuel needle slightly if necessary to obtain smooth acceleration.

Refer to Fig. CL11 for proper diaphragm lever setting. Always renew the copper sealing gasket if inlet needle seat is removed.

Brown CS Diaphragm Type Carburetor. Refer to Fig. CL12 for identification and exploded view of Brown CS-15 diaphragm type carburetor. Idle fuel mixture needle has slotted adjustment head and main fuel adjustment needle has a "T" head.

Initial adjustment of idle (22) and main (21) fuel mixture needles from a lightly seated position is 3/4 turn open for each screw.

Final adjustments are made with engine at operating temperature and running. Operate engine at rated speed and adjust main fuel mixture needle for smoothest engine operation, then open mixture needle slightly to provide an over-rich fuel mixture. Operate engine at idle speed and adjust idle mixture needle for smooth engine idle and proper accleration.

Refer to Fig. CL11 for proper diaphragm lever setting. Always renew the copper sealing gasket if inlet needle is removed.

Walbro Model SDC-34 Diaphragm Type Carburetor. Refer to Fig. CL13 for identification and exploded view of Walbro Model SDC-34 carburetor used on Model 502-0308-000 engine.

Initial adjustment of idle (20) and main (19) fuel mixture needles from a lightly seated position is 1 turn open for each needle.

Final adjustments are made with engine at operating temperature and running. Operate engine at rated speed and adjust main fuel mixture needle for smoothest engine performance. Operate engine at idle speed and adjust idle fuel mixture for smooth engine idle. Note that very slight adjustments to idle and main fuel mixture needles may be necessary to obtain satisfactory engine acceleration.

Model 502-0308-000 engines rely on throttle position changes to control speed only and are not equipped with governor.

Refer to Fig. CL14 for adjustment procedure for metering lever. Be sure metering lever spring is properly seated and hook on diaphragm and inlet valve needle are correctly engaged.

FUEL PUMP. Some models are equipped with a diaphragm type fuel pump as shown in the cross-sectional view in Fig. CL15. The pump diaphragm is actuated by pressure pulsations in the engine crankcase transmitted via an impulse tube connected between the fuel pump and crankcase. The pump is

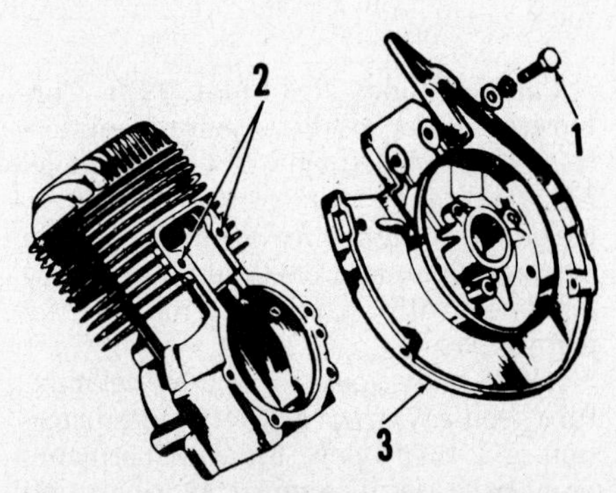

Fig. CL17—If cap screws (1) which thread into holes (2) are too long, they may bottom and damage cylinder walls.

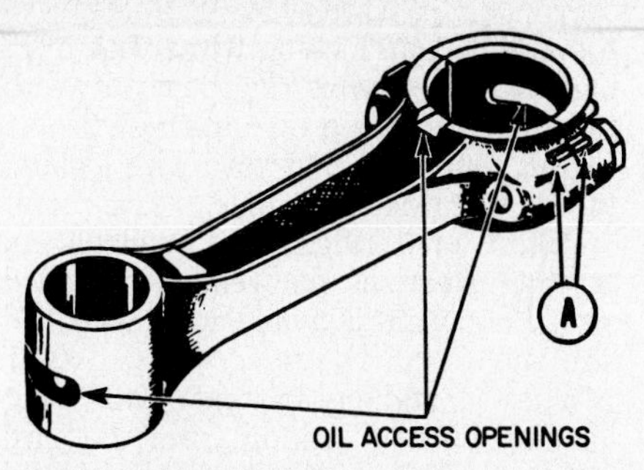

Fig. CL18—Be sure embossments (A) are aligned when assembling cap to connecting rod.

designed to lift fuel approximately 6 inches (152 mm). Service consists of renewing the complete fuel pump assembly.

GOVERNOR. Speed control is maintained by an air vane type governor. Renew bent or worn governor vane or linkage and be sure linkage does not bind in any position when moved through full range of travel.

On models with speed control lever, adjust lever stop so maximum speed does not exceed 3600 rpm. On models without speed control, renew or retension governor spring to limit maximum speed to 3600 rpm. On all models (so equipped), make sure blower housing is clean and free of dents that would cause binding of air vane.

MAGNETO AND TIMING. Ignition timing is 27° BTDC and is nonadjustable. Armature air gap should be 0.007-0.017 inch (0.18-0.43 mm). Adjust breaker point gap to 0.018-0.021 inch (0.46-0.53 mm) on all models.

Magento may be considered satisfactory if it will fire an 18 mm spark plug with gap set at 0.156-0.187 inch (3.97-4.76 mm). Refer to Fig. CL16.

LUBRICATION. Engine is lubricated by mixing a good quality two-stroke air-cooled engine oil with regular gasoline.

Oil to gasoline mixing ratio is determined by type of bearings installed in a particular two-stroke engine. Plain sleeve main and connecting rod bearings require 0.75 pint (0.36 L) of oil for each gallon of gasoline. Engines equipped with needle bearings require 0.50 pint (0.24 L) of oil for each gallon of gasoline. During break-in (first five hours of operation) a 50% increase of oil portion is advisable, especially if engine will be operated at or near maximum load or rpm.

CARBON. Power loss on two-stroke engines can often be corrected by cleaning the exhaust ports and muffler. To clean the exhaust ports, remove the muffler and turn engine until piston is below ports. Use a dull tool to scrape carbon from ports. Take care not to damage top of piston or cylinder walls.

REPAIRS

TIGHTENING TORQUES. Recommended tightening torque specifications are as follows:

Bearing plate
  to block . . . . 75-95 in.-lbs. (8-11 N·m)
Blower housing . . . . 65-70 in.-lbs. (7-8 N·m)
Carburetor mounting . . . . . 60-65 in.-lbs. (7 N·m)
Connecting rod:
  E65, GK590, 502, 503
    (aluminum) . . . . 70-80 in.-lbs. (8-9 N·m)
  E65, GK590 (steel) . . . . 90-100 in.-lbs. (10-11 N·m)
  All others
    (aluminum) . . . . 35-45 in.-lbs. (4-5 N·m)
Cylinder head
  (502 & 503) . . . . 140-160 in.-lbs. (16-18 N·m)
Engine base . . . . 125-150 in.-lbs. (14-17 N·m)
Flywheel:
  E65 & GK590 . . . . 250-300 in.-lbs. (28-34 N·m)
  All others . . . . 375-400 in.-lbs. (42-45 N·m)
Muffler . . . . 40-60 in.-lbs. (4-7 N·m)
Spark plug . . . . 275-300 in.-lbs. (31-34 N·m)
Stator plate to
  bearing plate . . . . 50-60 in.-lbs. (5-7 N·m)

BEARING PLATE. Because of the pressure and vacuum pulsations in crankcase, bearing plate and gasket must form an airtight seal when installed. Inspect bearing plate gasket surface for cracks, nicks or warpage. Ensure oil passages in crankcase, gasket and plate are aligned and that correct number and thickness of thrust washers are used. Also make certain cap screws of the correct length are installed when engine is reassembled.

CAUTION: Cap screws (1—Fig. C17) may bottom in threaded holes (2) if incorrect screws are used. When long screws are tightened, damage to cylinder walls can result.

FLYWHEEL. On vertical shaft engines, both a lightweight aluminum flywheel and a cast iron flywheel are available. The lightweight aluminum flywheel should be used only on rotary lawnmower engines where the blade is attached directly to the engine crankshaft. The cast iron flywheel should be used on engines with belt pulley drive, etc.

Some kart engine owners will prefer using the lightweight aluminum flywheel due to the increase in accelerating performance.

CONNECTING ROD. Connecting rod and piston unit can be removed after removing the reed plate and crankshaft.

On E-65 and GK590 models (karting engines), and on 502 and 503 models, needle roller bearings are used at crankpin end of connecting rod. Renew crankshaft and/or connecting rod if bearing surfaces are scored or rough. When reassembling, place the thirteen needle rollers between connecting rod and crankpin. Using a low melting point grease, stick the remaining twelve needle rollers to connecting rod cap. Install

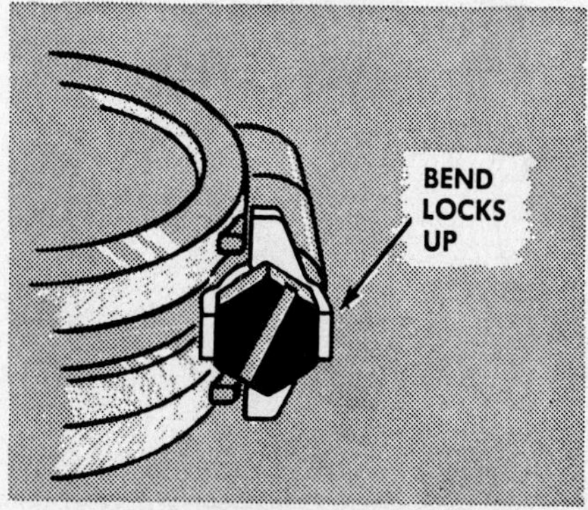

Fig. CL19—Bend connecting rod cap retaining screw locks up as shown.

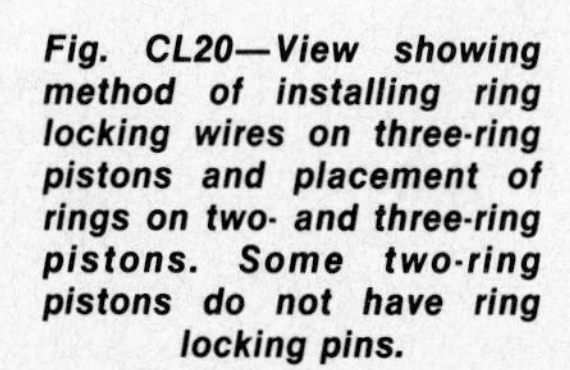

Fig. CL20—View showing method of installing ring locking wires on three-ring pistons and placement of rings on two- and three-ring pistons. Some two-ring pistons do not have ring locking pins.

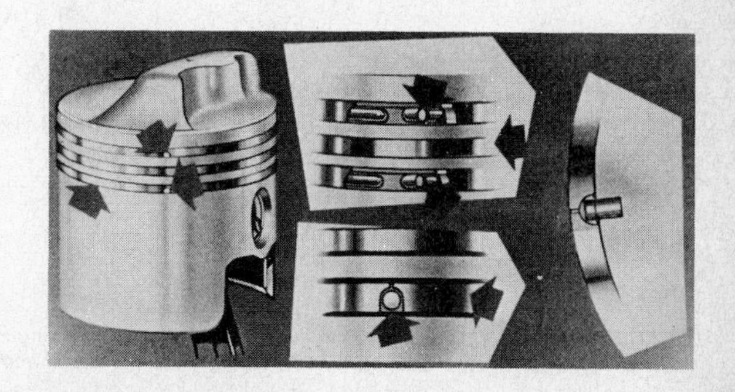

cap with the embossments (see A – Fig. CL18) on cap and rod properly aligned.

On all other models, an aluminum connecting rod with bronze bearing surfaces is used. Standard diameter of connecting rod crankpin bearing bore (plain bearing) is 0.7820-0.7827 inch (19.863-19.881 mm).

Clearance between connecting rod and crankpin journal should be 0.0026-0.0040 inch (0.07-0.10 mm). Renew connecting rod and/or crankshaft if clearance exceeds 0.0055 inch (0.14 mm).

Clearance between piston pin and connecting rod piston pin bore should be 0.0004-0.0011 inch (0.010-0.028 mm). Renew connecting and/or piston pin if clearance exceeds 0.002 inch (0.05 mm).

Reinstall cap to connecting rod with embossments (A – Fig. CL18) aligned and carefully tighten screws to recommended specification. Bend the locking tabs against screw heads as shown in Fig. CL19.

PISTON, PIN AND RINGS. Pistons may be equipped with either two or three compression rings. On three-ring pistons, a locking wire is fitted in a small groove behind each piston ring to prevent ring rotation on the piston. To install the three locking wires, refer to Fig. CL20 and hold piston with top up and intake side of piston to right. Install top and bottom locking wires with locking tab (end of wire ring bent outward) to right of locating hole in ring groove. Install center locking wire with locking tab to left of locating hole.

Piston and rings are available in several oversizes as well as standard size. Piston pin is available in standard size only.

When installing pistons with ring locating pins or locking wires, be sure end gaps of rings are properly located over the pins or the locking tabs and install piston using a ring compressor. Refer to Fig. CL20.

Be sure to reinstall piston and connecting rod assembly with exhaust side of

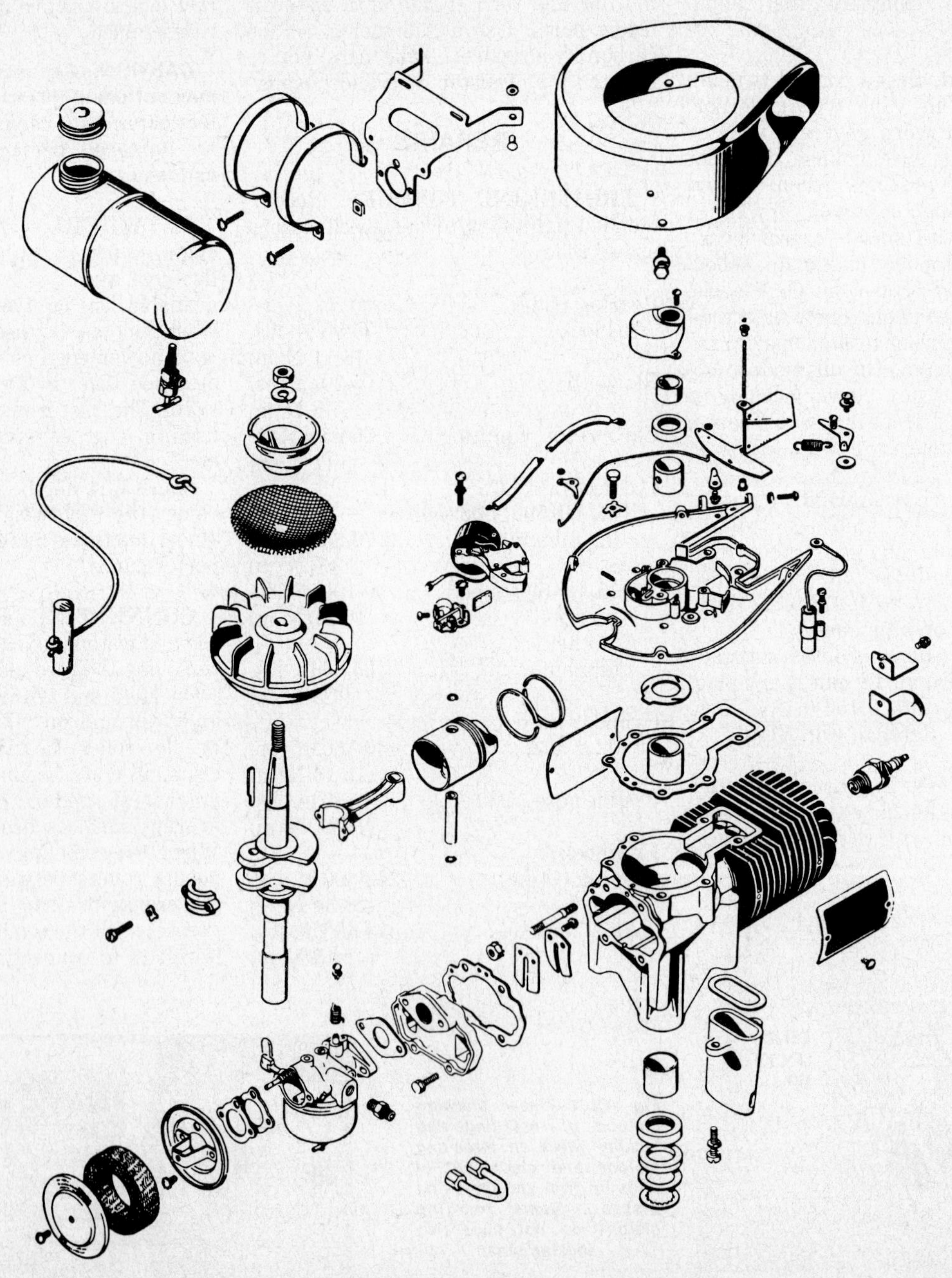

Fig. CL21—Exploded view of typical vertical shaft two-stroke engine. Model 503 differs in type of starter (see Fig. CL102—ACCESSORIES), use of an eight-petal rossette type reed valve, a detachable cylinder head and has ball and roller bearings to support crankshaft with needle bearing at crankpin.

piston (long sloping side of piston dome) towards exhaust ports in cylinder.

Check piston, pin and rings against the following specifications:

Specifications For 1.875 in. (47.6 mm) Pistons

Ring end gap:
Standard .... 0.005-0.013 in. (0.13-0.33 mm)
Maximum .... 0.020 in. (0.51 mm)
Ring side clearance:
Standard .... 0.0015-0.0040 in. (0.038-0.102 mm)
Maximum .... 0.006 in. (0.15 mm)
Piston skirt clearance:
Standard .... 0.0045-0.0065 in. (0.114-0.165 mm)
Maximum .... 0.008 in. (0.20 mm)
Piston pin diameter .... 0.4294-0.4296 in. (10.899-10.904 mm)
Piston pin bore diameter .... 0.4295-0.4298 in. (10.901-10.909 mm)

Specifications For 2.125 in. (53.2 mm) Piston

Ring end gap (two-ring piston):
Standard .... 0.007-0.017 in. (0.18-0.43 mm)
Maximum .... 0.025 in. (0.64 mm)
Ring end gap (three-ring piston):
Standard .... 0.010-0.015 in. (0.25-0.38 mm)
Maximum .... 0.017 in. (0.43 mm)
Ring side clearance (two-ring piston):
Standard .... 0.0015-0.0040 in. (0.038-0.102 mm)
Maximum .... 0.006 in. (0.15 mm)
Ring side clearance (three-ring piston):
Standard .... 0.002-0.004 in. (0.05-0.10 mm)
Maximum .... 0.0055 in. (0.140 mm)
Piston skirt clearance (two-ring piston):
Standard .... 0.005-0.007 in. (0.13-0.18 mm)
Maximum .... 0.008 in. (0.20 mm)
Piston skirt clearance (three-ring piston):
Standard .... 0.0045-0.0050 in. (0.114-0.127 mm)
Maximum .... 0.007 in. (0.18 mm)
Piston pin diameter .... 0.4999-0.5001 in. (12.698-12.703)
Piston pin bore diameter ...... 0.5000-0.5003 in. (0.127-0.135 mm)

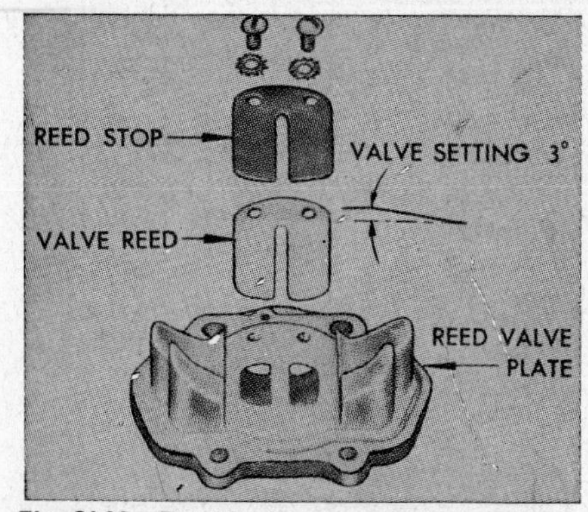

Fig. CL22—Details of two-petal reed inlet valve used on some models. Eight-petal rosette style reed valve is also used. See Fig. CL23.

Specifications For 2.375 in. (60.3 mm) & 2.500 in. (63.5 mm) Pistons

Ring end gap:
Standard .... 0.007-0.017 in. (0.18-0.43 mm)
Maximum .... 0.025 in. (0.64 mm)
Ring side clearance:
Standard .... 0.0015-0.0040 in. (0.038-0.102 mm)
Maximum .... 0.006 in. (0.15 mm)
Piston skirt clearance:
Standard .... 0.005-0.007 in. (0.13-0.18 mm)
Maximum .... 0.008 in. (0.20 mm)

CYLINDER AND CRANKCASE. The one-piece aluminum alloy cylinder and crankcase unit is integrally die-cast around a cast iron sleeve (cylinder liner). Cylinder can be rebored if worn beyond specifications. Standard cylinder bore diameter for models with 1.875 inch (47.63 mm) bore is 1.875-1.876 inches (47.63-57.65 mm); with 2.125 inch (53.98 mm) bore is 2.125-2.126 inches (53.98-54.00 mm); with 2.375 inch (60.33 mm) bore is 2.375-2.376 inches (60.33-60.35 mm) and with 2.500 inch (63.50 mm) bore is 2.500-2.501 inches (63.50-63.53 mm).

Refer to PISTON, PIN AND RINGS paragraphs for piston-to-cylinder clearance specifications.

Standard main bearing diameters (plain bushing type) should be 0.7517-0.7525 inch (19.093-19.114 mm) at flywheel end and either 0.877-0.878 inch (22.276-22.301 mm) or 1.002-1.003 inch (25.451-25.476 mm) at pto end. Standard main bearing clearance should be 0.0015-0.0030 inch (0.038-0.076 mm)

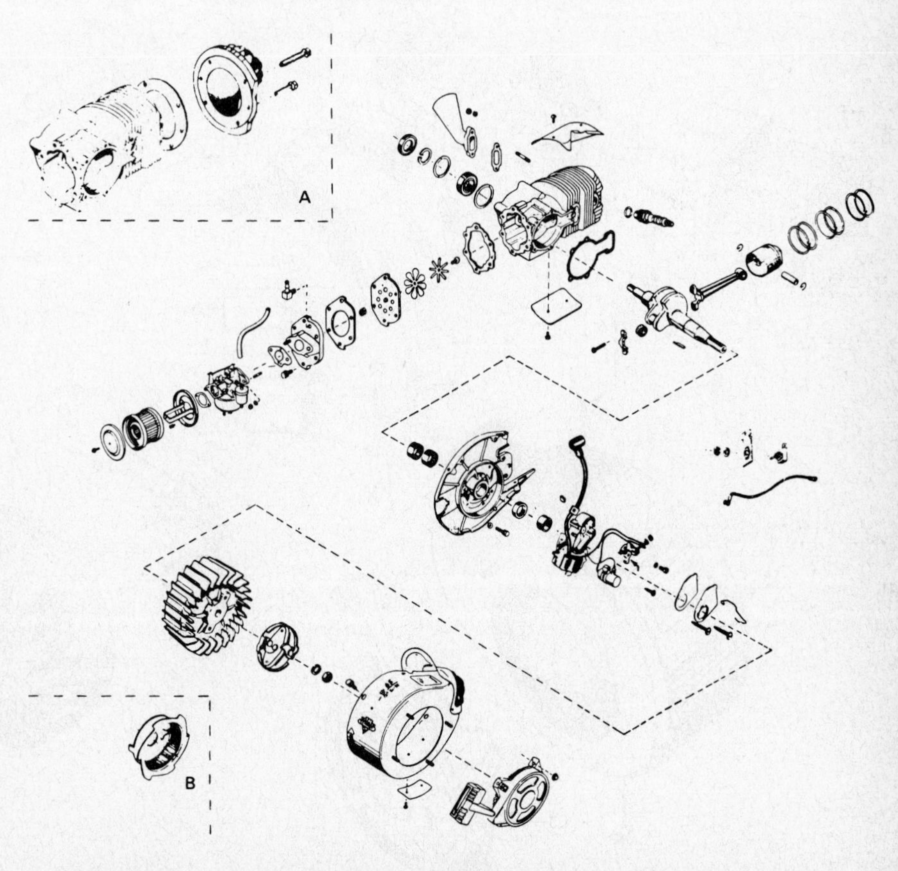

Fig. CL23—Exploded view of E-65 engine. Differences from Model 502 are minor: Cylinder head is detachable as in inset "A"; piston is fitted with two rings instead of three; muffler, starter cup (inset "B") and carburetor are of different design. See Fig. CL13 for carburetor used on 502-0308-000 engines. Model 502-0309-000 engines are equipped with float type carburetor.

and crankcase should be renewed if worn beyond specification or if out-of-round 0.0015 inch (0.038 mm) or more.

Main bearing bushings, bushing remover and driver, reamers and reamer alignment plate are available through Clinton parts sources for renewing bushings in crankcase and bearing plate. On vertical crankshaft engines having two bushings in the crankcase, the outer bushing must be removed towards inside of crankcase.

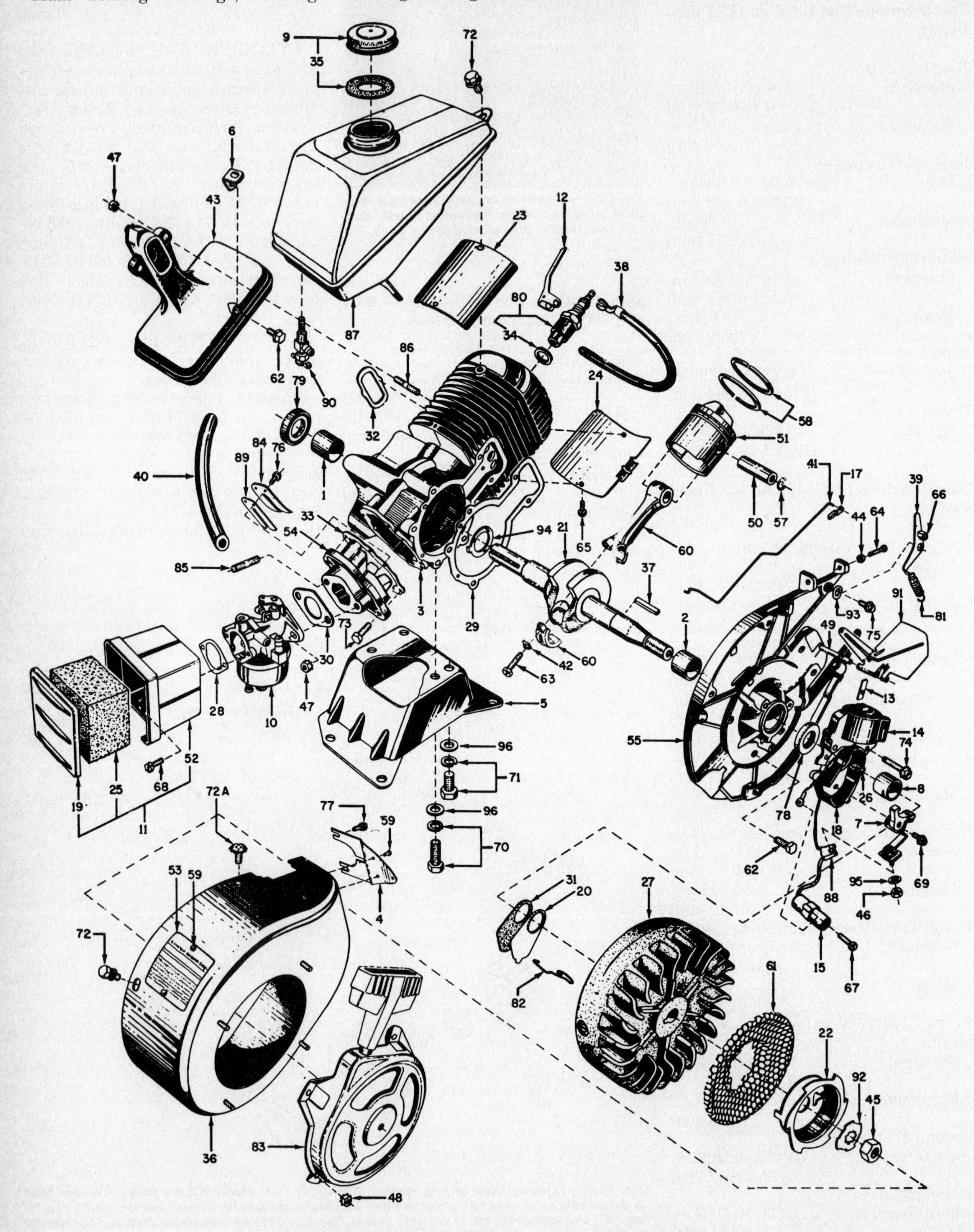

Fig. CL24—Exploded view of typical horizontal crankshaft engine.

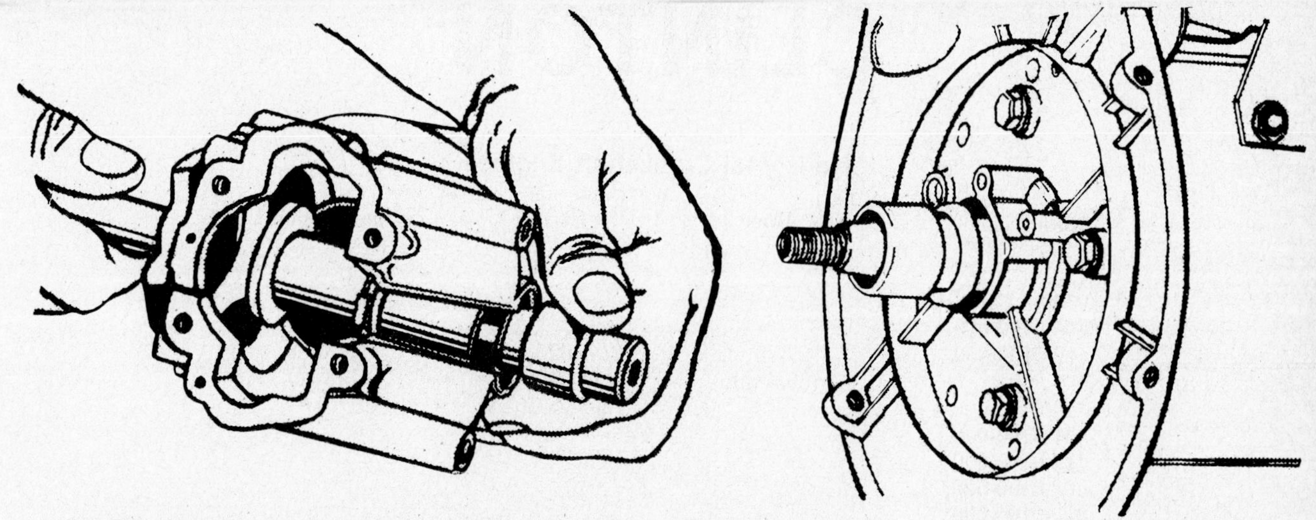

Fig. CL25—Examples of typical use of oil seal loaders for protection of seals during installation of crankshaft. Use ample lubrication.

Bushings must be recessed 1/32 inch (0.79 mm) from inside surface of bearing plate or crankcase except when thrust washer is used between crankcase or plate and thrust surface of crankshaft.

CRANKSHAFT. To remove crankshaft, remove blower housing and nut retaining flywheel to crankshaft. Thread an impact nut to within 1/8 inch (3.18 mm) of flywheel. Pull on flywheel and tap impact nut with hammer. After flywheel is removed, remove magneto assembly. Remove the engine base and inlet reed valve plate from crankcase and detach connecting rod from crankshaft. Be careful not to lose the 25 needle rollers on E-65, GK-590, 502 and 503 models. Push connecting rod and piston unit up against top of cylinder. On models with ball bearing main on output end of crankshaft, remove the snap ring retaining ball bearing in crankcase. Remove the magneto stator plate and withdraw crankshaft from engine taking care not to damage connecting rod.

On crankshaft used with needle roller main or crankpin bearings, renew crankshaft if bearing surface shows signs of wear, roughness or scoring. On mains and/or cast-in crankpin bearing, crankpin diameter should be 0.7788-0.7795 inch (19.782-19.799 mm). Main journal diameter at flywheel end should be 0.7495-0.7502 inch (19.037-19.055 mm) and main journal diameter at pto end should be 0.8745-0.8752 inch (22.212-22.230 mm). Crankshaft should be renewed if worn beyond specifications or if out-of-round 0.0015 inch (0.038 mm) or more.

Refer to CYLINDER AND CRANKCASE section for main bearing specifications.

Renew ball bearing type mains if bearing is rough or worn. Renew needle type main bearings if one or more needles show any defect or if needles can be separated the width of one roller.

When reinstalling crankshaft with bushing or needle mains, crankshaft end play should be 0.005-0.020 inch (0.13-0.51 mm). The gasket used between bearing (stator) plate and crankcase is available in several thicknesses. It may also be necessary to renew the crankshaft thrust washer.

CRANKSHAFT OIL SEALS. On all two-stroke engines, the crankshaft oil seals must be maintained in good condition to hold crankcase compression on the downward stroke of the piston. It is usually a good service practice to renew the seals whenever overhauling an engine. Apply a small amount of gasket sealer to the outer rim of the seal and install with lip towards inside of crankcase or bearing plate.

REED VALVE. Either a dual reed (Fig. CL21) or eight-petal rosette reed (Fig. CL23) is used.

The 3° setting of the dual reed shown in Fig. CL22 is the design of the reed as stamped in the manufacturing process. The 3° bend should be towards the reed seating surface. The reed stop should be adjusted to approximately 0.280 inch (7.11 mm) from tip of stop to reed seating surface.

Renew reed if any petal is cracked, rusted or does not lay flat against the reed plate. Renew the reed plate if rusted, pitted or worn.

CLINTON

Horizontal Crankshaft Engines

Model	Bore	Stroke	Displacement
100, 2100,A-2100, 400-0100-000, 402-0100-000	2.375 in. (60.3 mm)	1.625 in. (41.3 mm)	7.2 cu. in. (118 cc)
3100,4100, 404-0100-000, 406-0100-000, 408-0100-000, 424-0100-000, 426-0100-000, 492-0000-000	2.375 in. (60.3 mm)	1.875 in. (47.6 mm)	8.3 cu. in. (136 cc)
410-0000-000	2.500 in. (63.5 mm)	1.875 in. (47.6 mm)	9.2 cu. in. (151 cc)

Vertical Crankshaft Engines

Model	Bore	Stroke	Displacement
V-100,VS-100, VS-2100,VS300, 403-0000-000, 405-0000-000, 411-0002-000	2.375 in. (60.3 mm)	1.625 in. (41.3 mm)	7.2 cu. in. (118 cc)
AFV-3100, AV-3100, AVS-3100, FV-3100, V-3100, VS-3100, AVS-4100, VS-4100, 405-0000-000, 407-0000-000, 409-0000-000, 415-0000-000, 417-0000-000, 435-0000-000, 455-0000-000	2.375 in. (60.3 mm)	1.875 in. (47.6 mm)	8.3 cu. in. (136 mm)
419-0000-000, 429-0000-000, 431-0003-000	2.500 in. (63.5 mm)	1.875 in. (47.6 mm)	9.2 cu. in. (151 cc)

MAINTENANCE

SPARK PLUG. Recommended spark plug for all models is a Champion H10 or equivalent. Electrode gap is 0.025-0.028 inch (0.64-0.71 mm). Tighten spark plug to specified torque.

CARBURETOR. Either Clinton (Walbro) float type or Clinton suction type carburetor is used. Refer to the appropriate paragraph.

Clinton (Walbro) Float Type Carburetor. A variety of Clinton (Walbro) LMB, LMG and LMV series carburetors have been used. To identify carburetor, refer to Fig. CL61 for identification number.

Initial adjustment of idle (9) and main (22) fuel mixture needles from a lightly seated position is 1¼ turns open for each needle.

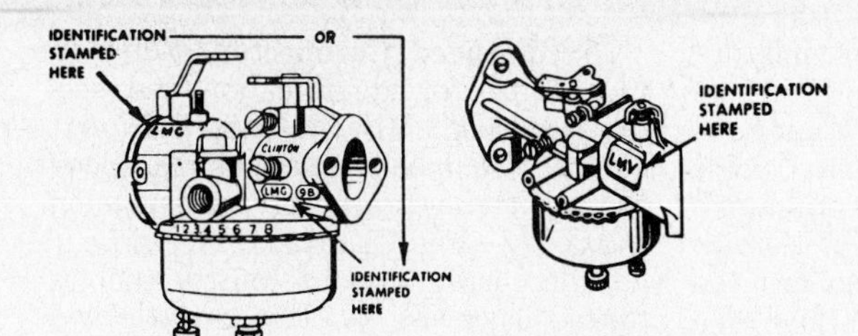

Fig. CL60—Views showing location of identification numbers on LMB, LMG and LMV series carburetors. Identification numbers must be used when ordering service parts.

Final adjustment is made with engine at operating temperature and running. Operate engine at rated speed and adjust main fuel mixture needle for smoothest engine operation. Operate engine at idle speed and adjust idle fuel mixture needle for smooth engine idle. If engine fails to accelerate smoothly it may be necessary to slightly richen main fuel needle adjustment.

When overhauling carburetor, do not remove the main fuel nozzle (13) unless necessary. If nozzle is removed, it must be discarded and a service type nozzle (Fig. CL61A) installed.

Choke plate (11 – Fig. CL61) should be installed with the "W" or part number to outside. Install throttle plate with the side marked "W" facing towards mounting flange and with the part number towards idle needle side of carburetor bore when plate is in closed position. When installing throttle plate, back idle speed adjustment screw out, turn plate and throttle shaft to closed position and seat plate by gently tapping wtih small screwdriver before tightening plate retaining screws.

If either the float valve or seat is damaged, install a new matched valve and seat assembly (12) and tighten seat to 40-50 in.-lbs. (5-6 N·m).

To check float level, invert carburetor throttle body and float assembly. Distance between free side of float and float bowl mating surface of carburetor surface should be 5/32 inch (3.97 mm). Carefully bend float lever tang which contacts fuel inlet needle if necessary to obtain correct float level.

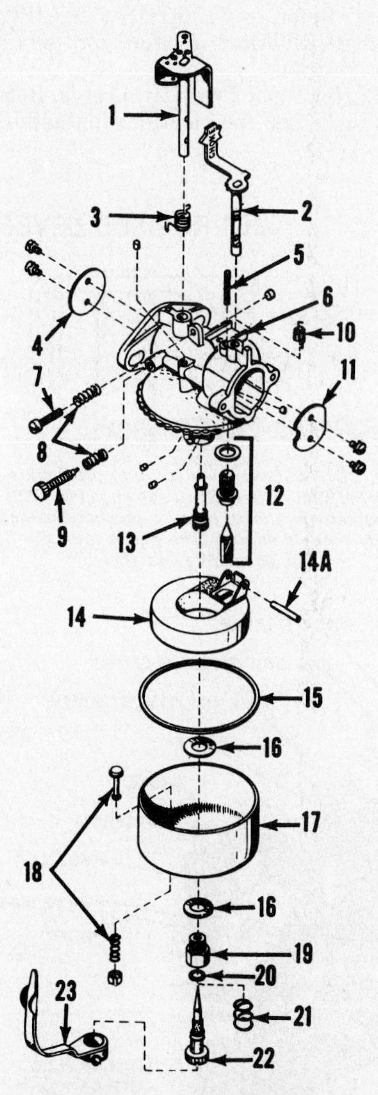

Fig. CL61 – Exploded view of typical LMG series carburetor. LMB and LMV series are similar.

1. Throttle shaft
2. Choke shaft
3. Spring
4. Throttle plate
5. Spring
6. Carburetor body
7. Idle stop screw
8. Springs
9. Idle fuel needle
10. Spring
11. Choke plate
12. Inlet needle
13. Main nozzle
14. Float
14A. Float pin
15. Gasket
16. Gaskets
17. Float bowl
18. Drain valve
19. Retainer
20. Seal
21. Spring
22. Main fuel needle
23. Lever (optional)

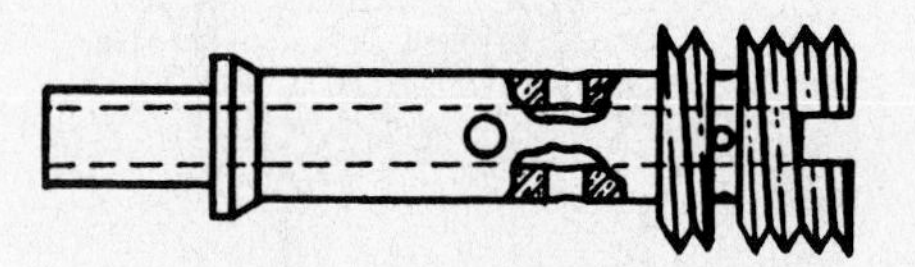

Fig. CL61A – When original main fuel nozzle is removed from LM series carburetors, it must be discarded and service type nozzle shown must be installed.

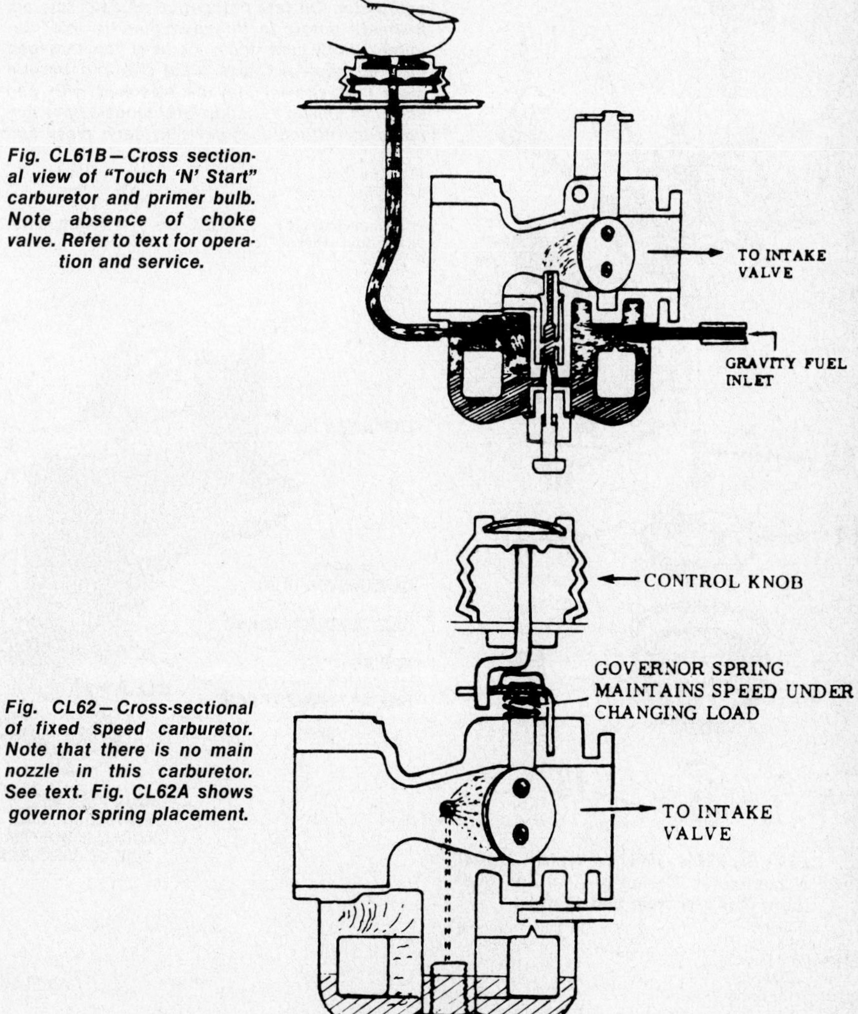

Fig. CL61B – Cross sectional view of "Touch 'N' Start" carburetor and primer bulb. Note absence of choke valve. Refer to text for operation and service.

Fig. CL62 – Cross-sectional of fixed speed carburetor. Note that there is no main nozzle in this carburetor. See text. Fig. CL62A shows governor spring placement.

Clinton "Touch 'N' Start" Carburetor. Refer to Fig. CL61B for identification and cross-sectional view of "Touch 'N' Start" carburetor. Instead of choke valve (11 – Fig. CL61) carburetor is furnished with a flexible primer bulb which provides a rich charging mixture to carburetor venturi and intake manifold for easier starting. Float bowl is vented through primer tube to flexible bulb and vent is closed when operator's finger depresses primer bulb. Service procedures and specifications are the same as for Clinton (Walbro) float type carburetor.

GOVERNOR LINK HOOKS IN THIS HOLE

GOVERNOR SPRING HOOKS ON THIS EDGE

Fig. CL62A – Top view of fixed speed carburetor showing correct installation of governor link and spring.

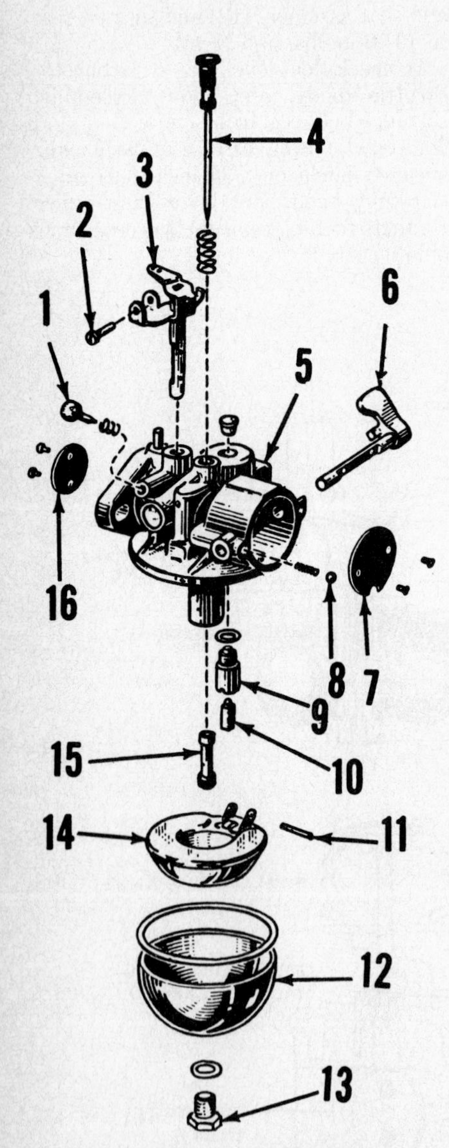

Fig. CL63 – Exploded view of typical Carter Model N carburetor. Design of float and float bowl may vary from that shown.

1. Idle fuel needle
2. Idle stop screw
3. Throttle shaft
4. Main fuel needle
5. Carburetor body
6. Choke shaft
7. Choke plate
8. Detent ball
9. Inlet valve seat
10. Inlet needle
11. Float pin
12. Float bowl
13. Retainer
14. Float
15. Main nozzle
16. Throttle plate

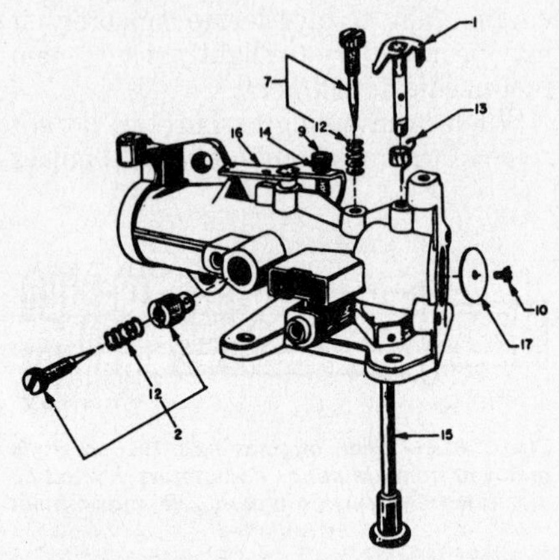

Fig. CL64—Exploded view of Clinton suction lift carburetor. On late production models, idle adjustment screw is threaded directly into carburetor body and does not have the threaded bushing shown. Choke plate (17) and throttle plate (not shown) may be attached with one screw as shown although later models have two holes for attaching screws in each plate and shaft.

1. Choke shaft
2. Idle fuel needle & bushing
7. Main fuel needle
9. Idle stop screw
10. Screw
12. Springs
13. Spring
14. Spring
16. Throttle shaft
17. Choke plate

Fixed Speed Carburetor. Refer to Fig. CL62 for identification and sectional view of LMB, LMG and LMV carburetors equipped with constant speed control. These carburetors have no main nozzle (13 – Fig. CL61) and are without an idle current. Speed control knob is turned clockwise to close throttle and stop engine.

Engine speed is controlled at 3000-3400 rpm (no load) by throttle governor spring. See Fig. CL62A for spring placement.

Initial adjustment of idle and main fuel adjustment needles are the same as for Clinton (Walbro) float type carburetor with the exception that main fuel needle initial adjustment is 1¼ to 1½ turns open. Refer to CLINTON (WALBRO) CARBURETORS section for service specifications.

Carter Float Type Carburetor. Refer to Fig. CL63 for identification and ex-

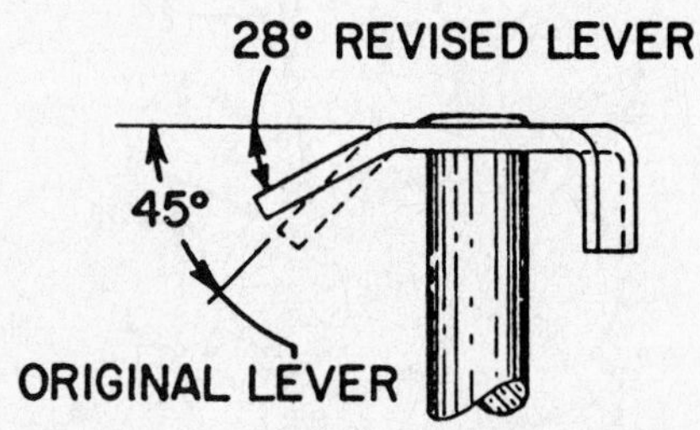

Fig. CL65—Choke lever for later production of Clinton suction carburetor shown in Fig. CL64 is modified as shown here to prevent breakage. Early models should have choke lever (1—Fig. CL64) modified as shown.

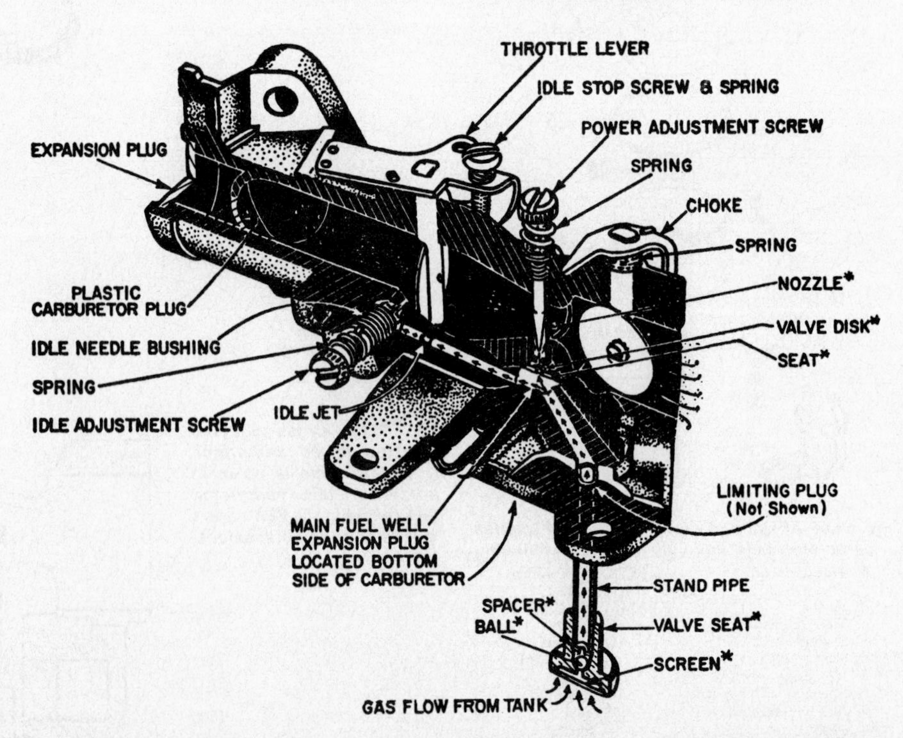

Fig. CL64A—Cut-away view of Clinton suction lift carburetor. Refer to text for disassembly procedure.

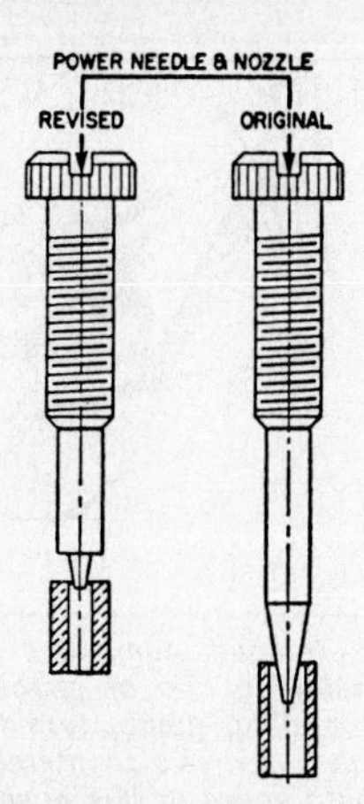

Fig. CL66—View showing early production and late production main fuel needles for Clinton suction lift carburetors. Original and revised needles are not interchangeable. Needle seats are nonrenewable; renew carburetor if seat is damaged.

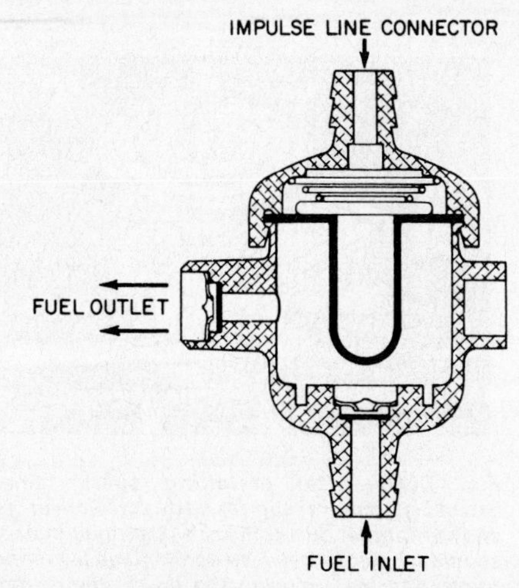

Fig. CL66A—Cross-sectional view of impulse type fuel pump used on four-stroke engines. Renew complete pump assembly if inoperative.

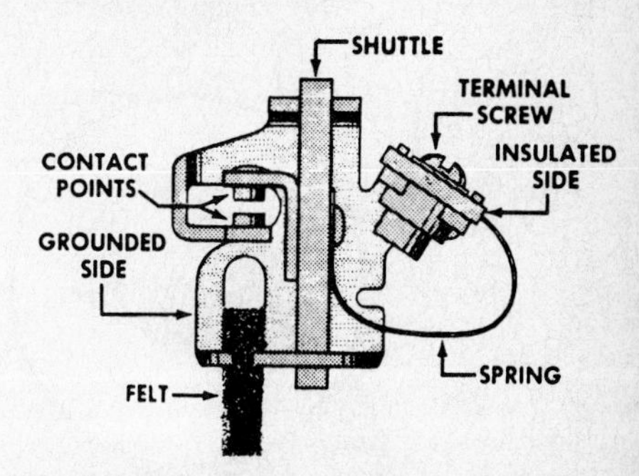

Fig. CL67—Shuttle type breaker points shown are used in early production models.

ploded view of Carter N series carburetors used on some models. Design of float and float bowl may vary.

Initial adjustment of idle (1) and main (4) fuel mixture needles from a lightly seated position is 1 turn open for idle mixture screw and 1½ turns open for main fuel mixture screw on N-2264S and N-2459S models or 1 turn open for idle fuel mixture needle and 2 turns open for main fuel mixture needles on N-2236S and N-2458S models.

Final adjustment is made with engine at operating temperature and running. Operate engine at rated speed and adjust main fuel mixture needle for smoothest engine operation. Operate engine at idle speed and adjust idle fuel mixture needle for smooth engine idle. If engine fails to accelerate smoothly it may be necessary to slightly richen main fuel needle adjustment.

To check float level, invert carburetor throttle body and float assembly. Distance between free side of float and float bowl mating surface on carburetor should be 11/64 inch (4.37 mm). Carefully bend float lever tang which contacts fuel inlet needle as necessary to obtain correct float level.

Clinton Suction Type Carburetors. Refer to Fig. CL64 for identification and exploded view of Clinton suction type carburetor used on vertical crankshaft engines. Carburetor used on horizontal crankshaft engines is similar except that expansion plug and plastic plug shown in cut-away view (Fig. CL64A) are not used.

Initial adjustment of idle fuel mixture needle (2–Fig. CL64) from a lightly seated position is 1½ turns open for mixture needle threaded directly into carburetor body or 4 to 4¼ turns open for mixture needle which thread into a bushing, and bushing is threaded into carburetor body.

Initial adjustment of main fuel mixture needle (7–Fig. CL64) adjustment needle from a lightly seated position is ¾ to 1 turn open for early style needle (refer to Fig. CL66 to identify fuel mixture needle used in early and late carburetors) or 1¼ to 1½ turns open for late style needle (Fig. CL64).

Final adjustments are made with engine at operating temperature and running with fuel tank approximately ½ full. Operate engine at rated speed (fully loaded) and adjust main fuel mixture needle for smoothest engine operation. Note that engine main fuel adjustments cannot be made without engine under load. Engine high speed operation must be adjusted for a richer fuel mixture which may not produce smooth engine operation until load is applied. Operate engine at idle speed and adjust idle fuel mixture needle for smoothest engine idle.

To remove throttle shaft on carburetors from vertical crankshaft engines, drill through the expansion plug at rear of carburetor body, insert punch in drilled hole and pry plug out. Remove plastic plug, throttle valve screws, throttle valve and the throttle shaft. When reassembling, use new expansion plug and seal plug with sealer.

Test check valve in the fuel stand pipe by alternately blowing and sucking air through pipe. Stand pipe can be removed by clamping pipe in vise and prying carburetor from pipe. Apply sealer to stem of new stand pipe and install so it projects 1.895-1.985 inches (48.13-50.32 mm) from carburetor body.

To remove idle jet (Fig. CL64A), remove expansion plug from bottom of carburetor, idle needle and if so equipped, idle needle bushing. Insert a 1/16 inch (1.59 mm) rod through fuel well and up through idle passage to push jet out into idle fuel reservoir. To install new jet, place mark on the rod exactly 1¼ inches (31.75 mm) from end and push new jet into passage until mark on rod is in exact center of fuel well.

Limiting plug (Fig. CL64A) located in fuel passageway from stand pipe should not be removed unless necessary for cleaning purposes. Use sealer on plug during installation.

FUEL PUMP. Some models are equipped with a diaphragm type fuel pump as shown in the cross-sectional view in Fig. CL66A. The pump diaphragm is actuated by pressure pulsations transmitted via an impulse tube connected between the fuel pump and intake manifold. The pump is designed to lift fuel approximately six inches (152 mm). Service consists of renewing the complete fuel pump assembly.

GOVERNOR. An air vane type governor is used on all models. The air vane, which is located in the blower housing, is linked to the throttle lever. The governor is actuated by air delivered by the flywheel fan and by governor spring. Any speed within the operating speed range of the engine can be obtained by adjusting tension on the governor spring. Use only the correct Clinton part for replacement of governor spring and do not adjust maximum governed speed above 3600 rpm.

Make sure the governor linkage does not bind when linkage is moved through full range of travel.

MAGNETO AND TIMING. Magneto coil and armature (laminations), breaker points and condenser are located under the engine flywheel. Two different types of breaker point assemblies are used. Early model engines were equipped with shuttle type breaker points as shown in Fig. CL67. Early type breaker points are enclosed in a box which is an integral part of the engine crankcase and are covered by a plate and gasket. Some shuttle type points may also be enclosed in a sealed unit within the breaker box. Magneto edge gap for models with shut-

Fig. CL67A—Magneto can be considered in satisfactory condition if it will fire an 18 mm spark plug with electrode gap set at 0.156-0.187 inch (3.96-4.75 mm).

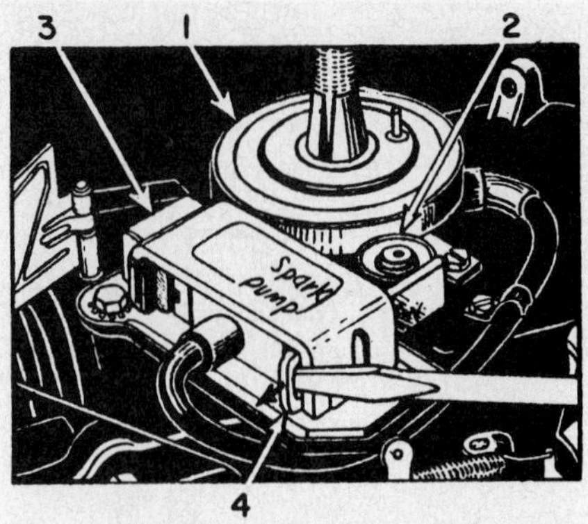

Fig. CL68—After installing spark pump assembly, remove clip (4) with screwdriver as shown. Clip (4) or a 0.10 inch (2.54 mm) spacer should be placed between spark pump lever and frame prior to removing the spark pump from engine. Refer to text.

1. Timing switch
2. Eccentric bearing
3. Spark pump
4. Spacer clip

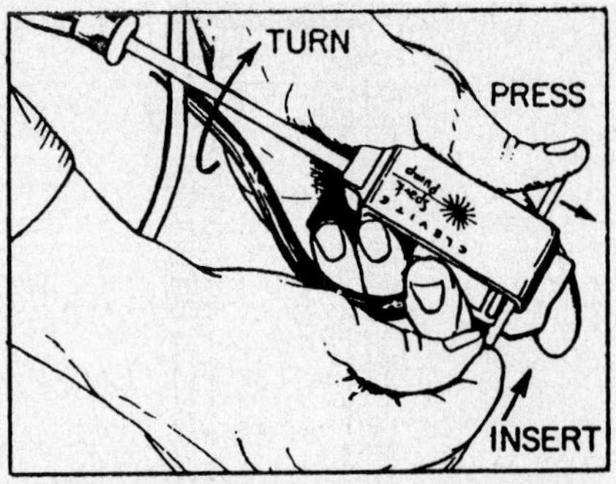

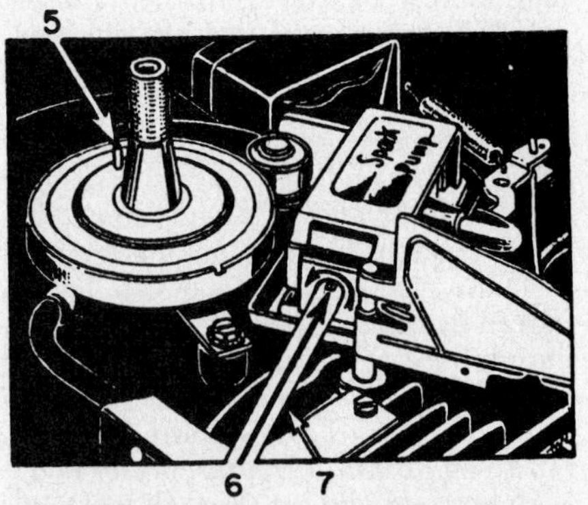

Fig. CL69—Turn load adjusting screw (6) counterclockwise with slotted tool (7) to remove pressure from spark generating cell; spring within spark pump will turn screw back clockwise to apply correct pressure. CAUTION: Never turn load adjusting screw in a clockwise direction. Timing switch is driven by pin (5) which engages engine flywheel.

tle type points is 0.156-0.187 inch (3.96-4.75 mm). Lubricate felt with a high melting point grease.

Later engine models are equipped with rocker type breaker points and the points are enclosed in a breaker box which is attached to the engine crankcase. Magneto edge gap for systems with rocker type breaker points is 0.094-0.250 inch (2.39-6.35 mm).

Breaker point gap for all models is 0.018-0.021 inch (0.46-0.53 mm). Ignition timing is 21° BTDC and is non-adjustable.

SPARK PUMP. On a limited number of engines, a Dyna-Spark ignition system (spark pump or "piezo-electric" ignition system) is used instead of the conventional flywheel type magneto. The system consists of a spark pump (3–Fig. CL68) and a timing switch (1).

The spark pump is actuated by an eccentric bearing (2) on the entended end of the engine camshaft and the timing switch is driven by the crankshaft. A spark is generated whenever the cam lever on the spark pump is moved.

As with the conventional magneto, the Dyna-Spark ignition system may be considered in satisfactory condition if it will fire a spark plug with electrode gap set at 0.156-0.187 inch (3.96-4.75 mm). See Fig. CL67A.

If system is inoperative, inspect wire from spark pump to timing switch and from timing switch to spark plug for shorts or breaks in wire. If inspection does not reveal open or shorted condition, the complete Dyna-Spark unit must be renewed.

To remove Dyna-Spark unit, disconnect spark plug and remove engine blower housing and flywheel. Turn engine slowly until eccentric bearing (2–Fig. CL68) on end of camshaft is at maximum lift position and install a 0.10 inch (2.54 mm) spacer between spark pump actuating lever and spark pump frame. Turn engine so the eccentric bearing is at minimum lift position and remove the spark pump and timing switch from engine. The eccentric bearing may be removed from the camshaft after removing the retaining snap ring.

To install Dyna-Spark unit, turn engine so eccentric bearing is at minimum lift position and install the spark pump and timing switch on engine. Turn engine so eccentric bearing is at maximum lift position and remove the clip (4) or spacer from spark pump. Turn the load screw (6–Fig. CL69) counterclockwise with slotted tool (7) or needle nose pliers to release tension on the spark generating element. Lever on spark pump should then follow eccentric bearing closely and smoothly as the engine is turned. Be sure pin (5) properly engages engine flywheel when reinstalling flywheel.

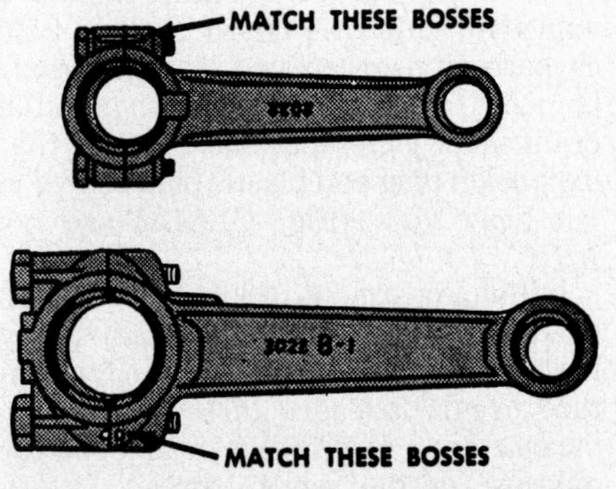

Fig. CL70—If spark pump has been removed without installing clip or spacer (Fig. CL68), before reinstalling pump, turn load adjusting screw about 1/2 turn in a counterclockwise direction and hold screw in this position. Press actuating lever in direction shown and insert a 0.10 inch (2.54 mm) spacer between inner end of lever and spark pump frame.

CAUTION: Never turn load screw (6) in a clockwise direction. A coil spring within the spark pump unit will return the load screw to correct tension.

If the spark pump has been removed without a 0.10 inch (2.54 mm) spacer (or the clip as provided in a new spark pump) installed, refer to Fig. CL70 prior to installation of spark pump on engine.

LUBRICATION. Manufacturer recommends oil with an API service classification SE or SF. Use SAE 30 oil for temperatures above 32° F (0° C), SAE 10W oil for temperatures between -10° F (-23° C) and 32° F (0° C) and SAE 5W oil for temperatures below -10° F (-23° C).

On models equipped with reduction gearing, use SAE 30 oil in gearbox.

CRANKCASE BREATHER. Crankcase breather assembly located in or behind valve chamber cover should be cleaned if difficulty is experienced with oil loss through breather. Be sure the breather is correctly reassembled and reinstalled.

Fig. CL71—When assembling cap to connecting rod, be sure embossments on rod and cap are aligned as shown.

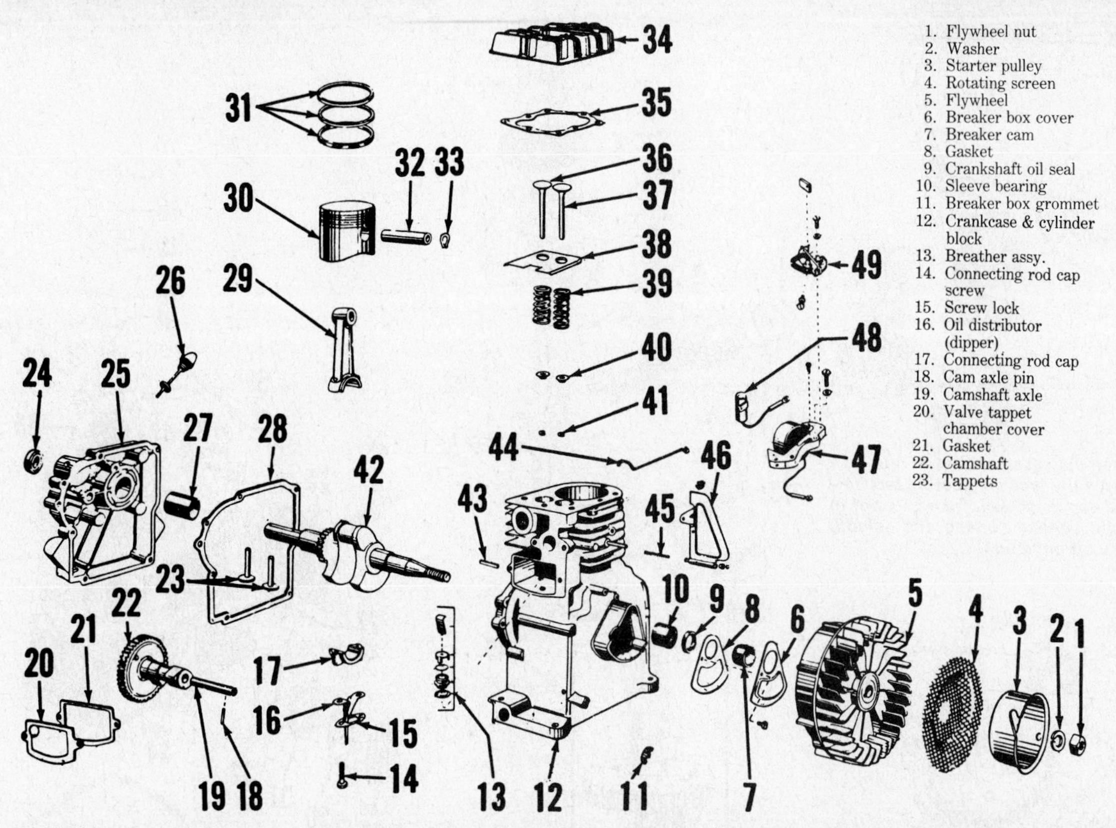

1. Flywheel nut
2. Washer
3. Starter pulley
4. Rotating screen
5. Flywheel
6. Breaker box cover
7. Breaker cam
8. Gasket
9. Crankshaft oil seal
10. Sleeve bearing
11. Breaker box grommet
12. Crankcase & cylinder block
13. Breather assy.
14. Connecting rod cap screw
15. Screw lock
16. Oil distributor (dipper)
17. Connecting rod cap
18. Cam axle pin
19. Camshaft axle
20. Valve tappet chamber cover
21. Gasket
22. Camshaft
23. Tappets
24. Crankshaft oil seal
25. Bearing plate
26. Oil dipstick
27. Sleeve bearing
28. Gasket
29. Connecting rod
30. Piston
31. Piston rings
32. Piston pin
33. Pin retainers
34. Cylinder head
35. Gasket
36. Exhaust valve
37. Intake valve
38. Valve spring retainer
39. Valve springs
40. Spring retainers
41. Valve keepers
42. Crankshaft
43. Flywheel key
44. Governor link
45. Air vane pin
46. Governor air vane
47. Armature & coil assy.
48. Condenser
49. Breaker points

Fig. CL71A—Exploded view of early production horizontal crankshaft model. Camshaft (22) rotates on cam axle (19). Breather assembly (13) is located inside valve chamber. Breaker box is integral part of crankcase and breaker points (49) are accessible after removing flywheel and cover (6). Refer to Fig. CL71B for exploded view of late production horizontal crankshaft model.

REPAIRS

TIGHTENING TORQUES. Recommended tightening torque specifications are as follows:

Base plate or side cover	75-80 in.-lbs. (8.5-9 N·m)
Blower housing	60-70 in.-lbs. (6.7-8 N·m)
Carburetor to manifold	35-50 in.-lbs. (4-6 N·m)
Carburetor (manifold) to block	60-65 in.-lbs. (6.7-7.3 N·m)

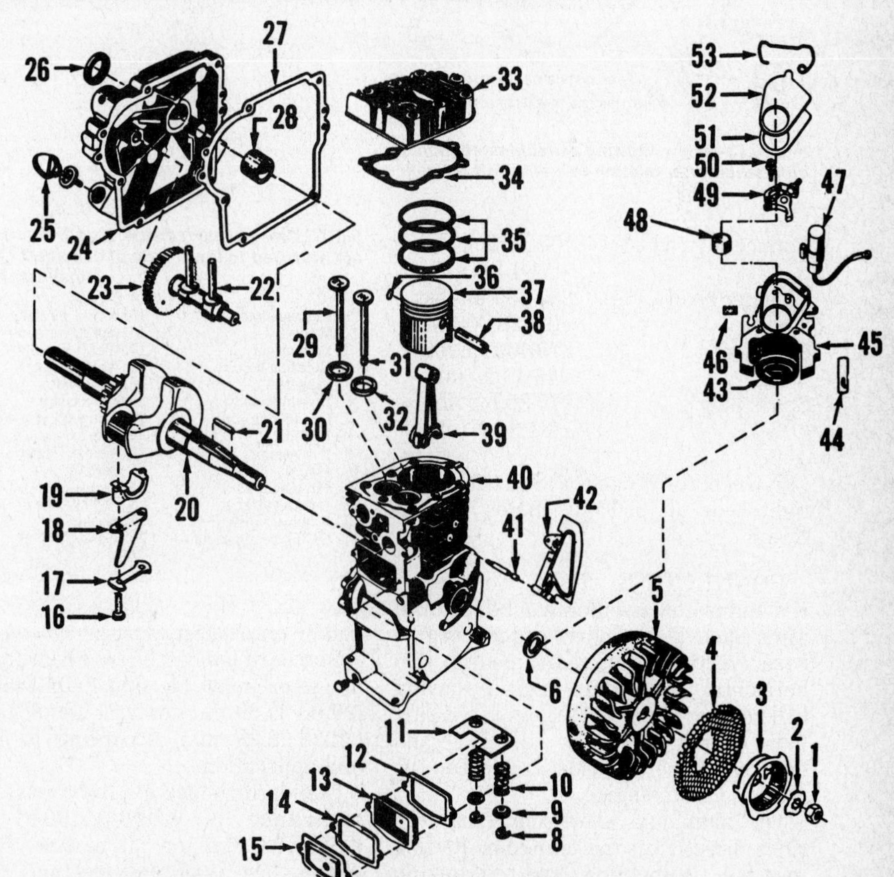

1. Flywheel nut
2. Washer
3. Starter cup
4. Rotating screen
5. Flywheel
6. Crankshaft oil seal
8. Valve keeper
9. Spring retainer
10. Valve springs
11. Spring retainer plate
12. Gasket
13. Breather assy.
14. Gasket
15. Tappet chamber cover
16. Connecting rod cap screws
17. Cap screw lock
18. Oil distributor
19. Connecting rod cap
20. Crankshaft
21. Flywheel key
22. Tappets
23. Camshaft
24. Bearing plate
25. Oil dipstick
26. Crankshaft oil seal
27. Gasket
28. Sleeve bearing
29. Exhaust valve
30. Exhaust valve seat
31. Intake valve
32. Intake valve seat
33. Cylinder head
34. Gasket
35. Piston rings
36. Piston pin retainers
37. Piston
38. Piston pin
39. Connecting rod
40. Crankcase & cylinder block
41. Air vane pin
42. Governor air vane
43. Ignition coil
44. Coil retaining clips
45. Armature & stator assy.
46. Cam wiper felt
47. Condenser
48. Breaker cam
49. Breaker points
50. Screw
51. Gasket
52. Breaker box cover
53. Retainer spring

Fig. CL71B—Exploded view of late production horizontal crankshaft model with die-cast aluminum crankcase and cylinder block assembly. Models with cast iron cylinder block are similar except for valve seat inserts (30 and 32).

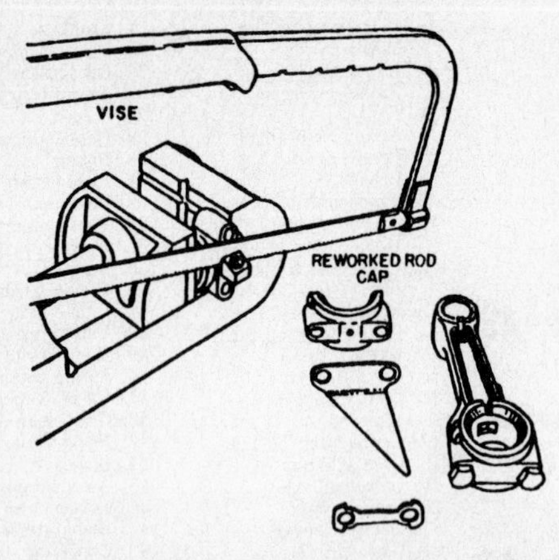

Fig. CL72—When necessary to install late type oil distributor on early type connecting rod, saw oil cup from rod cap as shown. Take care not to damage crankpin bearing surface and smooth off any burrs.

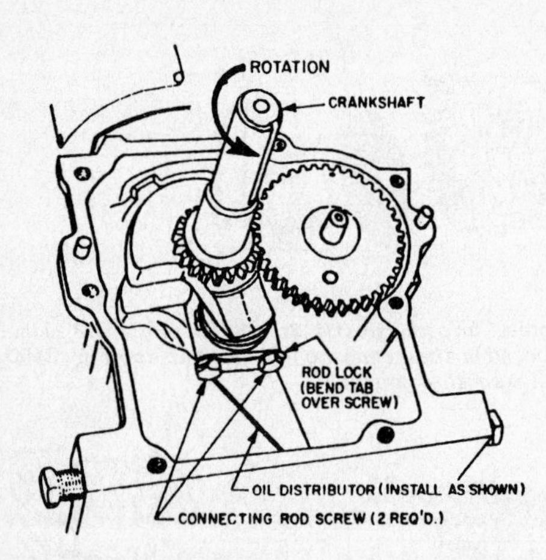

Fig. CL73—View showing correct installation of oil distributor in relation to crankshaft rotation.

Connecting rod	100-125 in. lbs. (11-14 N•m)
Cylinder head	225-250 in.-lbs. (25-28 N•m)
Flywheel	375-400 in.-lbs. (42-45 N•m)*
Spark plug	275-300 in.-lbs. (31-34 N•m)

*Flywheel for "Touch 'N' Start" brake is tightened to 650-700 in.-lbs. (73-79 N•m).

CONNECTING ROD. Connecting rod and piston assembly can be removed after removing cylinder head and engine base (vertical crankshaft models). On horizontal crankshaft engines having ball bearing mains, oil seal and snap ring must be removed from side cover and crankshaft before side cover can be removed from engine.

The aluminum alloy connecting rod rides directly on the crankpin. Recommended connecting rod-to-crankpin clearance is 0.0015-0.0030 inch (0.04-0.08 mm). If clearance is 0.004 inch (0.10 mm) or more, connecting rod and/or crankshaft must be renewed.

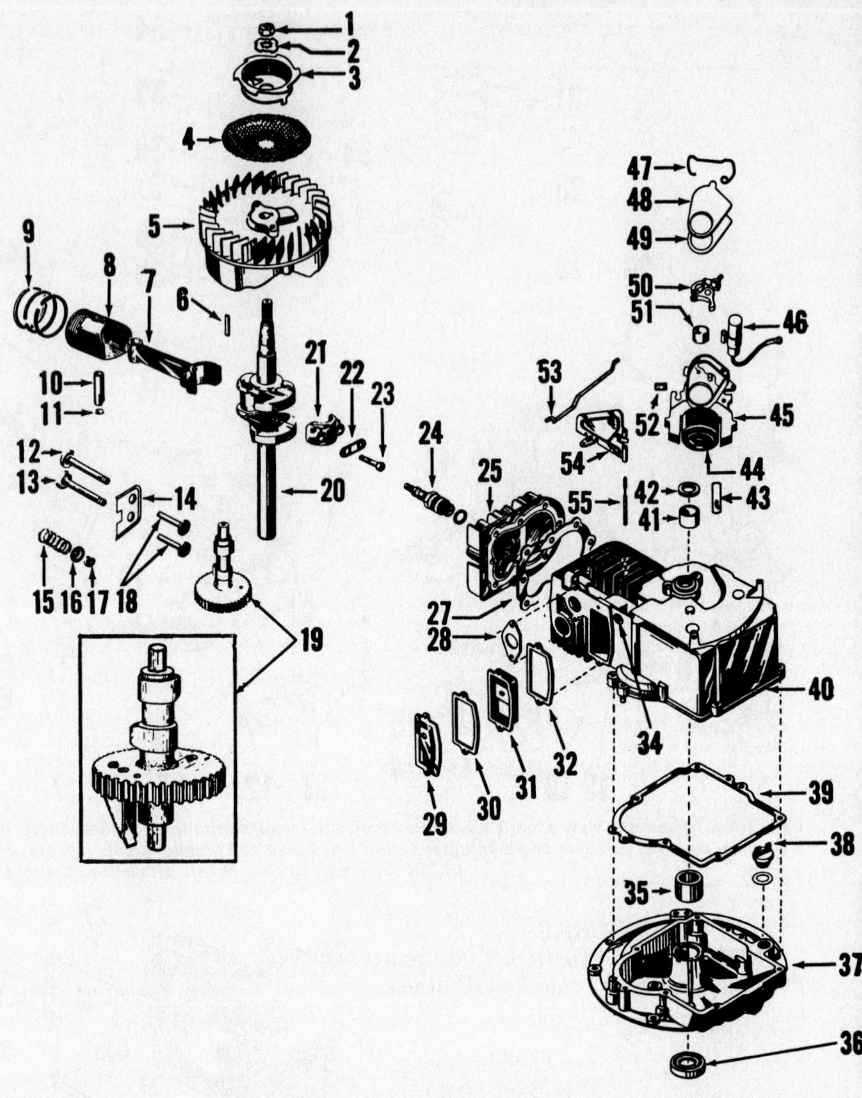

Fig. CL73A—Exploded view of vertical crankshaft model with splash lubrication system. Note oil dip per attached to lower side of camshaft gear (inset 19). Refer to Fig. CL76A for exploded view of ve tical crankshaft model with oil pump.

1. Flywheel nut
2. Washer
3. Starter cup
4. Rotating screen
5. Flywheel
6. Flywheel key
7. Connecting rod
8. Piston
9. Piston rings
10. Piston
11. Pin retainers
12. Intake valve
13. Exhaust valve
14. Spring retainer plate
15. Valve springs
16. Spring retainer
17. Valve keepers
18. Valve tappets
19. Camshaft
20. Crankshaft
21. Connecting rod cap
22. Cap screw lock
23. Connecting rod cap screws
24. Spark plug
25. Cylinder head
27. Gasket
28. Gasket
29. Tappet chamber cover
30. Gasket
31. Breather assy.
32. Gasket
34. Grommet
35. Sleeve bearing
36. Crankshaft oil seal
37. Engine base
38. Oil filler plug
39. Gasket
40. Crankcase & cylinder block
41. Sleeve bearing
42. Crankshaft oil seal
43. Coil retaining cli
44. Ignition coil
45. Armature & coil assy.
46. Condenser
47. Retainer spring
48. Breaker box cov
49. Gasket
50. Breaker points
51. Breaker cam
52. Felt cam wiper
53. Governor link
54. Governor air van
55. Air vane pin

Standard connecting rod bearing bore diameter may be 0.8140-0.8145 inch (20.68-20.69 mm) or 0.8770-0.8775 inch (22.28-22.29 mm) according to model and application.

Standard connecting rod-to-piston pin clearance is 0.0004-0.0011 inch (0.01-0.03 mm) on all models. Renew piston pin and/or connecting rod if clearance is 0.002 inch (0.05 mm) or more.

Standard piston pin bore diameter in connecting rod is 0.5630-0.5635 inc (14.30-14.31 mm).

Install connecting rod with oil ho towards flywheel side of engine and wit embossments on rod and cap aligned a shown in Fig. CL71.

Connecting rods on early horizont crankshaft models had an oil cup on ro cap which was designed for use with a oil distributor which is no longe available. If necessary to renew this o distributor, a new distributor, Clinto part 220-147, and two new rod lock

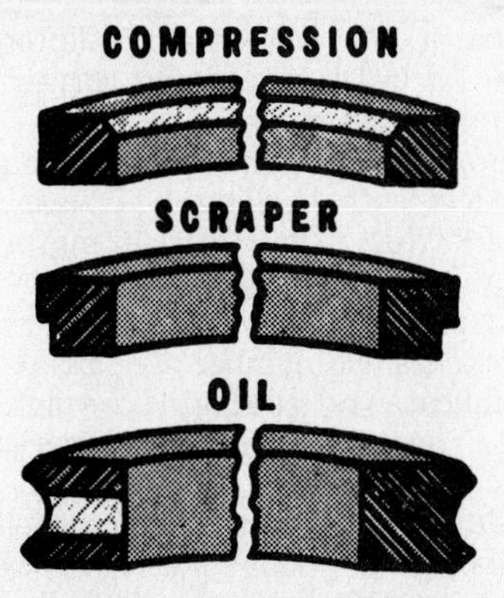

Fig. CL76—Drawing showing proper placement of piston rings. Note bevel at top of ring on inside diameter on top compression ring and notch in lower side of outside diameter of second compression (scraper) ring. Oil ring may be installed with either side up.

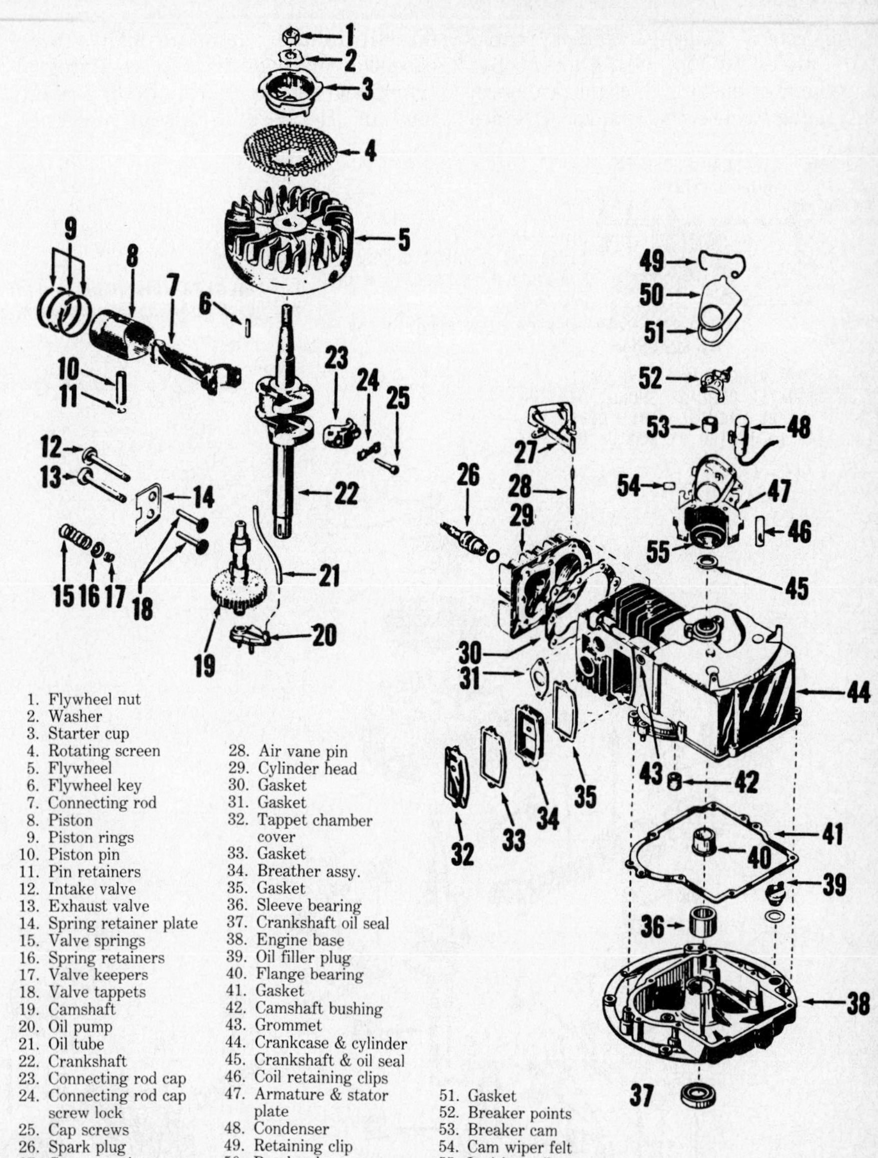

Fig. CL76A—Exploded view of vertical crankshaft model with lubricating oil pump (20). Refer to Fig. CL73A for vertical crankshaft model with splash lubrication.

part 31038 must be used. If connecting rod with oil cup is to be reused, it must be reworked as shown in Fig. CL72. Install oil distributor and rod locks as shown in Fig. CL73.

PISTON, PIN AND RINGS. Piston is equipped with two compression rings and one oil control ring.

Standard ring side clearance in groove is 0.002-0.005 inch (0.05-0.12 mm). If side clearance is 0.006 inch (0.15 mm), renew rings and/or piston.

Standard piston ring end gap is 0.007-0.017 inch (0.18-0.26 mm). If ring end gap is 0.025 inch (0.64 mm) or more, renew rings and/or recondition cylinder bore.

Standard piston skirt-to-cylinder bore clearance is 0.0045-0.0065 inch (0.114-0.165 mm). If clearance is 0.008 inch (0.203 mm) or more, renew piston and/or recondition cylinder bore.

Standard pin bore diameter in piston is 0.5625-0.5628 inch (14.29-14.30 mm). Standard piston pin diameter is 0.5624-0.5626 inch (14.288-14.295 mm). Piston pin is a 0.0001 inch (0.003 mm) interference to a 0.0004 inch (0.010 mm) loose fit in piston pin bore.

Piston and rings are available in a variety of oversizes as well as standard and a special chrome ring set is available for cylinders having up to 0.010 inch (0.25 mm) taper and/or out-of-round condition.

CLYINDER AND CRANKCASE. Cylinder and crankcase are an integral unit of either an aluminum alloy die-casting with a cast-in iron cylinder liner or a shell casting of cast iron.

Standard cylinder bore diameter for 419-0003-000, 429-0003-000 and 431-0003-000 models is 2.499-2.500 inch (63.48-63.50 mm). Standard cylinder bore diameter for all other models is 2.3745-2.3755 inch (60.31-60.34 mm). Refer to PISTON, PIN AND RINGS section for piston-to-cylinder specifications.

CRANKSHAFT, MAIN BEARINGS AND SEAL. Crankshaft may be supported in-bushing type, integral bushing type or ball bearing type main bearings.

Standard crankpin diameter is either 0.8119-0.8125 inch (20.62-20.64 mm) or 0.8745-0.8752 inch (22.21-22.23 mm) according to model and application. If crankshaft main journal is 0.001 inch (0.03 mm) or more out-or-round, or if crankpin journal is 0.0015 inch (0.04 mm) or more out-of-round, renew crankshaft.

Standard clearance between crankshaft main bearing journals and bushing or integral type main bearings is 0.0018-0.0035 inch (0.05-0.09 mm). If clearance is 0.005 inch (0.13 mm) or

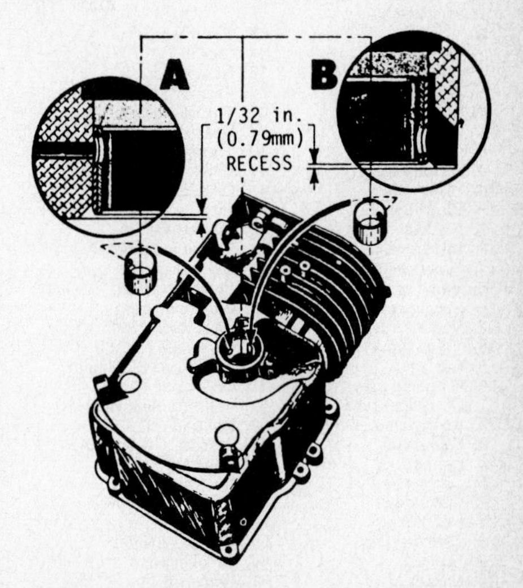

Fig. CL77—When installing crankshaft bushing in crankcase be sure oil holes are aligned as shown and that inner edge of bushing is 1/32 inch (0.79 mm) below thrust face of block.

more, renew bearings and/or crankshaft. Refer to Fig. CL77 for proper placement of bushing in crankcase bore. On models where the crankshaft rides directly in the aluminum alloy crankcase,side plate or base plate (integral type), bearing can be renewed by reaming out the bore to accept a service bushing. Contact nearest Clinton Central Parts Distributor for correct tools, reamers and parts.

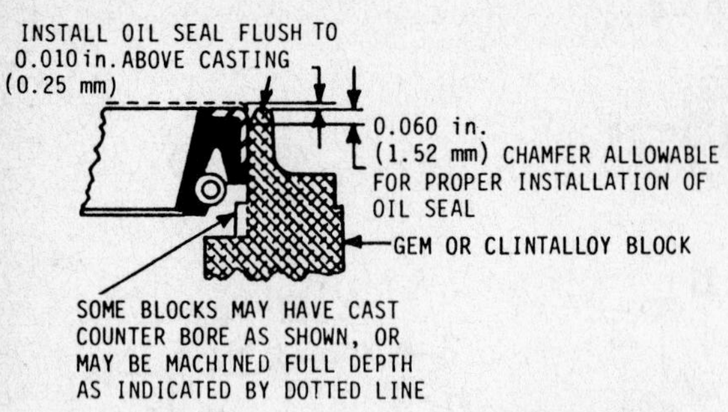

Fig. CL78—Install oil seal in block (crankcase) as shown.

Standard crankshaft end play for models with bushing type bearings is 0.008-0.018 inch (0.20-0.46 mm). If end play is 0.025 inch (0.64 mm) or more, end play must be adjusted by varying thickness and number of shims between crankcase and side plate/base plate. Gaskets are available in a variety of thicknesses.

On horizontal crankshaft models with ball bearing type main bearings, bearing on pto (side plate) end of crankshaft is retained in the side plate with a snap ring and is also retained on the crankshaft with a snap ring. To remove the side plate, first pry crankshaft seal from side plate and remove snap ring retaining bearing to crankshaft. Separate side plate with bearing from crankcase and

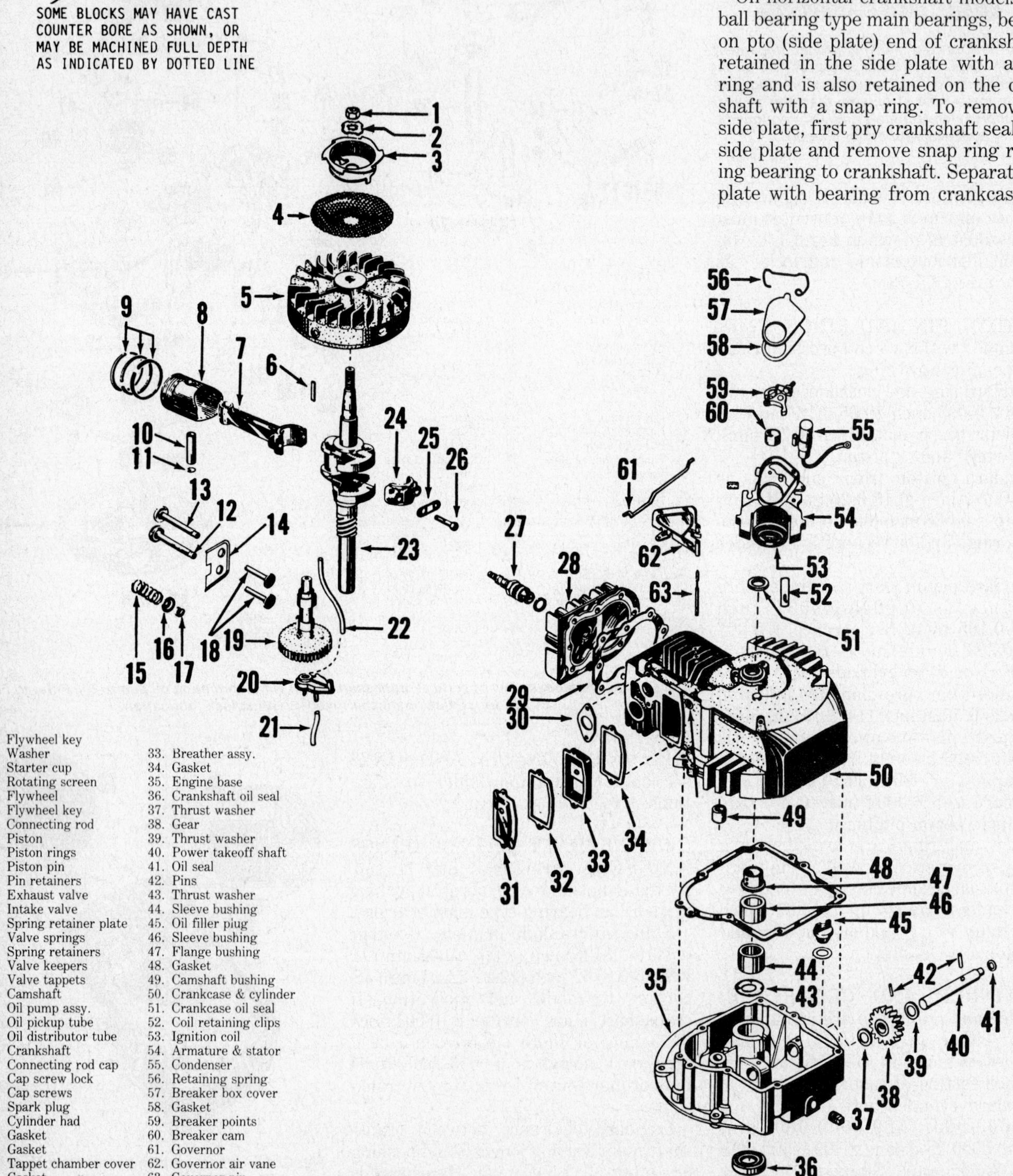

1. Flywheel key
2. Washer
3. Starter cup
4. Rotating screen
5. Flywheel
6. Flywheel key
7. Connecting rod
8. Piston
9. Piston rings
10. Piston pin
11. Pin retainers
12. Exhaust valve
13. Intake valve
14. Spring retainer plate
15. Valve springs
16. Spring retainers
17. Valve keepers
18. Valve tappets
19. Camshaft
20. Oil pump assy.
21. Oil pickup tube
22. Oil distributor tube
23. Crankshaft
24. Connecting rod cap
25. Cap screw lock
26. Cap screws
27. Spark plug
28. Cylinder had
29. Gasket
30. Gasket
31. Tappet chamber cover
32. Gasket
33. Breather assy.
34. Gasket
35. Engine base
36. Crankshaft oil seal
37. Thrust washer
38. Gear
39. Thrust washer
40. Power takeoff shaft
41. Oil seal
42. Pins
43. Thrust washer
44. Sleeve bushing
45. Oil filler plug
46. Sleeve bushing
47. Flange bushing
48. Gasket
49. Camshaft bushing
50. Crankcase & cylinder
51. Crankcase oil seal
52. Coil retaining clips
53. Ignition coil
54. Armature & stator
55. Condenser
56. Retaining spring
57. Breaker box cover
58. Gasket
59. Breaker points
60. Breaker cam
61. Governor
62. Governor air vane
63. Governor pin

Fig. CL78A—Exploded view of vertical crankshaft model with auxiliary power take-off; output shaft gear (38) is driven by worm gear on crankshaft. Sleeve bearing (47) is used in upper side of engine base and bearing (44) is installed in lower side.

crankshaft. Ball bearing and retaining snap ring can then be removed from the side plate. Bearing should be renewed if rough, loose or damaged.

When installing crankshaft seal in crankcase, refer to Fig. CL78 for proper placement of seal. When installing crankshaft, make certain crankshaft and camshaft gear timing marks are aligned as shown in Fig. CL79.

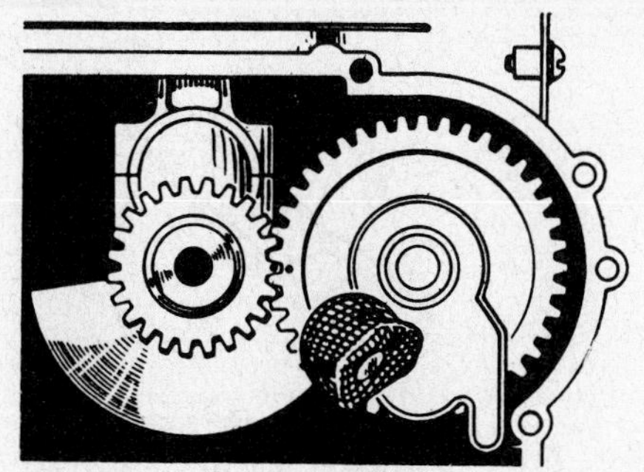

Fig. CL79—Valves are correctly timed when marks on camshaft gear and crankshaft gear are aligned as shown.

CAMSHAFT. On early models, integral camshaft and gear turned on a stationary axle. Later models are equipped with a solid integral camshaft and gear which is supported at each end in bearings which are an integral part of crankcase or side plate/baseplate assemblies.

On early models, clearance between camshaft and cam axle is 0.001-0.003 inch (0.03-0.08 mm). If clearance is 0.005 inch (0.13 mm) or more, camshaft and/or axle must be renewed.

Standard axle diameter is 0.3740-0.3744 inch (9.50-9.51 mm). Axle should be a 0.0009 inch (0.02 mm) tight to 0.001 inch (0.03 mm) loose fit in crankcase and side plate/base plate.

On models where camshaft turns in bores in crankcase and base or side plate, standard clearance between camshaft journals and bearing bore is 0.001-0.003 inch (0.03-0.08 mm). If clearance is 0.006 inch (0.15 mm) or more, renew camshaft and/or crankcase, side plate or base plate.

When installing camshaft, align camshaft and crankshaft gear timing marks as shown in Fig. CL79.

VALVE SYSTEM. Recommended valve tappet gap for intake and exhaust valves on all models is 0.009-0.011 inch (0.23-0.28 mm). Adjust valve clearance by grinding off end of valve stem to increase clearance or by renewing valve and/or grinding valve seat deeper into block to reduce clearance.

Valve face angle is 45° and valve seat angle is 44°. If valve face margin is 1/64 inch (0.4 mm) or less, renew valve. On models with aluminum alloy cylinder block, seat inserts are standard. On models with cast iron cylinder block, seats are ground directly into cylinder block surface. Standard valve seat width is 0.030-0.045 inch (0.76-1.29 mm). If seat width is 0.060 inch (1.52 mm) or more, seat must be narrowed.

Standard valve stem-to-guide clearance is 0.0015-0.0045 inch (0.04-0.11 mm). If clearance is 0.006 inch (0.15 mm) or more, guide may be reamed to 0.260 inch (6.60 mm) and a valve with a 0.010 inch (0.25 mm) oversize stem installed.

LUBRICATING SYSTEM. All horizontal crankshaft models and some vertical crankshaft models are splash lubricated. An oil distributor is attached to the connecting rod cap on horizontal crankshaft models and an oil scoop is riveted to the lower side of the camshaft gear on vertical crankshaft models. Refer to CONNECTING ROD section for additional information on oil distributor on connecting rod cap.

A gear type oil pump, driven by a pin on the lower end of the engine camshaft, is used on some vertical crankshaft models. When reassembling these engines, be sure the oil tube fits into the recesses in cylinder block and oil pump before installing base plate.

SERVICING CLINTON ACCESSORIES

IMPULSE STARTERS (EARLY PRODUCTION)

Difficulty in starting engine equipped with impulse type starter may be caused by improper starting procedure or adjustment. The throttle lever must be in the full choke position and left there until engine starts. After impulse starter is fully wound, move handle to start position and push handle down against stop until starter releases. Occasionally there is a hesitation because the engine is on the compression stroke. If engine does not start readily (within 5 releases of the starter), make the following checks:

1. Remove air cleaner and be sure choke is fully closed when throttle lever is in full choke position. If not, adjust controls so choke can be fully closed.
2. Check the idle and high speed fuel mixture adjustment needles. See engine servicing section for recommended initial adjustment for appropriate model being serviced.
3. Check magneto for spark.

To disassemble unit, make certain starter spring is released, remove the four phillips head screws (3–Fig. CL100) and invert the assembly. Holding the assembly at arms length, lightly tap the legs of the starter frame against work bench to remove bottom cover, power springs and cups, plunger assembly and the large gear. Carefully separate the spring and cup assemblies from the plunger. Hold plunger and unscrew ratchet using a 3/8-inch Allen wrench. Remove the snap ring (9) and disassemble plunger unit. Renew all damaged parts or assemblies.

Prior to reassembly, coat all internal parts with light grease. Install large gear with beveled edge of teeth to bottom (open) side of starter and engage the lock pawl as shown in Fig. CL100A. Assemble the plunger unit and install it through the large gear so release button protrudes through top of starter frame. Install power spring and cup assembly with closed side of cup towards large gear and carefully work inner end of spring over the plunger. If two power springs are used, install second spring and cup assembly with closed side of cup towards first spring and cup unit. In some starters, an empty spring cup is used as a spacer. Install cup with closed side toward power spring. Install the bottom cover (4–Fig. CL100) and spring (2); then, screw ratchet into plunger bushing. It is not necessary to tighten the ratchet as normal action of the starter will do this.

IMPULSE STARTER (LATE PRODUCTION)

Troubleshooting procedure for late type impulse starter unit shown in Fig. CL101 will be similar to that for early unit.

Service is limited to renewal of handle (3), pawl (6) and/or pawl spring (9). If power spring is broken, or if drive gear that engages starter cup (2) is worn or damaged, a complete new starter assembly must be installed. Starter cup

(2) is not included as part of the starter assembly.

REWIND STARTERS

Exploded view of rewind starter currently used on most Clinton engines is shown in Fig. CL102. Other starters used are shown in Figs. CL102A through CL110. Care should be taken when reassembling all starters to be sure the rewind spring is not wound too tightly. The spring should be wound tight enough to rewind the rope, but not so tight that the spring is fully wound before the rope is pulled out to full length. Coat spring and all internal parts with Lubriplate or equivalent grease when reassembling. Be sure spring, pulley and related parts are assembled for correct rotation.

12 VOLT STARTER AND LIGHTING COIL

Refer to Figs. CL113, CL114, CL115 and CL116. Service and/or parts for the 12 volt starter are available at American Bosch Service Centers.

The 12 volt lighting coils are mounted on the magneto armature core as shown in Fig. CL115 and CL116. Wiring diagram when unit is used for lighting circuit only is shown in Fig. CL115.

Fig. CL100 – Exploded view of early type Clinton impulse type starter. Some starters having only one power spring (5) used an empty power spring cup as a spacer.

1. Ratchet
2. Spring
3. Cap screws
4. Cover plate
5. Power spring & cup assy.
6. Plunger hub
7. Steel balls (2)
8. Plunger
9. Snap ring
10. Gear
11. Housing, crank & pinion assy.

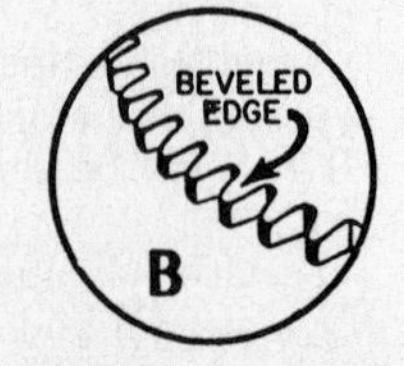

Fig. CL100A – When reinstalling large gear (10 – Fig. CL100), be sure beveled edge of gear is towards open side of housing and lock engages gear as shown.

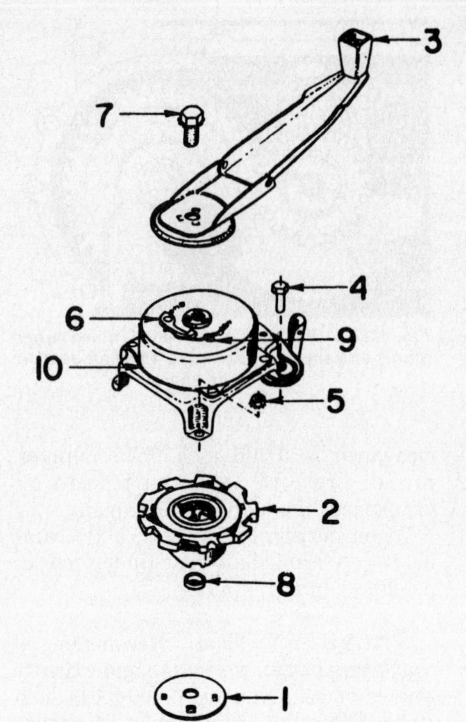

Fig. CL101 – Exploded view of late type Clinton impulse starter. Starter is serviced only in parts indicated by callouts (1) through (9).

1. Starter cup adapter
2. Starter cup
3. Handle assy.
4. Barrel nut
5. Nut
6. Starter pawl
7. Cap screw & washer
8. Starter cup spacer
9. Pawl spring
10. Housing, power spring & gear assy.

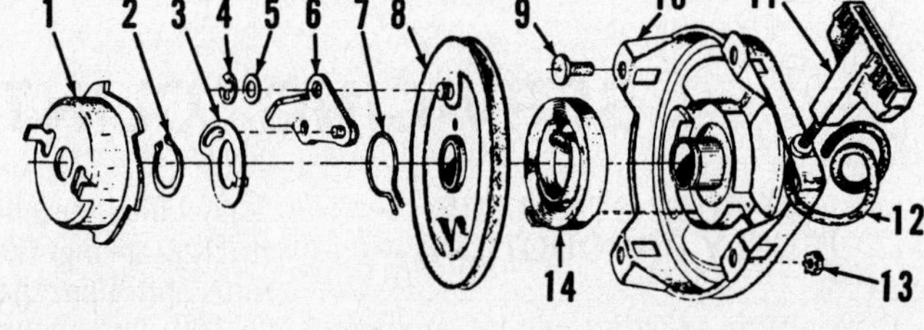

Fig. CL102 – Exploded view of late production Clinton rewind starter.

1. Starter cup
2. Snap ring
3. Actuator
4. Snap ring
5. Nut
6. Pawl & pin assy.
7. Actuator spring
8. Pulley
9. Pan head screw
10. Housing
11. Handle
12. Rope
13. Nut & washer assy.
14. Rewind spring

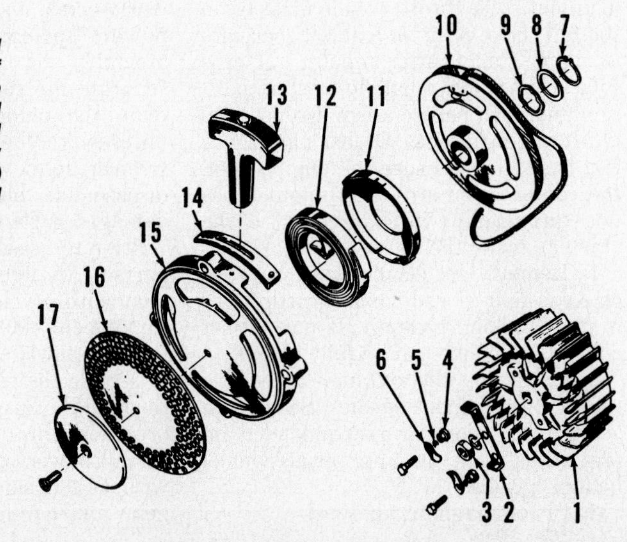

Fig. CL102A – Exploded view of Deluxe Model of Clinton rewind starter of the type used on chain saws and some engines. Starter will operate in either rotation by interchanging springs (5) and inverting pawls (6) and rewind spring (12).

1. Flywheel
2. Plate
3. Flywheel nut
4. Spacer
5. Spring
6. Pawl
7. Snap ring
8. Washer
9. Wave washer
10. Pulley
11. Cup
12. Rewind spring
13. Handle
14. Guide
15. Housing
16. Screen
17. Retainer

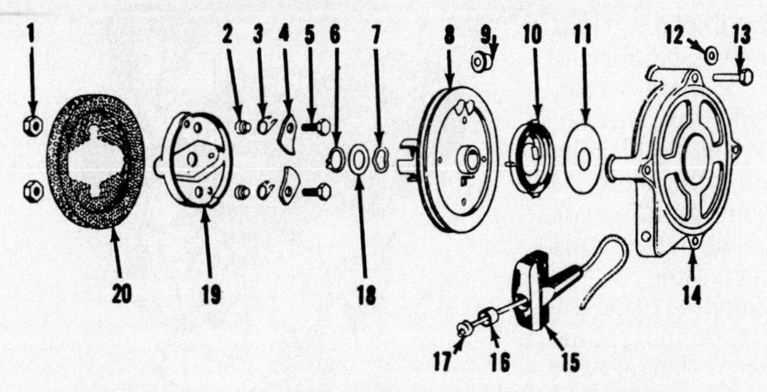

Fig. CL103—Exploded view of an earlier type Clinton rewind starter.

1. Locknut
2. Pawl spacer
3. Pawl spring
4. Pawl
5. Cap screw
6. Snap ring
7. Wave washer
8. Pulley
9. Rope bushing
10. Rewind spring & cup asy.
11. Washer
12. Lockwasher
13. Barrel nut
14. Housing
15. Handle
16. Retainer
17. Rope
18. Washer
19. Starer cup
20. Rotating screen

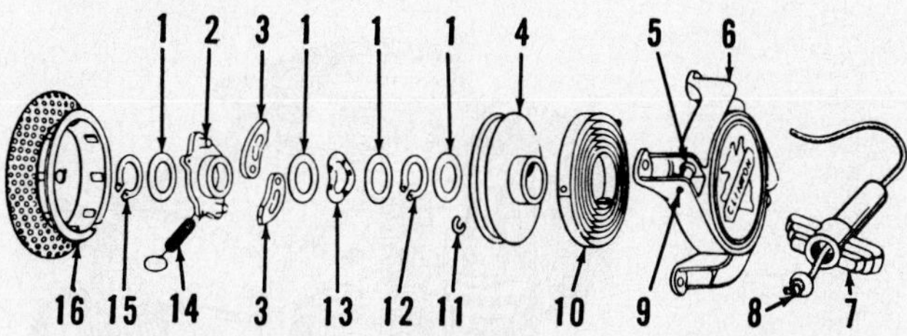

Fig. CL104—Exploded view of an earlier type of Clinton rewind starter.

1. Flat washer
2. Drive plate
3. Drive pawls
4. Pulley
5. Rope pulley
6. Housing
7. Handle
8. Rope
9. Roll pin
10. Rewind spring
11. Snap ring
12. Snap ring
13. Wave washer
14. Tension spring
15. Snap ring
16. Starter cup

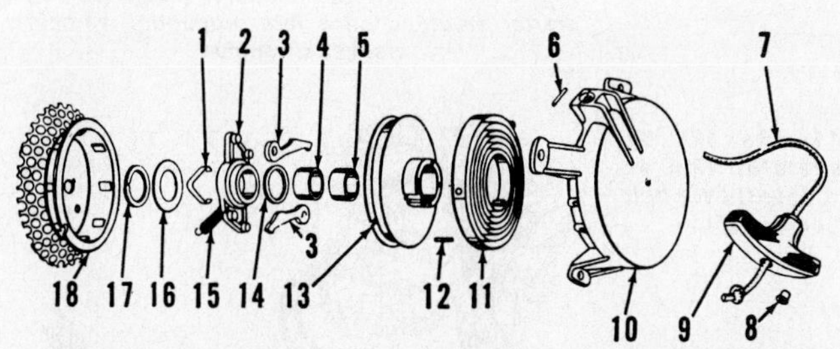

Fig. CL105—Exploded view of an earlier type Clinton rewind starter. Design of housing (10) may vary from that shown. Bushing (5) is not used in some starters.

1. Retaining spring
2. Pawl plate
3. Pawls
4. Bushing
5. Bushing
6. Roll pin
7. Rope
8. Washer
9. Handle
10. Housing
11. Rewind spring
12. Roll pin
13. Pulley
14. Spacer
15. Tension spring
16. Washer
17. Snap ring
18. Starter cup

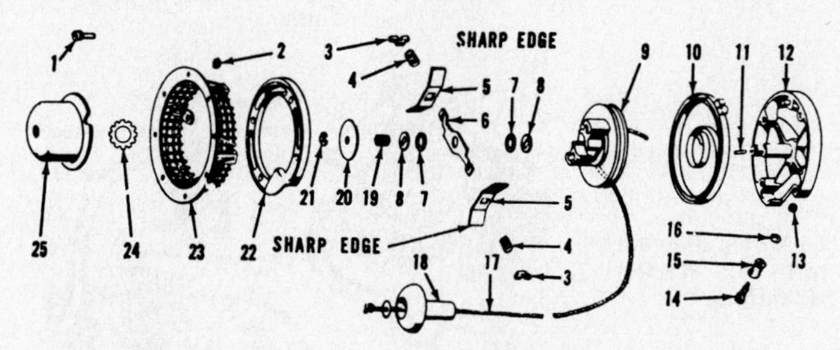

Fig. CL106—Exploded view of Bulldog starter used on some early production engines.

1. Screw & washer assy.
2. Nut & retainer assy.
3. Retainer
4. Spring
5. Friction shoe
6. Brake lever
7. Washer
8. Slotted washer
9. Pulley
10. Rewind spring
11. Centering pin
12. Housing
13. Nut
14. Shoulder screw
15. Roller
16. Washer
17. Rope
18. Handle
19. Spring
20. Washer
21. Snap ring
22. Flange
23. Base
24. Lockwasher
25. Pulley (cup)

When unit is used to provide a battery charging current, a rectifier must be installed in the circuit to convert the AC current into DC current as shown in Fig. CL116.

110 VOLT ELECTRIC STARTERS

Exploded views of the two types of 110 volt electric starters are shown in Figs. CL117 and CL118. Always connect starter cord at engine before connecting cord to 110 volt power source.

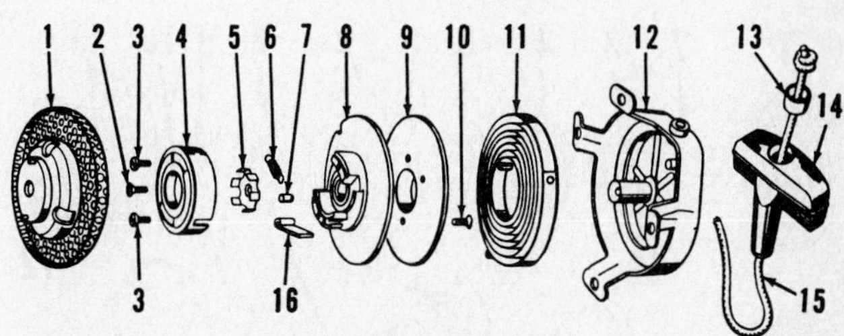

Fig. CL107—Exploded view of Eaton rewind starter used on some early production engines.

1. Starter pulley
2. Screws
3. Screws (2)
4. Retainer
5. Brake
6. Tension spring
7. Spring retainer
8. Hub
9. Plate
10. Screw
11. Rewind spring
12. Housing
13. Cup
14. Handle
15. Rope

Fig. CL108—Exploded view of Fairbanks-Morse rewind starter used on early production engines. Refer to Fig. CL109 for a second type of Fairbanks-Morse starter used on other Clinton models.

1. Centering pin
2. Washer
3. Fiber washer
4. Brake lever
5. Friction shoe
6. Spring
7. Retainer
8. Washer
9. Rope
10. Handle
11. Housing
12. Roller
13. Screw
14. Rewind spring
15. Pulley
16. Spring
17. Snap ring
18. Pulley (cup)

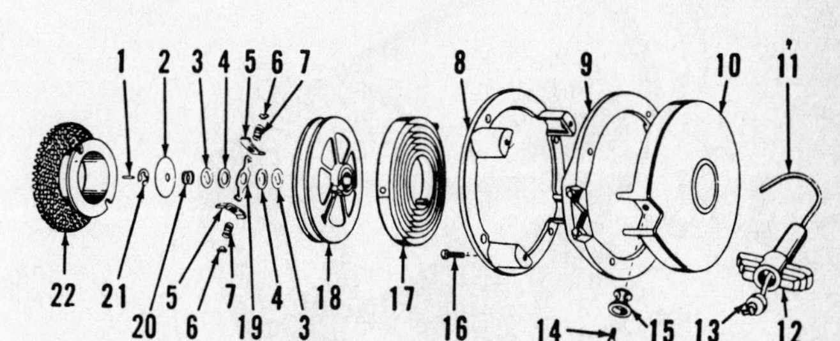

Fig. CL109—Exploded view of a Fairbanks-Morse rewind starter used on some Clinton engines. Refer to Fig. CL108 for another type of Fairbanks-Morse starter used.

1. Centering pin
2. Washer
3. Washer
4. Fiber washer
5. Friction shoe
6. Retainer
7. Spring
8. Mounting flange
9. Middle flange
10. Housing
11. Rope
12. Handle
13. Cup
14. Screw
15. Roller
16. Screws
17. Rewind spring
18. Pulley
19. Brake lever
20. Spring
21. Snap ring
22. Pulley (cup)

ELECTRA-START

Electra-Start models have a built-in battery, starter and flywheel mounted alternator. Early models used a wet cell 12 volt battey mounted on mower deck. Late models use a 12-cell, nickel-cadmium battery which attaches to crankcase, making the unit fully self-contained.

Electra-Start models are equipped with the bulb type fuel primer which pressurizes the carburetor float chamber when bulb is depressed, forcing a small amount of fuel out main nozzle.

Nominal voltage of the nickel-cadmium battery is 15 volts. Charging rate of the flywheel alternator is 0.2-0.25 DC amps. About 40-125 amp-seconds are normally required to start the engine and recovery time to full charge should be 3-10 minutes of operation. About 20 minutes running time should be allowed when battery or engine is first put into service.

Fig. CL113—View of American Bosch 12 volt starter mounted on a late production horizontal crankshaft engine. Square-shaped unit below starter motor is a rectifier to convet AC current from lighting coils on magneto armature core to DC current.

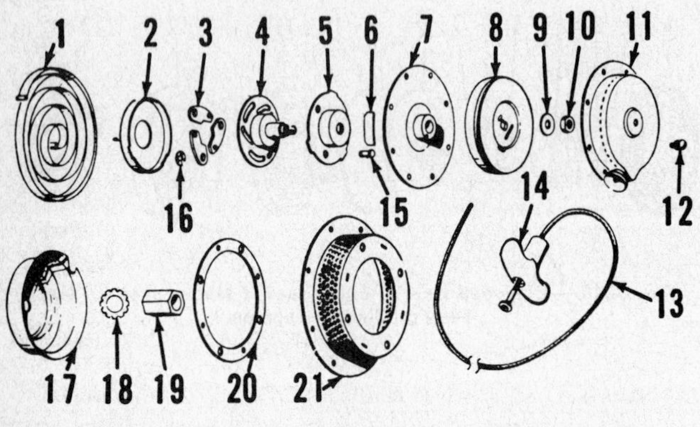

1. Rewind spring
2. Cover
3. Pawls
4. Cam plate & axle
5. Drive plate*
6. Spring*
7. Plate
8. Drum
9. Washer
10. Nut
11. Housing
12. Screws
13. Rope
14. Handle
15. Pin
16. Snap ring
17. Rope pulley
18. Lockwasher
19. Driven nut
20. Ring
21. Housing

Fig. CL110—Exploded view of typical Schnacke starter unit used on early production engines. Asterisk indicates part not used on all starters.

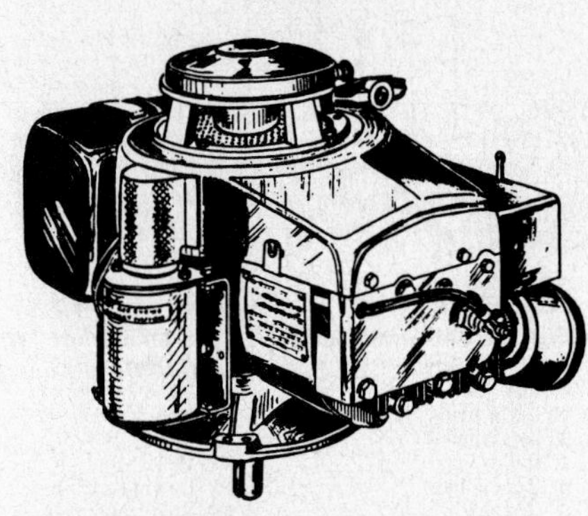

Fig. CL114—View of American Bosch 12 volt starter mounted on a late production vertical crankshaft engine.

Fig. CL111—The 12 volt starter-generator (28) is driven (and drives) by a belt and pulley (25).

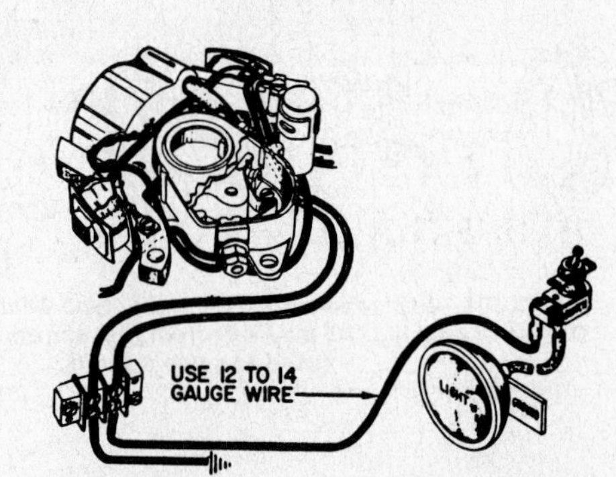

Fig. CL115—View showing wiring circuit from lighting coils when used for lighting purposes only.

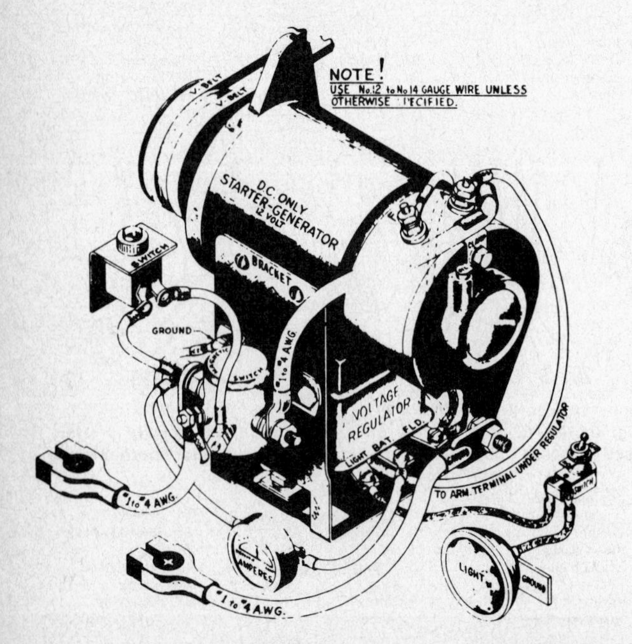

Fig. CL112—Wiring diagram for 12 volt DC starter-generator.

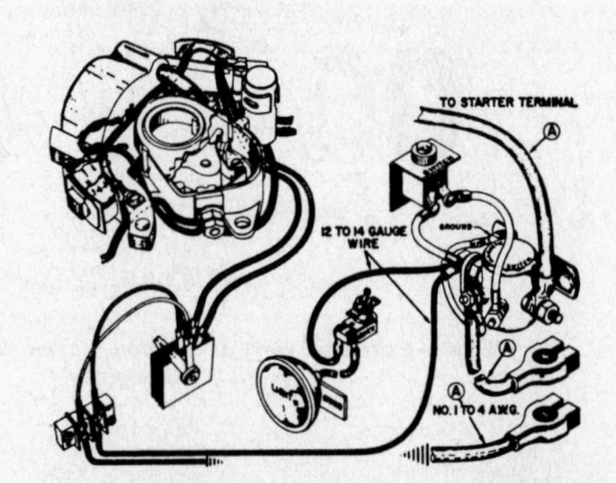

Fig. CL116—View showing wiring circuit from lighting coils. A rectifier is included in circuit to provide DC current for charging a 12 volt battery.

The alternator output (200-250 milliamperes) cannot be measured with regular shop equipment. If trouble is encountered, start the engine and operate at rated speed. Disconnect generator lead from red (positive) battery lead and check generator output using one of the following methods:

1. Connect one lead of a voltmeter or small test light to generator lead and ground the remaining lead. Voltmeter should indicate charging current or test light should glow, indicating charging current.

2. In the absence of test equipment, momentarily touch the disconnected generator lead to a suitable ground and watch for a spark.

If test equipment or spark indicated charging current, the generator should be considered satisfactory.

If a charging current is not indicated, remove the flywheel and renew the generator coil which is mounted on one leg of magneto pole shoe. The halfwave rectifier is built into magneto coil and is not renewable separately.

POLYURETHANE AIR CLEANER

Some Clinton models are equipped with a polyurethane air filter element. Element should be removed and washed in nonflammable solvent (mild solution of detergent and water) after every 10 hours of operation. Reoil element with SAE 10 motor oil and wring out excess oil. Insert the element evenly into air cleaner housing and snap cover in place. Refer to Fig. CL119.

REMOTE CONTROLS

Several different remote control options are available. Refer to Fig. CL122. Views A through G are for engines with following basic numbers: IMB basic numbers 494 and 498; old basic numbers 900, 960 990, B1260 and B1290.

View A shows standard fixed speed for hoizontal shaft engine. B, C and F show available remote control options. Typical cable and handle assembly is shown in view D. Detail of pivot pin to block is shown in E. View G shows method of securing control wire to lever on some applications.

The remote control allows the engine speed to be regulated by movement of the remove control lever some distance away from the engine. The engine speed will be maintained by the engine governor at any setting of the remote control lever within the prescribed speed range of the engine. Maximum engine speed should not exceed 3600 rpm.

Fig. CL117—Exploded view of the 110 volt below deck electric starter, which is available on some vertical shaft engines.

1. Engine base
3. Rotor clamp
6. Crankshaft
7. Cover
9. Insulator gasket
10. Insulator gasket
12. Rotor
17. Connector sleeves (3)
18. Insulating sleeve
19. Stator

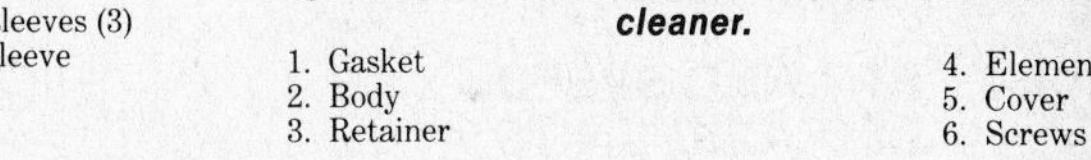

Fig. CL119—Exploded view of polyurethane air cleaner.

1. Gasket
2. Body
3. Retainer
4. Element
5. Cover
6. Screws

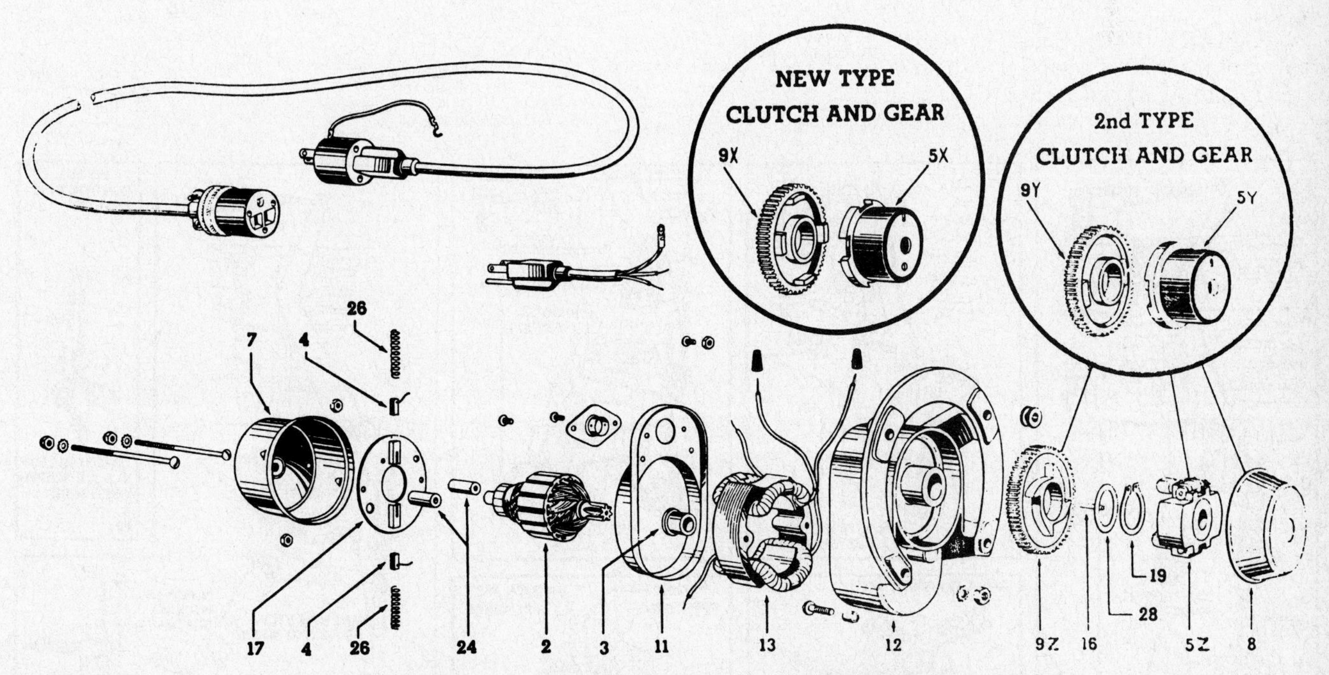

Fig. CL118—Exploded view of the 110 volt electric starter available on some models.

2. Armature
3. Bearing
4. Brushes
5X. Latest clutch assy.
5Y. Second clutch assy.
5Z. Earliest clutch assy.
7. Cover
8. Cup
9X. Latest clutch gear
9Y. Second clutch gear
9Z. Earliest clutch gear
11. Upper housing
12. Lower housing
13. Field winding
16. Centering pin
17. Brush mounting plate
24. Spacers
26. Springs
28. Fiber washer

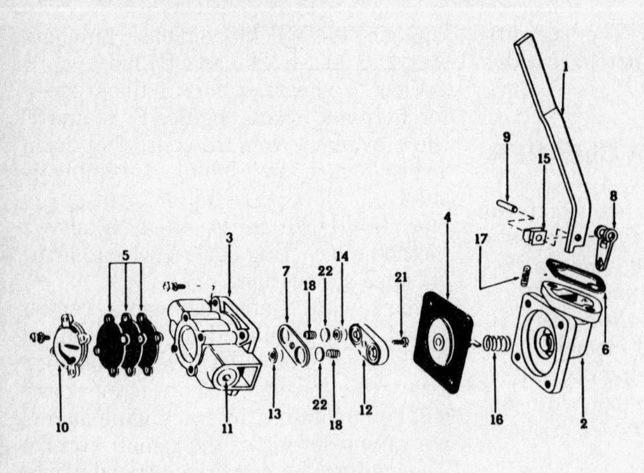

Fig. CL121—Exploded view of mechanical fuel pump used on some engines.

1. Rocker arm
2. Pump body
3. Top cover
4. Diaphragm
5. Pulsator
6. Gasket
7. Valve gasket
8. Lever link
9. Rocker arm pin
10. Cover plate
11. Pipe plug
12. Valve retainer
13. Valve seat
14. Valve seat
15. Rocker spring clip
16. Diaphragm spring
17. Rocker arm spring
18. Valve spring
21. Valve retainer
22. Valve (2)

To adjust remote control with the engine running:

1. Loosen screw in swivel nut on control lever.
2. Move control lever to high speed position (view B – Fig. CL122).
3. Move control wire through control casing and swivel nut until maximum desired engine speed is obtained. Do not exceed maximum engine speed of 3600 rpm. Lever should remain in high speed position.
4. Tighten screw in swivel nut to hold control wire.
5. Move control lever to slow speed position to be sure engine can slow down to idle speed.

PTO AND SPEED REDUCER UNITS

AUXILIARY PTO. Some vertical crankshaft models are equipped with an auxiliary pto as shown in Fig. CL123. Unit is lubricated by oil in engine crankcase. To disassemble unit, drive the pins (6, 7 and 22) partially out of shaft (11), turn shaft half-turn and pull pins. Remove shaft and gears (2 and 4). Remove output shaft (12) and gear (21) in similar manner.

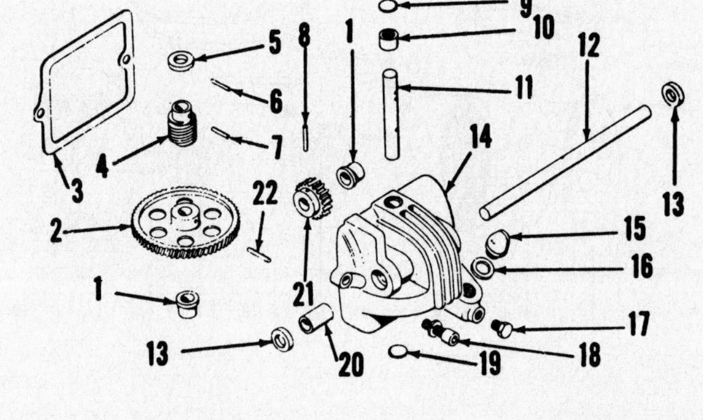

Fig. CL123—Exploded view of auxiliary pto unit used on some vertical crankshaft engines. Gear ratio is either 15:1 or 30:1.

1. Flange bearing
2. Drive gear
3. Gasket
4. Worm gear
5. Thrust washer
6. Pin
7. Pin
8. Pin
9. Expansion plug
10. Needle bearing
11. Worm shaft
12. Pto shaft
13. Oil seal
14. Housing
15. Oil filler cap
16. Gasket
17. Oil drain plug
18. Cap screw
19. Expansion plug
20. Sleeve bearing
21. Pto gear
22. Pin

SPEED REDUCERS. Speed reducers (gear reduction units) with ratios of 2:1, 4:1 and 6:1 are available on a number of horizontal shaft engines.

Refer to Fig. CL124 for exploded view of typical speed reducer for engines of less than 5 horsepower (3.7 kW). Unit should be lubricated by filling with SAE 30 oil to the level plug. Oil level should be checked after every 10 hours of operation. Drain and refill the reduction unit with clean SAE 30 oil after every 100 hours of operation.

Use Fig. CL124 as disassembly and reassembly guide. Unit can be mounted on engine in any of four positions; however, be sure outer housing is installed so filler plug is up.

An exploded view of typical speed reducer used on engines of 5 horsepower (3.7 kW) and larger is shown in Fig. CL125. Units used on early engines support output end of crankshaft in sleeve bearing (1) as shown. Late units support output end of crankshaft in a tapered roller bearing.

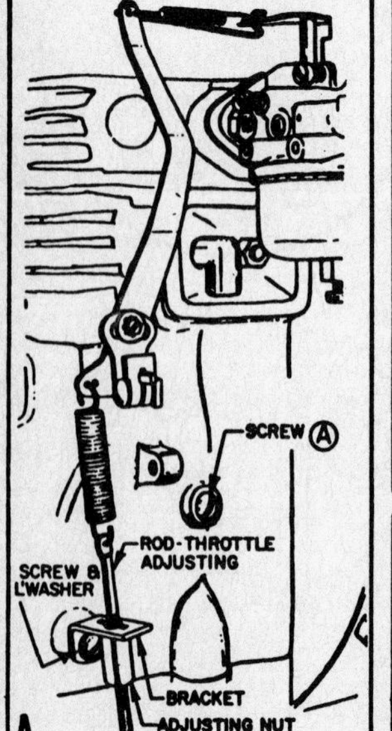

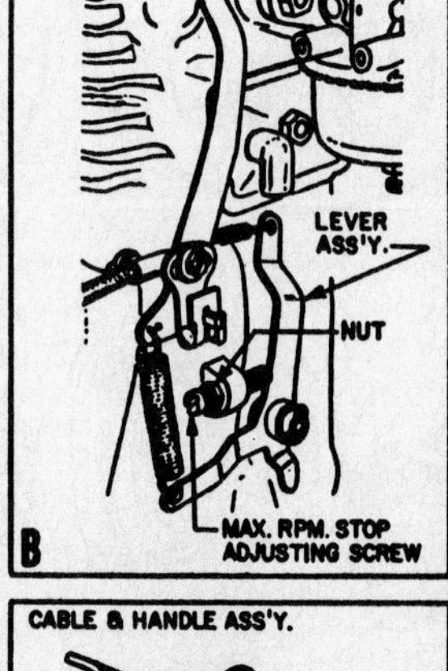

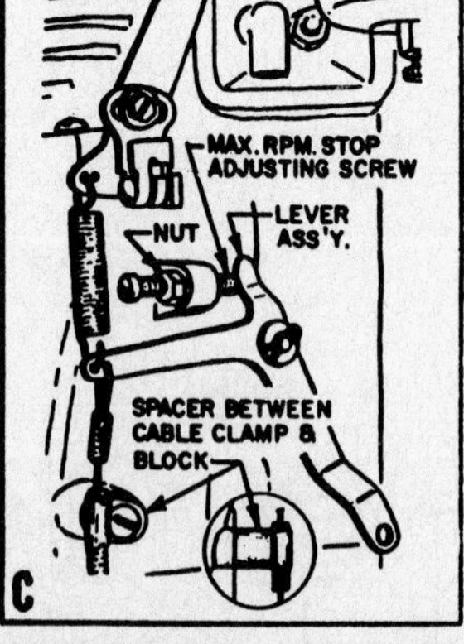

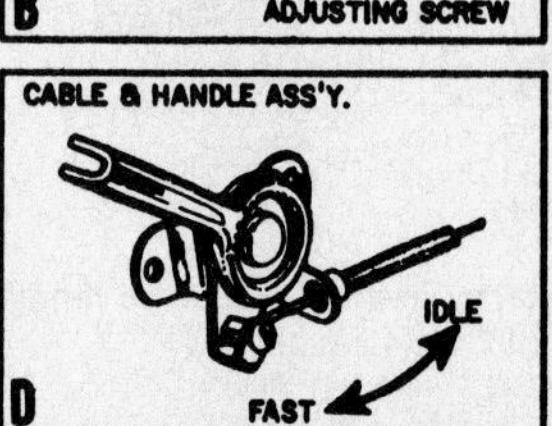

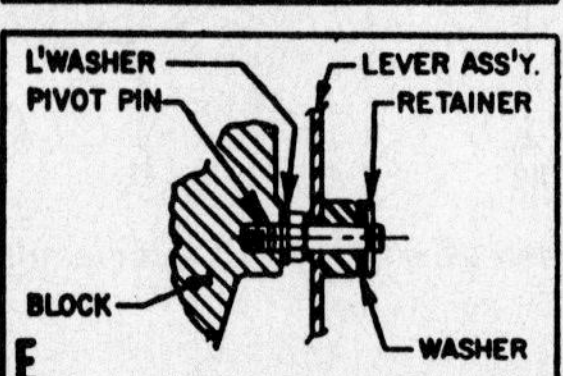

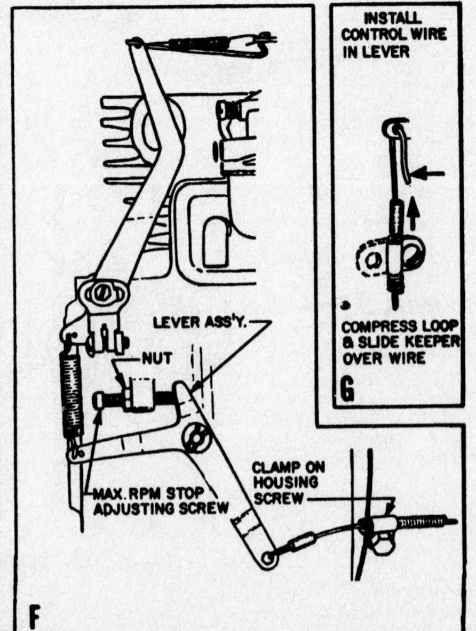

Fig. CL122—Remote control hook-ups for various Clinton engines. Refer to text for engine application and adjustment procedures.

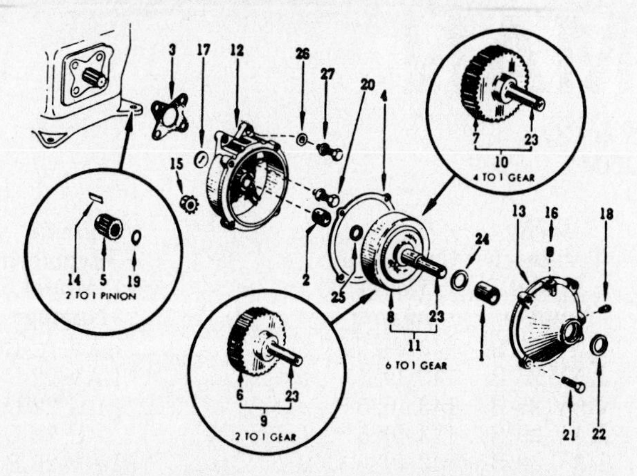

Fig. CL124—Exploded view of gear reduction unit used on engines of less than 5 horsepower (3.7 kW). Gear ratio may be 2:1, 4:1 or 6:1. Sleeve bearings must be reamed after installation.

1. Sleeve bearing
2. Sleeve bearing
3. Gasket
4. Gasket
5. 2:1 pinion
6. 2:1 gear
7. 4:1 gear
8. 6:1 gear
9. Gear & shaft
10. Gear & shaft
11. Gear & shaft
12. Inner housing
13. Outer housing
14. Key
15. Nut & washer assy.
16. Breather plug
17. Expansion plug
18. Oil level plug
19. Snap ring
20. Screw
21. Screw
22. Oil seal
23. Shaft
24. Thrust washer
25. Thrust washer
26. Flat washer
27. Lockwasher

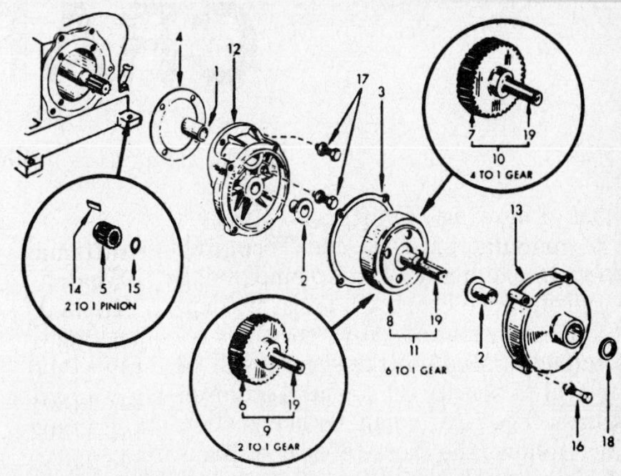

Fig. CL125—Exploded view of gear reduction unit used on engines larger than 5 horsepower (3.7 kW). Gear ratio may be 2:1, 4:1 or 6:1. Sleeve bearing (1) must be finish reamed after installation. Units used on late production engines have tapered roller engine main bearing instead of sleeve bearing (1) shown.

1. Sleeve bearing
2. Flange bearing
3. Gasket
4. Gasket
5. 2:1 pinion
6. 2:1 gear
7. 4:1 gear
8. 6:1 gear
9. Gear & shaft
10. Gear & shaft
11. Gear & shaft
12. Inner housing
13. Outer housing
14. Key
15. Snap ring
16. Screw & washer
17. Screw & washer
18. Oil seal
19. Shaft

After unit has been drained of oil or after reassembly, initially fill unit with same oil as used in engine. Lubricating oil supply is thereafter maintained from engine crankcase through passage drilled in inner housing (12). Gaskets (4) used between inner housing and engine crankcase or bearing plate are available in several thicknesses. Use proper thickness gasket to maintain specified crankshaft end play.

As unit can be mounted on engine in any of four different positions, be sure outer housing (13) is installed with filler plug to top and oil level drain plugs down. Note that after engine has been started, oil level will usually be below level of oil level plug.

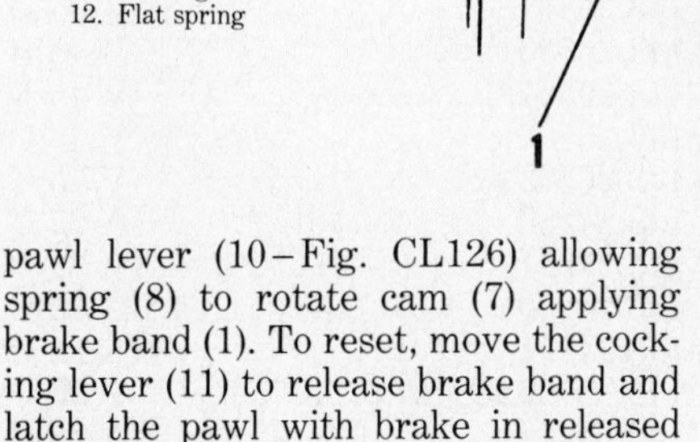

Fig. CL126—Schematic drawing of "Touch 'N' Stop" brake used on some vertical crankshaft rotary lawnmower engines.

1. Brake band
2. Flywheel nut
3. Flywheel
4. Pin
5. Pivot bolt
6. Screw
7. Cam
8. Spring
9. Spring
10. Pawl lever
11. Cocking lever
12. Flat spring

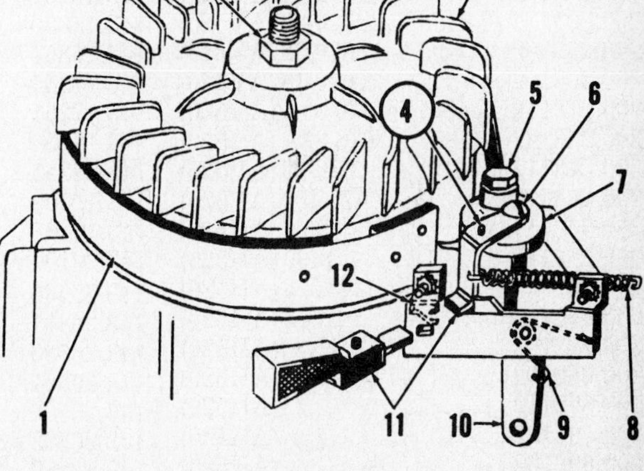

FLYWHEEL BRAKE

Some vertical crankshaft models used on rotary lawnmowers are equipped with a "Touch 'N' Stop" flywheel brake. Actuating the control assembly releases pawl lever (10–Fig. CL126) allowing spring (8) to rotate cam (7) applying brake band (1). To reset, move the cocking lever (11) to release brake band and latch the pawl with brake in released position.

Required service, such as renewal of worn or broken parts or broken springs, should be evident after inspection of unit and reference to Fig. CL126. Tighten flywheel nut (2) to 50-58 ft.-lbs. (68-79 N•m) on models so equipped.

CRAFTSMAN

SEARS ROEBUCK & CO.
Chicago, Ill. 60607

The following Craftsman engines were manufactured by the Tecumseh Products Company. The accompanying cross-reference chart will assist in identifying an engine for service requirements. Much of the service will be identical to service on similar Tecumseh engines. The Craftsman service section which follows the cross reference charts includes service which is different than repairing Tecumseh engines.

TWO-STROKE MODELS

Craftsman Model Number	Tecumseh Engine Type Number
143.17112	638-07 (AV520)
143.171112	638-07 (AV520)
143.171122	638-01A (AV520
143.171272	638-51 (AC520)
200.183112	638.07A (AV520)
200.183122	638-17A (AV520)
200.193132	638-07B (AV520)
200.193142	638-17B (AV520)
200.193152	650-44 (AV520)
200.193162	650-29A (AV520)
200.203172	670-39 (AV520)
200.203182	670-32 (AV520)
200.203192	670-18 (AV520)
200.213112	670-39A (AV520)
200.213122	670-32A (AV520)
200.243112	670-83 (AV520)
200.283012	670-100 (AV520)
200-503111	1525A (AH520)
200.583111	1499 (AH520)
200.593121	1525 (AH520)
200.613111	1553 (AH520)
200.672102	(AH520)
200.682102	(AH520)
200.692112	(AH520)
200.692122	(AH520)
200.692132	(AH520)

FOUR-STROKE ENGINES
Vertical Crankshaft

Craftsman Model Number	Basic Tecumseh Model Number
143.V20A-1-1WA	V 20 D
143.V25A-5-1-1WA	V 25 D
143.V27A-3-1-1WA	V 27 D
143.09300	LAV 25 B
143.09301	LAV 25 B
143.09302	LAV 25 B
143.10250	LAV 22 B
143.10251	LAV 22 B
143.10300	LAV 25 B
143.10301	LAV 25 B
143.11300	LAV 25 B
143.11301	LAV 25 B
143.11302	LAV 25 B
143.11350	LAV 30 B
143.11351	LAV 30 B
143.11352	LAV 30 B
143.12300	LAV 25 B
143.12301	LAV 25 B
143.12302	LAV 25 B
143.12303	LAV 25 B
143.12304	LAV 25 B
143.12350	LAV 30 B
143.12351	LAV 30 B
143.13250	LAV 22 B
143.13251	LAV 22 B
143.13300	LAV 25 B
143.13301	LAV 25 B
143.14350	LAV 30 B
143.14351	LAV 30 B
143.15300	LAV 25 B
143.15301	LAV 25 B
143.16350	LAV 30 B
143.16351	LAV 30 B
143.17350	LAV 30 B
143.17351	LAV 30 B
143.18350	LAV 30 B
143.18351	LAV 30 B
143.19400	LAV 35 B
143.19401	LAV 35 B
143.20100	V 20 D
143.20400	V 35 D
143.20401	V 35 D
143.20500	V 45 B
143:20501	V 45 B
143.20502	V 45 C
143.20503	V 45 C
143.23250	HA 22 C
143.23251	HA 22 C
143.24250	HA 22 C
143.24251	HA 22 C
143.25100	V 25 D
143.25250	HA 22 C
143.25251	HA 22 C
143.27100	V 27 D
143.27200	V 20 D
143.30250	LAV 22 B
143.30251	LAV 22 B
143.30350	LAV 30 B
143.30351	LAV 30 B
143.31600	V 55 B
143.31601	V 55 C
143.36250	HA 22 C
143.36251	HA 22 C
143.36252	H 22 D
143.36253	H 22 D
143.36254	H 22 H
143.36255	H 22 H
143.39250	LAV 22 B
143.39251	LAV 22 B
143.40250	H 22 D
143.40251	H 22 D
143.40300	LAV 25 B
143.40301	LAV 25 B
143.40350	LAV 30 B
143.40351	LAV 30 B
143.40500	V 45 B
143.40501	V 45 B
143.40502	V 45 C
143.40503	V 45 C
143.40600	V 55 B
143.41300	LAV 25 B
143.41301	LAV 25 B
143.41302	LAV 25 B
143.41350	LAV 30 B
143.41351	LAV 30 B
143.41352	LAV 30 B
143.42200	V 22 H-8
143.42201	V 22 H-8P
143.42202	V 22 H-37
143.42203	V 22 H-37P
143.42204	V 22 H-8
143.42205	V 22 H-8B
143.42500	V 22 H-8
143.42501	V 25 H-8P
143.42700	V 25 H-8
143.42701	V 25 H-8P
143.43004	V 30 H-3P
143.43200	V 30 H-8
143.43201	V 30 H-29
143.43202	V 30 H-40
143.43203	V 30 H-8P
143.43204	V 30 H-29P
143.43205	V 30 H-40P
143.43250	H 22 H
143.43251	H 22 H
143.43300	V 30 H-8
143.43301	V 30 H-8P
143.43500	V 45 B
143.43501	V 45 B
143.43700	V 32 A-28
143.43701	V 32 A-28P
143.44000	V 35 B-28
143.44400	H 35 D
143.44401	H 35 D
143.50020	V 20 G-8
143.50021	V 20 G-8P
143.50025	V 25 G-8
143.50026	V 25 G-8P
143.50030	V 30 G-14
143.50031	V 30 G-14P
143.50040	V 40 A-4
143.50045	V 40 B-4
143.50125	V 25 G-3
143.50126	V 25 G-3P
143.50130	V 30 G-8

Craftsman Model Number	Basic Tecumseh Model Number
143.50131	V 30 G-8P
143.50230	V 30 G-3
143.50231	V 30 G-3P
143.50250	HA 22 A
143.50251	HA 22 A
143.50300	H 30 D
143.50301	H 30 D
143.50400	H 35 D
143.50401	H 35 D
143.50402	H 35 H
143.50403	H 35 H
143.50700	V 25 H-8
143.52200	V 22 H-8
143.52201	V 22 H-8P
143.52701	V 25 H-8P
143.52702	V 25 H-8
143.52703	V 25 H-8P
143.53001	V 30 H-4P
143.53200	V 30 H-8
143.53201	V 30 H-8P
143.53300	V 30 H-8
143.53301	V 30 H-8P
143.54500	VX-45 A-24
143.54502	V 45 A-4
143.55250	HA 22 A
143.55251	HA 22 A
143.55300	H 30 D
143.56250	HA 22 A
143.56251	HA 22 A
143.56252	HA 22 A
143.56253	HA 22 A
143.58250	HA 22 C
143.58251	HA 22 C
143.59250	H 22 D
143.59251	H 22 D
143.60020	V 20 G-8
143.60021	V 20 G-8P
143.60025	V 25 G-14
143.60026	V 25 G-14P
143.60030	V 30 G-8
143.60031	V 30 G-8P
143.60040	VX 40 A-19
143.60125	V 25 G-8
143.60126	V 25 G-8P
143.60130	V 30 G-8
143.60131	V 30 G-8P
143.60140	VX 40 A-24
143.60225	V 25 G-8
143.60226	V 25 G-8P
143.60231	V 30 G-4P
143.60240	VX 40 A-19
143.60325	V 25 G-14
143.60326	V 25 G-14F
143.60327	V 27 G-16
143.60328	V 27 G-16P
143.60330	V 27 G-2
143.60331	V 27 G-2P
143.60340	VX 40 A-19
143.60351	LV 30 B
143.60401	LV 35 B
143.62200	V 22 H-8
143.62201	V 22 H-8P
143.62700	V 25 H-8
143.62701	V 25 H-14
143.62702	V 25 H-3
143.62703	V 25 H-8P

Craftsman Model Number	Basic Tecumseh Model Number
143.62704	V 25 H-14P
143.62705	V 25 H-3P
143.63200	V 30 H-8
143.63201	V 30 H-14
143.63202	V 30 H-3
143.63203	V 30 H-8P
143.63204	V 30 H-14P
143.63205	V 30 H-3P
143.64000	V 40 B-4
143.64500	V 45 A-4
143.65250	H 22 H
143.65251	H 22 H
143.66250	H 22 H
143.66251	H 22 H
143.66500	V 45 B
143.66501	V 45 B
143.67250	H 22 H
143.67251	H 22 H
143.71250	LAV 22 B
143.71251	LAV 22 B
143.75250	LAV 22 B
143.75251	LAV 22 B
143.76250	LAV 22 B
143.76251	LAV 22 B
143.76252	LAV 22 B
143.82000	H 20 B-15
143.82001	H 20 B-15P
143.82002	H 20 B-15
143.82003	H 20 B-15P
143.82004	H 20 B-15
143.82005	H 20 B-15P
143.82006	H 20 B-15
143.82007	H 20 G-15P
143.83000	H 30 B-15
143.83001	H 30 B-15P
143.83250	LAV 22 B
143.83251	LAV 22 B
143.83252	LAV 22 B
143.84300	LAV 25 B
143.84301	LAV 25 B
143.84302	LAV 25 B
143.86400	LAV 35 B
143.86401	LAV 36 B
143.86402	LAV 35 B
143.90020	H 20 A-15
143.90021	H 20 A-15P
143.91250	H 22 H
143.91251	H 22 H
143.93000	H 30 B-15P
143.97250	H 22 H
143.97251	H 22 H
143.101010	LAV 30 E
143.101011	LAV 30 E
143.101012	LAV 30 E
143.101020	LAV 30 E
143.101021	LAV 30 E
143.101022	LAV 30 E
143.101030	LAV 30 E
143.101031	LAV 30 E
143.101032	LAV 30 E
143.102010	LAV 30 E
143.102011	LAV 30 E
143.102012	LAV 30 E
143.102020	LAV 30 E
143.102021	LAV 30 E
143.102022	LAV 30 E

Craftsman Model Number	Basic Tecumseh Model Number
143.102030	LAV 30 E
143.102031	LAV 30 E
143.102040	LAV 30 E
143.102041	LAV 30 E
143.102041	LAV 30 E
143.102050	LAV 30 E
143.102051	LAV 30 E
143.102060	LAV 30 E
143.102061	LAV 30 E
143.102062	LAV 30 E
143.102070	LV 30 E
143.102071	LV 30 E
143.102072	LV 30 E
143.102100	LAV 30 E
143.102101	LAV 30 E
143.102102	LAV 30 E
143.102110	LAV 30 E
143.102111	LAV 30 E
143.102112	LAV 30 E
143.102120	LAV 30 E
143.102121	LAV 30 E
143.102130	LAV 30 E
143.102131	LAV 30 E
143.102132	LAV 30 E
143.102140	LAV 30 E
143.102141	LAV 30 E
143.102150	LAV 30 E
143.102151	LAV 30 E
143.102152	LAV 30 E
143.102160	LAV 25 E
143.102161	LAV 25 E
143.102162	LAV 25 E
143.102170	LAV 30 E
143.102171	LAV 30 E
143.102172	LAV 30 E
143.102190	LAV 30 E
143.102191	LAV 30 E
143.102200	LAV 30 E
143.102201	LAV 30 E
143.102202	LAV 30 E
143.102210	LV 30 E
143.102211	LV 30 E
143.102212	LV 30 E
143.102220	LAV 30 E
143.102221	LAV 30 E
143.102222	LAV 30 E
143.103010	LAV 30 E
143.103011	LAV 30 E
143.103012	LAV 30 E
143.103020	LAV 30 E
143.103021	LAV 30 E
143.103050	LAV 30 E
143.103051	LAV 30 E
143.103052	LAV 30 E
143.103060	LAV 30 E
143.103061	LAV 30 E
143.104010	LAV 35 E
143.104011	LAV 35 E
143.104020	LAV 35 E
143.104021	LAV 35 E
143.104030	LV 35 E
143.104031	LV 35 E
143.104040	LV 35 E
143.104041	LV 35 E
143.104050	LV 35 E
143.104051	LV 35 E

Craftsman Model Number	Basic Tecumseh Model Number
143.104080	LAV 35 E
143.104081	LAV 35 E
143.104090	LAV 35 E
143.104091	LAV 35 E
143.104100	LV 35 E
143.104101	LV 35 E
143.104110	LV 35 E
143.104111	LV 35 E
143.104120	LV 35 E
143.104121	LV 35 E
143.105010	V 45 D
143.105011	V 45 D
143.105020	V 45 D
143.105021	V 45 D
143.105030	V 45 D
143.105031	V 45 D
143.105040	V 55 D
143.105041	V 55 D
143.105050	V 45 D
143.105051	V 45 D
143.105060	V 45 D
143.105061	V 45 D
143.105070	V 45 D
143.105071	V 45 D
143.105090	V 45 D
143.105091	V 45 D
143.105100	V 45 D
143.105101	V 45 D
143.105110	V 55 D
143.105120	V 45 D
143.105121	V 45 D
143.106010	LAV 30 E
143.106011	LAV 30 E
143.106012	LAV 30 E
143.106020	LAV 30 E
143.106021	LAV 30 E
143.106030	LAV 30 E
143.106031	LAV 30 E
143.122011	LV 30 L
143.122012	LV 30 L
143.122021	LAV 30 L
143.122022	LAV 30 L
143.122031	LAV 30 L
143.122032	LAV 30 L
143.122041	LAV 30 L
143.122042	LAV 30 L
143.122051	LAV 30 L
143.122052	LAV 30 L
143.122061	LAV 30 L
143.122071	LAV 30 L
143.122081	LAV 30 L
143.122082	LAV 30 L
143.122091	LAV 30 L
143.122092	LAV 30 L
143.122101	LAV 25 L
143.122102	LAV 25 L
143.122201	LAV 30 L
143.122202	LAV 30 L
143.122211	LAV 30 M
143.122212	LAV 30 M
143.122221	LAV 30 M
143.122222	LAV 30 M
143.122231	LAV 30 M
143.122232	LAV 30 M
143.122242	LAV 35 M
143.122251	LV 30 M

Craftsman Model Number	Basic Tecumseh Model Number
143.122252	LV 30 M
143.122261	LAV 30 M
143.122262	LAV 30 M
143.122271	LAV 30 M
143.122272	LAV 30 M
143.122281	LAV 30 M
143.122282	LAV 30 M
143.122291	LAVT 30 C
143.122311	LV 30 N
143.122321	LAV 30 L
143.122322	LAV 30 L
143.123021	LAVR 30 L
143.123022	LAVR 30 L
143.123031	LAVR 30 L
143.123041	LAVR 30 L
143.123051	LAVR 30 L
143.123052	LAVR 30 L
143.123071	LAVR 30 L
143.123091	LAVR 30 L
143.123092	LAVR 30 L
143.124011	LV 35 L
143.124021	LAV 35 L
143.124031	LAV 35 L
143.124041	LV 35 M
143.124051	LV 35 N
143.124061	LAVT 35 C
143.124071	LV 35 N
143.125011	VT 45 A
143.125021	VT 45 C
143.125031	VT 45 A
143.125041	VT 45 C
143.125051	VT 45 C
143.125061	VT 45 C
143.126011	VT 55 A
143.126021	VT 55 B
143.126031	VT 55 B
143.126041	VT 55 C
143.126051	VT 55 C
143.126061	VT 55 C
143.131022	LAV 30
143.131032	LAV 30
143.131042	LAV 30
143.131052	LAV 30
143.131062	LAV 30
143.131072	LAV 30
143.131082	LAV 30
143.131092	LAV 30
143.131102	LAV 30
143.131112	LAV 22
143.131122	LAV 30
143.131132	LAV 30
143.131142	LAV 30
143.131152	LAV 30
143.131162	LAV 30
143.131172	LAV 30
143.131182	LAV 30
143.133012	LAV 30
143.133022	LAV 30
143.133032	LAV 30
143.133042	LV 35
143.133052	LAV 30
143.134012	LV 35
143.134022	LV 35
143.134032	LAV 35
143.134042	LV 35
143.134052	LV 35

Craftsman Model Number	Basic Tecumseh Model Number
143.135012	V 50
143.135022	V 50
143.135042	V 50
143.135052	V 50
143.135062	V 50
143.135072	V 50
143.135082	V 50
143.135092	V 50
143.136012	V 60
143.135112	V 50
143.136012	V 60
143.136052	V 60
143.136042	V 60
143.136052	V 60
143.137012	V 40
143.137032	V 40
143.141012	LAV 30
143.141022	LAV 30
143.141032	LAV 30
143.141042	LAV 30
143.141052	LAV 30
143.141062	LAV 30
143.141072	LAV 30
143.141082	LAV 30
143.141092	LAV 25
143.141102	LAV 30
143.141112	LAV 30
143.141122	LAV 30
143.141132	LAV 25
143.141142	LAV 30
143.141152	LAV 30
143.141162	LAV 30
143.141172	LAV 30
143.141182	LAV 30
143.141192	LAV 30
143.141202	LAV 30
143.141212	LAV 30
143.141222	LAV 30
143.141232	LAV 30
143.141242	LAV 30
143.141252	LAV 30
143.141262	LAV 30
143.141272	LAV 30
143.141282	LAV 30
143.141292	LAV 30
143.141302	LAV 30
143.143012	LAV 30
143.143022	LAV 30
143.143032	LAV 30
143.144012	LV 35
143.144022	LV 35
143.144032	VL 35
143.144042	VL 35
143.144052	VL 35
143.144062	LV 35
143.144072	LV 35
143.144082	LV 35
143.144092	LV 35
143.144102	LV 35
143.144112	LV 35
143.144122	LV 35
143.144132	LV 35
143.145012	V 50
143.145022	V 50
143.145032	V 50
143.145042	V 50

Craftsman Model Number	Basic Tecumseh Model Number
143.145052	V 50
143.145062	V 50
143.145072	V 50
143.146012	V 60
143.146022	V 60
143.147012	V 40
143.147022	V 40
143.147032	V 40
143.151012	LAV 30
143.151022	LAV 30
143.151032	LAV 30
143.151042	LAV 30
143.151052	LAV 30
143.151072	LAV 30
143.151082	LAV 30
143.151092	LAV 30
143.151102	LAV 30
143.151103	LAV 30
143.151104	LAV 30
143.151105	LAV 30
143.151142	LAV 30
143.151152	LAV 30
143.151162	LAV 30
143.153012	LAV 30
143.153022	LAV 30
443.153032	LAV 30
143.154012	LAV 35
143.154022	LAV 35
143.154032	LAV 35
143.154042	LAV 35
143.154052	LAV 35
143.154062	LAV 35
143.154072	LAV 35
143.154082	LAV 35
143.154092	LAV 35
143.154102	LAV 35
143.154112	LAV 35
143.154132	LAV 35
143.154142	LAV 35
143.155012	V 50
143.155022	V 50
143.155032	V 50
143.155042	V 50
143.155052	V 50
143.155062	V 50
143.156012	V 60
143.156022	V 60
143.156032	V 60
143.157012	V 40
143.157022	V 40
143.157032	V 40
143.161012	LAV 30
143.161022	LAV 30
143.161032	LAV 30
143.161042	LAV 30
143.161052	LAV 30
143.161062	LAV 30
143.161072	LAV 30
143.161082	LAV 30
143.161092	LAV 30
143.161102	LAV 30
143.161112	LAV 30
143.161132	LAV 30
143.161142	LAV 30
143.161152	LAV 30
143.161162	LAV 30
143.161172	LAV 30
143.161182	LAV 30
143.161192	LAV 30
143.161202	LAV 30
143.161212	LAV 30
143.161222	LAV 30
143.161232	LAV 30
143.161242	LAV 30
143.161262	LAV 30
143.162022	LAV 35
143.162032	LAV 35
143.162092	LAV 35
143.163012	LAV 30
143.163022	LAV 30
143.163032	LAV 30
143.163042	LAV 30
143.163052	LAV 30
143.163062	LAV 30
143.164012	LAV 35
143.164022	LAV 35
143.164032	LAV 35
143.164042	LAV 35
143.164052	LAV 35
143.164062	LAV 35
143.164072	LAV 35
143.174082	LAV 35
143.164102	LAV 35
143.164112	LAV 35
143.164122	LAV 35
143.164132	LAV 35
143.164142	LAV 35
143.164152	LAV 35
143.164162	LAV 35
143.164172	LAV 35
143.164182	LAV 35
143.164202	LAV 35
143.165012	V 50
143.165022	V 50
143.165032	V 50
143.165042	V 50
143.165052	V 50
143.166012	V 60
143.166022	V 60
143.166032	V 60
143.166042	V 60
143.166052	V 60
143.167012	V 40
143.167022	V 40
143.167032	V 40
143.167042	V 40
143.171012	LAV 30
143.171022	LAV 30
143.171032	LAV 30
143.171042	LAV 30
143.171052	LAV 30
143.171062	LAV 30
143.171072	LAV 30
143.171082	LAV 30
143.171092	LAV 30
143.171102	LAV 30
143.171132	LAV 30
143.171142	LAV 30
143.171152	LAV 30
143.171162	LAV 30
143.171172	LAV 30
143.171202	LAV 35
143.171212	LAV 30
143.171232	LAV 30
143.171242	LAV 30
143.171252	LAV 30
143.171262	LAV 30
143.171302	LAV 30
143.171312	LAV 30
143.171322	LAV 30
143.171332	LAV 30
143.173012	LAV 30
143.173042	LAV 30
143.174012	LAV 35
143.174022	LAV 35
143.174032	LAV 35
143.174042	LAV 35
143.174052	LAV 35
143.174062	LAV 35
143.174072	LAV 35
143.174082	LAV 35
143.174092	LAV 35
143.174102	LAV 35
143.174132	LAV 35
143.174142	LAV 35
143.174152	LAV 35
143.174162	LAV 35
143.174172	LAV 35
143.174182	LAV 35
143.174192	LAV 35
143.174232	LAV 35
143.174242	LAV 35
143.174252	LAV 35
143.174272	LAV 35
143.174292	LAV 35
143.175012	V 50
143.175022	V 50
143.175032	V 50
143.175042	V 50
143.175052	V 50
143.175062	V 50
143.175072	V 50
143.176012	V 60
143.176022	V 60
143.176032	V 60
143.176042	V 60
143.176052	V 60
143.176062	V 60
143.176072	V 60
143.176082	V 60
143.176092	V 60
143.176102	V 60
143.177012	V 40
143.177022	V 40
143.177032	V 40
143.177042	V 40
143.177062	V 40
143.177072	V 40
143.181042	LAV 30
143.181052	LAV 30
143.181062	LAV 30
143.181082	LAV 30
143.181092	LAV 30
143.181102	LAV 30
143.181112	LAV 30
143.181122	LAV 30
143.181132	LAV 30
143.183042	LAV 30

Craftsman Model Number	Basic Tecumseh Model Number
143.184012	LAV 35
143.184052	LAV 35
143.184082	LAV 35
143.184092	LAV 35
143.184102	LAV 35
143.184112	LAV 35
143.184122	LAV 35
143.184132	LAV 35
143.184142	LAV 35
143.184152	LAV 35
143.184162	LAV 35
143.184172	LAV 35
143.184182	LAV 35
143.184192	LAV 35
143.184202	LAV 35
143.184212	LAV 35
143.184232	ECV 100
143.184242	ECV 100
143.184252	ECV 100
143.184262	LAV 35
143.184272	LAV 35
143.184282	LAV 35
143.184292	LAV 35
143.184302	LAV 35
143.184402	LAV 35
143.185012	V 50
143.185022	V 50
143.185032	V 50
143.185042	V 50
143.185052	V 50
143.186012	V 60
143.186052	V 60
143.186062	V 60
143.186082	V 60
143.186092	V 60
143.186102	V 60
143.186112	V 60
143.186122	V 60
143.187022	LAV 40
143.187042	LAV 40
143.187052	LAV 40
143.187062	LAV 40
143.187072	LAV 40
143.187082	LAV 40
143.187094	LAV 40
143.187102	EVC 105
143.191012	LAV 30
143.191022	LAV 30
143.191032	LAV 30
143.191042	LAV 30
143.191052	LAV 30
143.194012	LAV 35
143.194022	LAV 35
143.194032	LAV 35
143.194042	LAV 35
143.194052	LAV 35
143.194062	ECV 100
143.194072	LAV 35
143.194082	LAV 35
143.194092	LAV 35
143.194102	ECV 100
143.194112	LAV 35
143.194122	LAV 35
143.194132	LAV 35
143.194142	LAV 35
143.195012	V 50
143.195022	V 50
143.196012	V 60
143.190622	V 60
143.196032	V 60
143.196082	V 60
143.17012	LAV 40
143.197022	ECV 105
143.197032	ECV 105
143.197042	LAV 40
143.197052	LAV 40
143.197062	LAV 40
143.197072	LAV 40
143.197082	ECV 105
143.201032	LAV 30
143.201042	LAV 30
143.203012	LAV 30
143.204022	ECV 100
143.204032	LAV 35
143.204042	LAV 35
143.204052	LAV 35
143.204062	ECV 100
143.204072	LAV 35
143.204082	LAV 35
143.204092	LAV 35
143.204102	EVC 100
143.204132	ECV 100
143.204142	LAC 35
143.204162	LAV 35
143.204172	LAV 35
143.204182	LAV 35
143.204192	LAV 35
143.204202	ECV 100
143.205022	V 50
143.206012	V 60
143.206032	V 60
143.207012	LAV 40
143.207022	LAV 40
143.207032	LAV 40
143.207052	LAV 40
143.207062	ECV 105
143.207072	LAV 40
143.207082	ECV 105
143.213012	LAV 30
143.213022	LAV 30
143.213042	LAV 30
143.214012	LAV 35
143.214022	LAV 35
143.214032	LAV 35
143.214042	ECV 100
143.214052	ECV 100
143.214062	ECV 100
143.214072	ECV 100
143.214082	LAV 35
143.214092	LAV 35
143.214102	LAV 35
143.214112	LAV 35
143.214122	LAV 35
143.214132	LAV 35
143.214192	LAV 35
143.214202	LAV 35
143.214212	LAV 35
143.214222	LAV 35
143.214232	LAV 35
143.214242	LAV 35
143.214252	LAV 35
143.214262	ECV 100
143.214272	ECV 100
143.214282	ECV 100
143.214292	LAV 35
143.214302	LAV 35
143.214312	ECV 100
143.214322	ECV 100
143.214332	LAV 35
143.214342	LAV 35
143.214352	ECV 100
143.216042	V 60
143.216052	V 60
143.216062	V 60
143.216122	V 60
143.216142	V 60
143.216182	V 60
143.217012	ECV 105
143.217022	ECV 105
143.217032	ECV 105
143.217042	LAV 40
143.217052	LAV 40
143.217062	LAV 40
143.217072	LAV 40
143.217092	ECV 105
143.217102	LAV 40
143.223012	LAV 30
143.223022	LAV 30
143.223032	LAV 30
143.223042	LAV 30
143.223052	LAV 30
143.224012	LAV 35
143.224022	LAV 35
143.224032	ECV 100
143.224062	LAV 35
143.224092	LAV 35
143.224102	LAV 35
143.224112	LAV 35
143.224122	LAV 35
143.224132	LAV 35
143.224142	LAV 30
143.224162	LAV 35
143.224172	LAV 35
143.224182	LAV 35
143.224192	LAV 35
143.224202	LAV 35
143.224212	LAV 35
143.224222	LAV 35
143.224232	ECV 100
143.224242	ECV 100
143.224252	LAV 35
143.224262	LAV 35
143.224272	LAV 35
143.224282	LAV 35
143.224292	ECV 100
143.224302	ECV 100
143.224312	LAV 35
143.224322	LAV 35
143.224332	LAV 35
143.224342	LAV 35
143.224352	ECV 100
143.224362	ECV 100
143.224372	LAV 35
143.224382	LAV 35
143.224392	LAV 35
143.224402	LAV 35
143.224412	LAV 35
143.224422	LAV 35

Craftsman Model Number	Basic Tecumseh Model Number
143.225012	ECV 120
143.225022	ECV 120
143.225032	V 50
143.225042	V 50
143.225052	V 50
143.225062	ECV 120
143.225072	ECV 120
143.225082	V 50
143.225092	V 50
143.225102	V 50
143.226012	V 60
143.226032	V 60
143.266132	V 60
143.226152	V 60
143.226162	V 60
143.226182	V 60
143.226222	V 60
143.226232	V 60
143.226242	V 60
143.226262	V 60
143.262322	V 60
143.226332	V 60
143.227012	ECV 110
143.227022	ECV 110
143.227062	ECV 110
143.227072	ECV 110
143.233012	LAV 30
143.233032	LAV 30
143.233042	LAV 30
143.234022	LAV 35
143.234042	LAV 35
143.234052	LAV 35
143.234062	ECV 100
143.234072	ECV 100
143.234082	ECV 100
143.234092	ECV 100
143.234102	LAV 35
143.234112	LAV 35
143.234122	LAV 35
143.234132	LAV 35
143.234142	LAV 35
143.234162	LAV 35
143.234182	ECV 100
143.234192	LAV 35
143.234202	LAV 35
143.234212	ECV 100
143.234222	ECV 100
143.234232	ECV 100
143.234242	LAV 35
143.234252	LAV 35
143.234262	LAV 35
143.235012	ECV 120
143.235022	ECV 120
143.235032	LAV 50
143.235042	ECV 120
143.235052	ECV 120
143.235062	V 50
143.235072	LAV 50
143.236012	V 60
143.236052	V 60
143.236082	V 60
143.236102	V 60
143.236112	V 60
143.236132	V 60
143.236152	V 60
143.237012	ECV 110
143.237022	ECV 110
143.237032	ECV 105
143.237042	LAV 40
143.243012	LAV 30
143.243022	LAV 30
143.242042	LAV 30
143.243052	LAV 30
143.243062	LAV 30
143.243072	LAV 30
143.243082	LAV 30
143.244012	LAV 35
143.244022	LAV 35
143.244032	LAV 35
143.244042	ECV 100
143.244052	ECV 100
143.244062	ECV 100
143.244072	LAV 35
143.244082	LAV 35
143.244092	LAV 35
143.244102	LAV 35
143.244112	LAV 35
143.244122	ECV 100
143.244132	ECV 100
143.244142	ECV 100
143.244202	LAV 35
143.244212	ECV 100
143.244222	LAV 35
143.244232	LAV 35
143.244242	ECV 100
143.244252	ECV 100
143.244262	LAV 35
143.244272	LAV 35
143.244282	LAV 35
143.244292	ECV 100
143.244302	ECV 100
143.244312	ECV 100
143.244322	ECV 100
143.244332	ECV 100
143.245012	LAV 50
143.245042	V 50
143.245052	ECV 120
143.245062	ECV 120
143.245072	ECV 120
143.245082	ECV 120
143.245092	LAV 50
143.245102	ECV 120
143.245112	ECV 120
143.245122	ECV 120
143.245132	ECV 120
143.245142	LAV 50
143.245152	LAV 50
143.245162	ECV 120
143.245172	LAV 50
143.245182	LAV 50
143.245192	ECV 120
143.246012	V 60
143.246042	V 60
143.246352	V 60
143.246392	V 60
143.254012	LAV 35
143.254022	LAV 35
143.254032	LAV 35
143.254042	LAV 35
143.254052	LAV 35
143.254062	ECV 100
143.254072	LAV 35
143.254082	LAV 35
143.254092	LAV 35
143.254102	LAV 35
143.254112	LAV 35
143.254122	LAV 35
143.254142	ECV 100
143.254152	ECV 100
143.264162	ECV 100
143.254172	ECV 100
143.254182	ECV 100
143.254192	ECV 100
143.254212	LAV 35
143.254222	LAV 35
143.254232	ECV 100
143.254242	ECV 100
143.254252	ECV 100
143.254262	ECV 100
143.254272	ECV 100
143.254282	ECV 100
143.254292	ECV 100
143.254302	LAV 35
143.254312	LAV 35
143.254322	ECV 100
143.254332	LAV 35
143.254342	ECV 100
143.254352	ECV 100
143.254362	LAV 35
143.254372	ECV 100
143.254382	ECV 100
143.254392	LAV 35
143.254402	ECV 100
143.254412	ECV 100
143.254432	LAV 30
143.254442	ECV 100
143.254452	LAV 35
143.254462	ECV 100
143.254472	LAV 35
143.254482	LAV 35
143.254492	ECV 100
143.254502	LAV 35
143.254512	LAV 35
143.254522	LAV 35
143.254532	LAV 35
143.255012	LAV 50
143.255022	LAV 50
143.255042	LAV 50
143.255052	LAV 50
143.255062	LAV 50
143.255072	LAV 50
143.255092	LAV 50
143.255112	LAV 50
143.256022	V 60
143.256052	V 60
143.256082	V 60
143.256092	V 60
143.256122	V 60
143.257012	LAV 40
143.257022	LAV 40
143.257032	LAV 40
143.257042	LAV 40
143.257052	LAV 40
143.257062	LAV 40
143.257072	LAV 40
143.264012	LAV 35
143.264022	LAV 35
143.264032	LAV 35

Craftsman Model Number	Basic Tecumseh Model Number
143.264042	LAV 35
143.264052	ECV 100
143.264062	ECV 100
143.264072	ECV 100
143.264082	ECV 100
143.264092	LAV 35
143.264102	ECV 100
143.264232	LAV 35
143.264242	LAV 35
143.264252	LAV 35
143.264262	LAV 35
143.264272	LAV 35
143.264282	LAV 35
143.264292	LAV 35
143.264302	LAV 35
143.264312	LAV 35
143.264322	LAV 35
143.264332	LAV 35
143.265352	ECV 100
143.264362	ECV 100
143.264372	ECV 100
143.264382	LAV 35
143.264392	ECV 100
143.264402	ECV 100
143.264412	ECV 100
143.264422	LAV 35
143.264432	ECV 100
143.264452	ECV 100
143.264462	ECV 100
143.264482	ECV 100
143.264492	LAV 35
143.264052	LAV 35
143.264512	ECV 100
143.264522	LAV 35
143.264542	LAV 35
143.264562	ECV 100
143.264572	ECV 100
143.264582	ECV 100
143.264592	ECV 100
143.264602	ECV 100
143.264612	ECV 100
143.264622	ECV 100
143.264632	ECV 100
143.264642	ECV 100
143.264652	ECV 100
143.264672	ECV 100
143.264682	LAV 35
143.265012	LAV 50
143.265032	LAV 50
143.265042	LAV 50
143.265052	LAV 50
143.265062	LAV 50
143.265072	LAV 50
143.265082	LAV 50
143.265092	LAV 50
143.265112	LAV 50
143.265122	LAV 50
143.265132	LAV 50
143.265142	LAV 50
143.265152	LAV 50
143.265162	LAV 50
143.265172	LAV 50
143.265192	LAV 50
143.266032	V 60
143.266062	V 60
143.266082	V 60
143.266252	V 60
143.266372	V 60
143.266382	V 60
143.266392	V 60
143.266402	V 60
143.266412	V 60
143.266432	V 60
143.266442	V 60
143.266452	V 60
143.267012	LAV 40
143.267022	LAV 40
143.267042	LAV 40
143.274022	ECV 100
143.274032	ECV 100
143.274042	ECV 100
143.274052	ECV 100
143.274062	ECV 100
143.274072	ECV 100
143.274092	LAV 35
143.274102	LAV 35
143.274112	LAV 35
143.274122	LAV 35
143.274132	LAV 35
143.274142	ECV 100
143.274162	LAV 35
143.274172	LAV 35
143.274182	LAV 35
143.274202	ECV 100
143.274212	ECV 100
143.274222	ECV 100
143.274232	ECV 100
143.274242	ECV 100
143.274252	LAV 35
143.274262	ECV 100
143.274272	LAV 35
143.274282	LAV 35
143.274292	LAV 35
143.274302	LAV 35
143.274312	LAV 35
143.274322	LAV 35
143.274332	LAV 35
143.274342	ECV 100
143.274352	LAV 35
143.274362	ECV 100
143.274372	LAV 35
143.274392	ECV 100
143.274402	ECV 100
143.274412	ECV 100
143.274422	ECV 100
143.274432	ECV 100
143.274442	ECV 100
143.274452	ECV 100
143.274462	ECV 100
143.274472	LAV 35
143.274482	ECV 100
143.274492	LAV 35
143.274502	ECV 100
143.274512	ECV 100
143.274522	ECV 100
143.274542	ECV 100
143.274552	LAV 35
143.274562	ECV 100
143.274582	ECV 100
143.274592	LAV 35
143.274602	ECV 100
143.274612	EVC 100
143.274622	EVC 100
143.274632	EVC 100
143.274642	LAV 35
143.274652	ECV 100
143.274662	LAV 35
143.274672	ECV 100
143.274682	LAV 35
143.274692	ECV 100
143.274702	LAV 35
143.274712	ECV 100
143.274722	ECV 100
143.274732	ECV 100
143.274742	ECV 100
143.274752	ECV 100
143.274762	ECV 100
143.274772	LAV 35
143.274782	ECV 100
143.274792	LAV 35
143.275012	LAV 50
143.275022	LAV 50
143.275042	LAV 50
143.275052	LAV 50
143.275062	LAV 50
143.275072	LAV 50
143.275082	LAV 50
143.276182	V 60
143.276202	V 60
143.276252	V 60
143.276412	V 60
143.277012	LAV 40
143.277022	LAV 40
143.284012	LAV 35
143.284022	LAV 35
143.284032	LAV 35
143.284042	ECV 100
143.284052	LAV 35
143.284062	LAV 35
143.284072	ECV 100
143.284082	LAV 35
143.284092	LAV 35
143.284102	ECV 100
143.284112	LAV 35
143.284142	LAV 35
143.284152	LAV 35
143.284162	LAV 35
143.284182	LAV 35
143.284212	ECV 100
143.284222	LAV 30
143.284232	TVS 90
143.284242	LAV 35
143.284252	LAV 35
143.284262	ECV 100
143.284272	TVS 90
143.284282	LAV 35
143.284292	LAV 35
143.284302	LAV 35
143.284312	LAV 35
143.284322	LAV 35
143.284332	ECV 100
143.284342	ECV 100
143.284352	ECV 100
143.284362	ECV 100
143.284372	ECV 100
143.284382	ECV 100
143.284392	LAV 35
143.284402	LAV 35

Craftsman Model Number	Basic Tecumseh Model Number
143.284412	LAV 35
143.284422	LAV 35
143.284432	ECV 100
143.284442	LAV 30
143.284452	ECV 100
143.284462	ECV 100
143.284472	ECV 100
143.284482	LAV 35
143.284492	ECV 100
143.284502	ECV 100
143.284512	LAV 35
143.284522	LAV 35
143.284532	ECV 100
143.284542	LAV 35
143.284552	LAV 35
143.284562	LAV 35
143.284572	LAV 35
143.284582	ECV 100
143.284592	LAV 35
143.284602	ECV 100
143.284612	ECV 100
143.284622	ECV 100
143.284632	LAV 35
143.284642	ECV 100
143.284652	LAV 35
143.284672	ECV 100
143.284682	ECV 100
143.284692	ECV 100
143.284702	ECV 100
143.284712	LAV 35
143.284722	LAV 35
143.284732	LAV 35
143.284742	ECV 100
143.284752	ECV 100
143.284762	LAV 35
143.284772	ECV 100
143.284782	ECV 100
143.285012	LAV 50
143.285022	LAV 50
143.285032	LAV 50
143.285042	LAV 50
143.285052	LAV 50
143.285062	LAV 50
143.285072	LAV 50
143.285082	LAV 50
143.285092	LAV 50
143.285102	LAV 50
143.286012	V 50
143.286022	V 60
143.286342	V 60
143.287012	LAV 40
143.293012	TVS 75
143.294012	TVS 90
143.294022	TVS 90
143.294032	TVS 90
143.294042	TVS 90
143.294052	TVS 90
143.294062	TVS 90
143.294072	TVS 90
143.294092	TVS 90
143.294102	TVS 90
143.294112	TVS 90
143.294122	TVS 90
143.294132	TVS 90
143.294142	ECV 100
143.294152	ECV 100
143.294162	ECV 100
143.294172	ECV 100
143.294182	TVS 90
143.294192	TVS 90
143.294202	TVS 90
143.294212	TVS 90
143.294222	ECV 100
143.294232	ECV 100
143.294242	TVS 90
143.294252	TVS 90
143.294262	TVS 90
143.294272	TVS 90
143.294282	TVS 90
143.294292	TVS 90
143.294302	TVS 90
143.294312	TVS 90
143.294322	TVS 90
143.294332	ECV 100
143.294342	TVS 90
143.294352	ECV 100
143.294362	ECV 100
143.294372	ECV 100
143.294382	ECV 100
143.294392	ECV 100
143.294402	ECV 100
143.294412	ECV 100
143.294422	ECV 100
143.294432	TVS 90
143.294442	TVS 90
143.294452	TVS 90
143.294462	TVS 90
143.294472	ECV 100
143.294482	ECV 100
143.294492	TVS 90
143.294502	TVS 90
143.294512	TVS 90
143.294522	TVS 90
143.294532	TVS 90
143.294542	ECV 100
143.294552	TVS 105
143.294562	TVS 105
143.294572	ECV 100
143.294582	ECV 100
143.294592	ECV 100
143.294602	TVS 90
143.294612	TVS 90
143.294632	TVS 105
143.294642	TVS 105
143.294652	TVS 90
143.294662	ECV 100
143.294672	ECV 100
143.294682	ECV 100
143.294692	TVS 90
143.294702	TVS 105
143.294712	TVS 90
143.294722	ECV 100
143.294732	ECV 100
143.294742	ECV 100
143.295012	LAV 50
143.295022	LAV 50
143.295032	LAV 50
143.295042	ECV 120
143.297012	TVS 105
143.304012	TVS 90
143.304032	ECV 100
143.304042	ECV 100
143.304052	TVS 90
143.304072	ECV 100
143.304092	TVS 90
143.304102	ECV 100
143.304112	ECV 100
143.304122	ECV 100
143.305012	ECV 120
143.305022	ECV 120
143.305032	ECV 120
143.305042	LAV 50
143.305052	ECV 120
143.305062	LAV 50
143.313012	TVS 75
143.314012	ECV 100
143.314022	ECV 100
143.314032	TVS 90
143.314042	TVS 90
143.314052	TVS 90
143.314062	TVS 90
143.314072	TVS 90
143.314082	TVS 90
143.314092	TVS 90
143.314102	TVS 90
143.314112	TVS 90
143.314122	ECV 100
143.314132	ECV 100
143.314142	ECV 100
143.314152	ECV 100
143.314162	ECV 100
143.314172	ECV 100
143.314182	TVS 90
143.314192	ECV 100
143.314202	ECV 100
143.314212	ECV 100
143.314222	ECV 100
143.314232	ECV 100
143.314242	ECV 100
143.314252	ECV 100
143.314262	TVS 90
143.314272	TVS 90
143.314282	TVS 90
143.314292	TVS 90
143.314302	TVS 90
143.314312	ECV 100
143.314322	TVS 90
143.314332	TVS 90
143.314342	TVS 90
143.314372	ECV 100
143.314382	TVS 90
143.314392	ECV 100
143.314402	TVS 90
143.314412	TVS 90
143.314422	ECV 100
143.314432	LAV 35
143.314442	ECV 100
143.314452	ECV 100
143.314462	ECV 100
143.314472	ECV 100
143.314152	ECV 100
143.314522	ECV 100
143.314532	LAV 35
143.314542	TVS 90
143.314552	TVS 90
143.314562	TVS 90
143.314572	TVS 90
143.314582	ECV 100

Craftsman Model Number	Basic Tecumseh Model Number
143.314592	ECV 100
143.314612	ECV 100
143.314622	ECV 100
143.314632	ECV 100
143.314642	ECV 100
143.314652	ECV 100
143.314662	ECV 100
143.314672	ECV 100
143.314682	ECV 100
143.314692	ECV 100
143.314702	LAV 35
143.315012	ECV 120
143.315022	LAV 50
143.315032	TVS 105
143.315042	TVS 105
143.315062	LAV 50
143.315072	TVS 105
143.315082	ECV 120
143.315092	LAV 50
143.315102	LAV 50
143.315112	LAV 50
143.315122	LAV 50
143.321012	TVS 75
143.321022	TVS 75
143.324012	ECV 100
143.324022	ECV 100
143.324042	ECV 100
143.324052	TVS 90
143.324062	ECV 100
143.324072	ECV 100
143.324082	ECV 100
143.324102	ECV 100
143.324112	TVS 90
143.324132	ECV 100
143.324142	TVS 90
143.324152	TVS 90
143.324162	TVS 90
143.324172	TVS 90
143.324182	TVXL 105
143.324192	TVS 90
143.324202	ECV 100
143.324212	ECV 100
143.324222	ECV 100
143.324232	ECV 100
143.326012	TVM 195
143.331012	TVS 75
143.331022	TVS 75
143.334032	TVS 90
143.334042	ECV 100
143.334052	TVXL 105
143.334062	VS 90
143.334072	TVS 90
143.334082	ECV 100
143.334102	ECV 100
143.334112	TVS 90
143.334122	TVS 90
143.334132	ECV 100
143.334142	TVS 90
143.334152	TVS 90
143.334162	TVS 90
143.334172	ECV 100
143.334182	ECV 100
143.334192	LAV 35
143.334202	TVS 90
143.334212	ECV 100
143.334222	ECV 100
143.334232	ECV 100
143.334242	ECV 100
143.334252	ECV 100
143.334262	TVS 90
143.334272	TVS 90
143.334282	TVS 90
143.334292	TVS 90
143.334302	TVS 90
143.334312	TVS 90
143.334322	ECV 100
143.334332	TVS 90
143.334342	ECV 100
143.334352	TVS 90
143.334362	TVS 90
143.334372	TVS 90
143.334382	TVS 90
143.335012	ECV 120
143.335022	ECV 120
143.335032	LAV 50
143.335042	LAV 50
143.335052	TVS 120
143.341012	TVS 75
143.344022	TVS 90
143.344032	TVS 90
143.344042	TVS 90
143.344052	ECV 100
143.344062	ECV 100
143.344072	TVS 90
143.344082	ECV 100
143.344092	ECV 100
143.344102	TVS 90
143.344112	TVXL 105
143.344122	ECV 100
143.344132	ECV 100
143.344142	TVS 90
143.344152	ECV 100
143.344162	TVS 90
143.344172	ECV 100
143.344182	TVS 90
143.344192	TVS 90
143.344202	TVS 90
143.344212	TVS 90
143.344222	TVS 90
143.344232	ECV 100
143.344242	ECV 100
143.344262	EVC 100
143.344272	EVC 100
143.344282	ECV 100
143.344292	ECV 100
143.344302	ECV 100
143.344312	ECV 100
143.344322	ECV 100
143.344332	ECV 100
143.344342	ECV 100
143.344352	ECV 100
143.344362	ECV 100
143.344372	ECV 100
143.344382	ECV 100
143.344392	ECV 100
143.344402	TVXL 105
143.344412	TVXL 105
143.344422	TVS 90
143.344432	TVS 90
143.344442	TVS 105
143.344452	ECV 100
143.344462	TVS 105
143.344472	ECV 100
143.345012	ECV 120
143.345022	ECV 120
143.345032	TVS 120
143.345042	LAV 50
143.345052	ECV 120
143.345062	ECV 120
143.346202	TVM 125
143.351012	TVS 75
143.354012	TVS 90
143.354022	ECV 100
143.354032	ECV 100
143.354042	ECV 100
143.354052	ECV 100
143.354062	TVS 90
143.354072	ECV 100
143.354082	ECV 100
143.354092	TVS 90
143.354102	TVS 90
143.354112	TVS 100
143.354122	TVS 90
143.354132	TVXL 105
143.354142	TVS 90
143.354152	EVC 100
143.354162	TVS 90
143.354172	TVS 90
143.354182	TVS 90
143.354192	TVS 90
143.354202	TVS 90
143.354212	TVS 90
143.354222	ECV 100
143.354232	TVS 90
143.354242	ECV 100
143.354252	ECV 100
143.354262	ECV 100
143.354272	EVC 100
143.354282	LAV 35
143.354292	TVS 90
143.354302	ECV 100
143.355012	ECV 120
143.355022	ECV 120

Horizontal Crankshaft

Craftsman Model Number	Basic Tecumseh Model Number
143.501011	H22K
143.501020	H22K
143.501041	H22K
143.501051	H22K
143.501061	H22K
143.501071	H22K
143.501081	H22K
143.501090	H22K
143.501091	H22K
143.501111	H22K
143.501121	H22K
143.501131	H22K
143.501141	H22K
143.501151	H22K
143.501161	H22K
143.501170	H22K
143.501171	H22K
143.501181	H22K
143.501190	H22K
143.501191	H22K
143.501201	H22K
143.501211	H22K

Craftsman Model Number	Basic Tecumseh Model Number
143.501221	H22K
143.501231	H22K
143.501241	H22K
143.501251	H22K
143.501261	H22K
143.501270	H22K
143.501271	H22K
143.502011	H35M
143.502021	HB35K
143.502031	HB35K
143.502041	H35P
143.504011	H35K
143.505010	H55D
143.505011	H55D
143.506011	H30P
143.521011	H22R
143.521021	H22R
143.521031	H22R
143.521051	H22R
143.521061	H22R
143.521071	H22R
143.521081	HT30A
143.521091	H22R
143.521101	H22R
143.521111	H22R
143.521121	H22R
143.521131	HT30B
143.524021	HT35A
143.524031	HT35A
143.524041	H35P
143.524051	HT35A
143.524061	HT35B
143.524071	HT35B
143.524081	HT35B
143.525021	HT45C
143.526011	HT45A
143.526021	HT55B
143.526031	HT55C
143.531011	HT30B
143.531022	H25
143.531032	H22
143.531042	H22
143.531052	H30
143.531062	H22
143.531072	H30
143.531082	H30
143.531092	H22
143.531112	H22
143.531122	H30
143.531132	H30
143.531142	H22
143.531152	H30
143.531162	H22
143.531172	H30
143.531182	H30
143.534012	H35
143.534022	H35
143.534032	H35
143.534042	H35
143.534052	H35
143.534062	H35
143.534072	H35
143.535012	H50
143.535022	H50
143.535062	H50
143.536012	H60
143.536022	H60
143.536032	H60
143.536042	H60
143.536052	H60
143.536062	H60
143.537012	H40
143.541012	H30
143.541022	H22
143.541032	H22
143.541042	H30
143.541052	H30
143.541062	H30
143.541072	H22
143.541082	H22
143.541102	H22
143.541112	H30
143.541122	H30
143.541132	H30
143.541142	H25
143.541152	H30
143.541162	H22
143.541172	H25
143.541182	H25
143.541192	H30
143.541202	H30
143.541212	H22
143.541222	H30
143.541232	H22
143.541252	H22
143.541262	H22
143.541282	H30
143.541292	H25
143.541302	H30
143.544012	H35
143.544022	H35
143.544032	H35
143.544042	H35
143.544052	H35
143.545012	H50
143.545022	H50
143.545032	H50
143.545042	H50
143.546012	H60
143.546022	H60
143.546032	H60
143.546042	H60
143.546052	H60
143.546062	H60
143.546072	H60
143.546082	H60
143.546092	H60
143.546102	H60
143.546112	H60
143.546142	H60
143.546142	H60
143.547012	H40
143.547022	H40
143.547032	H40
143.551012	H30
143.551032	H30
143.551042	H22
143.551052	H30
143.551062	H30
143.551072	H30
143.551082	H25
143.551092	H25
143.551102	H30
143.551152	H30
143.551162	H30
143.551172	H22
143.551192	H30
143.554012	H35
143.554022	H35
143.554032	H35
143.554042	H35
143.554052	H35
143.554072	H35
143.554082	H35
143.555012	H50
143.555022	H50
143.555032	H50
143.555042	H50
143.555052	H50
143.556012	H60
143.556022	H60
143.556032	H60
143.556042	H60
143.556052	H60
143.556062	H60
143.556072	H60
143.556082	H60
143.556092	H60
143.556102	H60
143.556112	H60
143.556122	H60
143.556132	H60
143.556142	H60
143.556152	H60
143.556162	H60
143.556172	H60
143.556182	H60
143.556192	H60
143.556202	H60
143.556212	H60
143.556222	H60
143.556232	H60
143.556242	H60
143.556252	H60
143.556262	H60
143.556272	H60
143.556282	H60
143.557012	H40
143.557022	H40
143.557032	H40
143.557042	H40
143.557052	H40
143.557062	H40
143.557072	H40
143.557082	H40
143.561012	H30
143.561022	H25
143.561032	H30
143.561042	H30
143.561052	H30
143.561062	H30
143.561072	H25
143.561082	H25
143.561092	H30
143.561102	H30
143.561112	H30
143.561122	H22
143.561132	H30

Craftsman Model Number	Basic Tecumseh Model Number	Craftsman Model Number	Basic Tecumseh Model Number	Craftsman Model Number	Basic Tecumseh Model Number
143.561142	H22	143.571182	H30	143.584112	H35
143.561152	H22	143.571202	H25	143.584122	H35
143.561162	H25	143.574022	H35	143.584132	H35
143.561172	H30	143.574032	H35	143.584142	H35
143.561182	H30	143.574042	H35	143.585012	H50
143.561202	H30	143.574052	H35	143.585032	H50
143.561212	H30	143.574062	H35	143.585042	H50
143.561222	H30	143.574072	H35	143.586012	H60
143.564012	H35	143.574082	H35	143.586022	H60
143.564022	H35	143.574092	H35	143.586032	H60
143.564032	H35	143.574102	H35	143.586042	H60
143.564042	H35	143.574112	H35	143.586072	H60
143.564052	H35	143.575012	H50	143.586082	H60
143.564072	H35	143.575022	H50	143.586152	H60
143.564082	H35	143.575032	H50	143.586172	H60
143.564092	H35	143.575042	H50	143.586182	HH60
143.564102	H35	143.576002	H60	143.586192	H60
143.564112	H35	143.576012	HH60	143.586202	HH60
143.565012	H50	143.576022	H60	143.586212	HH60
143.565022	H50	143.576032	H60	143.586222	H60
143.566002	H60	143.576042	H60	143.586232	H60
143.556012	H60	143.576052	H60	143.586242	H60
143.566022	H60	143.576062	H60	143.586262	H60
143.566032	H60	143.576072	H60	143.586272	H60
143.566042	H60	143.576082	H60	143.586282	H60
143.566052	H60	143.576102	H60	143.587012	HS40
143.566062	H60	143.576112	H60	143.587022	HS40
143.566072	H60	143.576122	H60	143.587032	HS40
143.566082	H60	143.576132	H60	143.587042	HS40
143.566092	H60	143.576142	H60	143.591012	H25
143.566102	H60	143.576172	H60	143.591022	H30
143.566112	H60	143.576182	H60	143.591032	H30
143.566122	H60	143.576192	H60	143.591042	H30
143.566132	H60	143.576202	H60	143.591052	H30
143.566142	H60	143.576212	H60	143.591062	H30
143.566152	HH60	143.576222	H60	143.591072	H30
143.556162	H60	143.576232	H60	143.591082	H25
143.566172	H60	143.576282	H60	143.591092	H30
143.566182	H60	143.576292	HH60	143.591102	H30
143.566192	H60	143.576302	HH60	143.591112	H25
143.566202	H60	143.576312	H60	143.591122	H30
143.566212	H50	143.576322	H60	143.591132	H25
143.566222	H60	143.576332	H60	143.591142	H30
143.566232	H60	143.577012	H40	143.594012	H35
143.566242	H60	143.577022	H40	143.594022	H35
143.566252	H60	143.577032	H40	143.594032	H35
143.567012	H40	143.577042	H40	143.594042	H35
143.567022	H40	143.577072	H40	143.594052	H35
143.567032	H40	143.577082	H40	143.594060	H35
143.567042	H40	143.581002	H30	143.594072	H35
143.571002	H30	143.581022	H25	143.594082	H35
143.571012	H30	143.581032	H30	143.594092	ECH90
143.571022	H30	143.581042	H25	143.594102	H35
143.571032	H30	143.581052	H30	143.595012	H50
143.571042	H30	143.581062	H30	143.595042	H50
143.571052	H30	143.581072	H30	143.596042	H60
143.571062	H25	143.581082	H30	143.596072	H60
143.571072	H30	143.581092	H30	143.596082	H60
143.571082	H25	143.581102	H25	143.596092	H60
143.571092	H30	143.584012	H35	143.596102	HH60
143.571102	H25	143.584032	H35	143.596112	HH60
143.571112	H25	143.584052	H35	143.596122	H60
143.571122	H30	143.584062	H35	143.597012	HS40
143.571152	H22	143.584072	H35	143.597022	HS40
143.571162	H30	143.584082	H35	143.597032	HS40
143.571172	H25	143.584102	H35	143.601022	H30

Craftsman Model Number	Basic Tecumseh Model Number
143.601032	H30
143.601062	H30
143.604012	H35
143.604022	ECH90
143.604032	H35
143.604042	H35
143.604052	ECH90
143.604062	H35
143.604072	H35
143.605012	H50
143.605022	H50
143.605052	H50
143.607012	HS40
143.607022	HS40
143.607032	HS40
143.607042	HS40
143.607052	HS40
143.607062	HS40
143.611012	H30
143.611022	H30
143.611032	H30
143.611042	H30
143.611052	H30
143.611062	H25
143.611072	H30
143.611082	H30
143.611092	H30
143.611102	H25
143.611112	H30
143.614012	ECH90
143.614022	ECH90
143.614032	ECH90
143.614042	H35
143.614052	ECH90
143.614062	H35
143.614072	H35
143.614082	H35
143.614092	H35
143.614102	H35
143.614112	H35
143.614122	H35
143.614132	H35
143.614142	H35
143.614152	H35
143.614162	H35
143.615012	H50
143.615022	H50
143.615052	H50
143.615062	H50
143.615072	H50
143.615082	H50
143.615092	H50
143.616022	H60
143.616042	H60
143.616052	HH60
143.616062	H60
143.616072	H60
143.616082	H60
143.616112	H60
143.616132	HH60
143.616142	HH60
143.617012	HS40
143.617022	HS40
143.617032	HS40
143.617042	HS40
143.617062	HS40
143.617082	HS40
143.617092	HS40
143.617112	HS40
143.617132	HS40
143.617152	HS40
143.617162	HS40
143.617182	HS40
143.621012	H25
143.621022	H25
143.621032	H30
143.621042	H25
143.621052	H30
143.621062	H30
143.621082	H25
143.621092	H25
143.624012	H35
143.624022	H35
143.624032	H35
143.624042	H35
143.624092	H35
143.624102	H35
143.624112	H35
143.625012	H50
143.625022	H50
143.625032	H50
143.625042	H50
143.625052	H50
143.625072	H50
143.625082	H50
143.625092	H50
143.625102	H50
143.625112	H50
143.625122	H50
143.625132	H50
143.626022	H60
143.626042	H60
143.626132	H60
143.626162	H60
143.626182	H60
143.626202	H60
143.626222	H60
143.626232	H60
143.626242	HH60
143.626252	H60
143.626262	H60
143.626272	HH60
143.626302	H60
143.627012	HS40
143.627032	HS40
143.627042	HS40
143.631012	H30
143.631022	H25
143.631032	H25
143.631042	H25
143.631052	H25
143.631062	H25
143.631072	H25
143.631082	H25
143.631092	H30
143.634012	H35
143.634022	H35
143.634032	H35
143.635012	H50
143.635022	H50
143.635032	HS50
143.635042	H50
143.635052	H50
143.636052	H60
143.637012	HS40
143.641012	H25
143.641022	H25
143.641032	H30
143,641042	H25
143.641052	H30
143.641062	H30
143.641072	H35
143.644012	H35
143.644022	H35
143.644032	H35
143.644052	H35
143.644072	H35
143.644082	H35
143.645012	HS50
143.645022	HS50
143.645032	HS50
143.646112	HH60
143.646192	H60
143.647012	HS40
143.647022	HS40
143.647032	HS40
143.647042	HS40
143.647052	HS40
143.647062	HS40
143.651012	H30
143.651022	H30
143.651032	H30
143.651042	H30
143.651052	H30
143.651062	H30
143.651072	H30
143.654022	H35
143.654032	H35
143.654042	H35
143.654052	H35
143.654062	H35
143.654072	H35
143.654082	H35
143.654092	H35
143.654102	H35
143.654112	H35
143.654122	H35
143.654132	H35
143.654142	H35
143.654152	H35
143.654162	H35
143.654172	H35
143.654182	H35
143.654192	H35
143.654202	H35
143.654212	H35
143.654222	H35
143.654232	H35
143.654242	H35
143.654252	H35
143.654262	H35
143.654272	H35
143.654282	H35
143.654292	H35
143.564302	H35
143.654312	H35
143.654322	H35
143.655012	HS50

Craftsman Model Number	Basic Tecumseh Model Number
143.655032	HS50
143.656012	H60
143.656022	H60
143.656032	H60
143.656042	H60
143.656052	H60
143.656092	H60
143.656112	HH60
143.656182	H60
143.656202	H60
143.656252	160
143.657012	HS40
143.657022	HS40
143.657032	HS40
143.657042	HS40
143.657052	HS40
143.661012	H30
143.661022	H30
143.661032	H30
143.661042	H30
143.661062	H30
143.664032	H35
143.664042	H35
143.664052	H35
143.664062	H35
143.664072	H35
143.664082	H35
143.664092	H35
143.664102	H35
143.664112	H35
143.664122	H35
143.664132	H35
143.664142	H35
143.664152	H35
143.664162	H35
143.664172	H35
143.664182	H35
143.664192	H35
143.664202	H35
143.664212	H35
143.664222	H35
143.664232	H35
143.664242	H35
143.664252	H35
143.664262	H35
143.664272	H35
143.664282	H35
143.664292	H35
143.664302	H35
143.664312	H35
143.664332	H35
143.665012	HS50
143.665022	HS50
143.665032	HS50
143.665042	HS50
143.665052	HS50
143.665062	HS50
143.665072	HS50
143.665082	HS50
143.666102	H60
143.666112	H60
143.666122	H60
143.666132	H60
143.666142	H60
143.666172	H60
143.666182	H60
143.666192	H60
143.666202	H60
143.666242	HH60
143.666272	H60
143.666292	H60
143.666372	H60
143.667012	HS40
143.667022	HS40
143.667032	HS50
143.667042	HS40
143.667052	HS40
143.667062	HS40
143.667072	HS40
143.667082	HS40
143.675012	HS50
143.675022	HS50
143.675032	H50
143.675042	HS50
143.675052	HS50
143.675062	H50
143.676112	H60
143.676132	H60
143.676232	H60
143.676242	H60
143.677012	HS40
143.677022	HS40
143.684012	H35
143.685012	HS50
143.685022	HS50
143.685032	HS50
143.686072	H50
143.686112	H50
143,686122	H60
143.686152	H60
143.686182	H60
143.687012	HS40
143.687022	HS50
143.687032	HS50
143.687042	HS40
143.694012	H35
143.694022	H35
143.694032	H35
143.696012	H60
143.696042	H60
143.696082	HS50
143.696122	H60
143.696142	H50
143.696152	H60
143.697012	HS50
143.697022	HS50
143.697032	HS40
143.697042	HS50
143.697052	HS50
143.701012	H30
143.706012	H60
143.707012	HS50
143.707042	HS40
143.707052	HS50
143.707062	HS40
143.707072	HS50
143.707082	HS50
143.707092	HS50
143.707102	HS40
143.707112	HS50
143.707122	HS40
143.707132	HS50
143.711012	H30
143.711022	H30
143.711032	H30
143.711042	H30
143.714012	H35
143.714022	H35
143.714032	H35
143.714042	H35
143.714052	H35
143.714062	H35
143.714072	H35
143.714082	H35
143.714092	H35
143.714102	H35
143.714112	H35
143.714122	H35
143.716032	H60
143.716042	H60
143.716162	H50
143.716232	H60
143.716242	H50
143.716272	H60
143.716282	H60
143.716292	H60
143.716352	H60
143.717012	HS40
143.717022	HS40
143.717032	HS40
143.717042	HS40
143.717052	HS50
143.717062	HS50
143.717072	HS50
143.717082	HS50
143.717092	HS50
143.717102	HS40
143.717112	HS50
143.721012	H30
143.721022	H30
143.721032	H30
143.724012	H35
143.724022	H35
143.724032	H35
143.724042	H35
143.724052	HS40
143.725012	HS50
143.726092	H50
143.726302	H50
143.731012	H30
143.734012	H35
143.734022	H35
143.734032	H35
143.734042	HS40
143.735022	HS50
143.741012	H30
143.741032	H30
143.741042	H30
143.741052	H30
143.741062	H30
143.741072	H30
143.742042	H50
143.744012	H35
143.744022	H35
143.744032	H35
143.744052	H35
143.744062	H35
143.744072	H35

Craftsman Model Number	Basic Tecumseh Model Number	Craftsman Model Number	Basic Tecumseh Model Number	Craftsman Model Number	Basic Tecumseh Model Number
143.744102	HS40	143.751022	H30	143.754012	H35
143.744112	HS50	143.751032	H30	143.754022	H35
143.751012	H30	143.751042	H30		

CRAFTSMAN

ENGINE IDENTIFICATION

Engines must be identified by the complete model number, including the serial number and type number in order to obtain correct repair parts. These numbers are located on the name plate as shown in Fig. C1A. If short block renewal is necessary, original identification plate must be transferred to replacement short block assemblies so unit can be identified when servicing at a later date.

PARTS PROCUREMENT

Parts for Craftsman engines are available from all Sears Retail Stores and Catalog Sales Offices. Be sure to give complete Craftsman model, serial and type number of engine when ordering parts or service material.

SERVICE NOTES (FOUR-STROKE)

CARBURETOR. The carburetors used on many Craftsman engines are the same as used on similar Tecumseh engines. Special Craftsman fuel systems are used on some models. The tank mounted suction carburetor is shown in view (A – Fig. C1).

The Craftsman float carburetor without speed control shown in view (B) is used without a mchanical or air vane governor.

Some engines use a mechanical governor (15, 16 and 17) which controls engine speed using a throttle valve (7 and 8) located in the intake manifold (6).

Refer to the appropriate following paragraphs for servicing Craftsman fuel systems.

Fuel Tank Mounted Carburetor. Early carburetors are equipped with a Bowden wire control as shown in Fig. C2. Later fuel tank mounted carburetors have a manual control knob and a positioning spring as shown in Fig. C3. Other differences are also noted and it is important to identify the type used before servicing.

On all tank mounted carburetors, turn the control valve clockwise to the position shown in Fig. C4, then withdraw the valve. There are no check valves or check balls in the carburetor pickup tube, but the tube can be removed for more thorough cleaning. To reinstall

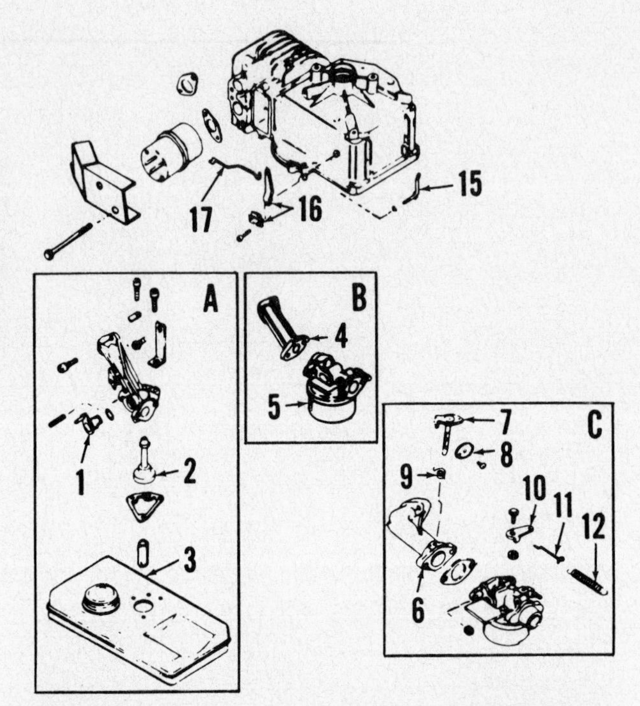

Fig. C1—Craftsman fuel systems are (A) tank mounted suction carburetor, (B) special float carburetor without governor or (C) the special float carburetor with mechanical governor linkage (15, 16 and 17) hooked up to a throttle (7 and 8) located in the intake manifold (6).

1. Control valve
2. Suction tube
3. Fuel tank
4. Intake manifold
5. Float carburetor
6. Intake manifold
7. Governor throttle shaft
8. Governor throttle plate
9. Return spring
10. Bellcrank
11. Linkage (10 to 12)
12. Governor spring
15. Internal lever
16. External lever
17. Linkage (16 to 7)

Fig. C1A—View of typical identification plate on Craftsman four-cycle engine.

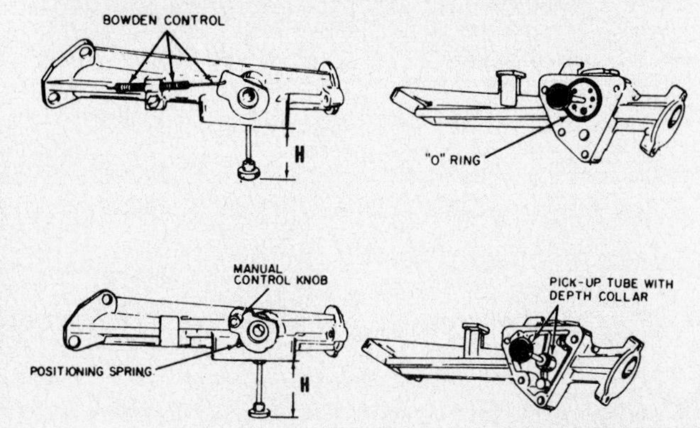

Fig. C2—View of early tank mounted suction carburetor. Notice the Bowden control cable and groove for "O" ring on lower surface. Refer to Fig. C3 for later type.

Fig. C3—View of later tank mounted suction carburetor. The unit is controlled by moving the manual control on carburetor. The lower surface is sealed with a gasket and the "O" ring used on early models is not used. The pickup tube has a collar to make sure it is installed at the correct depth.

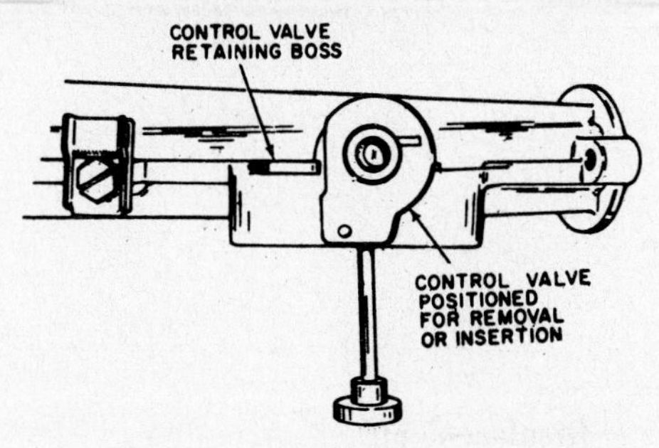

Fig. C4—The control valve must be turned clockwise to the position shown to clear the retaining rod.

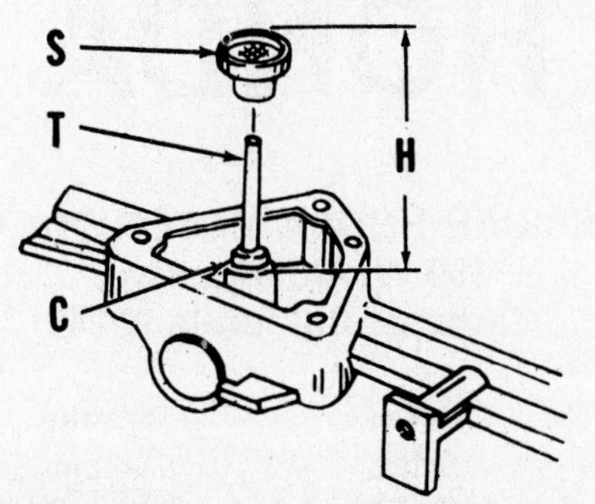

Fig. C6—Tube (T) should be pressed into bore until collar C) contacts body of late carburetor. Press strainer (S) onto tube until height (H) is correct.

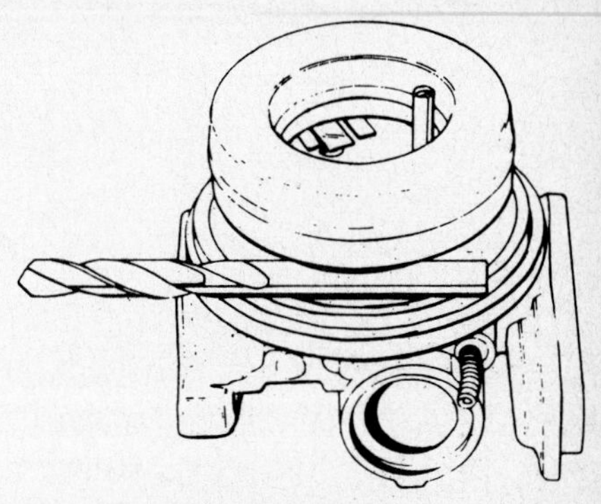

Fig. C11—Float height should be set using a number 4 drill bit positioned as shown. Drill size is 0.2090 inch (5.31 mm) diameter.

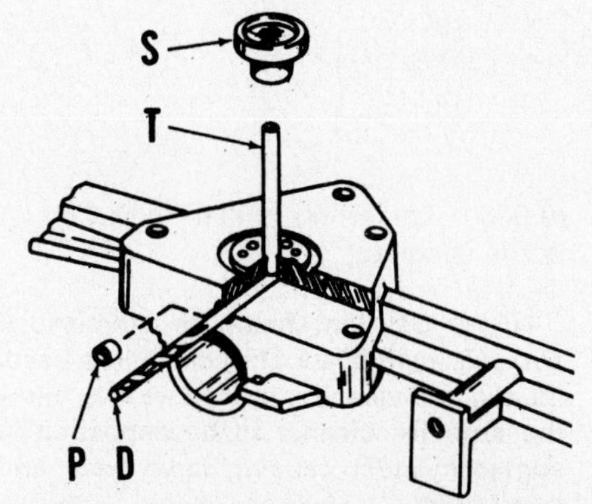

Fig. C5—A ¼-inch drill bit (D) should be used to gage the correct installed depth of tube (T). New plug (P) should be used when assembling.

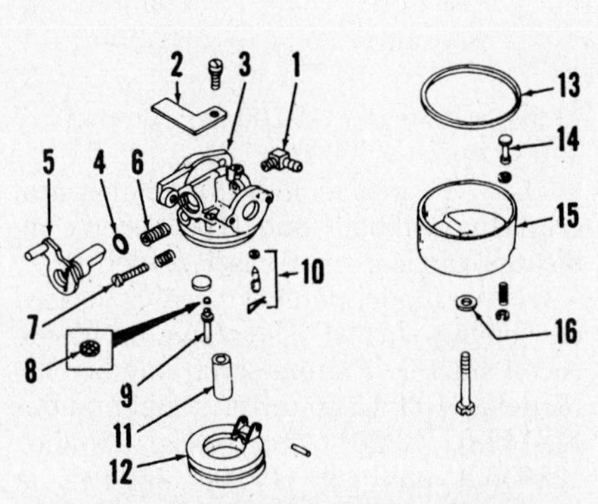

Fig. C10—Exploded view of Craftsman float type carburetor. Refer to text for service procedure.

1. Fuel inlet fitting
2. Retaining plate
3. Carburetor body
4. "O" ring
5. Control valve
6. Positioning spring
7. High speed stop
8. Screen
9. Fuel pickup tube
10. Fuel inlmt valve
11. Bowl spacer
12. Float
13. Gasket
14. Bowl drain
15. Float bowl
16. Gasket

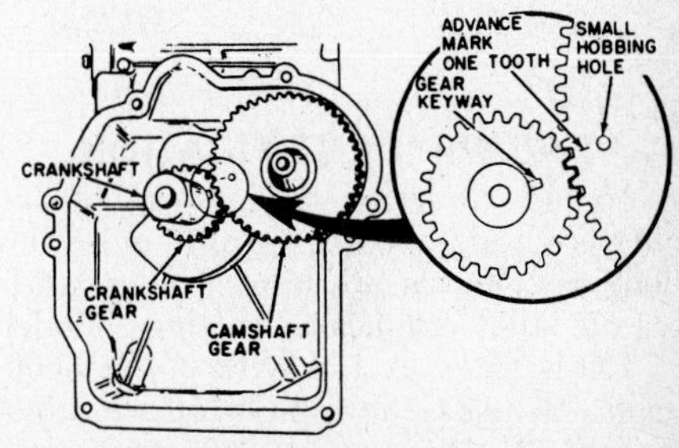

Fig. C12—Camshaft timing marks should be advanced one tooth when engine is not equipped with mechanical governor.

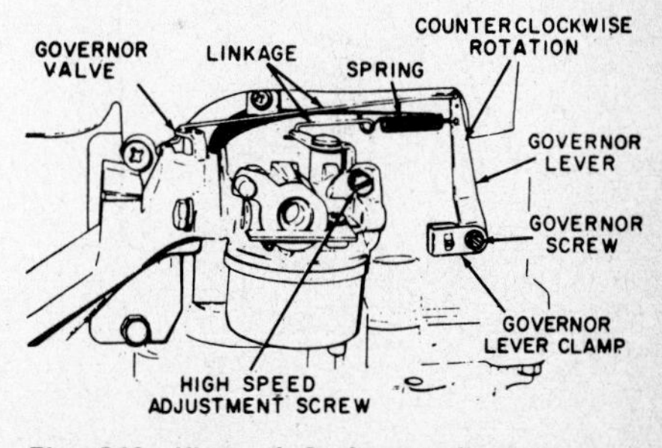

Fig. C13—View of Craftsman float type carburetor installed. The model shown uses mechanical governor controlling the governor throttle valve located in the intake manifold. This system is shown at (C—Fig. C1).

pickup tube in early models, it is necessary to remove plug (P–Fig. C5). Insert ⅛ inch drill into passage as shown and press tube into bore until seated against drill bit. Leave drill in place and press strainer (S) onto pickup tube until height (H–Fig. C2) is 1 27/64 to 1 7/16 inches (36.1-36.5 mm). Remove drill bit and install new cap plug.

On late carburetors, the tube (T–Fig. C6) has a collar near the top end. Press the tube into bore until the collar is against carburetor body casting; then press the strainer (S) onto tube until distance (H) is 1 15/32 inches (27.3 mm).

The reservoir tube in the fuel tank should have slotted end toward bottom of tank on all models.

Fuel mixture is nonadjustable, but the mixture will be changed by dirty or missing air filter. Be sure air filter is in good condition and clean.

NOTE: The camshaft timing should be advanced one tooth for engines with this carburetor. Refer to Fig. C12.

Craftsman Float Carburetor. The carburetor is serviced in manner similar to other float type carburetors. Refer to Fig. C10.

The carburetor is equipped with mixture adjustment. It is important that float height is correct. The air filter must be in good condition and clean. Refer to Fig. C11 for measuring float height with drill bit.

The fuel pickup tube (9) should be pressed into bore until collar is against carburetor body.

NOTE: Some engines equipped with this carburetor are not equipped with a variable speed governor. These models are equipped with a plain intake manifold as shown (B—Fig. C1). The camshaft timing should be advanced one tooth for engines without governor as shown in Fig. C12.

Camshaft timing marks should be aligned for models with variable speed governor. The governed high speed is adjusted by turning the adjustment screw shown in Fig. C13.

HONDA

AMERICAN HONDA MOTOR CO., INC.
100 W. Alondra Blvd.
Gardena, California 90247

Model	Bore	Stroke	Displacement
G100	46 mm (1.84 in.)	46 mm (1.84 in.)	76 cc (4.6 cu. in.)
G150	64 mm (2.5 in.)	45 mm (1.8 in.)	144 cc (8.8 cu. in.)
GV150	64 mm (2.5 in.)	45 mm (1.8 in.)	144 cc (8.8 cu. in.)
G200	67 mm (2.6 in.)	56 mm (2.2 in.)	197 cc (12.0 cu. in.)
GV200	67 mm (2.6 in.)	56 mm (2.2 in.)	197 cc (12.0 cu. in.)

ENGINE IDENTIFICATION

Honda G series engines are four-stroke, air-cooled, single-cylinder engines. Valves are located in cylinder block and crankcase casting. Model G100 is rated at 1.5kW (2 hp) at 3600 rpm, Models G150 and GV150 are rated at 2.6kW (3.5 hp) at 3600 rpm and Models G200 and GV200 are rated at 3.7 kW (5 hp) at 3600 rpm.

The "G" prefix indicates horizontal crankshaft model and "GV" prefix indicates vertical crankshaft model.

Engine model number decal is located on cooling shroud just above or beside recoil starter. Engine serial number for Models G100, G150 (after serial number 1181478), G200 (after serial number 1286400) and all GV200 engines is located on crankcase or crankcase cover near oil filler and dipstick opening. Engine serial number for all Model GV150 engines is located on oil pan just below cylinder head. Engine serial number for all other models is located on lower left edge (facing pto side) of crankcase. See Fig. HN1.

Always furnish engine model and serial number when ordering parts or service information.

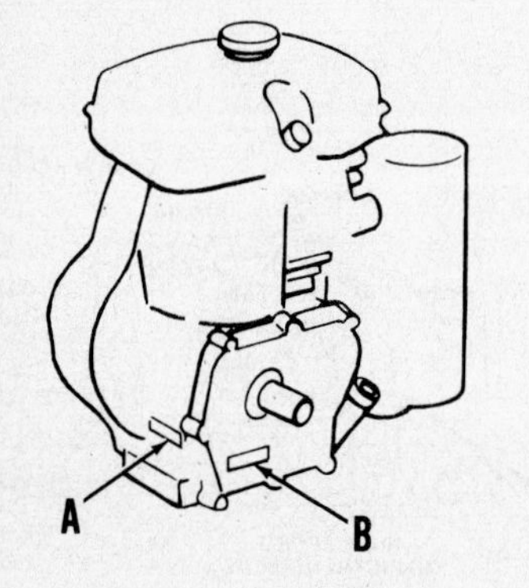

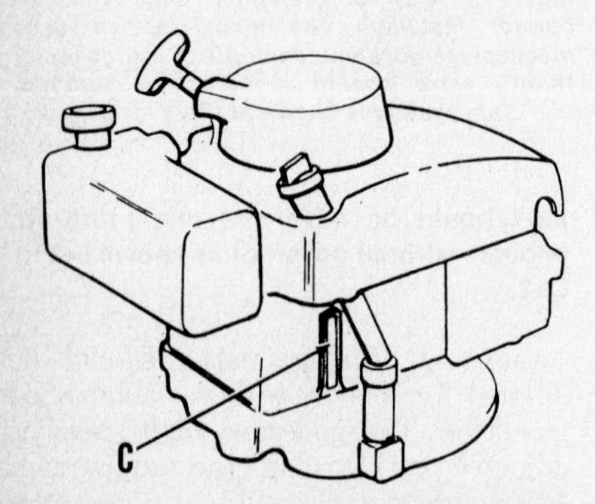

Fig. HN1—View showing serial number locations for all models. (A)—All Model G100, Models G150 (after serial number 1181478) and G200 (after serial number 1286400). (B)—Models G150 (prior to serial number 1181479) and G200 (prior to serial number 1286401). (C)—All model GV200.

MAINTENANCE

SPARK PLUG. Recommended spark plug is as follows:

Model	Standard	Resistor
G100	NGK BM4A	NGK BMR4A
G150*	NGK B4HS	NGK BR4HS
GV150*	NGK BM6A	NGK BMR6A
G200*	NGK B4HS	NGK BR4HS
G150**	NGK BP4HS	NGK BPR4HS
GV150**	NGK BPM6A	NGK BPMR6A
G200*	NGK BP4HS	NGK BPR4HS
GV200*	NGK BM6A	NGK BMR6A
GV200**	NGK BPM6A	NGK BPMR6A

*Breaker point ignition system
**CDI ignition system

Spark plug should be removed and cleaned after every 100 hours of operation. Set electrode gap at 0.6-0.7 mm (0.024-0.028 in.) for models with breaker point ignition system or 0.9-1.0 mm (0.035-0.039 in.) for models with CDI ignition system.

NOTE: Caution should be exercised if abrasive type spark plug cleaner is used. Inadequate cleaning procedure may allow the abrasive cleaner to be deposited in engine cylinder causing rapid wear and part failure.

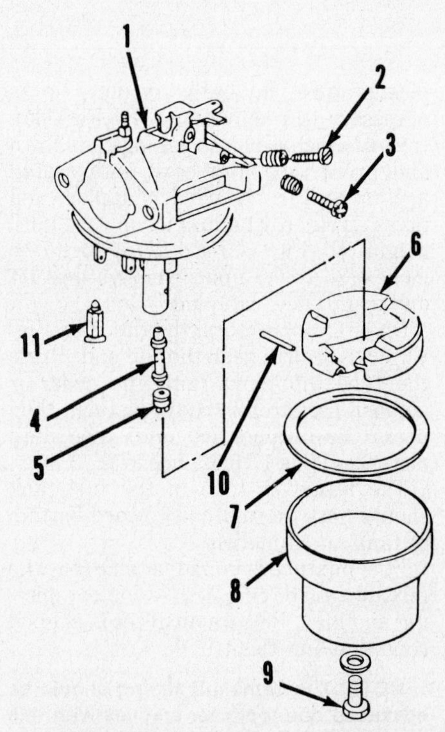

Fig. HN2—Exploded view of carburetor used on Models G150, G200, GV150 and GV200. Carburetor used on Model G100 is similar.

1. Carburetor throttle body
2. Idle mixture screw
3. Throttle stop screw
4. Nozzle
5. Main jet
6. Float
7. Gasket
8. Float bowl
9. Bolt
10. Float pin
11. Fuel inlet needle

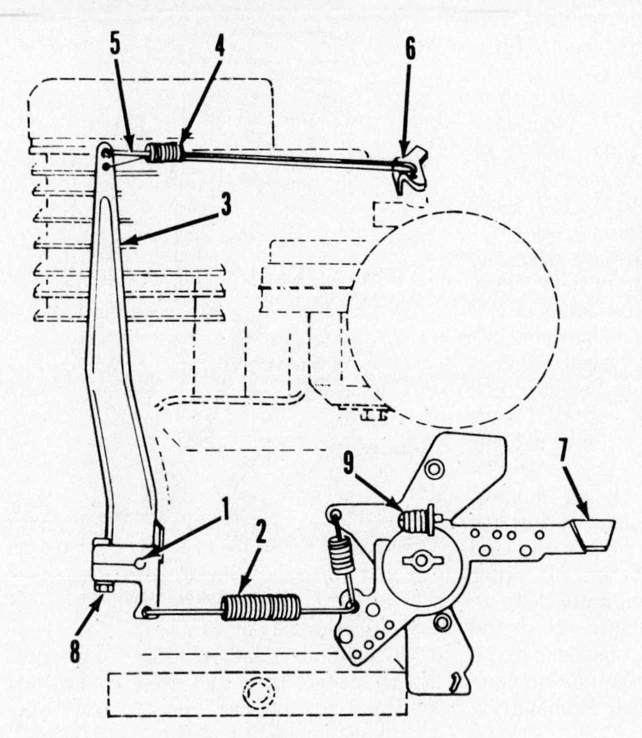

Fig. HN4—View of external governor linkage on Models G100, G150 and G200. Models GV150 and GV200 are similar.

1. Governor shaft
2. Tension spring
3. Governor lever
4. Spring
5. Carburetor-to-governor lever rod
6. Throttle pivot
7. Throttle lever
8. Clamp bolt
9. Maximum speed screw

CARBURETOR. All models are equipped with a float type side draft carburetor. Carburetor is equipped with idle fuel mixture screw. High speed fuel mixture is controlled by a fixed jet.

Engine idle speed is 1400 rpm for Models G100, G150 and G200 and 1700 rpm for Models GV150 and GV200. Idle speed is adjusted by turning throttle stop screw (3 – Fig. HN2). Initial adjustment of idle fuel mixture screw (2) from a lightly seated position is 1³/₈ turns open on Models G100 and G150, 3¼ turns open on Model G200 and 1¾ turns open on Models GV150 and GV200. On all models, final adjustment is made with engine at operating temperature and running. Adjust idle mixture screw to attain smoothest engine operation. Recheck engine idle speed and adjust if necessary.

Main jet (5) controls fuel mixture for high speed operation. Standard main jet size is #55 for Model G100, #65 for Models G150 and GV150, #72 for Model G200 and #75 for Model GV200.

Float level should be 6.7-9.7 mm (0.26-0.38 in.) on Models G100 and GV200 and 8.2 mm (0.32 in.) on Models G100, G150 and G200. To measure float level, invert carburetor throttle body and float assembly. Measure distance from top of float to float bowl mating surface. If dimension is not as specified, renew float.

FUEL FILTER. A fuel filter screen is located in sediment bowl below fuel shut-off valve. To remove sediment bowl, shut off fuel, unscrew threaded ring and remove ring, sediment bowl and gasket. Make certain gasket is in place before reassembly. To clean screen, fuel shut-off valve must be disconnected from fuel line and unscrewed from fuel tank.

AIR FILTER. Engines may be equipped with one of four different types of air cleaner (filter); single element type, dual element type, foam (semidry) type or oil bath type. On all models, air filter should be removed and serviced after every 20 hours of operation. Refer to appropriate paragraph for model being serviced.

Dry Type Air Filter. To remove element, loosen the two wing nuts and remove air cleaner cover. Remove element and separate foam element from paper element. Direct low pressure air from inside filter elements toward the outside to remove all loose dirt and foreign material. Reinstall elements.

Dual Element Type Air Filter. Remove wing nut, cover and elements. Separate foam outer element from paper element. Wash foam element in warm soapy water and thoroughly rinse. Allow element to air dry. Dip dry foam element in clean engine oil and gently squeeze our excess oil.

Direct low pressure air from inside paper element toward the outside to remove all loose dirt and foreign material. Reassemble elements and reinstall.

Foam (Semidry) Type Air Filter. Remove air cleaner cover and element. Clean element in nonflammable solvent and squeeze dry. Soak element in new engine oil and gently squeeze our excess oil. Reinstall element and cover.

Oil Bath Type Air Filter. Remove air cleaner assembly and separate cover, element and housing. Clean element in nonflammable solvent and air dry. Drain old oil and thoroughly clean housing. Fill housing to oil level mark with new engine oil and reassemble air cleaner.

GOVERNOR. The internal centrifugal flyweight governor assembly is located inside crankcase and is either gear or chain driven.

To adjust governor, first stop engine and make certain all linkage is in good condition and tension spring (2 – Fig. HN4) is not stretched or damaged. Spring (4) must pull governor lever (3) toward throttle pivot (6).

On all models except Models GV150 and GV200, loosen clamp bolt (8) and

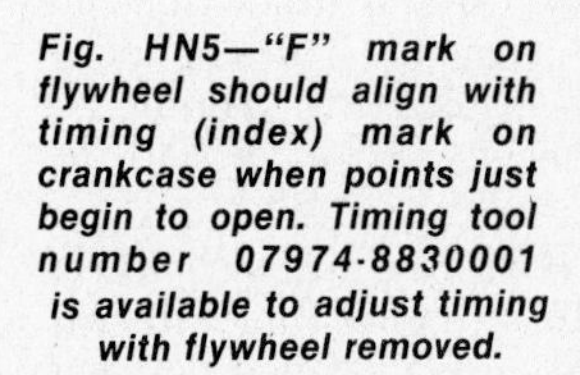
Fig. HN5—"F" mark on flywheel should align with timing (index) mark on crankcase when points just begin to open. Timing tool number 07974-8830001 is available to adjust timing with flywheel removed.

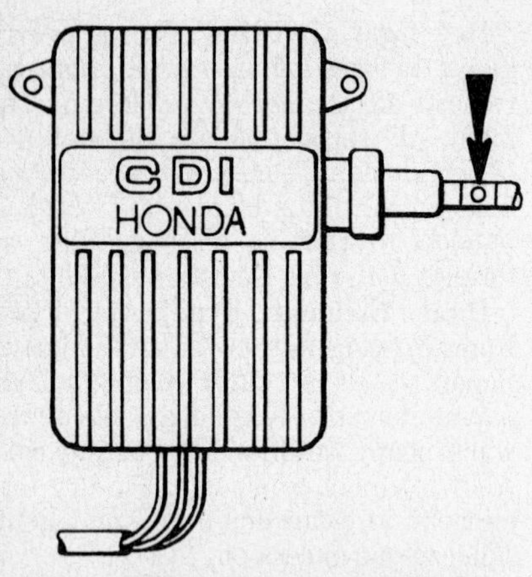

Fig. HN6—Exciter coil must be timed according to timing number on CDI unit.

turn governor shaft clockwise as far as it will go and move governor lever as far to the right as it will go. Tighten retaining bolt. Start and run engine until it reaches operating temperature. Adjust stop screw to obtain 4000 rpm for Models G150 and G200 or 4200 rpm for Model G100.

On Models GV150 and GV200, loosen clamp bolt, pull governor shaft all the way to the right and rotate governor shaft clockwise as far as possible. Tighten clamp bolt. With engine stopped, governor is assembled properly when governor lever is spring-loaded and moves freely. Maximum engine speed is 3600 rpm.

IGNITION SYSTEM. Engines may be equipped with a breaker point ignition system or a capacitor discharge ignition (CDI) system. Refer to appropriate paragraph for model being serviced.

Breaker Point Ignition System. Breaker points and ignition coil are located underneath the flywheel on all models. Breaker points should be checked after every 300 hours of operation. Initial breaker point gap should be 0.3-0.4 mm (0.012-0.016 in.) and can be varied to obtain 20° BTDC timing setting.

NOTE: Timing tool number 07974-8830001 is available from Honda Motor Company to allow timing adjustment with flywheel removed.

To check ignition timing, connect positive ohmmeter lead to engine stop switch wire and connect remaining lead to engine ground. Rotate flywheel until ohmmeter needle deflects. "F" mark on flywheel should align with index mark on crankcase (Fig. HN5). Remove flywheel and vary point gap to obtain correct timing setting. If timing tool is used, ignition timing can be checked with flywheel removed.

To check Model G100 ignition coil, connect positive ohmmeter lead to black wire and remaining lead to coil laminations. Ohmmeter should register 0.49 ohms. Disconnect positive lead and reconnect lead to spark plug wire. Ohmmeter should register 4 ohms.

To check Models G150, GV150, G200 and GV200 ignition coil, connect positive ohmmeter lead to spark plug wire and remaining lead to coil laminations. Ohmmeter should register 6.6 ohms.

CDI Ignition System. The capacitor discharge ignition (CDI) system consists of the flywheel magnets, exciter coil (located under flywheel) and CDI unit. CDI system does not require regular maintenance.

To test exciter coil on all models, disconnect the blue and black exciter coil leads on Model GV200 or the red and black exciter coil leads on all other models. Connect an ohmmeter lead to each exciter coil lead. Ohmmeter should register continuity, if not, renew coil. To renew exciter coil, remove engine cooling shrouds and flywheel. Remove exciter coil.

NOTE: Exciter coil must be installed and timed according to timing number on CDI unit. Refer to Fig. HN6.

Determine CDI unit timing number (Fig. HN6) and position coil so its inner edge aligns with scale position indicated by the CDI unit timing number (Fig. HN7). Distance between scale ridges indicates 2°. Tighten exciter coil mounting bolts and recheck alignment. Reinstall flywheel and cooling shrouds.

If timing, ignition stop switch and exciter coil check satisfactory, but an ignition system problem is still suspected, renew CDI unit. Make certain ground terminal on CDI unit is making an adequate connection.

VALVE ADJUSTMENT. Valves and seats should be refaced and stem clearance adjusted after every 300 hours of operation. Refer to REPAIRS section for service procedures and specifications.

CYLINDER HEAD AND COMBUSTION CHAMBER. Cylinder head, combustion chamber and piston should be cleaned and carbon and other deposits removed after every 300 hours of operation. Refer to REPAIRS section for service procedure.

LUBRICATION. Engine oil should be checked prior to each operating interval. Oil level should be maintained between reference marks on dipstick with dipstick just touching first threads. Do not screw dipstick in to check oil level.

Manufacturer recommends SAE 10W-40 oil with an API service

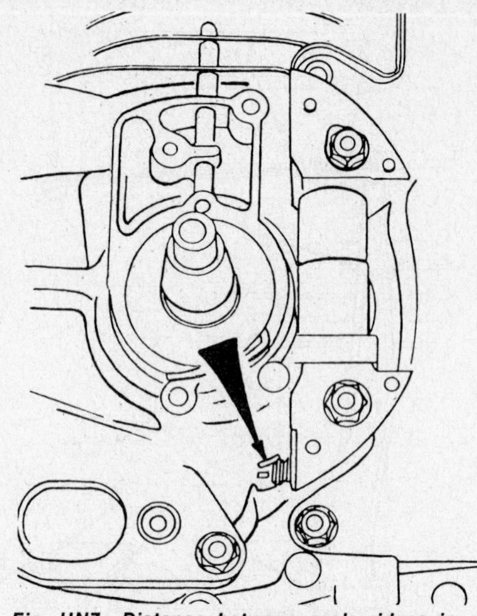

Fig. HN7—Distance between scale ridges indicate 2° ignition timing positions.

classification SE or SF.

Oil should be changed after the first 20 hours of operation and after every 100 hours of operation thereafter. Crankcase capacity is 0.45 L (0.95 pt.) for Model G100 and 0.7 L (1.48 pt.) for all other models.

GENERAL MAINTENANCE. Check and tighten all loose bolts, nuts and clamps prior to each operating interval. Check for fuel and oil leakage and repair if necessary.

Clean dust, dirt, grease and any foreign material from cylinder head and cylinder block cooling fins after every 100 hours of operation. Inspect fins for damage and repair if necessary.

REPAIRS

TIGHTENING TORQUES. Recommended tightening torque specifications are as follows:

Flywheel nut:
G100 .... 4.8 N·m (3.5 ft.-lbs.)
All others .... 73 N·m (54 ft.-lbs.)
Crankcase cover .... 8-12 N·m (6-8 ft.-lbs.)
Cylinder head bolts:
G100 .... 10 N·m (7 ft.-lbs.)
All others .... 24-26 N·m (18-19 ft.-lbs.)
Connecting rod bolts:
G100 .... 3 N·m (2.2 ft.-lbs.)
All others .... 9-11 N·m (6-9 ft.-lbs.)
Oil pump cover
(GV150 & GV200) .... 10 N·m (7 ft.-lbs.)

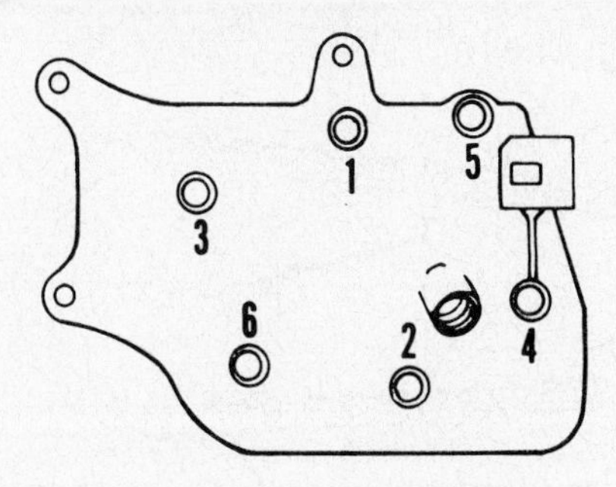

Fig. HN8—Tighten head bolts on Model G100 to specified torque following sequence shown.

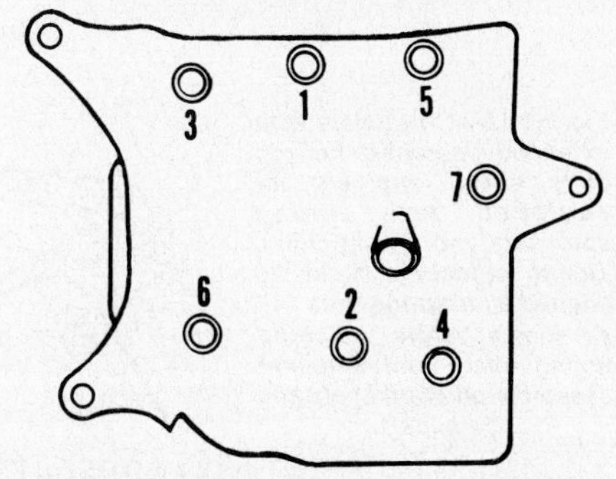

Fig. HN9—Tighten head bolts on all models except Model G100 to specified torque following sequence shown.

CYLINDER HEAD. To remove cylinder head, first remove cooling shrouds. Clean engine to prevent entrance of foreign material. Remove spark plug. Loosen cylinder head bolts in ¼ turn increments following the sequence shown in Fig. HN8 for Model G100 or Fig. HN9 for all other models until all bolts are loose enough to remove by hand. Remove cylinder head. Clean carbon and deposits from cylinder head.

Reinstall cylinder head and new gasket. Tighten head bolts to specified torque using the correct sequence shown (Fig. HN8 or HN9) according to model being serviced.

CONNECTING ROD. Connecting rod rides directly on crankshaft crankpin journal on all models. Piston and connecting rod are accessible after cylinder head removal and crankcase cover or oil pan is separated from crankcase. Remove the two connecting rod bolts, lock plate and connecting rod cap. Push piston and connecting rod assembly out through the top of cylinder block. Remove snap rings and piston pin to separate piston from connecting rod.

Standard diameter for piston pin bore in connecting rod small end is 10.006-10.017 mm (0.3939-0.3944 in.) for Model G100 or 15.005-15.020 mm (0.5907-0.5913 in.) for all other models. If dimension exceeds 10.050 mm (0.3957 in.) for Model G100 or 15.070 mm (0.5933 in.) for all other models, renew connecting rod.

Standard clearance between connecting rod bearing surface and crankpin journal is 0.016-0.033 mm (0.0006-0.0013 in.) for Model G100 and 0.040-0.066 mm (0.0016-0.0026 in.) for all other models. If clearance exceeds 0.1 mm (0.0004 in.) for Model G100 or 0.120 mm (0.0047 in.) for all other models, renew connecting rod and/or recondition crankshaft.

Standard connecting rod side play on crankpin journal is 0.2-0.9 mm (0.008-0.035 in.) for Model G100 or 0.10-0.80 mm (0.004-0.031 in.) for all other models. If side play exceeds 1.1 mm (0.043 in.) for Model G100 or 1.20 mm (0.047 in.) for all other models, renew connecting rod.

To install piston on connecting rod, refer to appropriate paragraph for model being serviced.

Model G100. Install piston on connecting rod so number 896 stamped on top of piston is toward long side of connecting rod (Fig. HN10). With match marks on connecting rod and cap aligned, long side of rod is installed toward valve side of engine.

Models G150, GV150 And G200. Piston may be installed on connecting rod either way. Install connecting rod and piston assembly in engine so marked side of piston top is toward valve side of engine. Align match marks on connecting rod and rod cap.

Model GV200. Install piston on connecting rod with mark on top of piston toward ribbed side of connecting rod. Install connecting rod so ribbed side is towards oil pan. Align connecting rod and rod cap match marks.

All models. Install oil dipper if equipped. Tighten connecting rod bolts to specified torque and lock bolts with lock plate if equipped.

PISTON, PIN AND RINGS. Piston and connecting rod are removed as an assembly. Refer to CONNECTING ROD section for removal and installation procedure.

After separating piston and connecting rod, carefully remove rings. Clean carbon and deposits from piston surface and ring lands.

CAUTION: Extreme care should be exercised when cleaning ring lands. Do not damage squared edges or widen ring grooves. If ring lands are damaged, piston must be renewed.

Measure piston diameter at piston thrust surfaces, 90° from piston pin. Standard piston diameters and service limits are as follows:

Model	Standard Diameter	Service Limit
G100	45.98-46.00 mm (1.810-1.811 in.)	45.92 mm (1.808 in.)
G150, GV150	63.98-64.00 mm (25.18-2.520 in.)	63.88 mm (2.515 in.)
G200, GV200	66.98-67.00 mm (2.637-2.638 in.)	66.88 mm (2.633 in.)

If piston diameter is less than service limit, renew piston.

Before installing rings, install piston in cylinder bore and use a suitable feeler gage to measure clearance between piston and cylinder bore. Standard clearance and service limits are as follows:

Model	Standard Clearance	Service Limit
G100	0.030 mm (0.001 in.)	0.130 mm (0.005 in.)
G150, GV150	0.060 mm (0.0024 in.)	0.285 mm (0.0112 in.)
G200, GV200	0.040 mm (0.0016 in.)	0.045 mm (0.0018 in.)

If clearance exceeds service limit dimension, renew piston and/or recondition cylinder bore.

Standard piston bore diameter in piston is 10.000-10.006 mm (0.3937-0.3939 in.) for Model G100 or 15.000-15.006 mm (0.5906-0.5908 in.) for all other models. Service limit for piston pin bore diameter is 10.046 mm (0.4096 in.) for Model G100 or 15.046 mm (0.5924 in.) for all other models. If diameter exceeds service limit, renew piston.

Standard piston pin outside diameter is 9.994-10.000 mm (0.3935-0.3937 in.) for Model G100 or 14.994-15.000 mm (0.5903-0.5906 in.) for all other models. Service limit for piston pin outside diameter is 9.950 mm (0.3917 in.) for Model G100 or 14.954 mm (0.5887 in.) for all other models. If diameter is less than service limit, renew piston pin.

Standard piston ring to piston groove side clearance for Model G100 is 0.025-0.055 mm (0.0009-0.0022 in.) for top ring and 0.010-0.040 mm (0.0004-0.0016 in.) for all remaining rings. Standard piston ring to piston groove side clearance for all other models is 0.01-0.05 mm (0.0004-0.0020 in.). If ring side clearance exceeds 0.10 mm (0.0039 in.) on Model G100 or 0.15 mm (0.0059 in.) on all other models, renew rings and/or piston.

On all models, if piston ring end gap exceeds 1.0 mm (0.039 in.) with ring

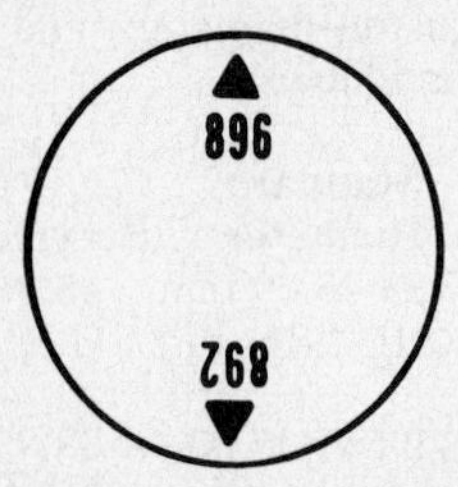

Fig. HN10—Note numbers on piston crown and install piston of Model G100 so "896" on piston is towards long side of rod.

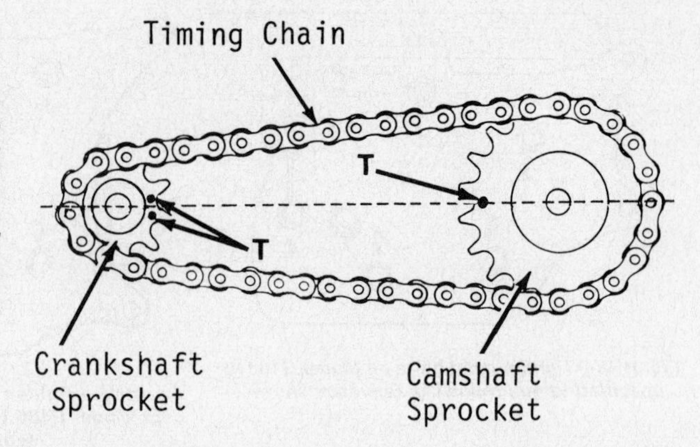

Fig. HN12—On models with extended camshaft for pto drive shaft, engines are equipped with timing sprockets and timing chain. Timing sprockets must be aligned with timing dots "T" as shown before installing timing chain and sprocket assembly on crankshaft and camshaft.

squarely installed in cylinder bore, renew ring and/or recondition cylinder bore.

Install piston rings with marked side towards top of piston. Stagger ring end gaps equally around circumference of piston.

CYLINDER AND CRANKCASE. Cylinder and crankcase are an integral casting. Standard cylinder bore diameters are as follows:

Model	Standard Diameter	Service Limit
G100	46.00-46.01 mm (1.811-1.812 in.)	46.05 mm (1.813 in.)
G150, GV150	64.00-64.02 mm (2.519-2.520 in.)	64.17 mm (2.526 in.)
G200, GV200	67.00-67.02 mm (2.638-2.639 in.)	67.17 mm (2.644 in.)

If cylinder bore diameter at any point in cylinder bore exceeds the service limit, recondition cylinder bore.

CRANKSHAFT, MAIN BEARINGS AND SEALS. Crankshaft is supported by ball bearing type main bearings at each end. To remove crankshaft, remove all cooling shrouds, flywheel, cylinder head and crankcase cover or oil pan. Remove piston and connecting rod assembly. Carefully remove crankshaft and camshaft. Remove main bearings and crankshaft oil seals if necessary.

Standard crankpin diameter is 17.973-17.984 mm (0.7076-0.7080 in.) for Model G100 or 25.967-25.980 mm (1.0223-1.0228 in.) for all other models. If crankpin diameter is less than 17.940 mm (0.7063 in.) for Model G100 or 25.197 mm (1.0204 in.) for all other models, renew or recondition crankshaft.

Main bearings are a light press fit on crankshaft and in bearing bores of crankcase and crankcase cover. It may be necessary to slightly heat crankcase or crankcase cover to reinstall bearings.

Inspect main bearings for roughness and looseness. Also check bearings for a loose fit on crankshaft journals or in crankcase and crankcase cover. Renew bearings if any of the previously described conditions are evident.

If crankshaft oil seals have been removed, use suitable seal driver to install new seals. Seals should be pressed in evenly until 4.5 mm (0.18 in.) below flush for seal in crankcase cover or oil pan and until 2.00 mm (0.08 in.) below flush for seal in crankcase.

Make certain crankshaft gear (sprocket) and camshaft gear (sprocket) timing marks are aligned (Fig. HN11 or Fig. HN12) during crankshaft installation.

CAMSHAFT, BEARINGS AND SEAL. Camshaft is supported at each end by bearings which are an integral part of crankcase or crankcase cover casting. An extended camshaft pto which utilizes a ball bearing for crankcase bearing, is available on some models. Refer to CRANKSHAFT, MAIN BEARINGS AND SEALS section for camshaft removal procedure.

Standard camshaft lobe height is 18.1-18.5 mm (0.71-0.73 in.) for intake and exhaust lobes on Model G100 or 33.4-33.6 mm (1.31-1.32 in.) for intake lobe, and 33.7-33.9 mm (1.33-1.34 in.) for exhaust lobe on all other models. If intake or exhaust lobe on Model G100 is less than 17.940 mm (0.7063 in.), renew camshaft. On all other models, if intake lobe is less than 33.25 mm (1.309 in.) or exhaust lobe is less than 33.55 mm (1.321 in.) renew camshaft.

Standard clearance between camshaft bearing journal and integral type bearings is 0.013-0.043 mm (0.0005-0.0017 in.). If clearance exceeds 0.1 mm (0.004 in.), renew camshaft and/or crankcase and crankcase cover.

On models with extended camshaft, ball bearing should be a light press fit on camshaft journal and in crankcase cover. Camshaft seal should be pressed into crankcase cover 2.0 mm (0.08 in.).

Make certain camshaft gear (sprocket) and crankshaft gear sprocket timing marks are aligned during installation.

GOVERNOR. The internal centrifugal flyweight governor is gear driven off of the camshaft gear on Model

Fig. HN11—When installing crankshaft or camshaft, make certain timing marks on gears are aligned as shown.

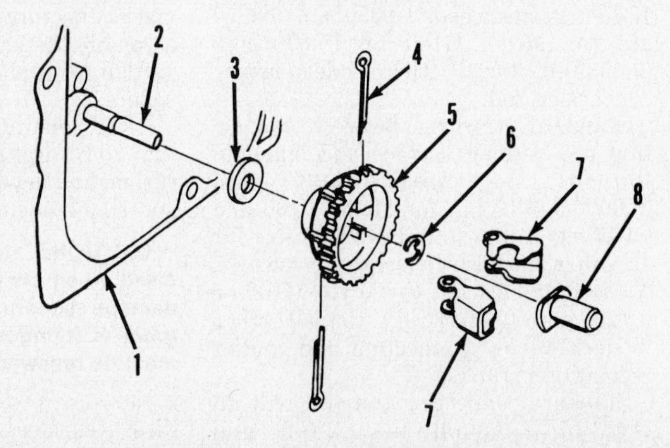

Fig. HN13—Exploded view of governor assembly used on Model G100. On all other models, governor weight assembly is mounted on camshaft gear or sprocket.

1. Crankcase cover
2. Governor stud
3. Thrust washer
4. Pin
5. Gear
6. "E" clip
7. Weights
8. Sleeve

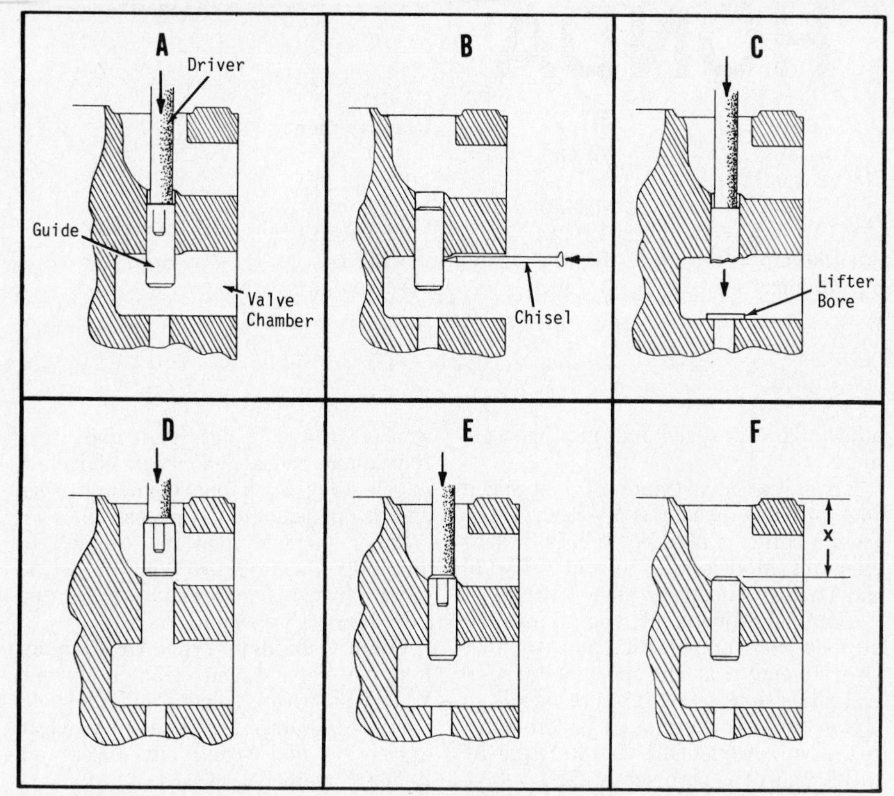

Fig. HN14—View showing valve guide removal and installation sequence on all models except Model G100. Refer to text.

G100 and is located on camshaft gear (sprocket) for all other models. Refer to GOVERNOR paragraphs in MAINTENANCE section for external governor adjustments.

To remove governor assembly, remove external linkage, metal cooling shrouds and crankcase cover or oil pan. On G100 models, remove "E" clip (Fig. HN13) and slide governor assembly off of shaft. On all other models, governor assembly is mounted on camshaft gear or sprocket.

When reassembling, make certain governor sliding sleeve and internal governor linkage is correctly positioned.

OIL PUMP. Models GV150 and GV200 are equipped with an internal oil pump located inside crankcase. All other models are splash lubricated.

Oil pump rotors may be removed and checked without disassembling the engine. Remove oil pump cover and withdraw outer and inner rotors. Remove "O" ring. Make certain oil passages are clear and pump body is thoroughly clean. Inner-to-outer rotor clearance should be 0.15 mm (0.006 in.). If clearance is 0.20 mm (0.008 in.) or more, renew rotors. Clearance between oil pump body and outer rotor diameter should be 0.15 mm (0.006 in.). If clearance is 0.26 mm (0.010 in.) or more, renew rotors and/or oil pan. Outside diameter of outer rotor should be 23.15-23.28 mm (0.911-0.917 in.). If diameter is 23.23 mm (0.915 in.) or less, renew rotors.

Reverse disassembly procedure for reassembly and tighten oil pump cover bolts to specified torque.

VALVE SYSTEM. Clearance between valve stem and valve tappet (cold) should be 0.04-0.10 mm (0.002-0.004 in.) for intake and exhaust valves on Model G100; 0.04-0.12 mm (0.002-0.005 in.) for intake and exhaust valves on Models G150 (prior to serial number 1100543) and G200 (prior to serial number 1168820); 0.05-0.11 mm (0.002-0.004 in.) for intake valve and 0.09-0.15 mm (0.004-0.006 in.) for exhaust valve on Models G150 (serial number 1100543 to 1380694) and G200 (serial number 1168820 to 1556007); 0.08-0.16 mm (0.003-0.006 in.) for intake valve and 0.16-0.24 mm (0.006-0.009 in.) for exhaust valve on Models G150 (after serial number 1380694) and G200 (after serial number 1556007); or 0.05-0.11 mm (0.002-0.004 in.) for intake valve and 0.09-0.015 mm (0.004-0.006 in.) for exhaust valve on all GV200 models.

On all models, valve clearance is adjusted as follows: To increase valve clearance, grind off end of stem. To reduce valve clearance, renew valve and/or grind valve seat deeper.

Valve face and seat angles are 45° for all models. Standard valve seat width is 0.42-0.78 mm (0.016-0.031 in.) for Model G100 while standard valve seat width is 0.7 mm (0.028 in.) for all other models. If valve seat width exceeds 1.0 mm (0.039 in.) for Model G100 or 2.0 mm (0.08 in.) for all other models, seats must be narrowed.

Standard valve stem diameters on Model G100 are 5.480-5.490 mm (0.2157-0.2161 in.) for intake valve stem and 5.435-5.445 mm (0.2140-0.2144 in.) for exhaust valve stem. If intake valve stem diameter is less than 5.450 mm (0.2146 in.) or exhaust valve stem diameter is less than 5.400 mm (0.2126 in.), renew valve.

Standard valve stem diameters on all models except Model G100 are 6.955-6.970 mm (0.2738-0.2744 in.) for intake valve stem and 6.910-6.925 mm (0.2720-0.2726 in.) for exhaust valve stem. If intake valve stem diameter is less than 6.805 mm (0.2679 in.) or exhaust valve stem diameter is less than 6.760 mm (0.2661 in.), renew valve.

Standard valve guide inside diameter for both intake and exhaust valve guides is 5.500-5.512 mm (0.2165-0.2170 in.) for Model G100 and 7.00-7.015 mm (0.2756-0.2762 in.) for all other models. If inside diameter of guide exceeds 5.560 mm (0.2189 in.) for Model G100 or 7.080 mm (0.2787 in.) for all other models, guides must be renewed.

To remove and install Model G100 valve guides, use Honda valve guide tool 07969-8960000 to pull guide out of guide bore and press new guide in. New guide is pressed in to a depth of 18 mm (0.7 in.) measured from end of guide to cylinder head surface as shown in section (F–Fig. HN14) at (X). Finish ream guide after installation with reamer 07984-2000000.

To remove and install valve guides on all models except Model G100, use the following procedure and refer to the sequence of illustrations in Fig. HN14. Use driver 07942-8230000 and drive valve guide down into valve chamber slightly (A). Use a suitable cold chisel and sever guide adjacent to guide bore (B). Cover tappet opening to prevent fragments from entering crankcase. Drive remaining piece of guide into valve chamber (C) and remove from chamber. Place new guide on driver and start guide into guide bore (D). Alternate between driving guide into bore and measuring guide depth below cylinder head surface (E). Guide is driven in to a depth of 27.5 mm (1.08 in.) measured from end of guide to cylinder head surface as shown in section (F) at (X). Finish ream guide after installation with reamer 07984-5900000.

HONDA

Model	Bore	Stroke	Displacement
GX110	57 mm	42 mm	107 cc
	(2.2 in.)	(1.7 in.)	(6.6 cu. in.)
GX140	64 mm	45 mm	144 cc
	(2.5 in.)	(1.8 in.)	(8.8 cu. in.)
GXV120	60 mm	42 mm	118 cc
	(2.4 in.)	(1.7 in.)	(7.2 cu. in.)

ENGINE INFORMATION

All models are four-stroke, overhead valve, single-cylinder air-cooled engines. Models GXV120 is a vertical crankshaft engine and all other models are horizontal crankshaft engines with cylinder inclined 25°.

Model GX110 is rated at 2.6kW (3.5 hp) at 3600 rpm, Model GX140 is rated at 3.8 kW (5.0 hp) at 3600 rpm and Model GXV120 is rated at 2.9 kW (4.0 hp) at 3600 rpm.

Engine model number is cast into side of crankcase (Fig. HN30) and engine serial number is stamped in crankcase (Fig. HN31). Always furnish engine model and serial number when ordering parts or service material.

MAINTENANCE

SPARK PLUG. Spark plug should be removed, cleaned and inspected after every 100 hours of operation.

Recommended spark plug for all models is a NGK BP6ES, or equivalent. Recommended spark plug electrode gap is 0.7 mm (0.028 in.).

When installing spark plug, manufacturer recommends installing spark plug finger tight, then, if new plug, tighten an additional ½ turn, if used plug, tighten an additional ¼ turn.

CARBURETOR. All models are equipped with a Keihin float type carburetor with a fixed main fuel jet and an adjustable low speed fuel mixture needle.

For initial adjustment of low speed fuel mixture needle (LS–Fig. HN32) from a lightly seated position is 3 turns open for Models GX110 and GXV120 and 1⅝ turns open for Model GX140.

For final adjustment, engine must be at operating temperature and running. Operate engine at idle speed (1400 rpm) and adjust low speed mixture needle to obtain smooth idle and satisfactory acceleration. Adjust idle to 1400 rpm by turning throttle stop screw (TS).

To check float level, remove float bowl and invert carburetor throttle body and float assembly. Measure from top edge of float to float bowl mating edge of carburetor throttle body. Measurement should be 12.2-15.2 mm (0.48-0.60 in.). Renew float if float height is incorrect.

Standard main jet on Models GX110 and GXV120 is a #65 and standard main jet on Model GX140 is a #68.

AIR CLEANER. Engine may be equipped with either a dry type, foam (semidry) type or oil bath air filter which should be cleaned and inspected after every 50 hours of operation. Refer to appropriate paragraph for type being serviced.

Dry Type Air Cleaner. Remove foam and paper air filter elements from air filter housing. Foam element should be washed in a mild detergent and water solution, rinsed in clean water and allowed to air dry. Soak foam element in clean engine oil. Squeeze out excess oil.

Paper element may be cleaned by directing low pressure compressed air stream from inside filter toward the outside. Reinstall elements.

Foam (Semidry) Type Air Cleaner. Remove element and clean in solvent. Wring out excess solvent and allow element to air dry. Dip element in clean engine oil and wring out excess oil. Reinstall element.

Oil Bath Type Air Cleaner. Remove elements and clean in suitable solvent. Foam element should be washed in a mild detergent and water solution and allowed to air dry. Discard old oil and clean oil reservoir with solvent. Refill oil reservoir with 60 mL (1.3 pt.) clean engine oil. Soak foam element in clean engine oil and wring out excess oil. Reassemble air cleaner.

GOVERNOR. The mechanical flyweight type governor is located inside engine crankcase. To adjust external linkage, stop engine and make certain all linkage is in good condition and tension

Fig. HN30—Engine model number (MN) is cast into side of engine crankcase.

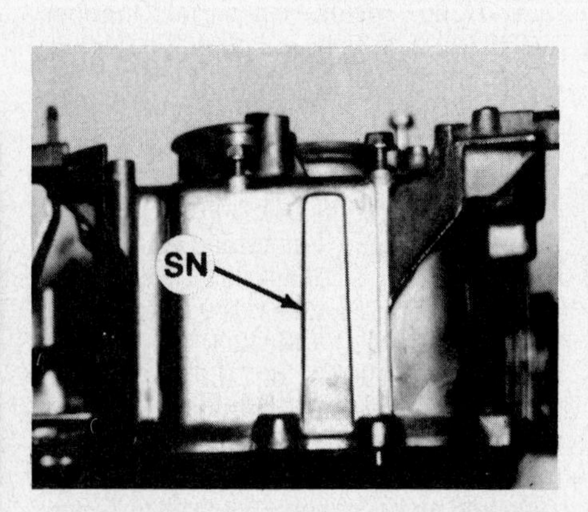

Fig. HN31—Engine serial number (SN) is stamped on raised portion of crankcase.

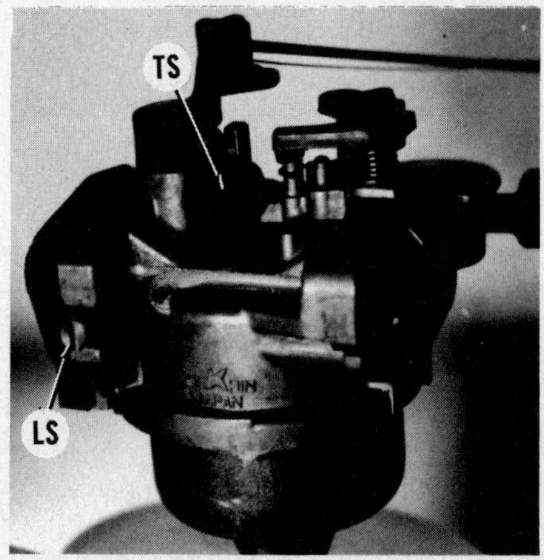

Fig. HN32—View of Keihin float type carburetor used on all models showing location of low speed mixture screw (LS) and throttle stop screw (TS).

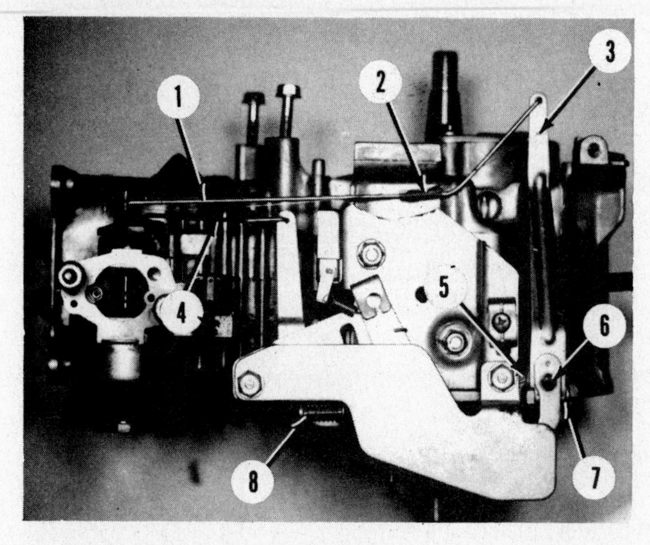

Fig. HN33—View of governor linkage.

1. Governor-to-carburetor rod
2. Spring
3. Governor lever
4. Choke rod
5. Tension spring (behind plate & lever)
6. Governor shaft
7. Clamp bolt
8. Throttle stop screw

spring (5–Fig. HN33) is not stretched or damaged. Spring (2) must pull governor lever (3) and throttle pivot toward each other. Loosen clamp bolt (7) and move governor lever (3) so throttle is completely open. Hold governor lever in this position and rotate governor shaft (6) in the same direction until it stops. Tighten clamp bolt.

Start engine and operate at an idle until operating temperature has been reached. Attach a tachometer to engine and move throttle so engine is operating at a maximum speed of 3800-4000 rpm. Adjust throttle stop screw (8) so throttle movement is limited to correct maximum engine rpm.

IGNITION SYSTEM. Breakerless ignition system requires no regular maintenance. Ignition coil unit is mounted outside the flywheel and air gap between flywheel and coil should be 0.2-0.6 mm (0.008-0.024 in.)

To check ignition coil primary side, connect one ohmmeter lead to primary (black) coil lead and touch iron coil laminations with remaining lead. Ohmmeter should register 0.7-0.9 ohms. To check ignition coil secondary side, connect one ohmmeter led to the spark plug lead wire and remaining lead to the iron coil laminations. Ohmmeter should register 6.3-7.7 ohms. If ohmmeter readings are not as specified, renew ignition coil.

VALVE ADJUSTMENT. Valve stem clearance should be checked and adjusted after every 300 hours of operation.

To adjust valve stem clearance, refer to Fig. HN34 and remove rocker arm cover. Rotate engine so piston is at top dead center on compression stroke. Insert a feeler gage between rocker arm (3) and end of valve stem. If necessary, loosen rocker arm jam nut (1) and turn adjusting nut (2) to obtain 0.10 mm (0.004 in.) clearance on intake valve and 0.15 mm (0.006 in.) clearance on exhaust valve on Model GXV120 or 0.15 mm (0.006 in.) clearance on intake valve and 0.20 mm (0.008 in.) clearance on exhaust valve for all other models. Tighten jam nut and recheck clearance. Install rocker arm cover.

CYLINDER HEAD AND COMBUSTION CHAMBER. After every 300 hours of operation, the manufacturer recommends carbon and other deposit removal from combustion chamber. The manufacturer also recommends refacing the valves and valve seats. Refer to CYLINDER HEAD paragraphs in REPAIRS section for cylinder head removal and disassembly procedures.

LUBRICATION. Engine oil level should be checked prior to each operating interval. Maintain oil level at top of reference marks (Fig. HN35) on gage. Oil level is checked with cap unscrewed and just touching first threads.

Oil should be changed after the first 20 hours of engine operation and after every 100 hours of operation thereafter.

Manufacturer recommends oil with an API service classification SE or SF. Use SAE 10W-30 or 10W-40 oil.

Crankcase capacity is 0.6 L (0.63 qt.) for all models.

GENERAL MAINTENANCE. Check and tighten all lose bolts, nuts or clamps prior to each operating interval. Check for fuel or oil leakage and repair if necessary.

Clean dust, dirt, grease or any foreign material from cylinder head and cylinder block cooling fins after every 100 hours of operation. Inspect fins for damage and repair if necessary.

REPAIRS

TIGHTENING TORQUES. Recommended tightening torque specifications are as follows:

Rocker arm cover	8-12 N·m (6-9 ft.-lbs.)
Cylinder head	22-26 N·m (16-19 ft.-lbs.)
Oil pan	10-14 N·m (7-10 ft.-lbs.)
Connecting rod bolts	10-14 N·m (7-10 ft.-lbs.)

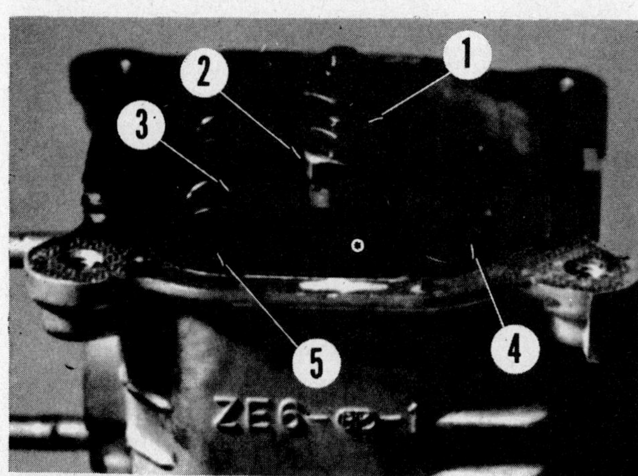

Fig. HN34—View of rocker arm and related parts.

1. Jam nut
2. Adjustment nut
3. Rocker arm
4. Valve stem clearance
5. Push rod

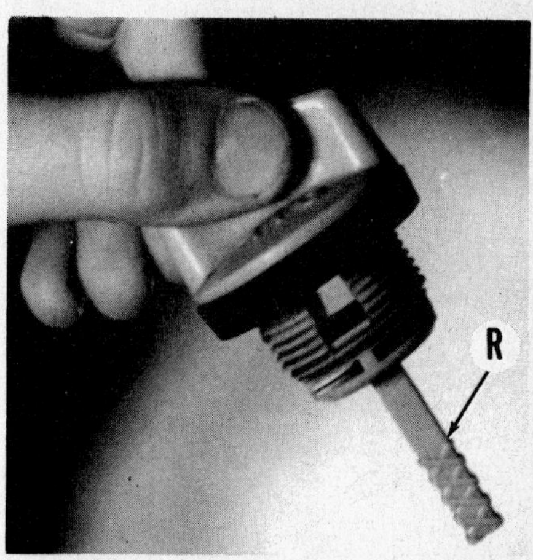

Fig. HN35—Do not screw in oil plug and gage when checking oil level. Maintain oil level at top edge of reference marks (R) on dipstick.

Oil drain plug . . . . 15-20 N•m (11-15 ft.-lbs.)
Rocker arm jam nut . . . . 8-12 N•m (6-9 ft.-lbs.)
Flywheel nut . . . . 70-80 N•m (51-58 ft.-lbs.)

CYLINDER HEAD. To remove cylinder head, remove cooling shroud, disconnect and remove carburetor linkage and carburetor. Remove muffler. Remove rocker arm cover and the four head bolts. Remove cylinder head. Use care not to lose push rods.

Remove rocker arms. Compress valve springs and remove valve retainers. Note exhaust valve on Models GX110 and GX140 are equipped with a valve rotator on valve stem. Remove valves and springs. Remove push rod guide plate if necessary.

Valve face and seat angles are 45° Standard valve seat width is 0.8 mm (0.032 in.). Narrow seat if seat width is 2.0 mm (0.079 in.) or more.

Standard valve spring free length is 34.0 mm (1.339 in.). Renew valve spring if free length is 32.5 mm (1.280 in.) or less.

Standard valve guide inside diameter is 5.50-5.51 mm (0.2165-0.2170 in.). Renew guide if inside diameter is 5.562 mm (0.219 in.) or more. Valve stem-to-guide clearance should be 0.02-0.04 mm (0.001-0.002 in.) for intake valve and 0.06-0.09 mm (0.002-0.003 in.) for exhaust valve. Renew valve and/or guide if clearance is 0.10 mm (0.004 in.) or more for intake valve or 0.12 mm (0.005 in.) or more for exhaust valve.

To renew valve guide, use valve guide driver number 07942-8920000 to remove and install guides. Remove old guides out toward combustion chamber side of head. Drive new guides into cylinder head from rocker arm side, until top of guide is 23.0 mm (0.905 in.) below cylinder head mating surface for Models GX110 and GXV120 or 25.5 mm (1.004 in.) below cylinder head mating surface for Model GX140.

New valve guides must be reamed after installation using reamer 07984-4600000 or 07984-2000000.

When installing cylinder head, tighten head bolts to specified torque in sequence shown in Fig. HN36. Adjust valves as outlined in VALVE ADJUSTMENT paragraphs in MAINTENANCE section.

Fig. HN36 – Tighten cylinder head bolts in sequence shown.

CONNECTING ROD. The aluminum alloy connecting rod rides directly on crankpin journal on all models. Connecting rod cap for all models except Model GXV120 is equipped with an oil dipper (Fig. HN38). Model GXV120 does not have dipper on connecting rod cap (Fig. HN37).

To remove connecting rod, remove cylinder head and crankcase cover or oil pan. Remove connecting rod cap screws and cap. Push connecting rod and piston assembly out of cylinder. Crankshaft and camshaft may also be removed if required. Remove piston pin retaining rings and piston pin. Separate piston from connecting rod.

Standard piston pin bore diameter in connecting rod is 13.005-13.020 mm (0.512-0.513 in.) for Models GX110 and GXV120 and 18.005-18.020 mm (0.7089-0.7094 in.) for Model GX140. Renew connecting rod if diameter is 13.07 mm (0.5 in.) or more for Models GX110 and GXV120 or 18.07 mm (0.711 in.) or more for Model GX140.

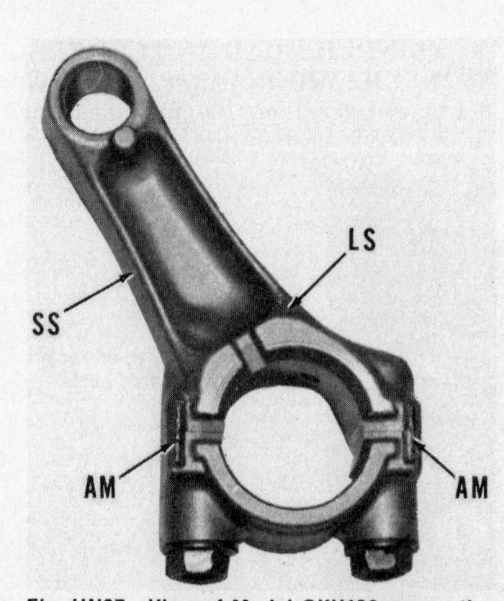

Fig. HN37—View of Model GXV120 connecting rod. Models GX110 and GX140 are equipped with an oil dipper as shown in Fig. HN38.

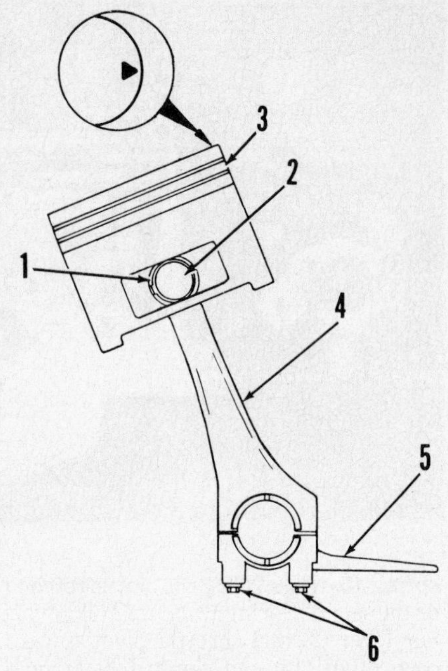

Fig. HN38 – Exploded view of rod and piston assembly used on Models GX110 and GX140. Note location of oil dipper on connecting rod cap.

1. Retaining rings
2. Piston pin
3. Piston
4. Connecting rod
5. Rod cap & dipper
6. Bolts

Standard piston pin-to-connecting rod clearance is 0.005-0.026 mm (0.0002-0.001 in.) for all models. Renew connecting rod and/or pin if clearance is 0.08 mm (0.0031 in.) or more.

Standard connecting rod bearing-to-crankpin clearance is 0.040-0.063 mm (0.0015-0.0025 in.) for all models. Renew connecting rod and/or crankshaft if clearance is 0.12 mm (0.0047 in.) or more.

Connecting rod side play on crankpin should be 0.1-0.7 mm (0.004-0.028 in.) for all models. Renew connecting rod if side play is 1.1 mm (0.043 in.) or more.

When reassembling piston on connecting rod, long side (LS – Fig. HN37) of connecting rod and arrowhead on piston top (Fig. HN39) must be on the same side.

When reinstalling piston and connecting rod assembly in cylinder, arrowhead on piston top must be on push rod side of engine. Align connecting rod cap and connecting rod match marks (AM-Fig. HN37), install connecting rod bolts and tighten to specified torque.

PISTON, PIN AND RINGS. Piston and connecting rod are removed as an assembly. Refer to CONNECTING ROD section for removal and installation procedure.

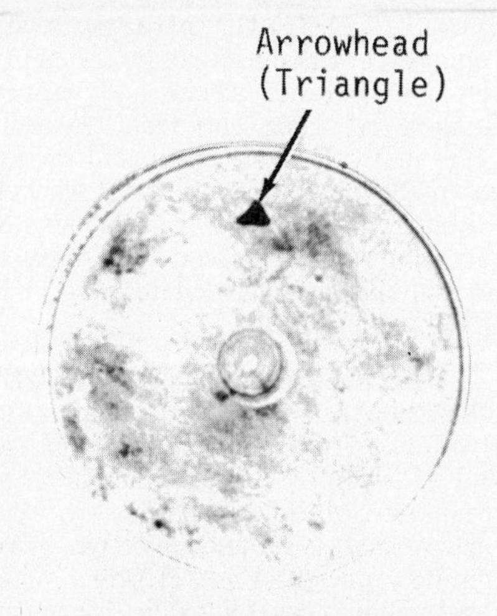

Fig. HN39—Arrowhead (triangle) on top of piston must be on push rod side of engine after installation. Refer to text.

Standard piston diameter measured at lower edge of skirt and 90° from piston pin bore, is 56.965-56.985 mm (2.2427-2.2434 in.) for Model GX110; 63.965-63.985 mm (2.518-2.519 in.) for Model GX140 and 59.965-59.985 mm (2.3567-2.3574 in.) for Model GXV120. Renew piston if diameter is 56.55 mm (2.226 in.) or less for Model GX110; 63.55 mm (2.502 in.) or less for Model GX140 or 59.55 mm (2.340 in.) or less for Model GXV120. Standard piston-to-cylinder clearance should be 0.015-0.050 mm (0.0006-0.002 in.) for all models. Renew piston and/or cylinder if clearance is 0.12 mm (0.0047 in.) or more.

Standard piston pin bore diameter is 13.002-13.008 mm (0.5118-0.5120 in.) for Models GX110 and GXV120 and 18.002-18.048 mm (0.709-0.711 in.) for Model GX140. Standard piston pin diameter is 12.994-13.000 mm (0.5115-0.5118 in.) for Models GX110 and GXV120 and 17.994-18.000 mm (0.7084-0.7087 in.) for Model GX140. If pin diameter is 12.954 mm (0.510 in.) or less for Models GX110 and GXV120, or 17.954 mm (0.707 in.) or less for Model GX140, renew piston pin. Standard clearance between piston pin and pin bore in piston should be 0.002-0.014 mm (0.0001-0.0006 in.) for all models. If clearance is 0.08 mm (0.0031 in.) or more, renew piston and/or pin.

Ring side clearance should be 0.015-0.045 mm (0.0006-0.0018 in.) for all models. Ring end gap for compression rings on all models should be 0.2-0.4 mm (0.008-0.016 in.). Ring end gap for oil control ring on all models should be 0.15-0.35 mm (0.006-0.014 in.). If ring end gap for any ring is 1.0 mm (0.039 in.) or more, renew ring and/or cylinder.

Install piston rings with marked side toward top of piston and stagger ring end gaps equally around circumference of piston.

CYLINDER AND CRANKCASE. Cylinder and crankcase are an integral casting. Standard cylinder bore diameter is 57.000-57.015 mm (2.244-2.245 in.) for Model GX110; 64.000-64.015 mm (2.519-2.520 in.) for Model GX140 or 60.000-60.015 mm (2.362-2.363 in.) for Model GXV120.

If cylinder diameter is 57.165 mm (2.251 in.) or more for Model GX110; 64.165 mm (2.526 in.) or more for Model GX140 or 60.165 mm (2.370 in.) or more for Model GXV120, renew cylinder.

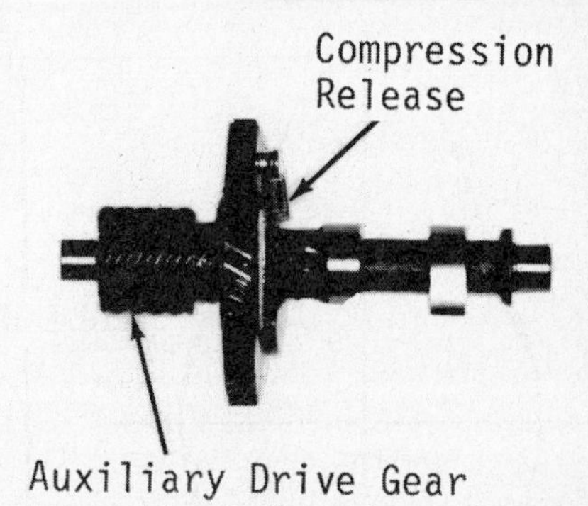

Fig. HN42—Camshaft and gear are an integral casting equipped with a compression release mechanism. Camshaft shown is for Model GXV 120 with auxiliary driveshaft.

CRANKSHAFT, MAIN BEARINGS AND SEALS. Crankshaft for Models GX110 and GX140 is supported at each end in ball bearing type main bearings. Crankshaft for Model GXV120 is supported at flywheel end in a ball bearing type main bearing and at pto end in a bushing type main bearing which is an integral part of the oil pan casting. To remove crankshaft, refer to CONNECTING ROD section.

Standard crankpin journal diameter is 26.0 mm (1.024 in.) for Model GX110; 30.0 mm (1.181 in.) for Model GX140 and 26.0 mm (1.024 in.) for Model GXV120. If crankpin diameter is 25.92 mm (1.020 in.) or less for Models GX110 and GXV120 or 29.92 mm (1.178 in.) or less for Model GX140, renew crankshaft.

Ball bearing type main bearings are a press fit on crankshaft journals and in bearing bores of crankcase and cover. Renew bearings if loose, rough or loose fit on crankshaft or in bearing bores. Bushing type bearing in oil pan of Model GXV120 is an integral part of oil pan.

Fig. HN41—Align crankshaft gear and camshaft gear timing marks (F) during installation.

Fig. HN43—Compression release mechanism spring (1) and weight (2) installed on camshaft gear.

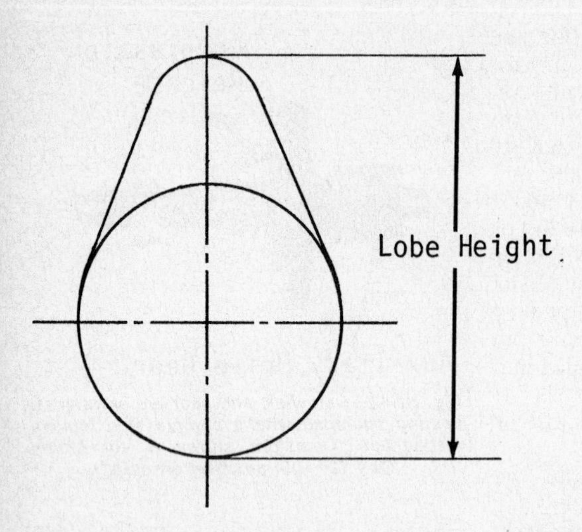

Fig. HN44—Drawing showing camshaft lobe height measurement.

Renew oil pan if bearing is worn, scored or damaged.

Seals should be pressed into seal bores until outer edge of seal is flush with seal bore. When installing crankshaft, make certain crankshaft gear and camshaft gear timing marks are aligned as shown in Fig. HN41.

CAMSHAFT. Camshaft and camshaft gear are an integral casting equipped with a compression release mechanism (Fig. HN42). To remove camshaft, refer to CONNECTING ROD section.

Standard camshaft bearing journal diameter is 14.0 mm (0.551 in.). Renew camshaft if journal diameter is 13.916 mm (0.548 in.) or less.

Standard camshaft lobe height is 27.7 mm (1.091 in.) for intake valve lobe and 27.75 mm (1.093 in.) for exhaust valve lobe (Fig. HN44). If intake lobe measures 27.45 mm (1.081 in.) or less or exhaust lobe measures 27.50 mm (1.083 in.) or less, renew camshaft.

Inspect compression release mechanism for damage. Spring must pull weight tightly against camshaft so decompressor lobe holds exhaust valve slightly open. Weight overcomes spring tension at 1000 rpm and moves decompressor lobe away from cam lobe to release exhaust valve.

When installing camshaft, make certain camshaft and crankshaft gear timing marks are aligned as shown in Fig. HN41.

GOVERNOR. Centrifugal flyweight type governor controls engine rpm via external linkage. Governor is located in oil pan on Model GXV120 and on flywheel side of crankcase on Models GX110 and GX140. Refer to GOVERNOR paragraphs in MAINTENANCE section for adjustment procedure.

To remove governor assembly on Model GXV120, remove oil pan. Remove governor assembly retaining bolt (1–Fig. HN45) and remove governor gear and weight assembly. Governor sleeve, thrust washer, retaining clip and gear may be removed from shaft.

To remove governor assembly on Models GX110 and GX140, refer to CONNECTING ROD paragraphs and remove crankshaft and camshaft. Remove governor sleeve and washer. Remove retaining clip from governor gear shaft. Remove gear and weight assembly and remaining thrust washer.

Reinstall governor assemblies by reversing removal procedure. Adjust external linkage as outlined in MAINTENANCE section.

AUXILIARY DRIVE (MODEL GXV120). Auxiliary drive shaft (Fig. HN45 and HN46) is mounted in oil pan and is driven by a gear which is an integral part of the camshaft.

Drive shaft is retained in oil pan by retaining pin (9–Fig. HN46). When reassembling, carefully slide drive shaft through oil pan seal, thrust washer, gear and remaining thrust washer. Insert retaining hair pin in hole in drive shaft so it is located between gear (8) and outer thrust washer (10).

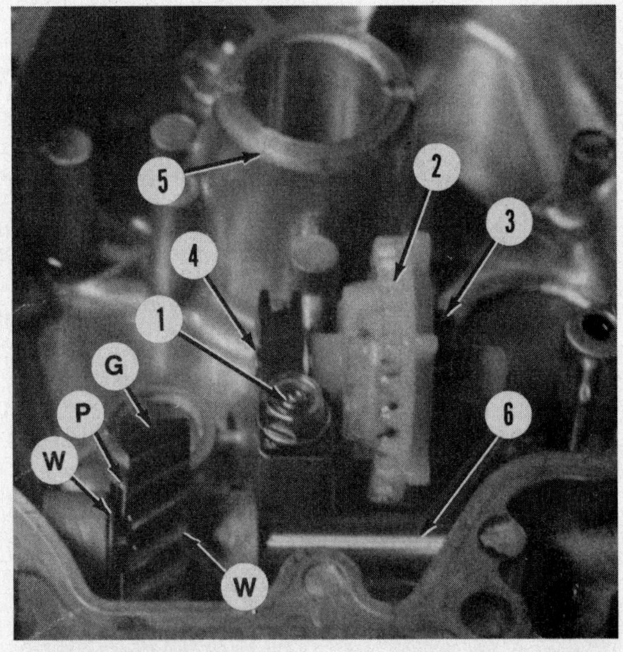

Fig. HN45—Governor assembly on Model GXV120 is mounted in oil pan. Refer to Fig. HN46 also. Governor assembly for all other models is similar except assembly is located in crankcase.

W. Thrust washers
G. Auxilliary drive gear
P. Retaining hair pin
1. Bolt
2. Governor gear
3. Weight
4. Governor gear shaft
5. Oil pan
6. Auxiliary drive shaft

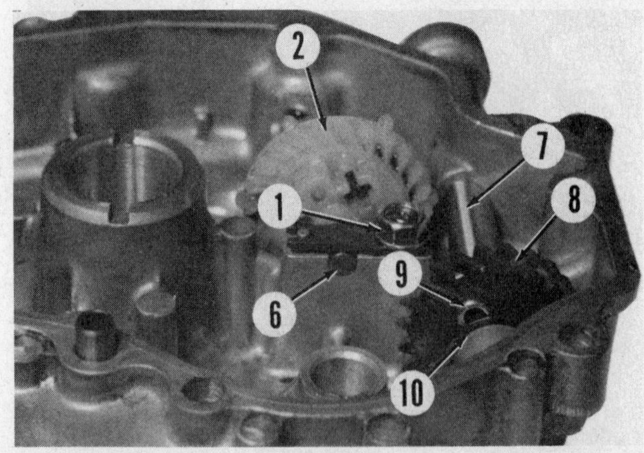

Fig. HN46—Auxiliary drive shaft is mounted in oil pan and driven by a gear which is an integral part of camshaft. See Fig. HN42 and HN43.

1. Bolt
2. Governor gear
6. Governor gear shaft
7. Auxiliary drive shaft
8. Auxiliary drive gear
9. Retainig hair pin
10. Thrust washer

JACOBSEN

Model	Bore	Stroke	Displacement
J-125	2.00 in. (50.8 mm)	1.50 in. (38.1 mm)	4.7 cu. in. (77 cc)
J-175	2.125 in. (54.0 mm)	1.75 in. (44.5 mm)	6.2 cu. in. (102 cc)
J-225	2.25 in. (57.2 mm)	2.00 in. (50.8 mm)	8.0 cu. in. (131 cc)
J-321	2.125 in. (54.0 mm)	1.75 in. (44.5 mm)	6.2 cu. in. (102 cc)
J-501	2.125 in. (54.0 mm)	1.75 in. (44.5 mm)	6.2 cu. in. (102 cc)

ENGINE INFORMATION

All models are two-stroke, air-cooled engines. Models with letter "V" suffix have vertical crankshafts. Models with letter "H" suffix have horizontal crankshafts.

MAINTENANCE

SPARK PLUG. Refer to the following chart for recommended Champion spark plug.

J-125, rotary mower J12J
J-125, reel mower UJ12
J-175, rotary mower J8J
J-225, rotary mower J8J
J-321 & J-501, rotary mower . . . J-17LM
J-321 & J-501, reel mower* UJ12
*For short spark plug, use TJ8.

Set electrode gap to 0.030 inch (0.76 mm) for all models and applications.

CARBURETOR. Tillotson MT58A carburetor is used on Model J-125, Tillotson MT59A carburetor is used on Model J-175, Tillotson MT54A carburetor is used on Model J-225 and either a Walbro LMB or LMG carburetor is used on Models J-321 and J-501. Refer to appropriate paragraph for model being serviced.

Tillotson Carburetors. Initial adjustment of fuel mixture screws from a lightly seated position, is 1 turn open for idle mixture screw (23 – Fig. JAC8) and 1¼ turns open for main fuel mixture screw (26).

Make final adjustments with engine at operating temperature and running. Operate engine at ½ throttle and turn main fuel adjustment screw in until engine loses speed. Turn main fuel screw counterclockwise until maximum rpm is obtained. Operate engine just above idle speed. Turn idle mixture screw in until engine loses speed and misses, then back screw out until engine is idling smoothly. Set idle speed as listed in chart (Fig. JAC7) for model and application by adjusting idle speed screw (18).

Carburetor inlet needle is spring loaded. Float setting is 1/16 to 3/32 inch

Engine Application	Engine RPM Idle Speed	Engine RPM Top Speed
9" Edge-R-Trim (32A9, 32B9 & 32C9)	1500-2000	3000-3300
9" Edge-R-Trim (50012)	1500-1800	Up to 3600
10" Trimo (3110-8610, 86A & 86B)	1500-2000	3400-3600
10" Trimo (50035)		3500-3700
18" Pacer (52C18, 42D18 & 42E-18)	1300-1600	3500 max.
18" Pacer (11814)	1500-1800	Up to 3600
18" Turbo-Cut (3418, 34B18, 34C18 & 34D18)	1300-1800	3400-3500
18" Turbo-Cut (7518, 75A18 & 75B18)	1500-2000	3400-3500
18" Turbo-Vac (31817)	1500-1800	3200-3400
18" Turbo-Vac (31819)*	1500-1800	3200-3400
18" Turbo-Vac (31819)	2400-2600	3200-3400
18" 4-Blade Rotary (31809)	2400-2600	3200-3400
18" Turbo Cone (117-18)	1500-1600	3200-3400
20" Scepter (8020 & 80A20)	1500-1800	3000-3200
20" Commercial Rotary (35025)	1500-1600	3200-3400
20" Commercial Rotary (32028)	2400-2600	3200-3400
20" Commercial Rotary (32028)*	1500-1600	3200-3400
20" Robust (32031)	1500-1700	3200-3400
20" Snow Jet (9620 & 96A20)	1500-1800	3600-3800
20" Snow Jet (52002, 52003)	1700-1900	3600-3800
21" 4-Blade Rotary (32114)	2400-2600	3200-3400
21" 4-Blade Rotary S.P. (42114, 42118, 42119)	2400-2600	3200-3400
21" Turbo-Cut (3921, 39B21 & 39C21)	1500-1800	3200-3400
21" Turbo Cone (119-21)	1500-1600	3200-3400
21" Turbo-Cut (3521, 35C21, 35D21, 35E21 & 35F21)	1500-1800	3500 max.
21" Turbo Cone (121-21)	1500-1600	3200-3400
21" Lawn Queen (2C21, 2D21 & 2E21)	1300-1600	3500 max.
21" Lawn Queen (12113)	1500-1800	Up to 3600
21" Manor (28F21 & 28G21)	1300-1600	3500 max.
21" Manor (22114, 32121-7B1)	1500-1800	Up to 3600
22" Putting Green (9A22 & 9B22)	1500-1800	3400 max.
22" Greensmower (62203, 62208)	1500-1800	Up to 3800
22" Scepter (8022 & 80A22)	1500-1800	3000-3200
24" Estate (8A24 & 8B24)	1300-1600	3800 max.
24" Rotary S. P. (40A24)	1900-2100	3000 max.
26" Estate (8A26, 8B26, 8C26 & 8D26)	1300-1600	3800 max.
26" Estate R.R. (22601, 22605-7B1)	1500-1800	Up to 3800
26" Estate F.R. (22611, 22615-7B1)	1500-1800	Up to 3800
26" Lawn King (12A26 & 12B26)	1300-1600	3200 max.
26" Lawn King (12601)	1500-1800	Up to 3800

*Speed control on handle.

Fig. JAC7—Chart showing engine idle and top speed for various engine applications. Refer to text.

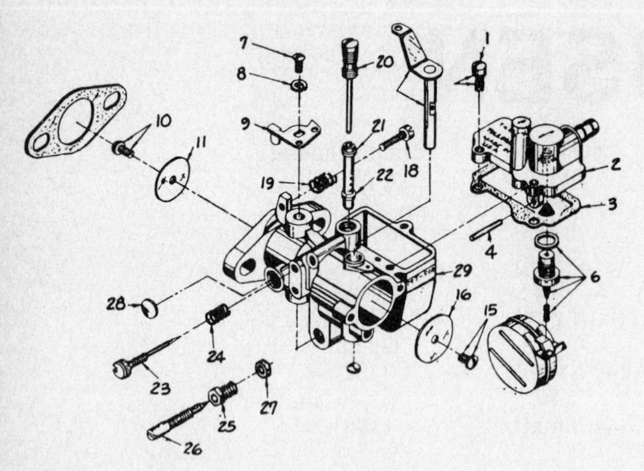

Fig. JAC8—Exploded view of typical Tillotson carburetor.

1. Screw & lockwasher
2. Bowl cover
3. Gasket
4. Lever pin
6. Inlet needle assy.
9. Throttle stop lever
10. Shutter screw
11. Throttle shutter
15. Shutter screw
16. Choke shutter
18. Idle speed screw
19. Spring
20. Idle tube
21. Main nozzle
22. Gasket
23. Idle mixture screw
24. Spring
25. Packing nut
26. Main adjusting screw
27. Packing
28. Welch plug
29. Carburetor body

engine begins to lose speed, then turn idle fuel screw counterclockwise 1/8 to 1/4 turn. Engine will have a slight "stutter" or intermittent exhaust sound at both idle and high speed when fuel mixture adjustments are correct.

Idle speed is controlled by proper adjustment of the governor rather than by adjustment of the idle speed stop screw on carburetor throttle. This is to prevent engine stalling when traction and/or reel clutch is engaged with engine at idle speed. Refer to GOVERNOR section.

To check float setting on Walbro carburetor, invert the body casting and float level should be 5/32 inch (3.97mm) (1.59-2.38 mm) from bowl cover flange to top of float with bowl cover assembly inverted and float resting lightly on inlet needle. Refer to Fig. JAC9. Float level should be 5/16 inch (7.94 mm) if a plastic float is used. Bend tab on float if necessary to obtain correct setting.

Walbro Carburetor. Refer to Figs. JAC10, JAC11 and JAC12 for exploded view of typical Walbro carburetors used on J-321 and J-501 models.

Walbro carburetor shown in Fig. JAC12 does not have adjustable idle or main fuel orifices. Note some carburetors shown in Fig. JAC10 or JAC11 are not equipped with an idle mixture adjusting screw. Refer to the following paragraphs for carburetor adjustments on carburetors equipped with idle adjusting screw and/or main fuel mixture screw.

For initial adjustment, open the idle and/or main fuel adjustment screw 1 to 1¼ turns from a lightly seated position. Make final adjustments with engine at operating temperature and running at "FAST" throttle position. Slowly turn main fuel mixture screw clockwise until engine begins to lose speed, then turn the main screw counterclockwise 1/8 to 1/4 turn. Move throttle to "IDLE" position and slowly turn idle fuel screw on carburetors so equipped, clockwise until

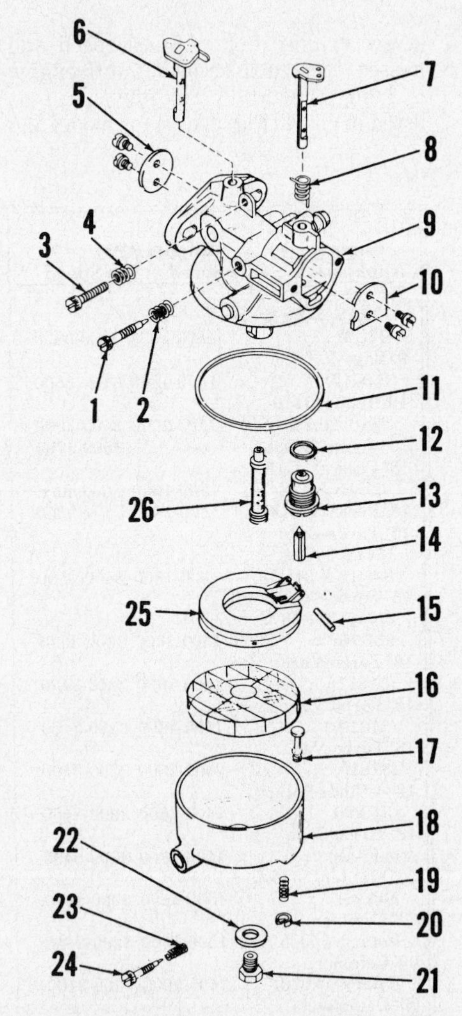

Fig. JAC10—Exploded view of Walbro series LMB float carburetor used on some J-321 and J-501 models. Note the fuel bowl screen (16) used on this typed carburetor.

1. Idle fuel needle
2. Spring
3. Idle speed screw
4. Spring
5. Throttle plate
6. Throttle shaft
7. Choke shaft
8. Choke return spring
9. Carburetor body
10. Choke plate
11. Gasket
12. Gasket
13. Inlet valve seat
14. Inlet valve
15. Float pin
16. Fuel bowl screen
17. Drain valve
18. Fuel bowl
19. Spring
20. Retainer
21. Bowl retainer
22. Gasket
23. Spring
24. Main fuel nozzle
25. Float
26. Main nozzle

Fig. JAC11—Exploded view of typical Walbro series LMG carburetor used on some Model J-321 and J-501 engines.

2. Main fuel nozzle
3. Spring
4. Seals
5. Bowl retainer
6. Washer
7. Adapter
8. Seal
9. Washer
10. Float bowl
11. Gasket
12. Retainer
13. Spring
14. Drain valve
15. Gasket
16. Float pin
17. Float
18. Main nozzle
19. Inlet valve
21. Gasket
22. Idle fuel needle
23. Spring
24. Idle stop screw
25. Spring
27. Throttle plate
28. Throttle shaft
30. Choke plate
31. Choke shaft
32. Choke spring
33. Carburetor body

Fig. JAC9—Distance (D) between bowl cover flange and nearest edge of float with bowl cover assembly inverted should be 1/16 to 3/32 inch (1.59-2.38 mm).

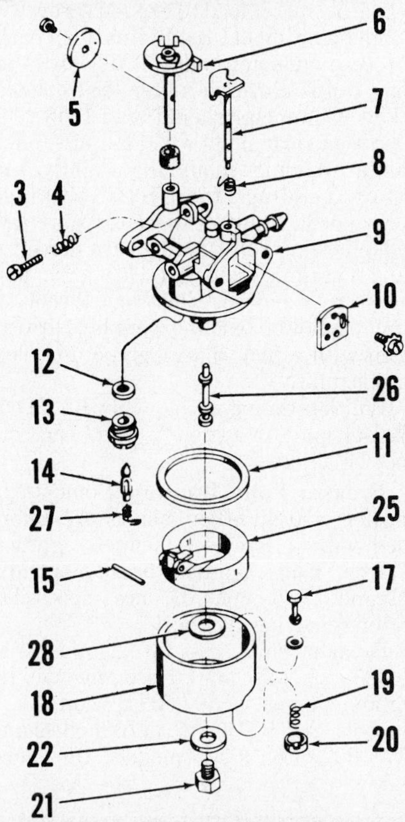

Fig. JAC12—Exploded view of Walbro carburetor with fixed idle and main fuel jets. Refer to Fig. JAC10 for parts identification except for retainer (27) and gasket (28).

measured between carburetor body casting and free side of float. On carburetor with fixed main fuel orifice, float level should be 1/16-3/32 inch (1.59-2.38 mm). Adjust clearance by bending tab on float which contacts fuel inlet needle on all models.

Check float free travel. Float should have 3/16 inch (4.76 mm) of free movement. Adjust free travel by bending tab on float which contacts float stop.

Fixed main fuel orifice on carburetors so equipped, may be cleaned with a #63 drill bit. Be careful not to damage orifice during cleaning.

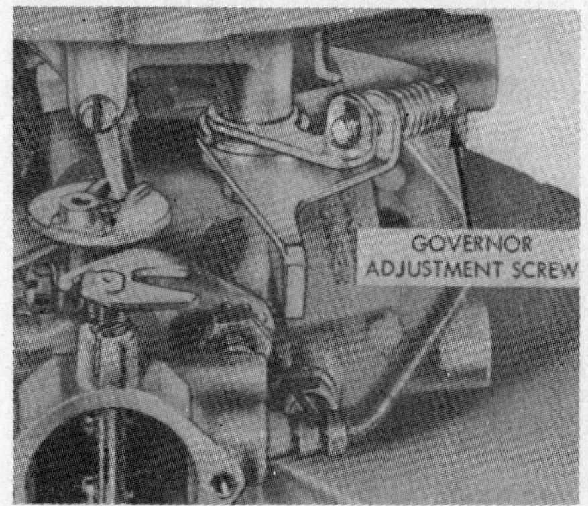

Fig. JAC13—View of governor adjustment screw on Model J-321 or J-501 engine.

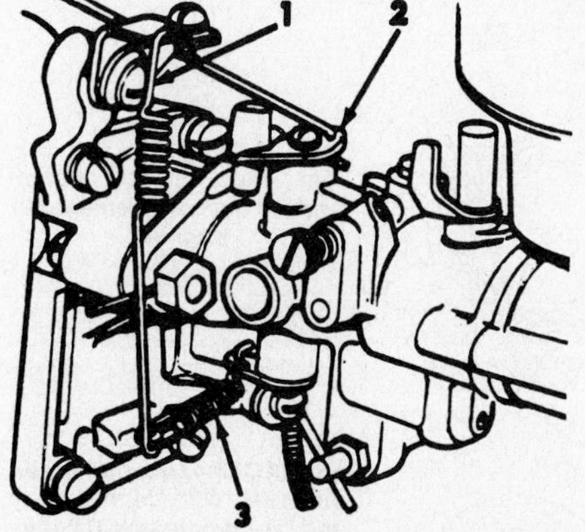

Fig. JAC14—Drawing showing correct installation of Tillotson carburetor controls. Refer to text.

GOVERNOR. Speed control on all models is maintained by a pneumatic (air vane) type governor. The air vane which is linked directly to the carburetor upper throttle lever is actuated by air from the flywheel fan. Make certain the linkage does not bind in any position when moved through the full range of travel.

To adjust governor linkage, refer to chart (Fig. JAC7) for recommended idle and top speed for various engine applications and refer to appropriate paragraphs for model being serviced.

Models J-321 And J-501 With Fixed Speed. Governor adjustment on models with fixed governed speed is accomplished by turning governor adjustment screw shown in Fig. JAC13. Turning adjustment screw provides adjustment range of approximately 400 rpm. Refer to recommended governor speeds.

Governor Adjustments With Tillotson Carburetor. With engine stopped, close the throttle control lever. The upper throttle shaft lever (2–Fig. JAC14) should be in the ½ open position. If not in this position, bend the balance spring (1) at upper end until the desired position is obtained. The lower throttle shaft lever is connected to the governor balance spring (1) by a small antisurge spring (3). Correct spring (3) installation is accomplished by hooking one end into the inner hole of the lower throttle shaft lever from the top side. Hook the other end into the governor balance spring loop so it is underneath the throttle lever link.

Governor Adjustment With Remote Control Walbro Carburetor. Set the throttle control to "FAST" position. The hook in the choke lever (link) should just touch the first loop in choke spring, or clear the first loop by 1/16 inch (1.59 mm) as shown in Fig. JAC15. If choke is partially closed, or clearance between hook in link and loop in spring exceeds 1/16 inch (1.59 mm), loosen the engine control cable clamp. Slide cable back or forward until choke control lever (link) just touches the loop in choke spring, then tighten the control cable clamp. Test setting by moving control to "STOP" position. The carburetor control lever should touch the contact point of "STOP" switch. Move the control to "CHOKE (START)" position. Carburetor choke should be completely closed.

MAGNETO AND TIMING. Model J-321 may be equipped with a breakerless ignition system. All other models are equipped with a flywheel magneto ignition system with breaker points. Refer to appropriate section for model being serviced.

Breakerless Ignition System. A Blaser ignition system is used on some engines and is identified by use of a

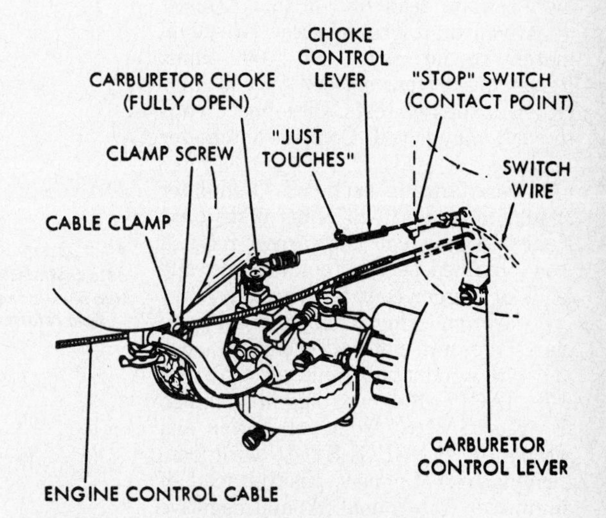

Fig. JAC15—Drawing showing correct installation of Walbro carburetor controls. Refer to text.

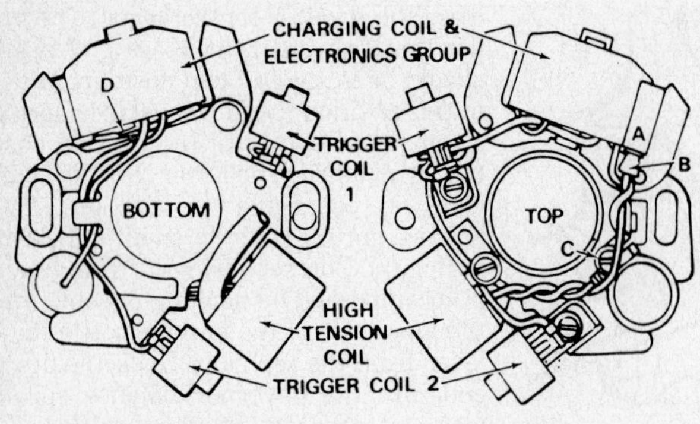

Fig. JAC16—Top and bottom views of CD ignition system used on some J-321 engines.

Fig. JAC17—Location of terminal (D) will identify early and late production CD ignition systems. Refer also to Fig. JAC16 for location of terminal (D).

single trigger coil instead of two trigger coils used in CD ignition system shown in Fig. JAC16. Individual components and complete assembly are not available for Blaser unit. Blaser ignition must be replaced with either a standard breaker type ignition system or with other CD ignition system if renewal is necessary. Flywheel must be renewed when installing different ignition system.

Some J-321 engines are equipped with the CD ignition system shown in Fig. JAC16. Components are available for this ignition. If service is required, the following troubleshooting procedure may be used. Be sure ignition is defective by using Wico test plug #14281 and rotating engine with starter. If test plug does not fire, ignition should be tested.

Remove ignition system from engine. To test high tension coil, connect positive lead of an ohmmeter to coil terminal (B–JAC16) and connect negative lead to ground screw (C). Ohmmeter should read 0.1-0.5 ohms. Disconnect ohmmeter and connect positive lead to high tension lead of coil and connect negative lead to ground screw (C). Ohmmeter should read 900-1100 ohms. Renew high tension coil if coil fails either of the two tests. Disconnect leads to each trigger coil. Connect ohmmeter leads to the leads of a trigger coil and read resistance of each coil. Ohmmeter should indicate 15-25 ohms resistance. Renew trigger coil if incorrect reading was obtained. Check connections and leads of trigger coils for continuity.

Before checking charging coil and electronic unit, note difference in early and late production units as shown in Fig. JAC17. Remove plug-in terminal (B–Fig. JAC16) from coil tower and disconnect "ENGINE STOP" wire from terminal (D). Connect positive lead of ohmmeter to terminal (A) and negative lead to terminal (D). Ohmmeter should read 5-25 ohms on early units and 1 megohm to infinity on late units. Reverse leads of ohmmeter. Ohmmeter should read infinity on early units and 1 megohm to infinity on late units. Disconnect ohmmeter leads and connect positive meter lead to terminal (D) and negative meter lead to ground screw (C). Ohmmeter should read infinity on early production units and 560-760 ohms on late units. Reverse meter lead connections. Ohmmeter should read 1500-2000 ohms on early units. On late units, ohmmeter reading should be slightly less than reading (560-760) obtained previously. If any of these tests are failed, charging coil windings are defective and charging coil and electronic unit must be renewed. Electronic circuit of unit cannot be tested except by substitution with a new charging coil and electronic unit.

Ignition timing on models with CD or Blaser ignition system is fixed and cannot be adjusted.

Breaker Point Ignition. Some J-321 models and all other models are equipped with a flywheel magneto ignition system using ignition breaker points. Magneto components are accessible after removing flywheel.

Breaker point gap for all models is 0.020 inch (0.51 mm). Recommended ignition timing is 30° BTDC on J-125 models, 28° BTDC on J-175 models and 27° BTDC on J-225 models. On Model

Fig. JAC18—View showing use of special Wico Test Plug number S14281 to check magneto output. Refer to text.

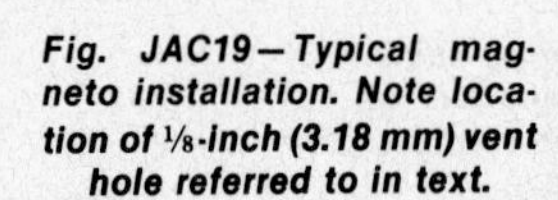

Fig. JAC19—Typical magneto installation. Note location of 1/8-inch (3.18 mm) vent hole referred to in text.

J-321 engine used on Snow Jet, rotate stator to full counterclockwise (advanced) position. On J-501 models and all other J-321 models with breaker point ignitions, rotate stator plate to full clockwise (retarded) position.

On J-125, J-175 and J-225 models, ignition timing is satisfactory if breaker points just begin to open when piston is 0.125 inch (3.18 mm) BTDC. Piston position can be determined by inserting a suitable dial indicator through spark plug hole. Reposition magneto stator plate if necessary to obtain desired timing.

Magneto output at starting speeds can be checked without removing magneto from engine using a special Wico Test Plug S14281. Refer to Fig. JAC18. Note engine spark plug remains in place when using test plug.

On early models, the manufacturer recommends drilling an 1/8-inch (3.18 mm) vent hole in the lower left-hand corner of magneto breaker box to aid in ventilating the contact points (refer to Fig. JAC19).

AIR CLEANER. If the engine is operated under dry or dusty conditions, the manufacturer recommends servicing the air cleaner after every 25 hours of operation. Refer to the appropriate paragraphs for type being serviced.

Foil Type Cleaner. Wash and rinse thoroughly in a suitable nonflammable solvent. Shake off solvent and immerse in clean SAE 30 oil. Allow excess oil to drain off before reinstallation.

Paper Filter. Brush or wipe outside of cleaner. Tap gently to loosen dirt from inside of filter. Do not oil or wash filter. Filter can also be cleaned by gently blowing compressed air from the inside. If, after extended use, filter is too dirty to clean properly, renew the filter.

Oil Bath Air Cleaner. Remove cover from oil reservoir and discard old oil. Wash cleaner in a suitable nonflammable solvent. Dry cleaner thoroughly. Refill to level indicated by arrow with clean SAE 30 motor oil.

Foam Type Filter. Remove filter and wash and rinse thoroughly in a nonflammable solvent. Soak in clean SAE 30 engine oil and carefully compress element to remove excess oil.

LUBRICATION. A good quality two-stroke, air-cooled engine oil should be mixed with regular grade gasoline at a ratio of 50:1 for Model J-501; 30:1 for Model J-321 and 16:1 for all other models.

Use SAE 10 oil in the reduction gear box on models so equipped. Gearbox is fitted with an oil level plug.

CLEANING CARBON. Power loss can often be corrected by cleaning carbon from exhaust ports and muffler. To clean, remove spark plug and muffler. Turn engine so piston is below bottom of exhaust ports and remove carbon from ports with a dull knife or similar tool. Clean out the muffler openings, then replace muffler and spark plug.

BEND LOCK TAB
MATCH MARKS
MATCH MARKS

Fig. JAC20—Be sure match marks are aligned when reassembling rod cap to connecting rod. Bend lock tab against screw heads on models so equipped.

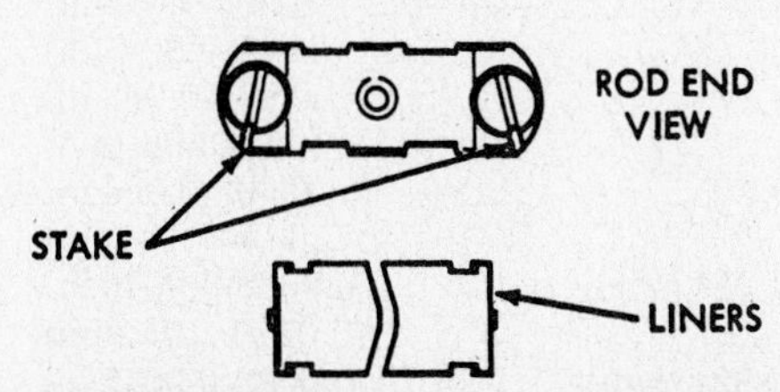

Fig. JAC21—On Model J-321 and J-501 engines with aluminum connecting rod, be sure bearing liners fit together as shown. Stake cap retaining screws as shown.

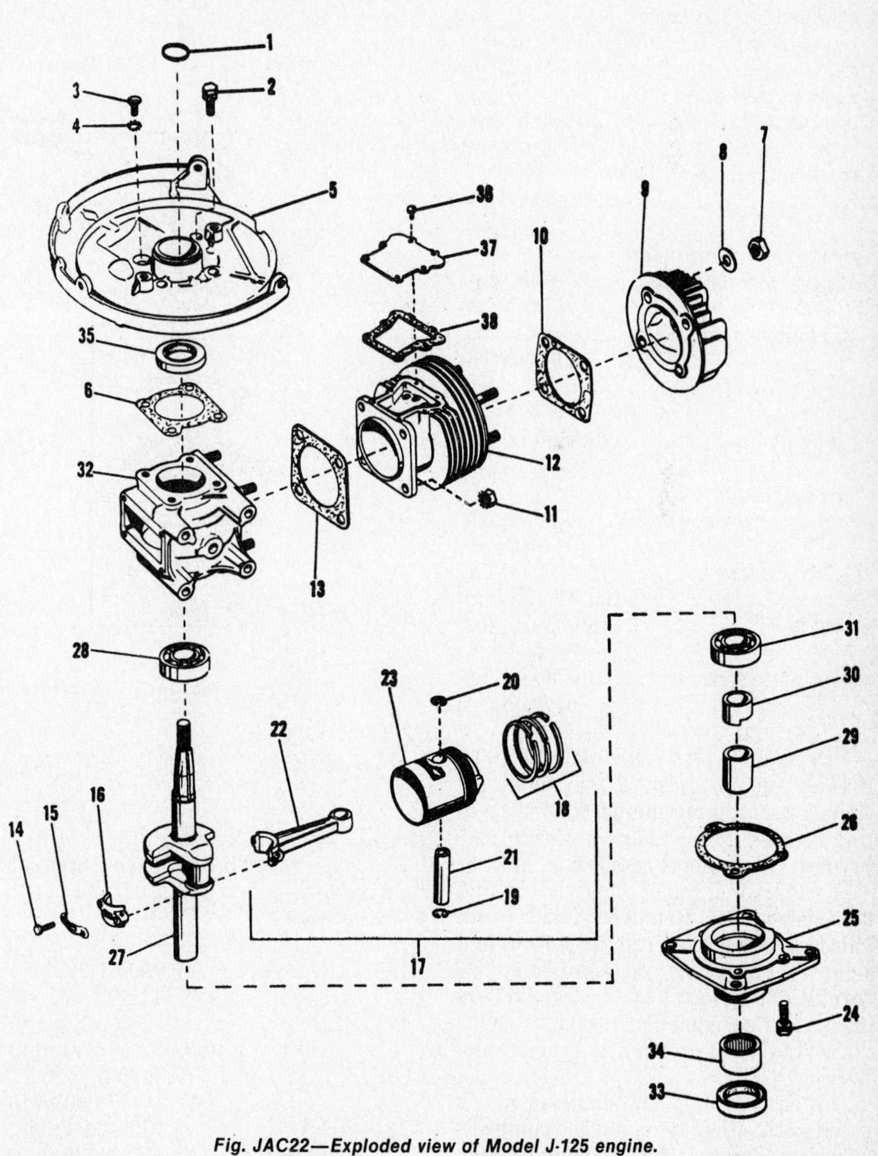

Fig. JAC22—Exploded view of Model J-125 engine.

1. Snap ring
5. Back plate
6. Gasket
9. Cylinder head
10. Gasket
12. Cylinder
13. Gasket
15. Lock tab
16. Rod cap
18. Piston rings
19. Snap ring
20. Snap ring
21. Piston pin
22. Connecting rod
23. Piston
25. Crankcase head
26. Gasket
27. Crankshaft
28. Ball bearing
29. Bearing race
30. Spacer
31. Ball bearing
32. Crankcase
33. Crankshaft seal
34. Needle bearing
35. Crankshaft seal
37. Transfer cover
38. Gasket

REPAIRS

TIGHTENING TORQUES. Recommended tightening torques are as follows:

Back plate screws . . . . 60-80 in.-lbs. (7-9 N·m)

Carburetor adapter screws:
- J-321 & J-501 with needle bearing connecting rod . . 60-80 in.-lbs. (7-9 N·m)
- All other models . . . . 30-50 in.-lbs. (3-6 N·m)

Connecting rod screws:
- J-321 & J-501 with needle bearing connecting rod . . . . 45-50 in.-lbs. (5-6 N·m)
- All other models . . . . 65-75 in.-lbs. (7-9 N·m)

Crankcase bearing plate bolts . . . . 80-100 in.-lbs. (9-11 N·m)

Cylinder to crankcase nuts (J-125) . . . . 150-180 in.-lbs. (17-20 N·m)

Cylinder head bolts or nuts . . . . 150-180 in.-lbs. (17-20 N·m)

Engine mounting bolts . . . . 120-150 in.-lbs. (13-17 N·m)

Muffler mounting bolts:
- Rotary mower engine . . . 60-80 in.-lbs. (7-9 N·m)
- Reel mower engine . . . . 80-100 in.-lbs. (9-11 N·m)

Fan housing screws . . . . . . 60-80 in.-lbs. (7-9 N·m)

Flywheel nut . . . . 300-360 in.-lbs. (34-41 N·m)

Gearbox cover:
- ¼ inch screws . . . . 60-80 in.-lbs. (7-9 N·m)
- 5/16 inch screws . . . . 80-100 in.-lbs. (9-11 N·m)

Spark plug . . . . 180-200 in.-lbs. (20-23 N·m)

Stator plate screws . . . . 60-80 in.-lbs. (7-9 N·m)

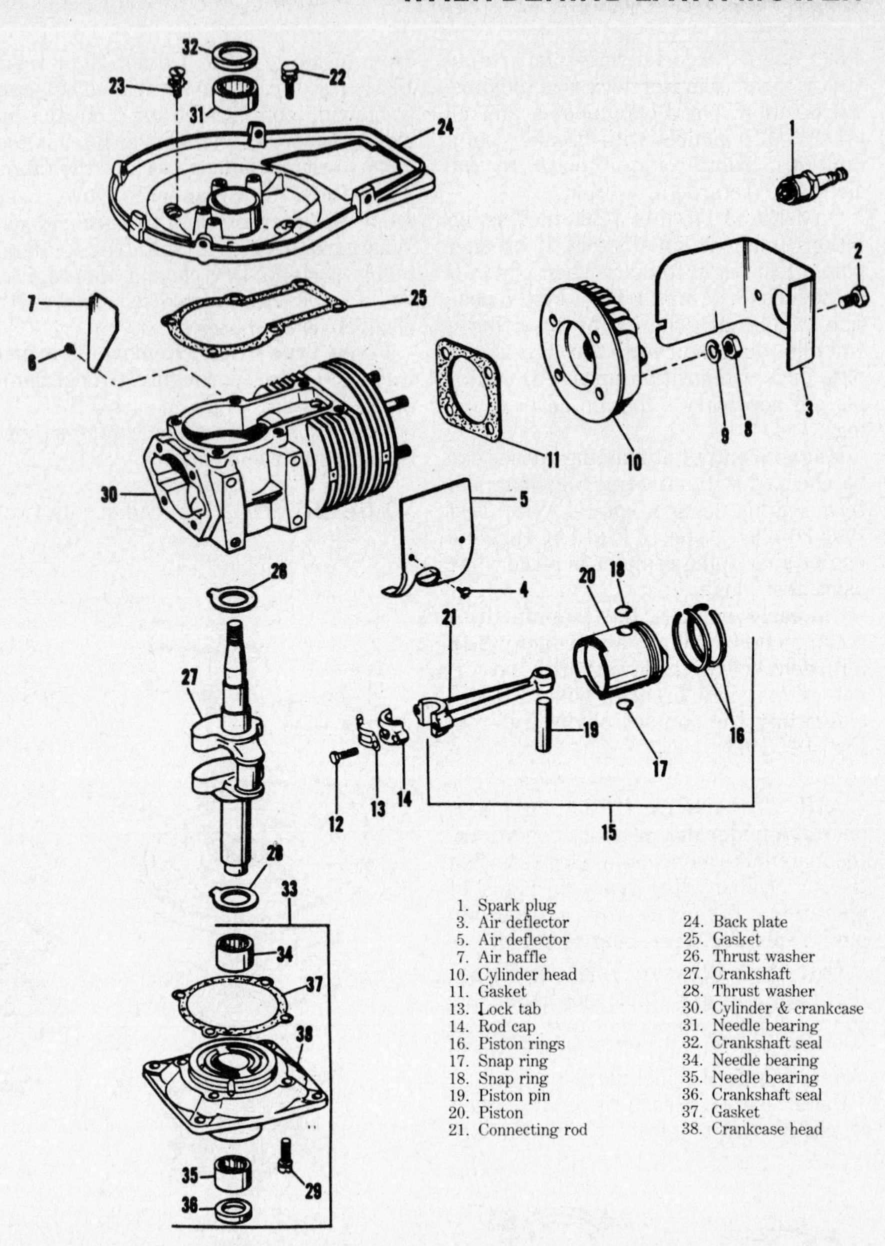

Fig. JAC23—Exploded view of Model J-175 engine.

PISTON, PIN AND RINGS. To remove piston, remove gas tank, air cleaner and interfering shrouds preventing access to cylinder head or carburetor. Disconnect governor link to carburetor and remove carburetor and reed valve plate. Remove cylinder head. Unscrew connecting rod cap screws and remove connecting rod and piston. Be careful not to lose loose bearing rollers on models so equipped. Refer to CONNECTING ROD section to service connecting rod.

The cam ground aluminum piston is fitted with either two or three compression rings depending upon engine model. Renew piston if badly worn or scored, or if side clearance of new ring in top ring groove is 0.010 inch (0.25 mm) or more. Install top ring with chamfer down and second ring with chamfer up.

Piston skirt diameters (measured at right angle to piston pin) are listed in the following table.

Model	Diameter
J-125	1.9965-1.9970 in. (50.711-50.724 mm)
J-175	2.1235-2.1240 in. (53.937-53.950 mm)
J-225	2.2475-2.2480 in. (57.087-57.099 mm)
J-321 (early)	2.1225-2.2480 in. (53.912-53.929 mm)
J-321 (late)	2.1220-2.1227 in. (53.899-53.917 mm)
J-501	2.1220-2.1227 in. (53.899-53.917 mm)

Piston skirt-to-cylinder wall clearances (new, measured at right angle to piston pin) are shown in the following table.

Model	Clearance
J-125	0.004-0.005 in. (0.10-0.30 mm)
J-175	0.002-0.003 in. (0.05-0.08 mm)
J-225	0.003-0.004 in. (0.08-0.10 mm)
J-321 (early)	0.0028-0.0040 in. (0.071-0.101 mm)
J-321 (late)	0.0033-0.0045 in. (0.084-0.114 mm)
J-501	0.0033-0.0045 in. (0.084-0.114 mm)

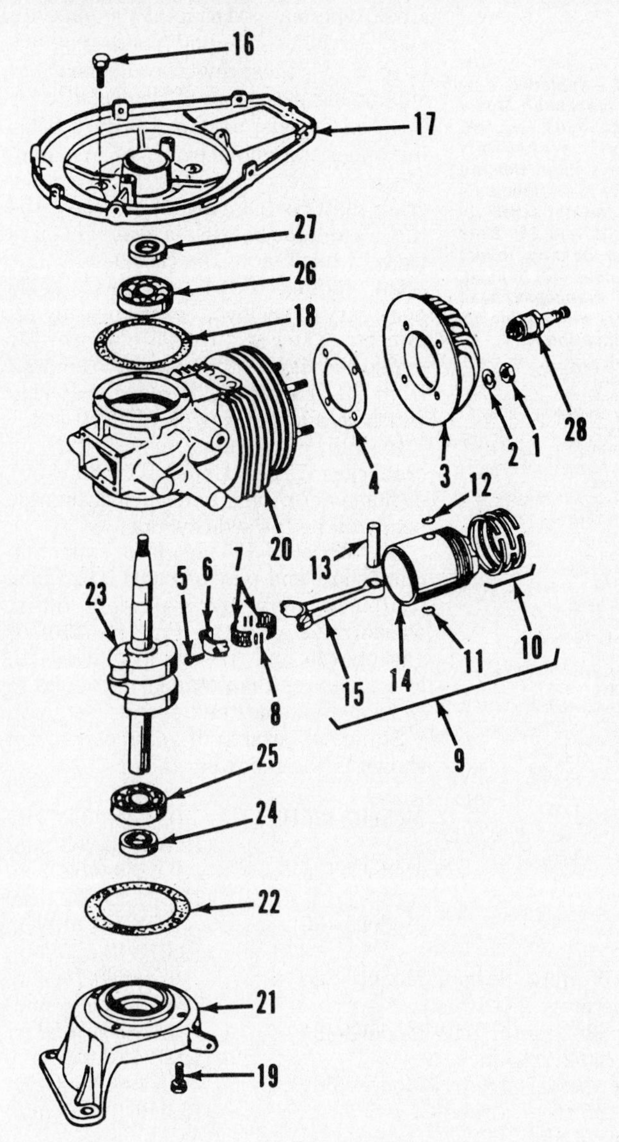

Fig. JAC24—Exploded view of Model J-225 engine.

3. Cylinder head
4. Gasket
6. Connecting rod cap
7. Bearing cage
8. Needle rollers
10. Piston rings
11. Snap ring
12. Snap ring
13. Piston pin
14. Piston
15. Connecting rod
17. Back plate
18. Gasket
21. Crankcase head
22. Gasket
23. Crankshaft
24. Crankshaft seal
25. Ball bearing
26. Ball bearing
27. Crankshaft seal

Piston ring end gap specifications are shown in the following table.

Model	End gap
J-125	0.005-0.010 in. (0.13-0.25 mm)
J-175 (un-pinned ring)	0.005-0.013 in. (0.13-0.33 mm)
J-175 (pinned ring)	0.052-0.060 in. (1.32-1.52 mm)
J-225, J-321, J-501	0.005-0.013 in. (0.13-0.33 mm)

Piston pin diameter for Model J-321 and J-501 engines is 0.4999-0.5001 inch (12.697-12.726 mm). Piston pin diameter for all other models is 0.49975-0.50025 in. (12.693-12.705 mm).

On Model J-225 engines (needle bearing in connecting rod) renew piston pin if scoring or excessive wear evident.

Piston pin-to-connecting rod clearances for all models without needle bearings are listed in the following table.

Model	Clearance
J-125	0.0005-0.0015 in. (0.013-0.038 mm)
J-175	0.00055-0.00175 in. (0.0140-0.0445 mm)
J-321, J-501	0.0007-0.0016 in. (0.018-0.041 mm)

On Models J-321 and J-501, piston pin should be 0.0003 inch (0.008 mm) tight to 0.0002 inch (0.005 mm) loose fit in pin bore of piston. On all other models, piston pin should be 0.00025 inch (0.006 mm) tight to 0.00055 inch (0.013 mm) loose fit in pin bore of piston.

Piston pins are available in several oversizes as well as standard size. If oversize pin is used, refer to the previous paragraphs for the appropriate pin to piston and rod bore specifications and ream the bores accordingly.

CONNECTING ROD. Connecting rod and piston unit can be removed from above after removing cylinder head and carburetor adapter.

Several types of connecting rods have been used according to engine model and application. Refer to appropriate paragraphs for model being serviced.

MODEL J-225. Model J-225 engine is equipped with a connecting rod which has needle roller crankpin and piston pin bearings. The crankpin needle rollers are separated by a split type cage and use crankpin and connecting rod surfaces as bearing races. The caged piston pin needle bearing is not renewable except by renewing connecting rod assembly.

Renew connecting rod is crankpin bearing surface is worn or scored and if any of the piston pin needle rollers have a flat spot or if two needles can be separated the width of one needle. Connecting rod crankpin bearing bore diameter is 0.9399-0.9403 inch (23.874-23.884 mm). Connecting rod side play on crankpin is 0.008-0.018 inch (0.20-0.46 mm).

MODELS J-321 AND J-501. Models J-321 and J-501 engines may be equipped with either a plain bearing bronze connecting rod (see All Other Models paragraphs) or an aluminum connecting rod with needle roller crankpin bearing and plain piston pin bearing. Model J-501 is equipped with an aluminum connecting rod.

The aluminum connecting rod has renewable steel bearing inserts for needle roller outer race. Twenty eight loose needle bearing rollers are used. Bearing guides (10–Fig. JAC25) are used to center bearing rollers on crankpin. Bearing rollers and guides may be held on crankpin with grease before installation of connecting rod.

Piston pin-to-connecting rod bearing bore clearance should be 0.0007-0.0016 inch (0.018-0.041 mm) on all models. On bronze connecting rod, crankpin-to-connecting rod clearance is 0.0035-0.0045 inch (0.09-0.11 mm) and connecting rod side play on crankpin should be 0.004-0.017 inch (0.10-0.43 mm).

On aluminum connecting rods, rod should have a crankpin bore diameter of 0.9819-0.9824 inch (24.940-24.953 mm). Connecting rod side play on crankpin should be 0.005-0.008 inch (0.13-0.46 mm).

ALL OTHER MODELS. A bronze connecting rod with plain crankpin and piston pin bearings is used in all models except J-225, J-501 and some J-321 engines.

Piston pin-to-connecting rod clearance should be 0.0005-0.0015 inch (0.013-0.038 mm) on Models J-125H and

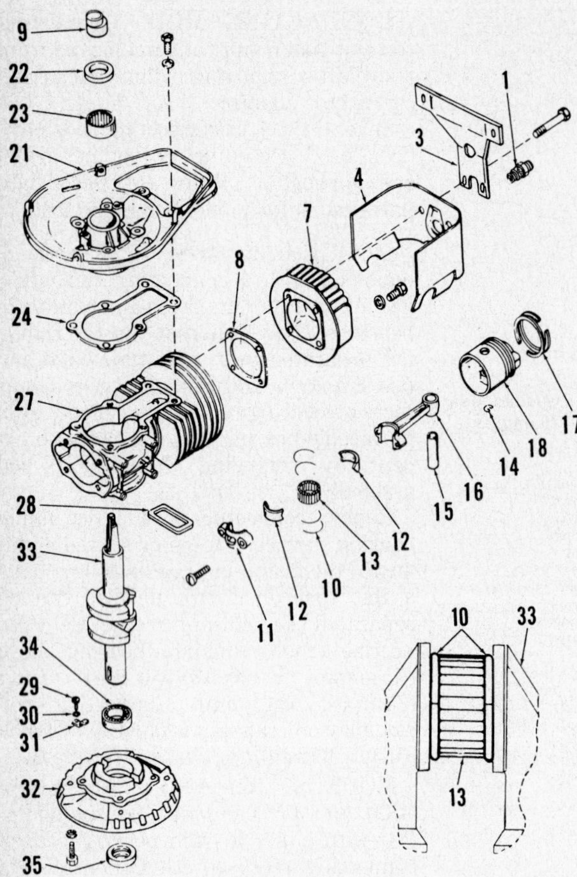

Fig. JAC25—Exploded view of vertical crankshaft Model J-321 and J-501 engine. Horizontal crankshaft models are similar. Bearing (34) on J-501 is retained by a snap ring and not screw retainer (29, 30 and 31). Note location of bearina rollers (13) and guides (10) in inset. Design of crankcase head (32) will vary with engine application.

1. Spark plug
3. Bracket
4. Air deflector
7. Cylinder head
8. Gasket
9. Breaker cam
10. Bearing roller guides
11. Rod cap
13. Bearing rollers (28)
14. Snap rings
15. Piston pin
16. Connecting rod
17. Piston rings
18. Piston
21. Back plate
22. Crankshaft seal
23. Needle bearing
24. Gasket
27. Cylinder & crankcase
28. Gasket
31. Bearing retainer
32. Crankcase head
33. Crankshaft
34. Ball bearing
35. Crankshaft seal

J-125V and 0.00055-0.00175 inch (0.0140-0.0445 mm) on Model J-175.

Crankpin-to-connecting rod bearing clearance should be 0.0015-0.0030 inch (0.038-0.076 mm) for Model J-125H, 0.0025-0.0040 inch (0.064-0.102 mm) for Model J-125V and 0.0035-0.0045 inch (0.090-0.114 mm) for Model J-175.

Connecting rod side play on crankpin for Models J-125H, J-125V and J-175 should be 0.004-0.017 inch (0.10-0.43 mm).

On all models, make certain match marks on connecting rod and cap are aligned as shown in Fig. JAC20 during installation. Bend lock tabs, if equipped, against cap screw heads. On Model J-321 and J-501 with aluminum connecting rod, steel bearing inserts must be installed correctly and screw heads staked as shown in Fig. JAC21.

CYLINDER. Cylinder bore should be resized if scored, out-of-round more than 0.0015 inch (0.038 mm) or worn more than 0.002 inch (0.05 mm). Standard cylinder bore diameter is 2.0010-2.0015 inch (50.825-50.838 mm) for Model J-125 engines, 2.1260-2.1265 inch (54.000-54.013 mm) for Model J-175 engines, 2.2510-2.2515 inch (57.175-57.188 mm) for Model J-225 engines and 2.1260-2.1265 inch (54.000-54.013 mm) for Models J-321 and J-501 engines.

Several oversize pistons as well as standard size are available. If reboring is not necessary, the manufacturer recommends deglazing the bore to aid in seating new rings.

CRANKSHAFT AND SEALS. On all models except Models J-175, J-321 and J-501, crankshaft is supported at each end by ball type main bearings which are pressed onto crankshaft. On Models J-175, crankshaft is supported by needle type main bearings and on Models J-321 and J-501, crankshaft is supported by a needle type main bearing at one end and a ball type main bearing at the opposite end. Model J-125 and J-175 engines have a third bearing (needle bearing) supporting outer end of crankshaft.

On all models, use an approved bearing puller to remove bearings. When installing new bearings, be sure crankshaft is supported between the throws and the bearing is heated in oil to prevent bending of the crankshaft.

On Model J-321 engines, it is necessary to remove the two bearing retainers (31–Fig. JAC25) before the crankcase head (32) can be removed from the crankshaft. Remove snap ring which retains bearing on Model J-501.

Install crankcase head (32) on crankcase (27) of Model J-321 or J-501 so flat spot on inner portion of crankcase head will be towards cylinder.

On Model J-175 engines, maintain crankshaft end play of 0.004-0.018 inch (0.10-0.46 mm) by renewing thrust washers (26 and 28–Fig. JAC23); or gaskets (25 and 37) are available with less thickness than standard gaskets to reduce crankshaft end play.

Standard crankpin diameters are shown in the following table.

Model J-125H	0.6240-0.6245 in. (15.850-15.862 mm)
Model J-125V	0.6230-0.6235 in. (15.824-15-837 mm)
Model J-175	0.7485-0.7490 in. (19.012-19.025 mm)
Model J-225	0.7500-0.7503 in. (19.050-19.058 mm)
Model J-321	0.7485-0.7490 in. (19.012-19.025 mm)
Model J-501	0.7496-0.7501 in. (19.040-19.053 mm)

On all models with direct drive, install crankcase seals so lip faces toward inside of engine. On models with speed reducer, the crankcase oil seal on the speed reducer side should be installed with lip facing the reducer.

REED VALVE. On all models the carburetor is mounted on an adapter plate. The plate carries a small spring steel leaf or reed, on the engine side, which acts as an inlet valve. The reed has a bend in it and must be installed so reed is forced firmly against the mounting plate. Blow-back through the carburetor may be caused by foreign matter holding reed valve open or by a damaged or improperly installed reed.

SERVICING JACOBSEN ACCESSORIES

REWIND STARTER

All models may be equipped with one of the rewind starters shown in Fig. JAC26, Fig. JAC27 or Fig. JAC28. Refer to the appropriate following paragraph and exploded view when servicing rewind starter.

To disassemble dog type starter shown in Fig. JAC26, remove starter from engine and pull starter rope out of starter approximately 12 inches (305 mm). Prevent pulley (3) from rewinding and pull slack rope back through rope outlet. Hold rope away from pulley and allow rope pulley to unwind. Unscrew retaining screw (11) and remove dog assembly and rope pulley being careful not to disturb rewind spring (2) in cover. Note direction of spring winding and carefully remove spring from cover.

To reassemble starter, reverse disassembly procedure. Rewind spring must be preloaded by installing rope through rope outlet of housing and hole in rope pulley, tie a knot at each end of rope and turn rope pulley approximately four turns in direction that places load on rewind spring. Hold pulley and pull remainder of rope through rope pulley. Attach handle to rope end and allow rope to rewind into starter. If spring is properly preloaded, rope will fully rewind.

Fig. JAC26—Exploded view of dog type rewind starter.

1. Cover
2. Rewind spring
3. Rope pulley
4. Dog
5. Washer
6. Brake
7. Spring
8. Retainer
9. Washer
10. Washer
11. Screw

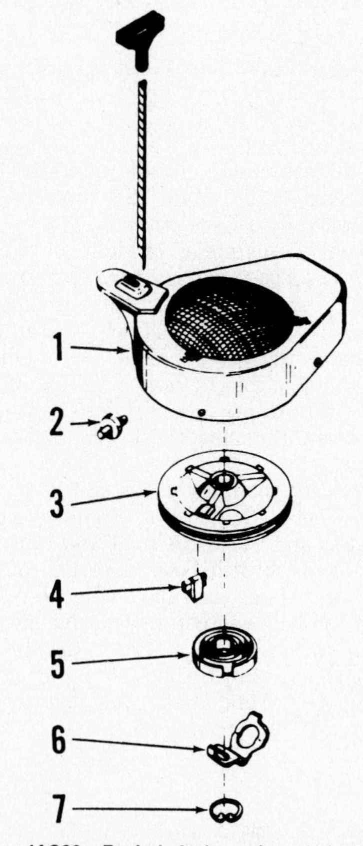

Fig. JAC28—Exploded view of rewind starter used on some later models.

1. Starter housing
2. Rope guide
3. Rope pulley
4. Pawl
5. Rewind spring & Cup
6. Pawl plate
7. Snap ring

To disassemble starter shown in Fig. JAC27, remove rope handle and allow rope to wind into starter. Remove roll pin (1) and washers (2 and 3). Remove rope pulley and remainder of starter components. When reassembling starter, turn rope pulley approximately 4 turns counterclockwise (viewed from open side) before connecting rope to rope pulley. Clip at rope end should have open side facing outward. After assembling starter, check operation and note if rope is fully rewound into starter.

Rewind starter shown in Fig. JAC28 is a pawl type starter with pawl (4) engaging a flywheel fin when the starter rope is pulled. To disassemble starter, remove rope handle and allow rope to rewind into starter. Remove snap ring (7) and remove starter components from starter housing. Inspect components for wear and damage and install in reverse order of disassembly. Spring cup (5) should be installed so spring attachment point on cup is 180° from rope entry hole in starter housing (1). Turn rope pulley approximately 3½ turns against spring tension before passing end of rope through hole in starter housing. Note that four embossed projections on starter housing adjacent to rope pulley are friction devices and should not be lubricated.

GEAR REDUCTION UNIT

Before the housing can be removed from the engine, the reduction gear cover and gear must be removed. Crankcase oil seal on the speed reducer side should be installed with the lip facing the reducer. On the power takeoff side of the reducer unit, the oil seal lip faces the reduction unit. Refer to Fig. JAC29.

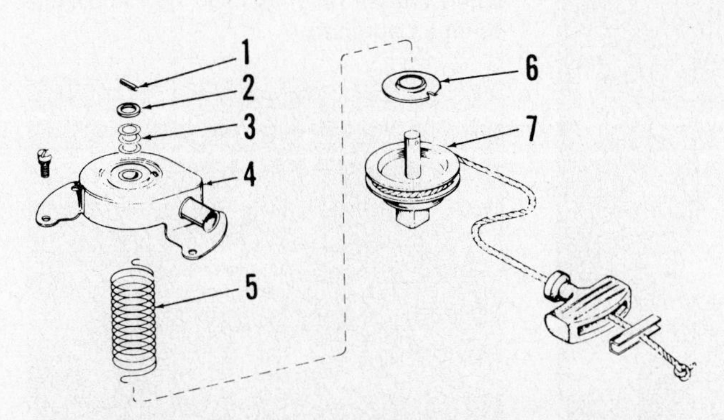

Fig. JAC27—Exploded view of rewind starter used on some engines. Some models use a snap ring in place of pin (1).

1. Pin
2. Washer
3. Washer
4. Cover
5. Spring
6. Washer
7. Rope pulley

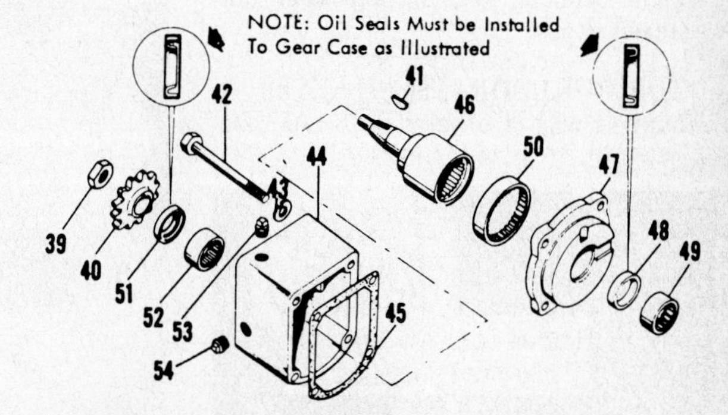

Fig. JAC29—Exploded view of reduction gear assembly used on horizontal crankshaft engines.

40. Drive sprocket
41. Woodruff key
44. Housing
45. Gasket
46. Drive gear
47. Crankcase head
48. Crankshaft seal
49. Needle bearing
50. Needle bearing
51. Oil seal
52. Needle bearing
53. Oil filler plug
54. Oil level plug

KAWASAKI

KAWASAKI MOTORS CORP. USA
P.O. Box 504
Shakopee, Minnesota 55379

Model	Bore	Stroke	Displacement
KT43	54 mm (2.13 in.)	48 mm (1.89 in.)	110 cc (6.7 cu. in.)

ENGINE INFORMATION

KT43 models are two-stroke, air-cooled engines. Cylinder head and cylinder bore are bolted to an aluminum die-cast crankcase. The three-piece crankshaft is supported at each end by ball bearing type main bearings.

Model KT43 is rated at 3.2 kW (4.3 hp) at 5000 rpm.

Engine model identification decal is located on cooling shroud and serial number is stamped in crankcase (Fig. KW1). Always furnish engine model and serial number when ordering parts or service material.

MAINTENANCE

SPARK PLUG. Recommended spark plug is NGK BM7, or equivalent.

Spark plug should be removed and cleaned and electrode gap set at 0.6-0.7 mm (0.024-0.027 in.) after every 25 hours of operation. Renew spark plug if electrode is severely burnt or damaged. Tighten spark plug to 27 N·m (20 ft.-lbs.).

NOTE: Caution should be exercised if abrasive type spark plug cleaner is used. Inadequate cleaning procedure may allow the abrasive cleaner to be deposited in engine cylinder accelerating wear and part failures.

CARBURETOR. Model KT43 is equipped with a Mikuni float type carburetor equipped with idle mixture adjustment needle. Main fuel mixture is controlled by a fixed jet.

Initial adjustment of idle mixture screw (2 – Fig. KW2) is 1 turn open from a lightly seated position. Make final adjustment with engine at operating temperature and running. Set engine idle at 1400 rpm by adjusting idle stop screw (1). Adjust idle mixture screw (2) to obtain the smoothest idle and acceleration.

Standard main jet size is #82.5.

To check float level, remove carburetor and separate float bowl from carburetor throttle body. Invert throttle body and float assembly (Fig. KW3). Float should contact float valve lever when bottom edge of float is just level with edge of throttle body. Float lever is spring-loaded and float should just touch lever, but not compress fuel inlet valve spring.

GOVERNOR. Model KT43 is equipped with a flyweight type governor. To adjust external linkage, stop engine. Make certain all linkage is in good condition and tension spring (3 – Fig. KW4) is not stretched. Spring (2) around governor-to-carburetor rod (1) must pull governor lever (4) and throttle arm toward each other. Loosen clamp bolt (5) and push governor lever until throttle is in full open position. Hold lever in this position and use a screwdriver in governor shaft slot to rotate governor shaft (6) clockwise until it stops (range is very slight). Tighten governor clamp bolt (5).

To set minimum speed, engine should be at operating temperature and running, no-load. Close throttle valve by hand until it hits throttle stop screw (1 – Fig. KW2). Adjust throttle stop screw so engine is running at 1400 rpm.

To set maximum speed, engine should be at operating temperature and running, no-load. Attach a tachometer to engine. Slowly move throttle lever increasing engine speed while observing tachometer.

CAUTION: At no time, even during adjustment, should engine be allowed to run above 5000 rpm. Stop throttle movement when engine reaches 4950 rpm and continue as outlined.

Fig. KW1—View showing location of engine serial number.

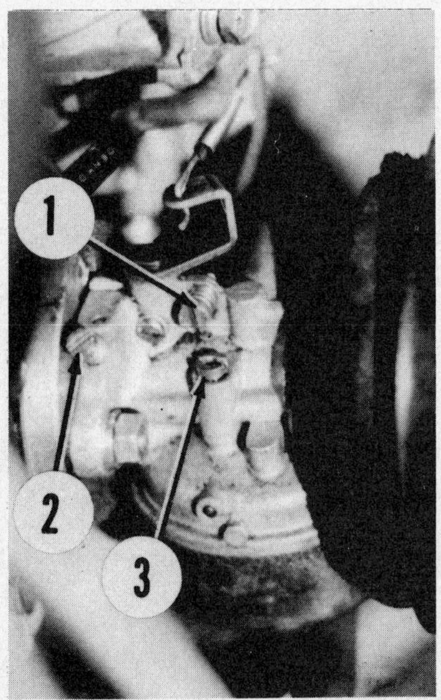

Fig. KW2—Model KT43 engines are equipped with Mikuni float type carburetor.

1. Throttle stop screw
2. Idle mixture screw
3. Pilot air jet

Fig. KW3—Carburetor float must be level with lower edge of carburetor throttle body without compressing fuel inlet needle spring.

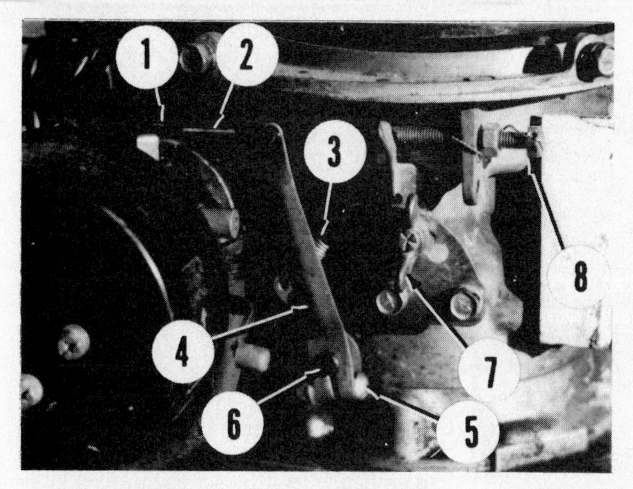

Fig. KW4—View of external governor linkage.
1. Carburetor-to-governor rod
2. Spring
3. Tension spring
4. Governor lever
5. Clamp bolt
6. Governor shaft
7. Throttle & lock
8. Maximum speed limit bolt

Fig. KW8—Loosen or tighten cylinder head bolts in sequence shown. Refer to text for torque specifications.

Adjust maximum speed limit screw (8–Fig. KW4) to obtain 4850-5000 rpm and lock it securely in place with jam nut.

IGNITION SYSTEM. A breaker point ignition system is standard. Ignition coil, breaker points and condenser are located behind flywheel (Fig. KW5). Flywheel must be removed to perform service or adjustment. Breaker point gap should be 0.3-0.5 mm (0.012-0.020 in.).

To check timing, first remove flywheel, then gently place flywheel on crankshaft. Rotate engine so "P" mark on flywheel is aligned with timing mark on crankcase (Fig. KW7). Very carefully remove flywheel without rotating engine. Breaker points should just begin to open. Loosen breaker plate screw and move points if necessary. This sets ignition timing at 25° BTDC. When the "T" mark on flywheel is aligned with timing mark, piston is at top dead center.

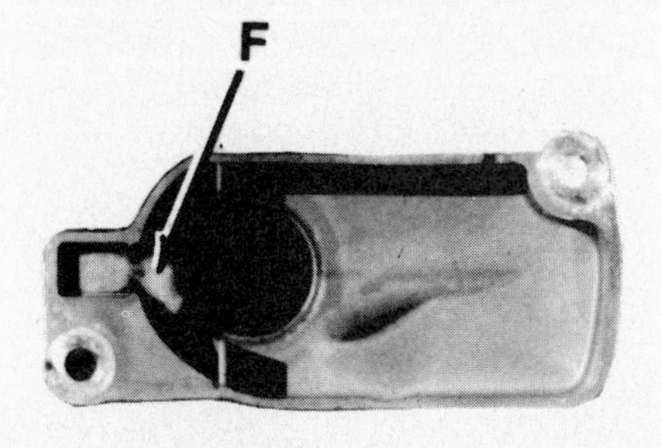

Fig. KW5—View of ignition coil (1), breaker points (2) and crankshaft (4). Condenser is not shown but is attached to wire leads (3).

Fig. KW6—Lubricating felt (F) in point cover should receive a drop of oil whenever points are renewed.

CYLINDER HEAD AND COMBUSTION CHAMBER. Cylinder head, combustion chamber and piston should be cleaned and carbon and other deposits removed after every 300 hours of operation. Refer to REPAIR section of this manual for service procedure.

LUBRICATION. Manufacturer recommends mixing a good quality two-stroke, air-cooled engine oil with regular grade gasoline at a 25:1 ratio. Always mix fuel in a separate container and add only mixed fuel to engine fuel tank.

GENERAL MAINTENANCE. Check and tighten all loose bolts, nuts or clamps prior to each day of operation. Check for fuel or oil leakage and repair if necessary.

Clean dust, dirt, grease or any foreign material from cylinder head and cylinder block cooling fins after every 100 hours of operation. Inspect fins for damage and repair if necessary.

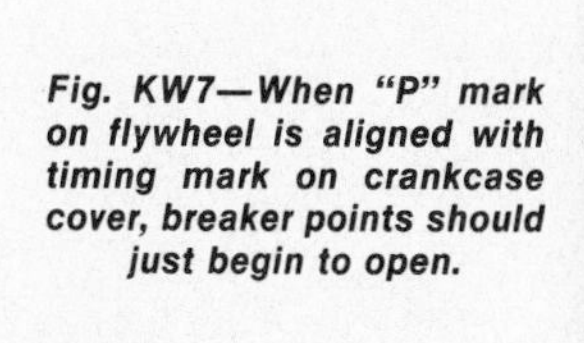

Fig. KW7—When "P" mark on flywheel is aligned with timing mark on crankcase cover, breaker points should just begin to open.

REPAIRS

TIGHTENING TORQUES. Recommended tightening torques are as follows:

Spark plug . . . . 27 N·m (20 ft.-lbs.)
Flywheel nut . . . . 39-44 N·m (30-33 ft.-lbs.)
Cylinder head . . . . 20-22 N·m (15-16 ft.-lbs.)
Crankcase . . . . 7-9 N·m (5-6 ft.-lbs.)

CYLINDER HEAD. To remove cylinder head, disconnect spark plug. Remove rewind starter and cooling shroud assembly. Unbolt cylinder head retaining bolts in sequence shown in Fig. KW8.

Reinstall by reversing removal procedure and tighten head retaining bolts to specified torque in sequence shown in Fig. KW8.

CONNECTING ROD. Connecting rod is forged steel with needle bearings in piston pin bore and crankpin bearing bore. To remove connecting rod, remove cylinder head, cylinder and flywheel. Remove piston pin retainers and piston pin. Separate piston from connecting rod. Use care not to lose needle bearings in connecting rod piston pin bore. Loosen upper crankcase cover bolts in sequence shown in fig. KW9. Loosen lower crankcase cover bolts in a criss-cross pattern and separate crankcase from crankcase covers. Remove connecting rod and crankshaft assembly.

CAUTION: Do not separate crankshaft from connecting rod. Crankshaft and connecting rod are available only as an assembly. Kawasaki does not recommend crankshaft disassembly.

If connecting rod side play at crankpin bearing is 0.5 mm (0.02 in.) or more, or if radial play at piston pin bearing or crankpin bearing is 0.05 mm (0.002 in.) or more, renew crankshaft and connecting rod assembly.

Use light grease to retain needle bearings in their bore during reassembly. Tighten crankcase cover bolts evenly, in graduated steps, to specified torque in sequence shown in Fig. KW9 for upper crankcase bolts. Use a criss-cross pattern when tightening lower crankcase cover bolts.

PISTON, PIN AND RINGS. Piston may be removed after removing cylinder head and cylinder. Refer to CONNECTING ROD paragraphs.

Check clearance between piston and cylinder. If clearance is 0.2 mm (0.008 in.) or more, renew piston and/or cylinder. Maximum side clearance for new ring in ring groove is 0.15 mm (0.006 in.). Maximum ring end gap is 1.0 mm (0.04 in.). Maximum clearance between piston pin and piston is 0.03 mm (0.001 in.). Renew piston if specifications are not as specified.

Fig. KW9—Upper crankcase bolts must be tightened evenly, in graduated steps, in sequence shown. Refer to text for torque specifications.

When installing piston rings, marked side of ring is towards top of piston. Align ring end gaps with pins in piston ring grooves.

Piston should be installed on connecting rod with the arrow mark on piston top toward the exhaust port side of engine. Lubricate and install piston pin and retaining rings. Use a suitable ring compressor to avoid damaging rings and very carefully install cylinder over piston and ring assembly.

CYLINDER. The cylinder is a low pressure aluminum alloy with iron sleeve. To remove cylinder, remove cylinder head and carefully work cylinder up off of piston.

Standard cylinder bore diameter is 54 mm (2.126 in.) with a service limit of 0.1 mm (0.004 in.). If cylinder bore wear exceeds service limit, renew cylinder.

Refer to PISTON, PIN AND RINGS section for piston-to-cylinder clearance and reassembly procedure.

CRANKSHAFT, MAIN BEARINGS AND SEALS. Refer to CONNECTING ROD section for crankshaft removal. Note that Kawasaki does not recommend separating connecting rod from crankshaft and that crankshaft and connecting rod are available as an assembly only. Ball bearing type main bearings are a slight press fit in crankcase covers. It may be necessary to slightly heat crankcase covers for removal and installation. Maximum axial play for main bearings is 0.15 mm (0.006 in.). Renew main bearings if loose, rough or a loose fit on crankshaft.

GOVERNOR. Refer to MAINTENANCE section for adjustment procedure of external governor linkage.

To remove governor, remove crankcase cover, loosening bolts in sequence shown in Fig. KW9. Separate cover from crankcase. Slide governor holder off the crankshaft and remove the pin which will allow governor sleeve and thrust washer removal.

Check governor tip contact surface and governor weight contact surface of the thrust washer for wear and roughness.

Remove governor lever from external end of governor shaft and governor tip from internal end of governor shaft. Pull governor shaft from crankcase bore. Check governor sleeve contact surface of governor tip for wear. Check governor shaft for wear where it is supported by crankcase bushing bore.

When reassembling governor, lubricate all parts with engine oil and make certain pin is correctly positioned in notch of governor sleeve. Tighten lower crankcase cover bolts equally in sequence shown in Fig. KW9 to specified torque.

KAWASAKI

Model	Bore	Stroke	Displacement
FA76D	52 mm (2.05 in.)	36 mm (1.42 in.)	76 cc (4.8 cu. in.)
FA130D	62 mm (2.44 in.)	43 mm (1.69 in.)	129 cc (7.9 cu. in.)
FA210D	72 mm (2.83 in.)	51 mm (2.01 in.)	207 cc (12.7 cu. in.)

ENGINE IDENTIFICATION

Kawasaki FA series engines are four-stroke, single-cylinder, air-cooled horizontal crankshaft engines. Model FA76D develops 1.3 kW (1.7 hp) at 4000 rpm, Model FA130D develops 2.3 kW (3.1 hp) at 4000 rpm and Model FA210D developes 3.9 kW (5.2 hp) at 4000 rpm.

Engine model number is located on the cooling shroud just above the rewind starter. Engine model number and serial number are both stamped on crankcase cover (Fig. KW20). Always furnish engine model and serial numbers when ordering parts or service information.

MAINTENANCE

SPARK PLUG. Recommended spark plug for all models is a NGK BM-6A, or equivalent.

Spark plug should be removed and cleaned and electrode gap set at 0.6-0.7 mm (0.024-0.027 in.) after every 100 hours of operation. Renew spark plug if electrode is severely burnt or damaged. Tighten spark plug to 27 N·m (20 ft.-lbs.).

NOTE: Caution should be exercised if abrasive type spark plug cleaner is used. Inadequate cleaning procedure may allow the abrasive cleaner to be deposited in engine cylinder accelerating wear and part failures.

CARBURETOR. Engines may be equipped with either a pulse type (Fig. KW21) or a float type (Fig. KW22) carburetor. Engine idle speed on all models should first be adjusted to 1500 rpm using idle stop screw (IS–Fig. KW21 or KW22) on carburetor, then, adjust idle limit screw (IL–Fig. KW25) on throttle plate assembly so engine idles at 1600 rpm. Adjust maximum speed limit screw (ML) so maximum engine speed is 4000 rpm.

CAUTION: Maximum engine speed must not exceed 4000 rpm even during adjustment procedure. Running engine at speeds in excess of 4000 rpm may result in engine damage and possible injury to operator.

For initial carburetor adjustment of pulse type carburetor, lightly seat pilot air screw (P–Fig. KW21) and then open pilot screw 1¼ turns on FA130D and FA210D models, or 1 turn on FA76D models. Make final adjustments with engine at operating temperature and running. Back out idle limit screw (IL–Fig. KW25) so it does not limit idle speed, then, adjust idle stop screw (IS–Fig. KW21) on carburetor so engines runs at lowest speed possible. Adjust pilot air screw (P) to obtain highest engine idle speed. Readjust idle stop screw (IS) and idle limit screw (IL–Fig. KW25) for correct speeds.

For initial carburetor adjustment of float type carburetor, lightly seat pilot

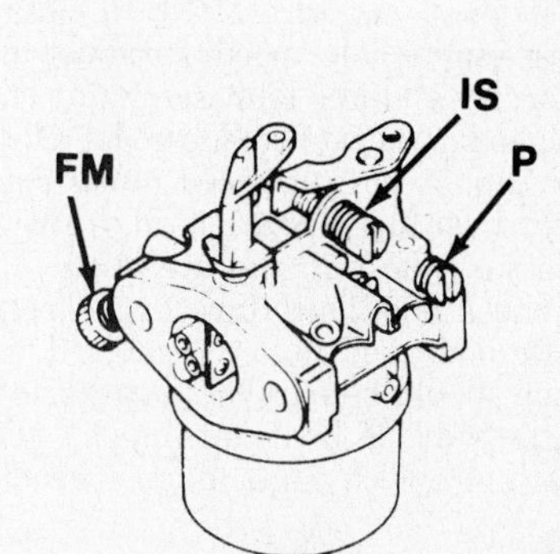

Fig. KW22—View of float type carburetor used on some models. Fuel mixture needle (FM) has a slotted head on Models FA76D and FA210D as shown. Model FA130D has a larger wheel type knob.

P. Pilot air screw
IS. Throttle stop screw
FM. Main fuel mixture screw

Fig. KW20—View showing location of engine model and serial number stamped on engine crankcase cover.

Fig. KW21—View of pulse type carburetor used on some models showing location of throttle stop screw (IS) and pilot air screw (P). Refer to text for adjustment procedure.

air screw (P–Fig. KW22) and fuel mixture needle (FM). Open pilot air screw (P) 1½ turns on FA130D and FA210D models and 1¼ turns on FA76D models. Open fuel mixture screw (FM) 1½ turns on all models. Make final adjustments with engine at operating temperature and running. Back out idle limit screw (IL–Fig. KW25) so it does not limit idle speed, then, adjust idle stop screw (IS–Fig. KW22) on carburetor so engine runs at lowest speed possible. Adjust pilot air screw (P) to obtain highest engine idle speed. Readjust idle stop screw and idle limit screw for correct speeds. Operate engine under a load at maximum engine speed (4000 rpm) and readjust fuel mixture needle (FM) to obtain smoothest engine operation.

To check float level, invert carburetor throttle body and float assembly. Float surface should be parallel to carburetor throttle body. If float is equipped with metal tang which contacts inlet needle,

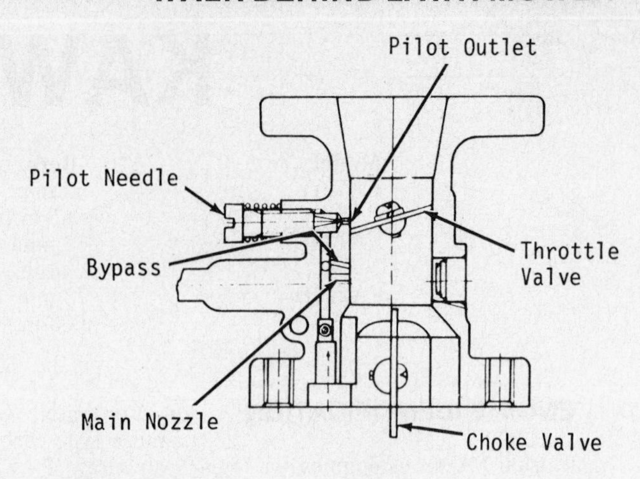

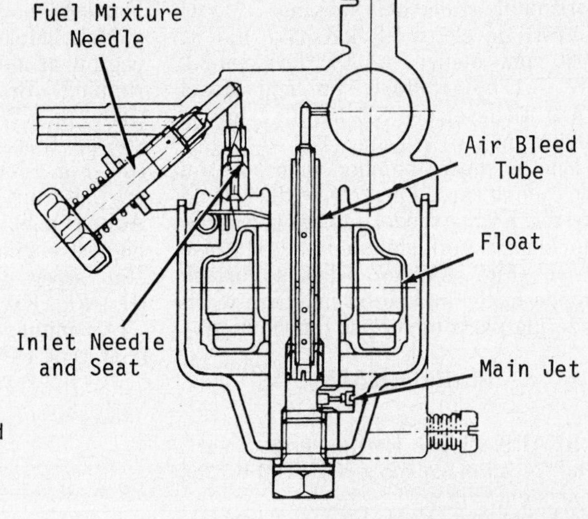

Fig. KW24—Cross-sectional view of float type carburetor used on Model FA130D. Carburetor on Models FA76D and FA210D is similar.

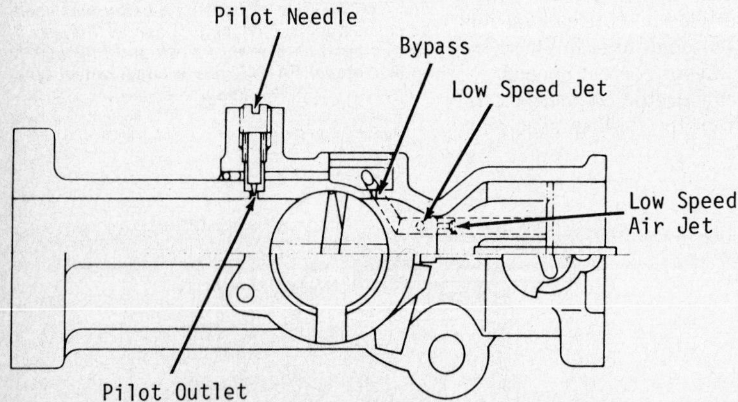

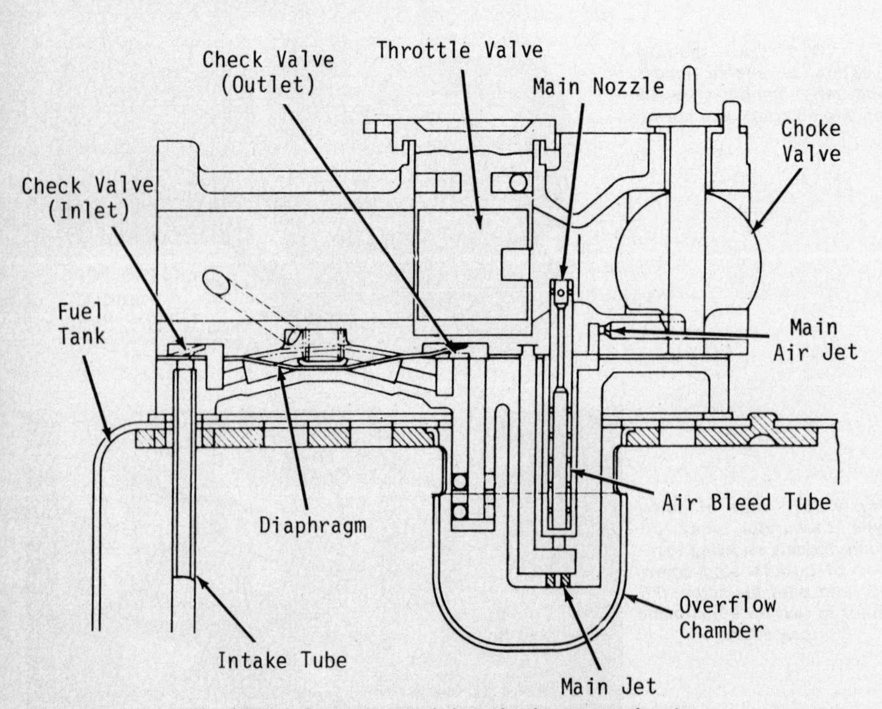

Fig. KW23—Cross-sectional view of pulse type carburetor.

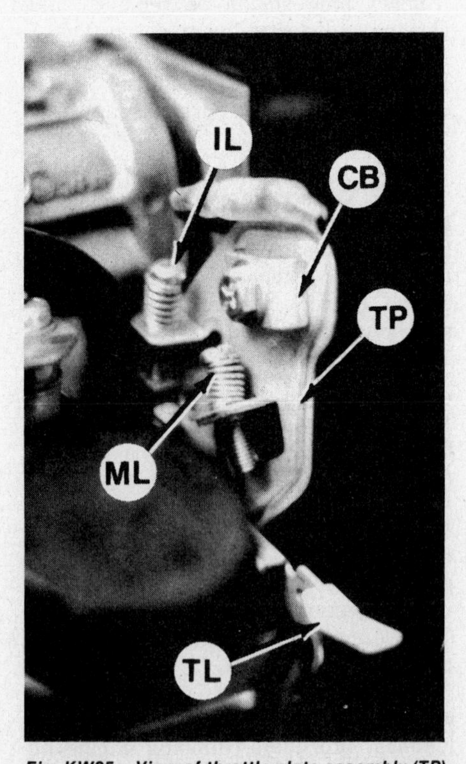

Fig. KW25—View of throttle plate assembly (TP) showing idle limit screw (IL) and maximum speed limit screw (ML). Throttle lever (TL) and remote throttle cable clamp (CB) are also shown.

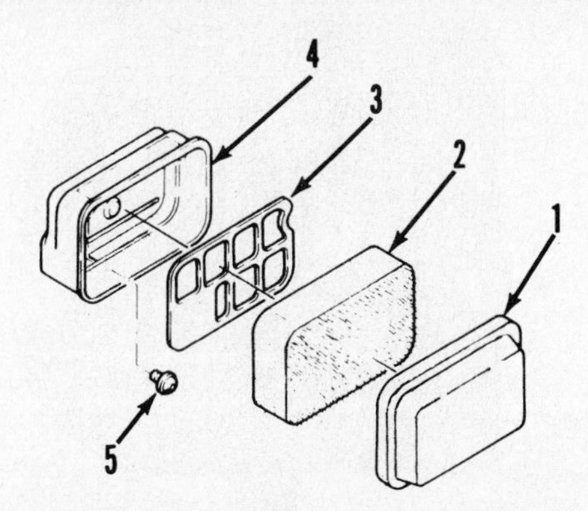

Fig. KW26—Exploded view of foam type air cleaner. Element (2) is installed with filament (hair like protrusions) away from carburetor.

1. Cover
2. Element
3. Plate
4. Element case
5. Screw

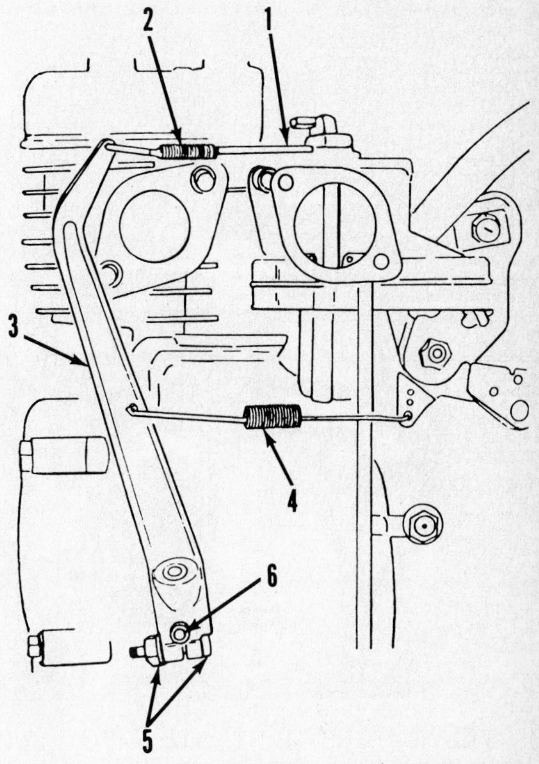

Fig. KW28—View of external governor linkage. Refer to text for adjustment procedure.

1. Governor-to-carburetor rod
2. Spring
3. Governor lever
4. Tension spring
5. Clamp bolt & nut
6. Governor shaft

carefully bend tang to adjust float. If float is a plastic assembly, float level is nonadjustable and float must be renewed if float level is incorrect.

AIR FILTER. Engine may be equipped with either a foam type (Fig. KW26) or a paper (dry) type (Fig. KW27) air cleaner. Refer to appropriate paragraph for model being serviced.

Foam Type Air Cleaner. Foam type air cleaner (Fig. KW26) should be removed and cleaned after every 25 hours of operation under normal operating conditions. Remove foam element and clean in a suitable nonflammable solvent. Carefully squeeze out solvent and allow element to air dry. Soak element in clean engine oil and squeeze out the excess. Reinstall element making certain filament (hair like protrusions) side is toward cover (Fig. KW26).

Paper (Dry) Type Air Cleaner. Paper type air cleaner (Fig. KW27) should be removed and cleaned after every 100 hours of operation under normal operating conditions. Remove foam and paper element, separate foam precleaner from paper element, wash element and foam precleaner in water and detergent and rinse in clean water. Allow to air dry. Soak foam precleaner in clean engine oil and carefully squeeze out excess oil. Slide precleaner over paper element and reinstall elements in air cleaner case.

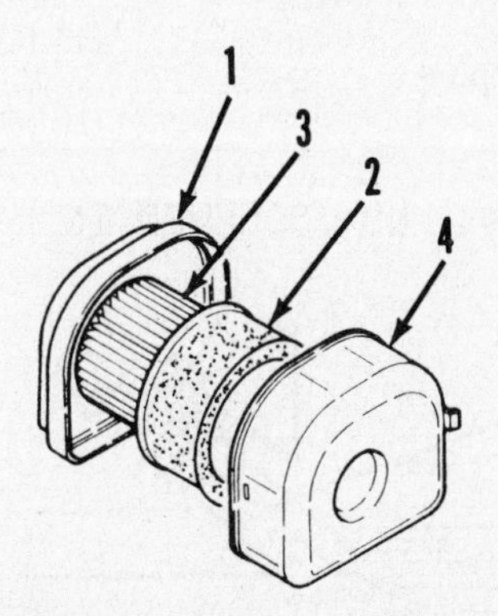

Fig. KW27—Exploded view of paper (dry) type air cleaner.

1. Cover
2. Precleaner (foam)
3. Paper element (filter)
4. Element case

GOVERNOR. A gear driven flyweight governor assembly is located inside engine crankcase on crankcase cover. To adjust external linkage, place engine throttle control in idle position. Make certain all linkage is in good condition and tension spring (4–Fig. KW28) is not stretched. Spring (2–Fig. KW29) around governor-to-carburetor rod must pull governor lever (3) and throttle pivot (4) toward each other. Loosen clamp bolt (5–Fig. KW28) and rotate governor shaft (6) clockwise as far as possible. Hold shaft in this position and pull governor lever (3) as far right as possible. Tighten clamp bolt (5). Install tension spring (4) in correct hole for speed desired.

IGNITION SYSTEM. Ignition coil (1–Fig. KW30) is located just above flywheel and breaker point set and condenser are located behind flywheel. Ignition timing should be set at 23° BTDC in the following manner. Remove flywheel and then reinstall it loosely on crankshaft. Rotate flywheel until edge of coil laminations are aligned with edge of notch in flywheel as shown in Fig. KW31. Remove flywheel making certain crankshaft does not turn. Slowly move breaker point plate assembly until points just begin to open. Tighten breaker point plate screw. Turn crankshaft clockwise until point gap is at widest open position and adjust gap to 0.3-0.5 mm (0.012-0.020 in.). Place a small amount of point cam lube on lubricating felt (4–Fig. FW30) and reinstall flywheel. Coil edge gap should be 0.05 mm (0.020 inch).

Fig. KW29—Spring (2) around governor-to-carburetor rod (1) must pull governor lever (3) toward throttle pivot (4).

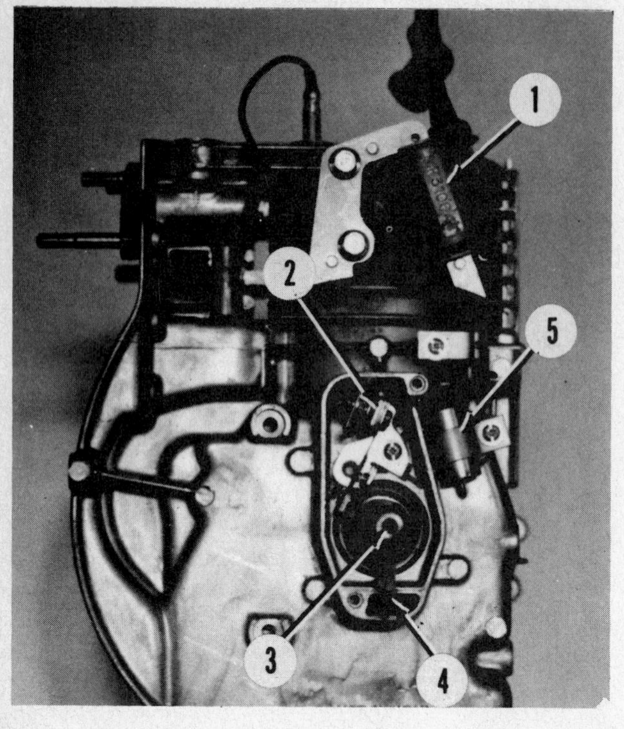

Fig. KW30—View showing location of breaker point ignition system components.
1. Ignition coil
2. Breaker points
3. Crankshaft
4. Lubricating felt
5. Condenser

Fig. KW33—On all models, loosen or tighten cylinder head bolts using the sequence shown.

VALVE ADJUSTMENT. Valves and seats should be refaced and stem clearance adjusted after every 300 hours of operation. Refer to REPAIRS section for service procedure and specifications.

CYLINDER HEAD AND COMBUSTION CHAMBER. Cylinder head, combustion chamber and piston should be cleaned and carbon and other deposits removed after every 300 hours of operation. Refer to REPAIRS section for service procedure.

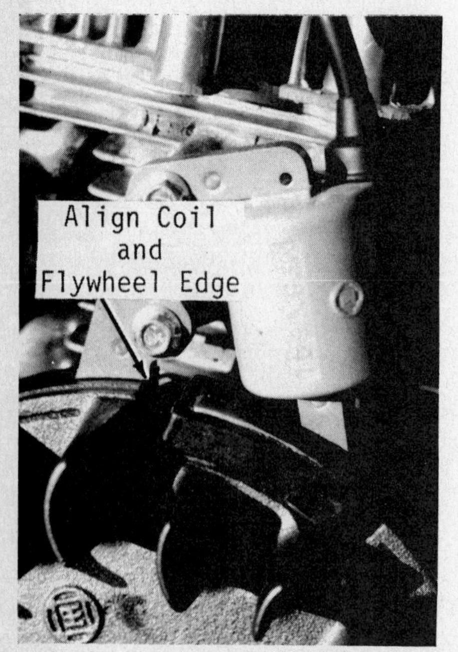

Fig. KW31—Align edge of flywheel notch with coil core laminations as shown to adjust timing. Refer to text.

LUBRICATION. All models are splash lubricated. Engine oil level should be checked prior to each operating interval. Oil level should be maintained between reference marks on gage with oil fill plug just touching first threads. Do not screw oil fill plug in to check oil level (Fig. KW32).

Manufacturer recommends oil with an API service classification SC, SD or SE. Use SAE 20 oil in winter and SAE 30 oil in summer.

Oil should be changed after the first 20 hours of operation and every 100 hours of operation thereafter. Crankcase capacity is 320 cc (0.68 pt.) on FA76D models, 500 cc (1.06 pt.) on FA130D models and 600 cc (1.27 pt.) on FA210D models.

GENERAL MAINTENANCE. Check and tighten all loose bolts, nuts or clamps prior to each day of operation. Check for fuel or oil leakage and repair if necessary.

Clean dirt, dust, grease or any foreign materal from cylinder head and cylinder block cooling fins after every 100 hours of operation. Inspect fins for damage and repair if necessary.

REPAIRS

TIGHTENING TORQUES. Recommended tightening torques are as follows:

Spark plug .... 16 N·m (12 ft.-lbs.)
Head bolts:
FA76D .... 7 N·m (5 ft.-lbs.)
All other models .... 19 N·m (14 ft.-lbs.)
Connecting rod bolts:
FA76D .... 7 N·m) (5 ft.-lbs.)
All other models .... 11 N·m (8 ft.-lbs.)
Crankcase cover bolts .... 5 N·m (4 ft.-lbs.)
Flywheel nut:
FA76D .... 34 N·m (25 ft.-lbs.)
All other models .... 60 N·m (44 ft.-lbs.)
Drain plug .... 13 N·m (10 ft.-lbs.)

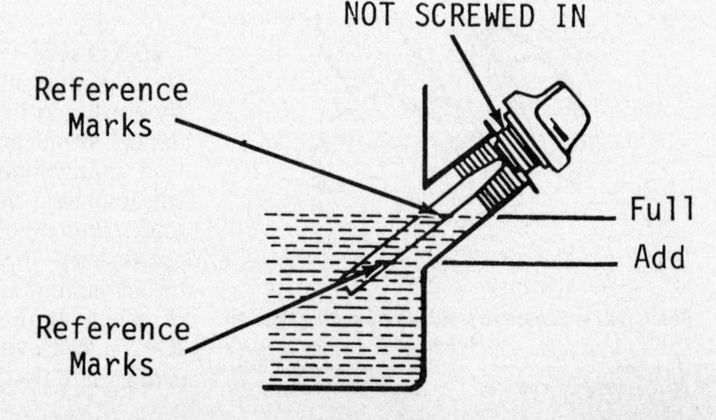

Fig. KW32—View showing oil fill plug and oil gage. Plug is not screwed in, but is just touching threads when checking oil level.

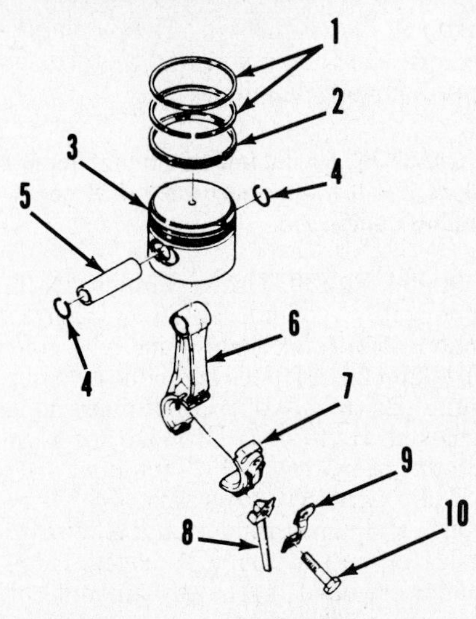

Fig. KW34—Exploded view of piston and connecting rod assembly.

1. Compression rings
2. Oil control ring
3. Piston
4. Retaining rings
5. Pin
6. Connecting rod
7. Connecting rod cap
8. Oil dipper
9. Lock plate
10. Bolts (2)

CYLINDER HEAD. To remove cylinder head, first remove cylinder head shroud. Clean engine to prevent entrance of foreign material. Loosen cylinder bolts in 1/4 turn increments in sequence shown in Fig. KW33 until all bolts are loose enough to remove by hand.

Remove spark plug and clean carbon and other deposits from cylinder head. Place cylinder head on a flat surface and check entire sealing surface for warpage. If warpage exceeds 0.25 mm (0.010 in.), cylinder head must be renewed. Slight warpage may be repaired by lapping cylinder head. In a figure eight pattern, lap head against 200 grit and then 400 grit emery paper on a flat surface.

Reinstall cylinder head and tighten bolts evenly to specified torque in sequence shown in Fig. KW33.

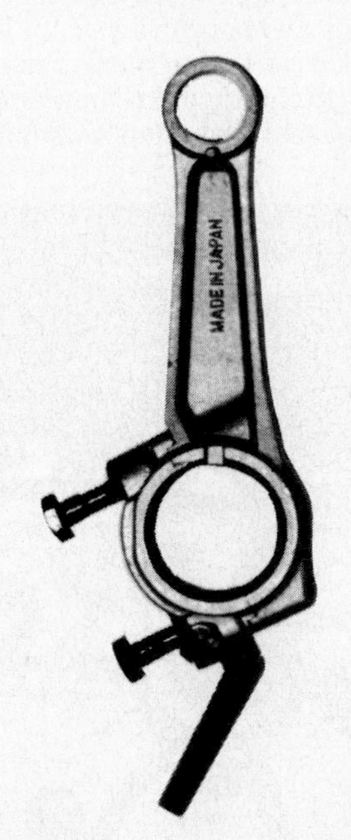

Fig. KW35—Connecting rod is assembled on piston so the side marked "MADE IN JAPAN" is toward "M" stamped on piston pin boss (Fig. KW36).

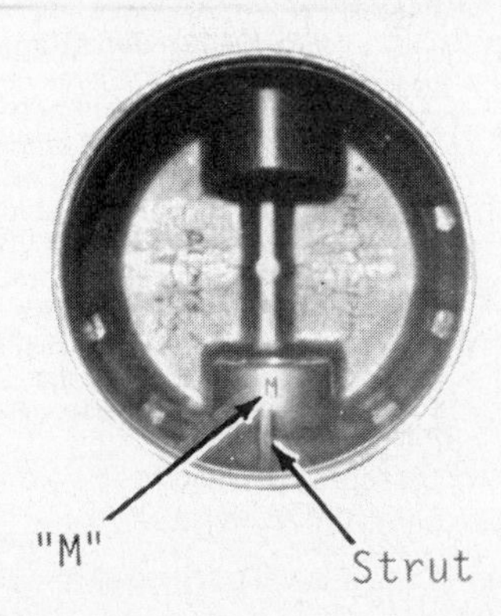

Fig. KW36—Piston pin boss is stamped with an "M". Note "M" is on boss with the strengthening strut.

CONNECTING ROD. Connecting rod rides directly on crankshaft journal on all models. Piston and connecting rod are removed as an assembly after cylinder head and crankcase cover have been removed. Bend lock plate up to release connecting rod bolts and remove the two bolts, lock plate, oil dipper and connecting rod cap. Rotate crankshaft so piston can be pushed out of cylinder and remove piston and rod assembly. Carefully remove piston rings (1 and 2–Fig. KW34). Remove the two wire retaining rings (4) and push pin (5) out to separate piston (3) from connecting rod (6).

Fig. KW37—On some models, piston top has a mark which must be toward flywheel side of engine after installation.

Inspect connecting rod and renew if piston pin bore or crankpin bearing bore are scored or damaged. If clearance between piston pin and connecting rod piston pin bore exceeds 0.05 mm (0.002 in.), renew connecting rod.

If clearance between crankpin journal and connecting rod bearing bore exceeds 0.07 mm (0.003 in.) on FA76D models or 0.10 mm (0.004 in.) on all other models, renew connecting rod and/or crankshaft. Correct connecting rod side play on crankpin should be 0.70 mm (0.028 in.).

Install piston on connecting rod so the side of connecting rod which is marked "MADE IN JAPAN" (Fig. KW35) is toward "M" stamped on piston pin boss (Fig. KW36). Secure piston pin with wire retaining rings. Install piston and connecting rod assembly so side of connecting rod marked "MADE IN JAPAN" is toward flywheel side of engine. Lubricate crankpin journal and install connecting rod cap and oil dipper as shown in Fig. KW38. Install new lock plate and bolts. Tighten bolts to specified torque and lock by bending lock plate.

PISTON, PIN AND RINGS. Piston and connecting rod are removed as an assembly. Refer to CONNECTING ROD section for removal and installation procedure.

After carefully removing piston rings and separating piston and connecting rod, carefully clean carbon and other deposits from piston surface and ring lands.

CAUTION: Extreme care should be exercised when cleaning ring lands. Do not damage squared edges or widen ring grooves. If ring lands are damaged, piston must be renewed.

Fig. KW38—Install connecting rod and piston assembly with oil dipper in position shown.

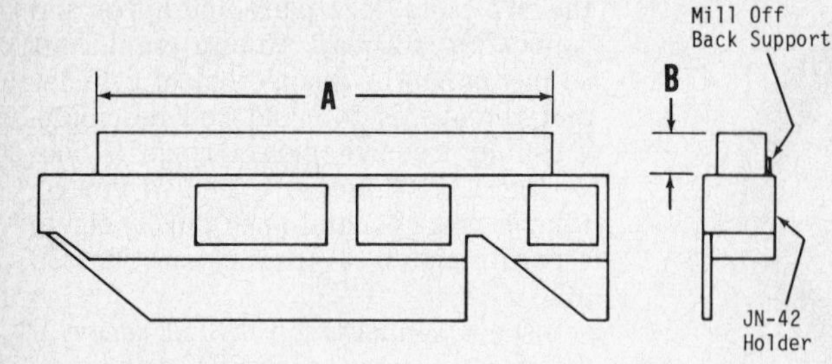

Fig. KW39—Felt strips must be altered for each model. Dimension A is 51.003 mm (2.80 in.) for FA76 models or 76.251 mm (3.20 in.) for FA130 models. Modification of felt strip length (A) for FA210 models is not required. Dimension B is 0.205 inch (0.18 mm) for FA76 and FA130 models or 0.305 inch (0.889 mm) for FA210 models.

Before installing rings on piston, install piston in cylinder and use a suitable feeler gage to measure clearance between piston and cylinder bore. If clearance exceeds 0.20 mm (0.008 in.) on FA76D models or 0.25 mm (0.010 in.) on all other models, renew piston and/or renew cylinder bore.

If ring side clearance between new ring and piston groove exceeds 0.15 mm (0.006 in.), renew piston. If ring end gap on new rings exceeds 0.10 mm (0.004 in.) when installed, renew cylinder. If clearance between piston pin and piston pin bore exceeds 0.05 mm (0.002 in.), renew piston.

Refer to CONNECTING ROD section to properly assemble piston to rod. Piston is marked "M" on piston pin boss (Fig. KW36) and on top of piston (Fig. KW37). Mark on top of piston must be toward flywheel side of engine after piston installation.

CYLINDER AND CRANKCASE. Cylinder and crankcase are an integral casting. Standard cylinder bore diameter is 52 mm (2.05 in.) on FA76D models, 62 mm (2.44 in.) on FA130D models and 72 mm (2.83 in.) on FA210D models. If cylinder bore diameter exceeds these dimensions by 0.07 mm (0.003 in.), renew cylinder.

Because of the high silicon content of the die-cast aluminum block, special honing procedures, stones and conditioning compounds are required to hone FA series engine cylinders. Procedures outlined and material prescribed are recommended to produce a cylinder bore finish comparable to the factory finish.

The following procedures outlined are designed to remove 0.258 mm (0.0102 in.) total material from cylinder bore diameter upon completion. Use 38.1 mm (1½ in.) strokes throughout the honing process. Honing time should be varied to remove only the amount of material specified to obtain desired bore size.

The cylinder bore is silicone impregnated. Refacing a cylinder bore removes the thin silicone contact surface and exposes both aluminum and silicon particles. If the aluminum particles are not removed, damage to the cylinder and piston may result. To erode the aluminum particles so only a silicone surface is present in bore, the last honing procedure must be performed exactly as outlined.

Model FA76. Use a Sunnen JN-95 hone with a spindle speed of approximately 150 rpm. Equip hone with four T20-J85-75xxx stones, without guides, and hone cylinder until 0.229 mm (0.009 in.) of material is removed. Hone should remove the specified amount of material in approximately 50 seconds.

Equip hone with four T20-C05-75xxx stones, without guides, and hone cylinder until 0.025 mm (0.001 in.) of material is removed. Hone should remove specified amount of material in approximately 20 seconds.

Alter T20-F05-75xxx felt set as shown in Fig. KW39. Saturate altered felt set with MB-30 honing oil. Mix AN-30 conditioning compound thoroughly and coat felts and entire cylinder wall surface with compound. Equip hone with altered felt set and insert hone into cylinder. Tighten wing nut finger tight. Hone cylinder, periodically tightening feed pinion, until 0.004 mm (0.0002 in.) of material is removed. Hone should remove specified amount of material in approximately 1½ minutes.

CAUTION: Do not use additional honing oil as it will wash away the AN-30 conditioning compound.

Model FA130. Use a Sunnen JN-95 hone with a spindle speed of approximately 150 rpm. Equip hone with four V24-J85-85xxx stones and hone cylinder until 0.229 mm (0.009 in.) of material is removed. Hone should remove specified amount of material in 1½ minutes.

Equip hone with four V24-J85-85xxx stones and hone cylinder until 0.025 mm (0.001 in.) of material is removed. Hone should remove specified amount of material in 20 seconds.

Alter V24-J85-85xxx felt set as shown in Fig. KW39. Saturate altered felt set with MB-30 honing oil. Mix AN-30 conditioning compounds thoroughly and coat felts and entire cylinder wall surface with compound. Equip hone with altered felt set and insert hone into cylinder. Tighten wing nut finger tight. Hone cylinder, periodically tightening feed pinion, until 0.004 mm (0.0002 in.) of material is removed. Hone should remove specified amount of material in approximately 1½ minutes.

CAUTION: Do not use additional honing oil as it will wash away the AN-30 conditioning compound.

Model FA210. Use Sunnen AN-112 hone with a spindle speed of approximately 150 rpm. Equip hone with GG25-J65 stones and hone cylinder until 0.152 mm (0.006 in.) of material has been removed. Hone should remove specified amount of material in approximately 45 seconds.

Fig. KW40—Loosen or tighten crankcase cover bolts in sequence shown. Refer to text for torque specifications.

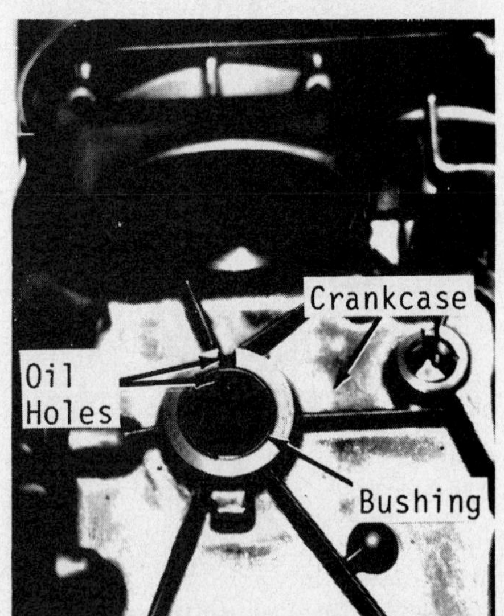

Fig. KW41—View of crankcase bushing. Oil holes must be aligned during installation. Refer to text.

Equip hone with GG25-J85 stones and hone cylinder until 0.076 mm (0.003 in.) of material is removed. Hone should remove specified amount of material in approximately 35 seconds.

Equip hone with GG25-C05 stones and hone cylinder until 0.025 mm (0.001 in.) material is removed. Hone should remove specified amount of material in approximately 20 seconds.

Alter GG25-F05 felt set as shown in Fig. KW39. Saturate altered felt set with MB-30 honing oil. Mix AN-30 conditioning compound thoroughly and coat felts and entire cylinder wall surface with compound. Equip hone with altered felt set and insert hone into cylinder. Tighten wing nut finger tight. Hone cylinder, periodically tightening feed pinion, until 0.005 mm (0.0002 in.) of material is removed. Hone should remove specified amount of material in approximately 1½ minutes.

CAUTION: Do not use additional honing oil as it will wash away the AN-30 conditioning compound.

All Models. Thoroughly clean cylinder and block after honing procedure. Lightly oil and assemble engine.

If crankcase cover has been removed, refer to CRANKSHAFT, MAIN BEARINGS AND SEALS section for bearing, bushing and seal installation. Tighten crankcase cover bolts to specified torque in sequence shown in Fig. KW40.

CRANKSHAFT, MAIN BEARINGS AND SEALS. Crankshaft on FA76D models is supported at each end in bushing type main bearings which are pressed into crankcase and crankcase cover. Crankshaft on all other models is supported in a ball bearing at crankcase cover (pto) end and in a bushing type main bearing which is pressed into crankcase at flywheel end.

Fig. KW42—Align crankshaft and camshaft gear timng marks as shown during assembly.

To remove crankshaft, remove all metal shrouds, flywheel, cylinder head and crankcase cover. Remove connecting rod and piston assembly and remove crankshaft and camshaft. Remove valve tappets.

Minimum crankpin diameter is 19.95 mm (0.785 in.) for FA76D models, 23.92 mm (0.942 in.) for FA130D models and 26.92 mm (1.060 in.) for FA210D models. If dimensions are not as specified, renew crankshaft.

Maximum clearance between bushing type main bearing and crankshaft journal is 0.13 mm (0.005 in.). When renewing bushing type main bearing, make sure oil hole in bushing is aligned with oil hole in crankcase or cover. Press bushing in 1 mm (0.039 in.) below inside edge of bushing bore. See Fig. KW41.

Ball bearing type main bearing is a light press fit on crankshaft and maximum bearing play is 0.30 mm (0.001 in.). It may be necessary to slightly heat crankcase cover to remove or install ball bearing.

Crankcase cover oil seals should be pressed into seal bores until seal is flush with outer edge of seal bore on FA76D models or 4 mm (0.158 in.) below outer edge of seal bore on all other models. Crankcase crankshaft seal should be pressed into seal bore until flush with outer edge of seal bore on all models.

To reinstall crankshaft, reverse removal procedure making sure crankshaft and camshaft gear timing marks are aligned during installation. Refer to Fig. KW42.

CAMSHAFT AND BEARINGS. Camshaft is supported at each end in bushing type bearings which are an integral part of crankcase and crankcase cover. To remove camshaft, refer to CRANKSHAFT, MAIN BEARINGS AND SEALS section.

Maximum clearance between camshaft and bushing type bearings is 0.10 mm (0.004 in.). If clearance exceeds specifications, renew camshaft and/or crankcase cover or crankcase.

Make certain camshaft lobes are smooth and free of scoring, scratches and other damage. Make certain camshaft and crankshaft gear timing marks are aligned during installation. Refer to Fig. KW42.

GOVERNOR. The internal centrifugal flyweight governor is gear driven off of the camshaft gear. Refer to GOVERNOR paragraph in MAINTENANCE section for external governor adjustments.

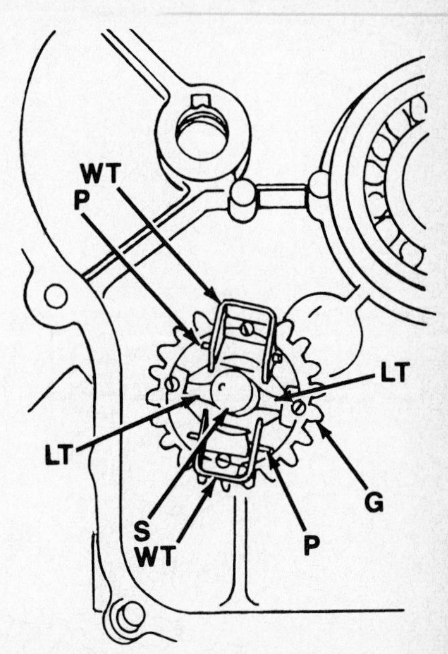

Fig. KW43—View of governor assembly. Lock tabs (LT) must be pressed in to remove sleeve and governor assembly. Renew sleeve prior to reassembly to ensure lock tabs will retain sleeve and governor assembly.

G. Gear
P. Pins
S. Sleeve
LT. Lock tabs
WT. Flyweights

To remove governor assembly, remove external linkage, metal cooling shrouds and crankcase cover. Governor gear and flyweights are attached to crankcase cover. Remove governor sleeve (S–Fig. KW43), gear (G) and flyweight assembly only if renewal is necessary. Use a screwdriver to press lock tabs (LT) in to remove sleeve. Flyweights are pinned to gear. Remove thrust washer located between gear and crankcase cover.

Reverse removal procedure for reinstallation. Adjust external linkage as outlined in MAINTENANCE section.

VALVE SYSTEM. Clearance between valve stem and valve tappet (cold) on FA76D models should be 0.1-0.2 mm (0.004-0.008 in.) for intake valve and 0.1-0.3 mm (0.004-0.012 in.) for exhaust valve. Clearance between valve stem and valve tappet (cold) on all other models should be 0.12-0.18 mm (0.005-0.007 in.) for intake valve and 0.10-0.034 mm (0.004-0.013 in.) for exhaust valve. If clearance is not as specified, valves must be removed and end of stems ground off to increase clearance or seats ground deeper to reduce clearance.

Valve seats are ground directly into crankcase casting and valve face and seat angles are 45°. Correct valve seat width for FA76D models is 0.5-1.1 mm (0.020-0.043 in.) and correct valve seat width for all other models is 1.0-1.6 mm (0.039-0.063 in.).

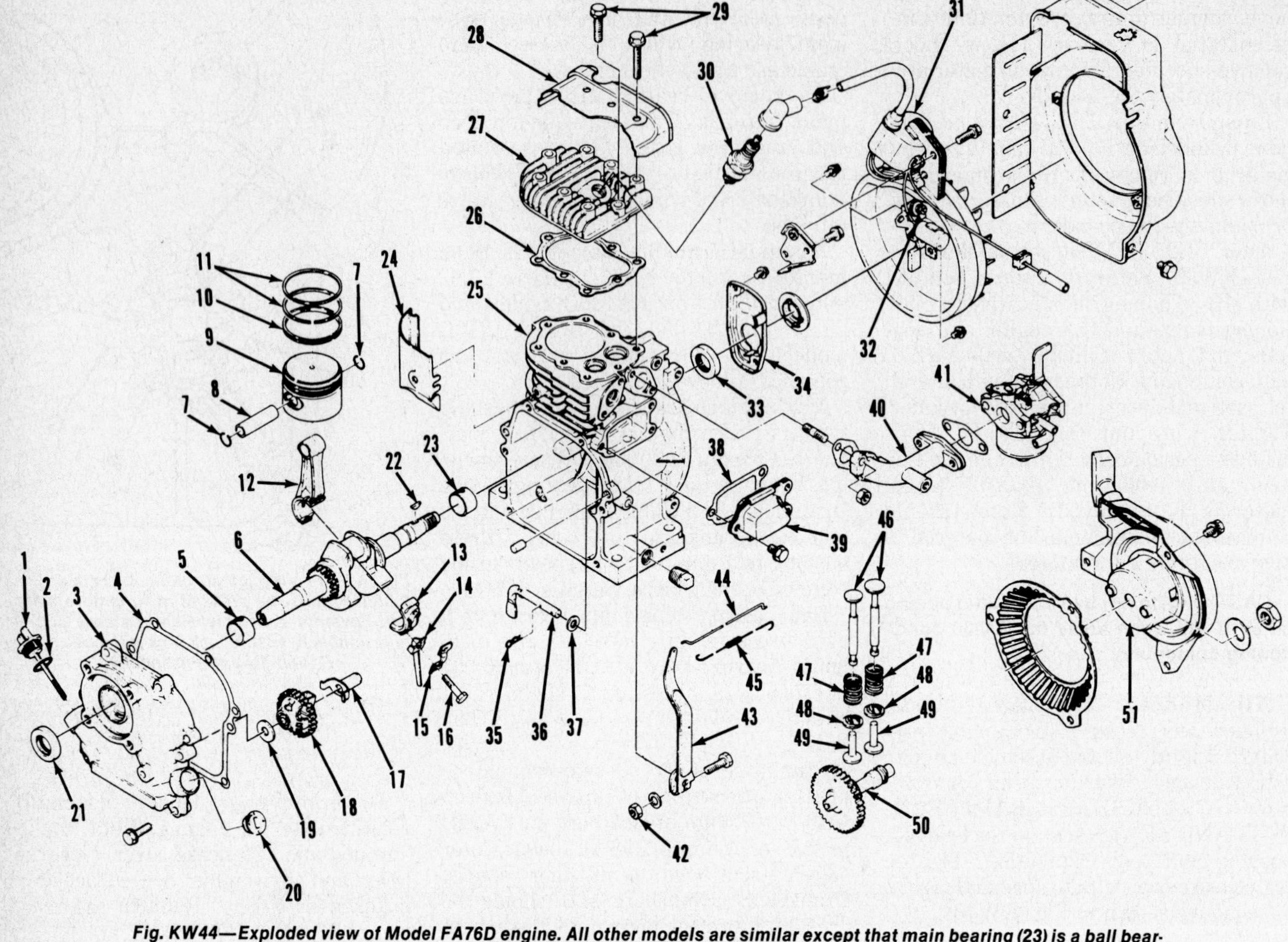

Fig. KW44—Exploded view of Model FA76D engine. All other models are similar except that main bearing (23) is a ball bearing on Models FA130D and FA210D.

1. Oil plug & gage
2. "O" ring
3. Crankcase cover
4. Gasket
5. Main bearing bushing
6. Crankshaft
7. Retaining ring
8. Piston pin
9. Piston
10. Oil control ring
11. Compression rings
12. Compression rod
13. Connecting rod cap
14. Oil plate
15. Lock plate
16. Connecting rod bolts
17. Governor sleeve
18. Governor gear & weight assy.
19. Thrust washer
20. Oil fill plug
21. Crankcase cover seal
22. Crankshaft key
23. Main bearing bushing (FA76D)
24. Cooling shroud
25. Cylinder block (crankcase)
26. Gasket
27. Cylinder head
28. Cooling shroud
29. Cylinder head cap screws
30. Spark plug
31. Ignition coil
32. Points & condenser
33. Seal
34. Point & condenser cover
35. Retaining clip
36. Governor crank
37. Washer
38. Gasket
39. Valve chamber cover
40. Manifold
41. Carburetor
42. Clamp bolt nut
43. Governor lever
44. Governor-to-carburetor rod
45. Spring
46. Valves
47. Valve springs
48. Valve spring retainers
49. Valve lifters
50. Camshaft & gear assy.
51. Rewind starter assy.

LAWN BOY

OUTBOARD MARINE CORPORATION
Post Office Box 82409
Lincoln, Nebraska 68501

Model	Bore	Stroke	Displacement
C-10, C-12, C-20, C-21, C-22, C-70, C-71, C-72, C-73, C-75	1.94 in. (49.2 mm)	1.50 in. (38.1 mm)	4.43 cu. in. (72.4 cc)
C-13 through C-17, C-40, C-41, C-42, C-50, C-51, C-60, C-61, C-74, C-76, C-77, C-78, C-80, C-81	2.13 in. (54.1 mm)	1.50 in. (38.1 mm)	5.22 cu. in. (86.9 cc)
C-18, C-44, D-400 through D-408, D405E through D-408E, D-409, D-430, D-431, D-432, D-433, D-440, through D-448, D-445E through D-448E, D-449, D-450, D-451, D-452, D-460, D-461, D-462, D-475, D-476, D-480, D-481, D-570, D-600, D-600E, D-601, D-640, D-640E, D-641, D-641E	2.38 in. (60.4 mm)	1.50 in. (38.1 mm)	6.65 cu. in. (108.8 cc)

ENGINE INFORMATION

On 1960 and earlier models, engine model and serial numbers will be found on engine shroud or gas tank strap (Fig. LB1). On later models, engine model and serial number are located on either starter spring cap or cylinder (Fig. LB1).

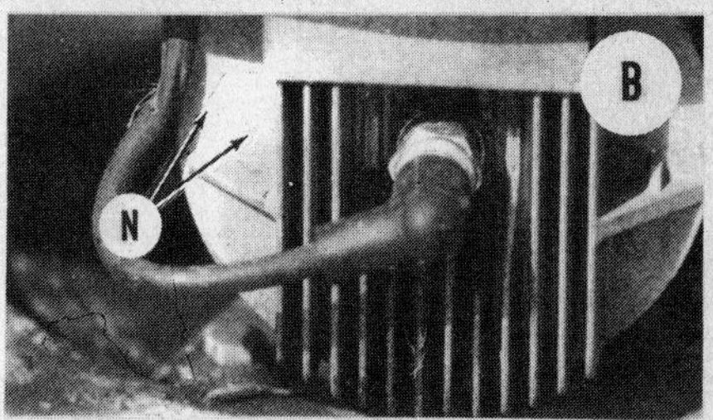

Fig. LB1—Engine model and serial number locations (N) on 1960 and earlier models are shown in views "A" and "B". On newer models, engine model number is located as shown in view "C".

MAINTENANCE

SPARK PLUG. Recommended spark plug for 4.43 cu. in. (72.4 cc) and 5.22 cu. in. (86.9 cc) engines is a Champion J14J, or equivalent. Recommended spark plug for 6.65 cu. in. (108.8 cc) engine is a Champion CJ14, or equivalent.

Electrode gap should be set at 0.025 inch (0.6 mm) for all models except D-600 series (solid state ignition) which calls for a gap of 0.035 inch (0.9 mm). Tighten plugs on all models to 12-15 ft.-lbs. (16-20 N•m).

CARBURETOR. Lawn Boy float type carburetors are used on all engine models. Figure LB2 shows an exploded view of carburetor used on C-series engines. Figure LB3 shows an exploded view of carburetor used on early D-series engines. Later D-series do not have a low speed mixture needle (6 – Fig. LB3), but are equipped instead with a fixed jet requiring no adjustments. D-600 series and Modular D-series carburetors are fully automatic with no external adjustments to regulate normal fuel intake. However, these carburetors are equipped with an altitude compensation needle which is covered later in this section.

Removal of C and early D-series carburetors require removal of reed plate. D-series engines equipped with plastic modular carburetors are secured by externally mounted screws.

C AND D-SERIES CARBURETORS. On C and D-series carburetors equipped with idle (low speed) needle, initial adjustment is approximately 3/4-turn out from seated position. Initial adjustment on main fuel needle is approximately 2½ turns out from seated position. After engine is started, adjust main fuel needle so engine runs smoothly until properly warmed up (about five minutes). Turn main fuel needle in (clockwise) until engine begins to lose speed, then turn needle out 1/8-1/4 turn. Adjust idle needle when engine is at normal operating temperature for smoothest idle performance. If idle mixture is too lean, engine is likely to surge at low speed.

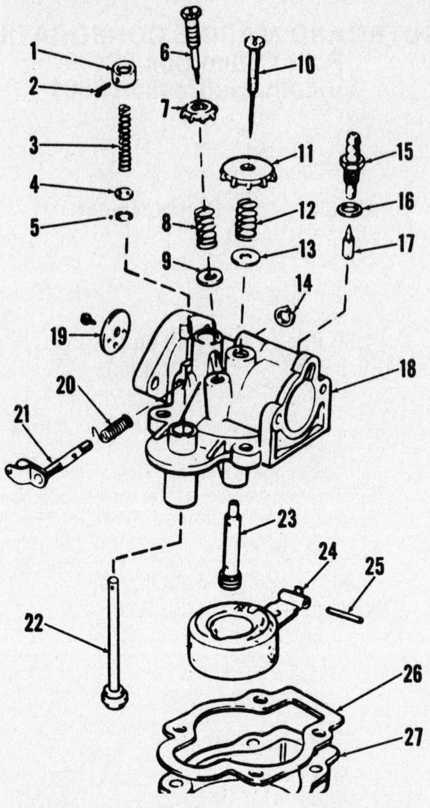

Fig. LB3—Exploded view of carburetor typical of type used on D-series engines. Late D models are not equipped with idle fuel needle (6).

1. Knob
2. Cotter pin
3. Spring
4. Washer
5. "O" ring
6. Idle fuel needle
7. Knob
8. Spring
9. Washer
10. Main fuel needle
11. Knob
12. Spring
13. Washer
14. Snap ring
15. Float valve seat
16. Washer
17. Float valve
18. Carburetor body
19. Throttle disc
20. Throttle spring
21. Throttle shaft
22. Primer plunger
23. Nozzle
24. Float
25. Float pin
26. Gasket
27. Float bowl

Float setting (H – Fig. LB3A) for C and D-series carburetors is 15/32 inch (12 mm) above carburetor body flange. Adjust float level by bending float arm using needle nose pliers. Do not apply strain or pressure to cork float. If float shows signs of damage or if epoxy varnish is flaked off or chipped, renew float assembly.

Do not use commercial carburetor cleaners. Use a mild solvent to clean and clear carburetor parts and passages. Do not dry carburetor parts with a cloth. Loose lint may be the cause of a later problem.

When reassembling carburetor, reinstall main fuel nozzle (94 – Fig. LB2 or 23 – Fig. LB3) before installing fuel needle (104 – Fig. LB2 or 10 – Fig. LB3) to prevent jamming needle into nozzle seat. Be sure to set a light preload on throttle spring (103 – Fig. LB2 or 20 – Fig. LB3) by adding about ½-turn tension when attaching spring to throttle shaft lever.

SERIES D-600 CARBURETOR. See Fig. LB4 for exploded view of Series D-600 carburetor. Difference between this and preceding models is elimination of external adjustments for main and idle circuits.

Altitude adjustment needle (15 – Fig. LB4) is designed to compensate for differences in atmospheric pressure if engine is to be operated at elevations which are higher or lower in relation to sea level. To adjust, preset altitude needle 1½ turns (counterclockwise) from fully seated position. Start engine and

Fig. LB2—Exploded view of typical Lawn Boy carburetor used on C-series engines. Refer to Fig. LB3 for carburetor used on D-series engines.

87. Bowl retaining screws
88. Float bowl
89. Gasket
90. Float pin
91. Float
92. Inlet valve assy.
93. Gasket
94. Nozzle
95. Choke detent
96. Spring
97.
98. Choke disc
99. Choke shaft
100. Throttle disc
101. Throttle shaft
103. Throttle spring
104. Fuel adjustment needle
105. Knob
106. Spring
107. Washer
108. Carburetor body

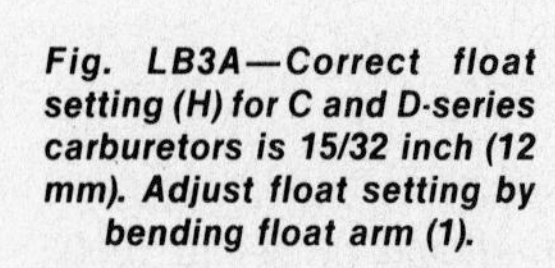

Fig. LB3A—Correct float setting (H) for C and D-series carburetors is 15/32 inch (12 mm). Adjust float setting by bending float arm (1).

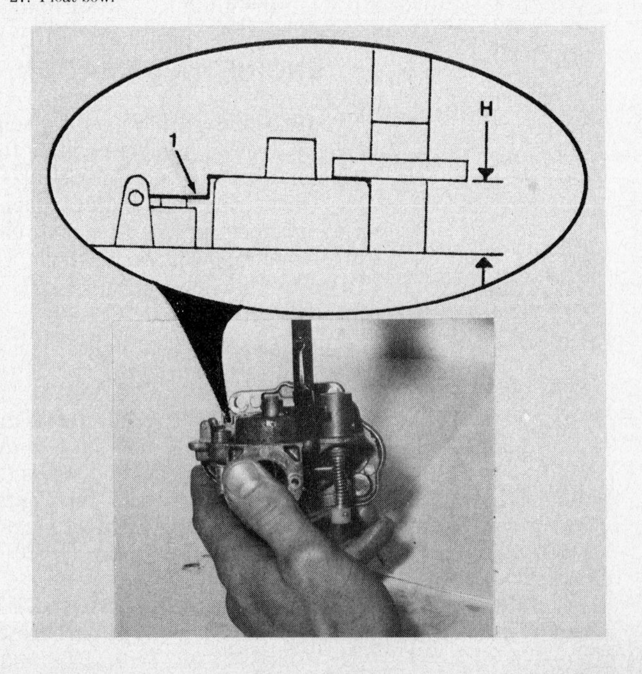

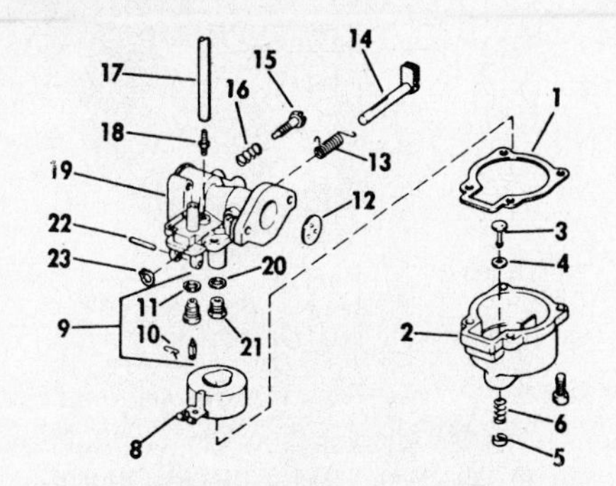

Fig. LB4—Exploded view of D-600 series carburetor.

1. Gasket
2. Float bowl
3. Dump valve stem
4. Valve seat
5. Valve keeper
6. Valve spring
8. Float
9. Float valve assy.
10. Clip
11. Washer
12. Throttle disc
13. Throttle spring
14. Throttle shaft
15. Altitude screw
16. Spring
17. Fuel tube
18. Connector
19. Body assy.
20. Washer
21. Plug
22. Float pin
23. Throttle shaft retainer

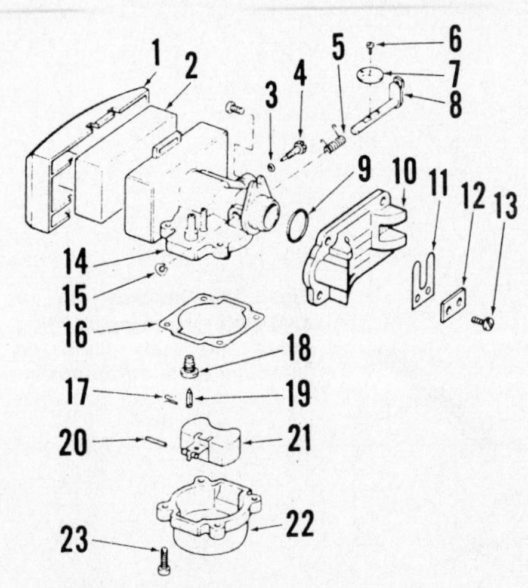

Fig. LB6—Exploded view of "Hinged Float" style modular carburetor used on some later Model D-series engines.

1. Filter cover
2. Filter element
3. "O" ring
4. Air screw
5. Return spring
6. Screw
7. Throttle disc
8. Throttle shaft
9. "O" ring
10. Reed plate
11. Reed
12. Reed retainer
13. Screw
14. Main body
15. Retainer
16. Bowl gasket
17. Clip
18. Valve seat
19. Needle
20. Pin
21. Float
22. Bowl
23. Screw

Fig. LB7—Correct float height setting for hinged float style modular carburetor.

run 3 to 5 minutes to warm up. Place speed control lever in low speed position. Slowly turn altitude needle in (clockwise) until engine begins to surge or slow down. Turn altitude needle out slowly until engine runs smoothly, then allow engine to run a few minutes while observing engine performance. If engine does not run smoothly repeat altitude needle adjustment procedure. Set speed control to high speed position. If engine is not running smoothly, turn altitude needle out (counterclockwise) in 1/4-turn increments until engine runs smoothly.

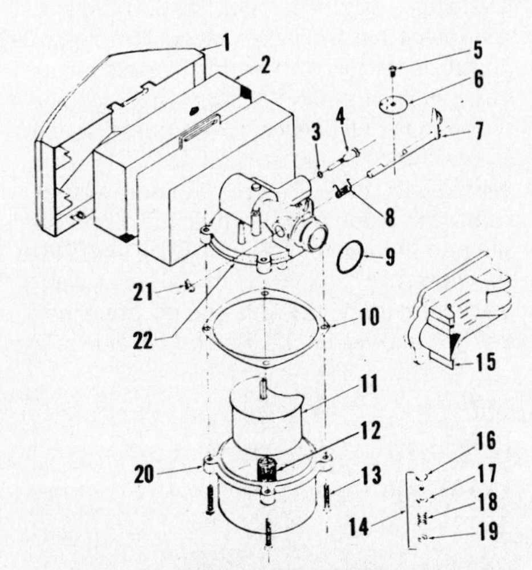

Fig. LB5—Exploded view of "Free Float" style modular carburetor used on some later Model D-series engines.

1. Filter cover
2. Filter element
3. "O" ring
4. Air screw
5. Screw
6. Disc
7. Throttle shaft
8. Return spring
9. "O" ring
10. Bowl gasket
11. Float
12. Spring
13. Screw
14. Dump valve assy.
15. Reed plate
16. Valve stem
17. Valve seat
18. Valve spring
19. Valve keeper
20. Float bowl
21. Retainer
22. Main body

Shut off engine and restart immediately. If difficulty in restarting is encountered, turn altitude needle out an additional 1/8-1/4 turn to richen mixture.

Carburetor servicing is limited to visual inspection, cleaning with a mild solvent (**DO NOT** use a standard carburetor cleaner) and adjustment of float height. Adjust float height to 15/32 inch (12 mm) above carburetor body flange as shown in Fig. LB3. Adjust float lever only by bending float arm using needle nose pliers. Do not apply strain or pressure to cork float. If float shows signs of damage or if epoxy varnish is flaked off or chipped, renew float assembly.

Early 1972 production was fitted with a plastic throttle shaft and disc. If renewal becomes necessary, use assembly number 681008 which contains bronze parts.

If throttle shaft and disc are removed for servicing or renewal, be sure to set a light preload on throttle spring by adding about 1/2-turn tension when attaching spring.

MODULAR CARBURETOR. Engine Models D-409, D-481, D-601, D-641 and D-641E may be equipped with a modular-type carburetor. There are two styles, the "Free Float" style as shown in Fig. LB5 and the "Hinged Float" style as shown in Fig. LB6. These carburetors are fully automatic and require no external adjustments except for the altitude adjustment needle.

On "Free Float" style carburetor, the specific free length of float spring (12–Fig. LB5) should be 5/8 inch (16 mm). If length is incorrect, renew spring.

On "Hinged Float" style carburetors, the correct float height should be 11/16 inch (17 mm) measured above carburetor body flange as shown in Fig. LB7. Adjust float level by bending float arm using needle nose pliers. Do not bend arm by applying pressure on cork float.

Altitude adjustment needle should be opened 3/4-turn from fully seated for elevations from 500-1500 feet above sea level.

Check all parts for excessive wear or damage and renew as needed. **DO NOT** use standard carburetor cleaners, use only a mild grease solvent to clean carburetor parts and blow dry with compressed air. **DO NOT** dry carburetor parts with a cloth, as lint from cloth could plug fuel passage and cause a performance problem.

PRIMER. D-series carburetors are equipped with priming devices to aid in starting. C-series carburetors are equipped with a standard manual choke butterfly. D-series carburetor uses a mechanical primer as shown in Fig. LB8. Primer plunger lifts fuel into car-

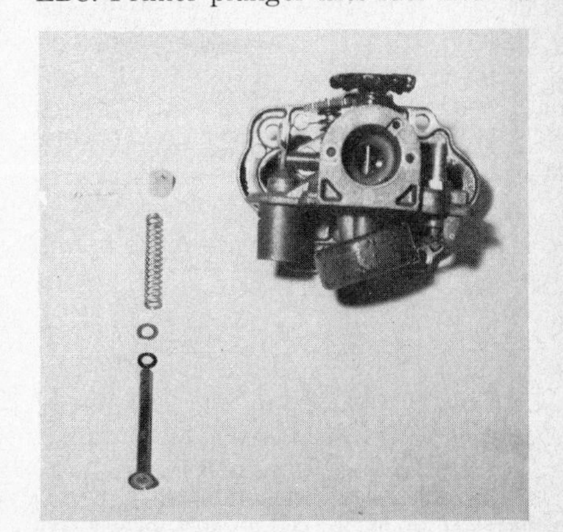

Fig. LB8—View showing primer plunger assembly used on D-series carburetors.

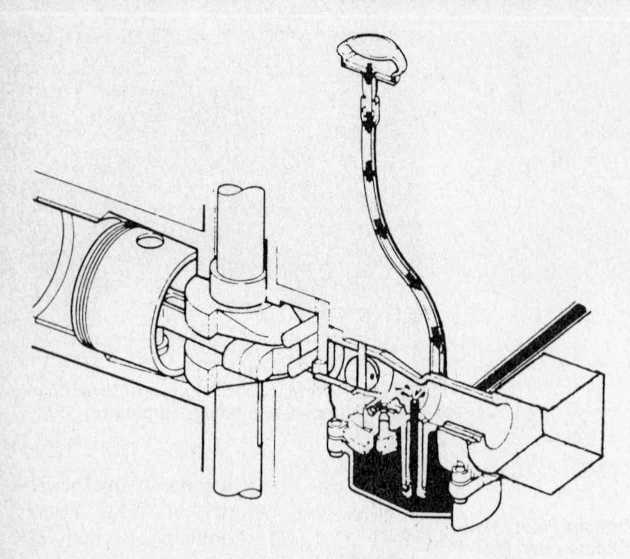

Fig. LB9—Cutaway view of carburetor showing priming pump assembly used on modular style carburetors.

Fig. LB12—View showing installation of governor yoke, weights and collar as a unit.

buretor chamber. Modular carburetors use a pneumatic bulb type primer as shown in Fig. LB9. Depressing primer bulb forces air in float chamber which forces fuel through main jet into carburetor venturi. **DO NOT** over-prime; one priming stroke is normally sufficient.

REED VALVE. Reed valves permit fuel-air mixture to enter crankcase on the compression stroke and seal fuel-air mixture in crankcase on the power stroke. Reeds are attached to reed plate on C and D-series engines and can be cleaned and serviced after removal of reed plate and carburetor. Reeds can be cleaned with solvent or carburetor cleaner. **DO NOT** use compressed air on reed assemblies or distortion may occur resulting in hard starting or loss of power. Bent or distorted reeds cannot be repaired. Rough edge of reed must be positioned away from machined surface of mounting plate. Check reed-to-plate clearance as shown in Fig. LB10. Clearance of 0.015 inch (0.38 mm) between reed tip and machined surface is allowable. Renew if clearance is excessive. Coat reed mounting screws with Lawn Boy Nut & Screw Lock (Part 682301) or a suitable equivalent and tighten to 10-13 in.-lbs. (1 N•m).

GOVERNOR. All C and D-series engines are equipped with a mechanical governor. The governor weight unit is located under the flywheel. View of a typical engine governor assembly is shown in Fig. LB11. Most D-series engines are also equipped with a variable speed spring and governor lever which will be explained later. Engine speed is controlled by compression of coil spring under governor weight unit as shown in Fig. LB12.

Two different governor springs are available on C-series engines providing governed speeds of 2800 rpm and 3200 rpm. Data on governor springs is as follows:

C-Series Engines

Rpm	Free Length	Color
2800	0.750-0.780 in. (19-20 mm)	Red
3200	0.735-0.765 in. (18.67-19.44 mm)	....

On D-series engines, all models except D-430 and D-460 are equipped with a variable speed spring located between speed control lever and governor lever (See Fig. LB16) to provide a variable speed governor. Governor spring color is green. Models D-430 and D-460 are equipped with a unpainted (Natural) governor spring to provide a governed speed of 3200 rpm. Data on governor springs is as follows:

D-Series Engines

Rpm	Free Length	Color
Variable (2500-3200)	0.735-0.765 in. (18.67-19.44 mm)	Green
3200	0.735-0.765 in. (18.67-19.44 mm)	Natural

Shown in Fig. LB13 is procedure for checking governor adjustment on D-series engines. On D-600 series engines turn variable speed control to "light" position and rotate thrust collar clockwise against its stop on governor lever. On all D-series engines, place Lawn Boy special tool 604541 (1) over crankshaft end. Push governor weight unit as far down on crankshaft as possible and hold carburetor throttle shaft (4) in closed position. Gap (G) between governor lever (2) and end of governor rod (3) should be 1/16-inch (1.6 mm). If

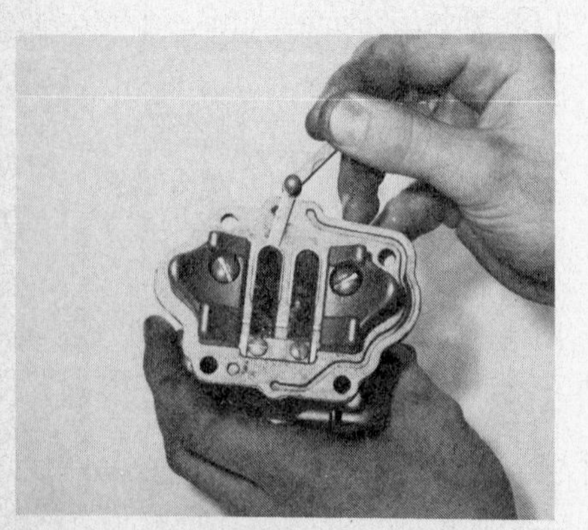

Fig. LB10—View showing a typical reed plate assembly. Clearance of 0.015 inch (0.38 mm) measured as shown is allowable. Renew reed valve if clearance is excessive or if bent or distorted so as to prevent sealing under pressure.

Fig. LB11—View showing components of a typical engine governor assembly used on all C and D-series engines.

1. Governor yoke
2. Weights
3. Collar
4. Wear block
5. Lever
6. Rod

Fig. LB13—View showing procedure for checking governor adjustment on D-series engine. C-series engine is similar. Measure gap (G) and refer to text.

1. Special tool
2. Governor lever
3. Governor rod
4. Throttle shaft

not, bend governor lever at crease in lever near point where it contacts governor rod.

Governor adjustment procedures for C-series engine are the same as for D-series engine except for the following: Lawn Boy special tool 602885 should be used over crankshaft end. Gap between governor lever and end of governor rod should be 1/8 inch (3.2 mm).

Governor lever thrust collars on D-400 and D-600 series engines are different as shown in Fig. LB14. D-600 collar is designed to prevent engine from starting and running backward due to preignition, combustion chamber hot spot or other malfunction. D-400 and D-600 thrust collars are **NOT** interchangeable.

When reassembling governor on D-series engines with variable speed governor, place web in groove of plastic thrust collar (See Fig. LB15) between arms on governor lever with lugs on arms inserted in groove of collar. Engage tabs on governor lever in slots on breaker point dust cover. Hook governor spring into lever as shown in Fig. LB16. Apply a light coat of Lawn Boy "A" grease or a suitable equivalent to both faces of steel thrust washer and place washer on top of plastic collar.

On all C and D-series engines place governor spring over crankshaft. Install governor yoke, weights and collar assembly over end of crankshaft and governor spring. Turn governor yoke to position word "Key" located on top of yoke under flywheel key, then install key and flywheel.

Fig. LB14—D-400 and D-600 governor thrust collars shown for comparison. They cannot be interchanged. Refer to text.

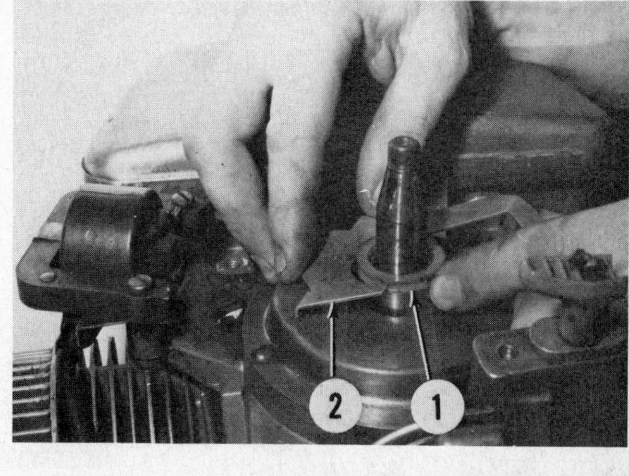

Fig. LB15—View showing governor lever (2) installed in thrust collar (1) on D-series engines equipped with variable speed governor.

Fig. LB16—Hooking variable speed spring (1) into governor lever (2) on so equipped D-series engines.

Fig. LB17—Lug (1) on flywheel drives governor weight unit on C-series and early D-400 series engines. Late D-400 and D-600 series engines are equipped with two "dimples" on top of governor yoke.

NOTE: Bore in flywheel and end of crankshaft should be clean and dry before installing flywheel.

Lug on flywheel (1 – Fig. LB17) on C and early D-400 engines drives governor weight unit. On late D-400 and D-600 series engines, governor yokes have two "dimples" located on top which engage bottom side of flywheel to drive governor.

MAGNETO AND TIMING. Refer to the following paragraphs for magneto service information on each Lawn Boy engine series.

C-SERIES MAGNETOS. All C-series engines use similar magnetos, refer to

Fig. LB18. On all models, timing is fixed and nonadjustable.

Armature air gap should be 0.010 inch (0.25 mm). Air gap can be considered correct if heels of armature core are flush with machined rim on armature plate as shown in Fig. LB19. Breaker point gap should be 0.020 inch (0.51 mm). When installing breaker points and/or condenser, wires from coil and stop switch should be installed under breaker point spring.

For reference in disassembling ignition grounding switch refer to Fig. LB20. Figure LB21 shows grounding switch correctly assembled and installed.

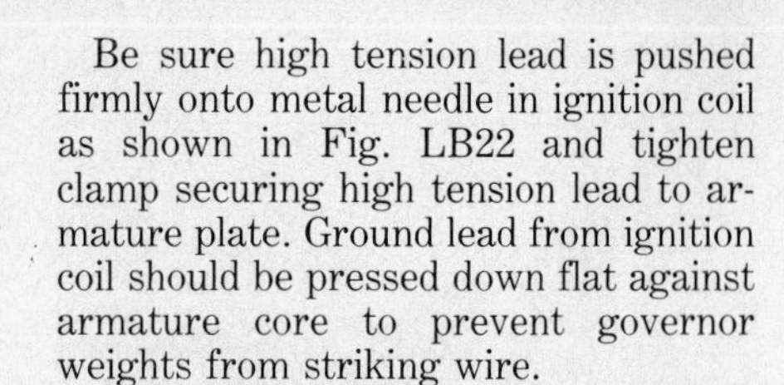

Be sure high tension lead is pushed firmly onto metal needle in ignition coil as shown in Fig. LB22 and tighten clamp securing high tension lead to armature plate. Ground lead from ignition coil should be pressed down flat against armature core to prevent governor weights from striking wire.

D-SERIES MAGNETOS. The magneto breaker can on D-series engines is driven by a flyweight (1 – Fig. LB23) to provide an automatic timing advance. The flyweight is retained in position by a pin through a hole in crankshaft which engages small end of flyweight. A coil spring surrounds the pin and forces small end of flyweight away from crankshaft when engine is not running as shown in Fig. LB23. When engine speed reaches approximately 1000 rpm, centrifugal force moves the flyweight against spring pressure which advances breaker cam to running position as shown in Fig. LB24. At starting position, ignition occurs at 6° BTDC; at running position, the timing is advanced to 26° BTDC.

The breaker points, condenser, cam and flyweight are enclosed by a dust cover under the flywheel and governor assembly.

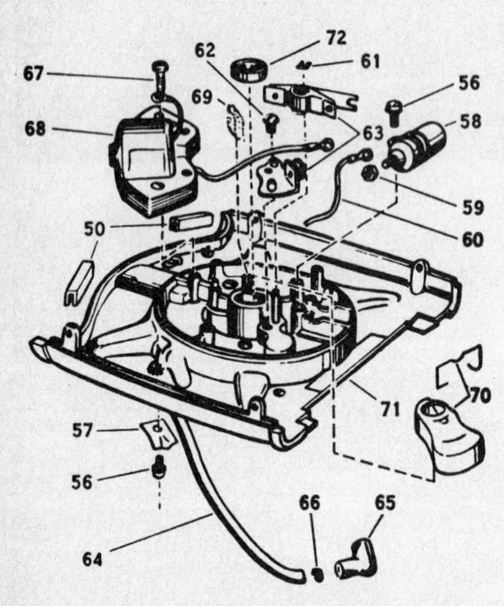

Fig. LB18—Exploded view of magneto used on C-series engines. Design of armature plate (71) is different on some models of C engines.

58. Condenser
60. Stop switch wire
63. Breaker points
68. Coil
69. Cam oiler felt
70. Dust cover
71. Armature plate

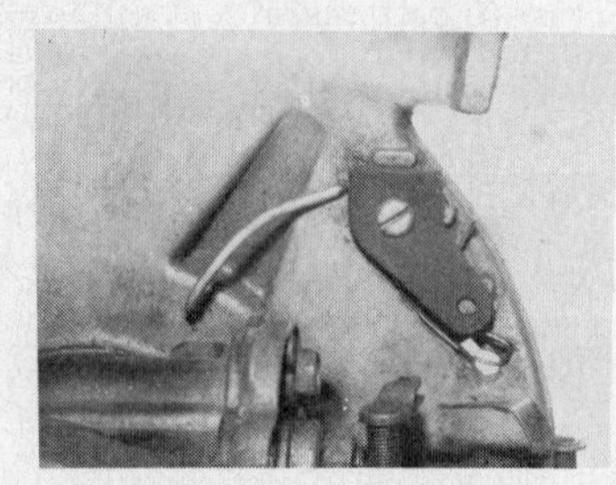

Fig. LB21—View showing grounding switch correctly assembled and installed on C-series engines.

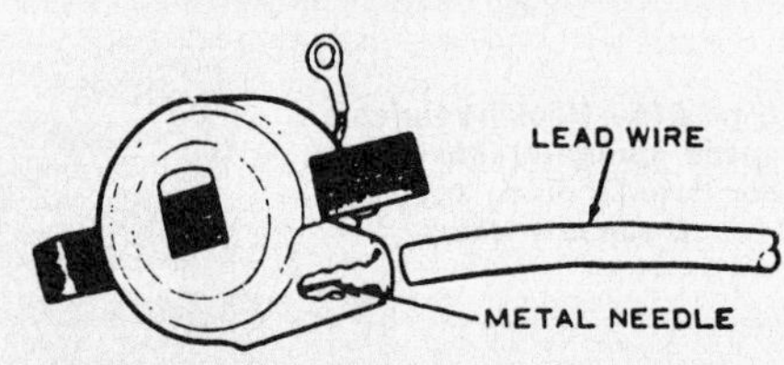

Fig. LB22—Push high tension lead firmly onto metal needle in ignition coil and be sure lead retaining clamp is tight.

Fig. LB19—To set armature air gap on C-series engines, loosen mounting screws and set coil heels flush with rim of armature plate as shown.

Fig. LB23—Flyweight (1) is shown in starting position.

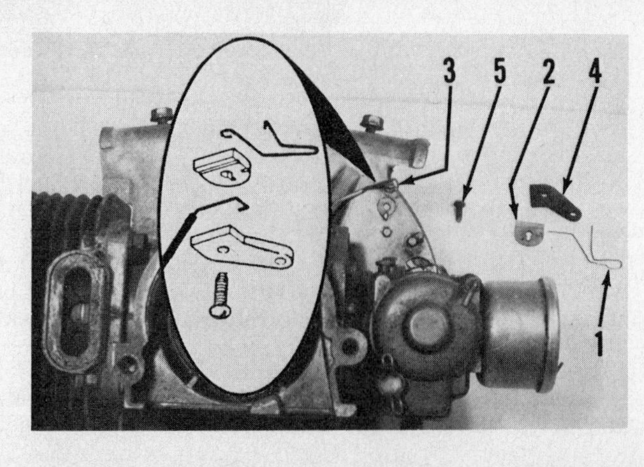

Fig. LB20—View showing ignition grounding switch components used on C-series engines.

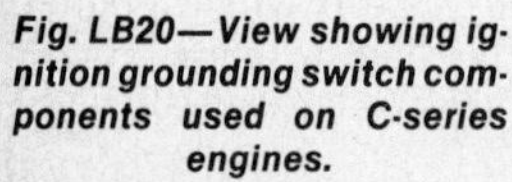

1. Spring
2. Upper insulator
3. Switch lead
4. Lower insulator
5. Mounting screw

Fig. LB24—View showing movement of flyweight as centrifugal force moves flyweight against spring pressure as engine rpm increases.

Fig. LB25—Remove flyweight and breaker cam as outlined in text.

To remove flyweight and breaker cam, withdraw flywheel, governor assembly and dust cover. Refer to Fig. LB25 and proceed as follows: Push small end of flyweight towards crankshaft, hold pin and spring in this position, and disengage small end of flyweight from pin and withdraw pin and spring from crankshaft. Lift flyweight and cam from crankshaft.

To adjust breaker points, install breaker cam on crankshaft and turn cam to position of widest breaker point gap. Pull upper end of crankshaft toward carburetor. Adjust breaker point gap to 0.020 inch (0.51 mm) on all later models with flywheel part number 678355 and 0.016 inch (0.41 mm) on early models with flywheel part number 678103. Flywheel part number is located on flywheel. Later flywheel is used in servicing early engines.

NOTE: When servicing magneto, always reinstall wire leads on condenser terminal first, then install breaker point spring and secure with nut. Wire from coil to condenser terminal should be under breaker point base.

When reinstalling cam and flyweight assembly, hold breaker points open and slip cam and flyweight as a unit on crankshaft. Turn cam and flyweight to position small end of flyweight on keyway side of crankshaft. Install spring on pin, push small end of flyweight down and insert inner end of pin in crankshaft hole. Push pin in against spring and engage outer end of pin in small end of flyweight.

When reinstalling flywheel on crankshaft, be sure Woodruff key is installed correctly in crankshaft groove as shown in Fig. LB26.

Armature air gap should be 0.010 inch (0.25 mm). With flywheel installed, place Lawn Boy air gap gage 604659 or a suitable piece of 0.010 inch (0.25 mm) nonmetallic shim stock as shown in Fig. LB27 between armature core and flywheel magnets. Loosen armature core mounting screws and allow magnets to pull core against flywheel. Tighten armature core mounting screws and remove shim stock.

NOTE: Excessive clearance between crankshaft and top main bearing will affect armature air gap and breaker point gap resulting in faulty ignition.

Top main bearing on 1965 and later models is a renewable needle bearing. The bushing, integral with armature plate on earlier production models, is serviced by installing late type armature plate and needle bearing.

D-600 SERIES CAPACITIVE DISCHARGE IGNITION SYSTEM. Lawn Boy's solid-state (capacitive discharge) ignition design eliminates all moving parts which are normally associated with magneto ignition. On these models, there is no spark advance assembly, breaker cam, breaker point set, condenser or coil. Complete system is electronic, contained in a single, sealed module which is dust and moisture proofed and serviced only as an assembly.

Maintenance is confined to setting air gap between flywheel magnets and CD module, checking for spark output by use of a test spark plug grounded to cylinder and isolation test of ignition switch for continuity. No procedures for troubleshooting or testing ignition pack are offered by manufacturer. If unit will not produce a spark at electrodes of test spark plug and ignition switch and spark plug cable check out to be good, then renewal of module assembly is necessary. Substitution of a spare ignition pack known to be good is an ideal test procedure.

Ignition timing is fixed. System design is set up for 9° retarded spark timing for ease of starting. After engine starts, spark is advanced electronically to 29° BTDC as crankshaft speed reaches 800 rpm.

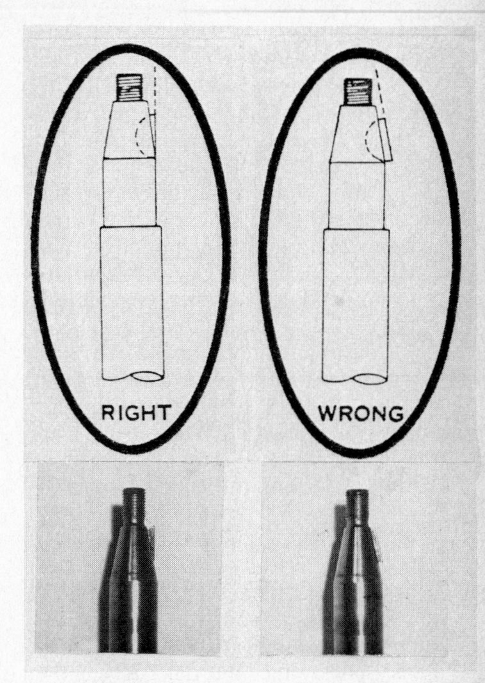

Fig. LB26—When installing flywheel, be sure Woodruff key is installed correctly in crankshaft groove as shown.

To service ignition module, remove three screws holding air baffle to armature plate and separate kill switch lead from ignition switch. Primer and starter need not be removed. With CD ignition pack exposed, 0.010 inch (0.25 mm) air gap may be checked. To do so, rotate flywheel to align flywheel magnets with laminated core heels of ignition module, then insert Lawn Boy Air Gap Gage 604659 or a suitable piece of 0.010 inch (0.25 mm) nonmetallic shim stock as shown in Fig. LB27. With gage in position, when module mounting screws are loosened, magnets will pull module snug against gage. Retighten screws and remove gage.

Operators of this engine should be cautioned to allow 10-15 seconds after shutdown before removing spark plug lead. This is to allow high voltage (approximately 30,000 volts) from CD pack secondary to leak off.

LUBRICATION. Engine is lubricated by premixing oil and gasoline in a separate container before filling engine

Fig. LB27—Adjust air gap using Air Gap Gage between flywheel magnets and coil heels on D-400 series engines. D-600 series engines with capacitive discharge ignition are adjusted following the same procedure. Air gap should be 0.010 inch (0.25 mm). Refer to text.

tank. Manufacturer recommends using Special Lawn Boy lubricant or a suitable good quality two-stroke oil. **DO NOT** use oil additives or standard automotive oils.

The use of regular gasoline (minimum octane rating of 89), unleaded or no-lead (minimum octane rating of 86) is recommended. **DO NOT** use gasoline additives except OMC 2+4 fuel conditioner. When using Special Lawn Boy lubricant mix fuel on 1971 and earlier engines at a ratio of 16:1 or one gallon of gasoline to eight ounces of lubricant. On 1972 and later engines mix fuel at a ratio of 32:1 or two gallons of gasoline to eight ounces of lubricant. When using a different brand of two-stroke oil mix fuel at a ratio of 16:1 on all models.

CARBON. If ignition, fuel supply to combustion chamber and compression check out as satisfactory, but engine will not run or runs poorly, it is likely that there is heavy carbon build-up in exhaust ports and muffler. A common symptom is "four-cycling" or firing every other power stroke.

Manufacturer recommends cleaning out carbon every 35 operating hours. Carbon cleaning procedure is as follows: On all models except D-600, remove two retaining nuts and muffler cover. On D-600 engines, muffler cover is secured by four nuts and mower blade and stiffener must first be removed from lower end of crankshaft. On all models, rotate crankshaft so piston wall blocks exhaust ports, then use a 3/8-inch wooden dowel to break carbon away from port openings. Clean carbon from muffler chamber and cover plate before reassembling.

REPAIRS

TIGHTENING TORQUES. Recommended tightening torques are as follows:

Spark plug............144-180 in.-lbs. (16-20 N·m)
Carburetor to reed plate..63-75 in.-lbs. (7-8 N·m)
Reed to reed plate.........10-13 in.-lbs. (1-1.5 N·m)
Reed plate to crankcase...63-75 in.-lbs. (7-8 N·m)
Breaker point base........20-25 in.-lbs. (2-3 N·m)
Coil to armature..........20-25 in.-lbs. (2-3 N·m)
Condenser to armature...20-25 in.-lbs. (2-3 N·m)
High tension cable clamp screw...........20-25 in.-lbs. (2-3 N·m)
Armature to crankcase...63-75 in.-lbs. (7-8 N·m)
Shroud to armature (front)................25-30 in.-lbs. (3-3.4 N·m)
Shroud to armature (side)..................10-15 in.-lbs. (1-1.7 N·m)

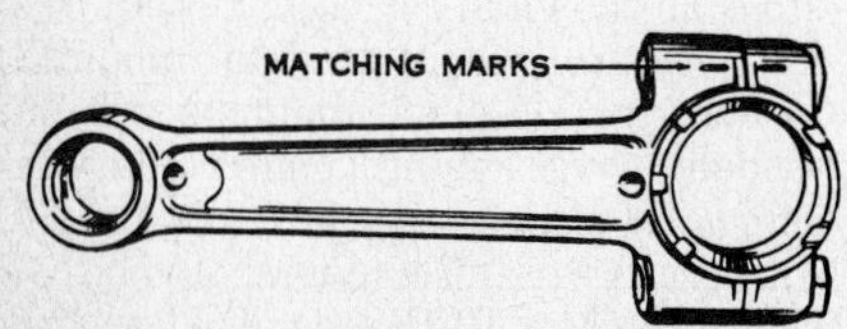

Fig. LB28—When reassembling cap to connecting rod, be sure match marks are aligned as shown.

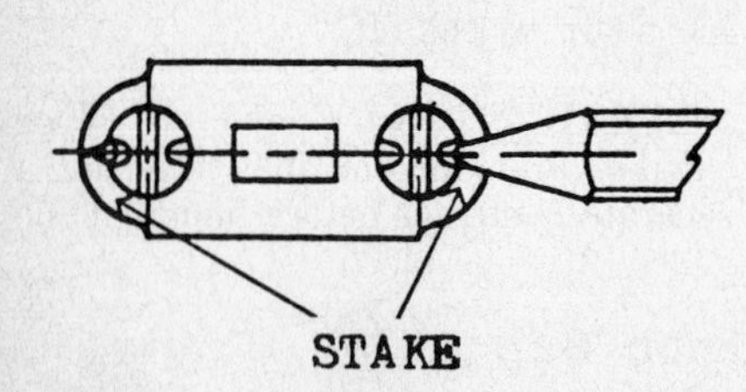

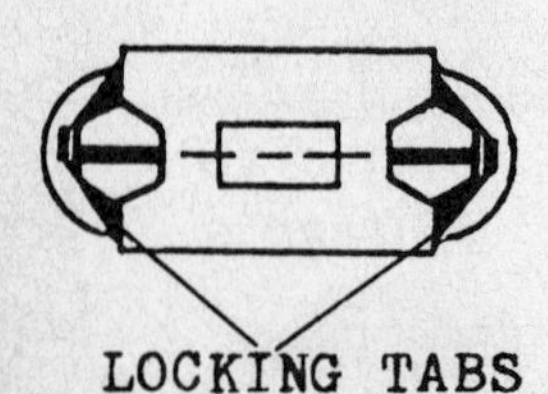

Fig. LB29—Connecting rod screws are locked into place by either staking screw heads or bending locking tabs.

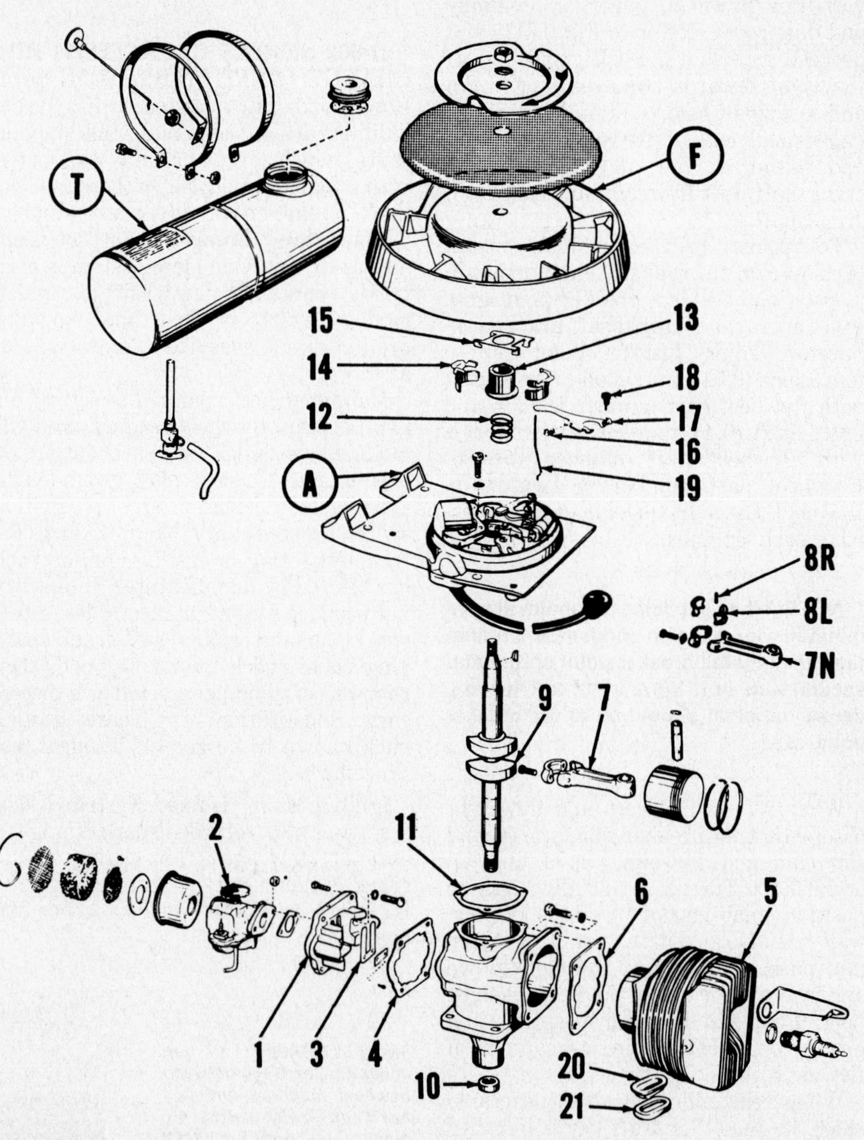

Fig. LB30—Exploded view of C-20, C-21 and C-22 engine. Armature plate (A) is different on other C-series engines. Some models use needle bearings in connecting rod as shown at 7N, 8L and 8R.

A. Armature plate
F. Flywheel
T. Fuel tank
1. Reed plate
2. Carburetor
3. Valve reeds
4. Gasket
5. Cylinder
6. Gasket
7. Connecting rod (plain bearing)
7N. Connecting rod
8L. Bearing liners
8R. Bearing rollers (33)
9. Crankshaft
10. Seal
11. Shim
12. Governor spring
13. Governor collar
14. Governor weight
15. Governor yoke
16. Wear block
17. Governor lever
18. Screws
19. Carburetor link rod
20. Sleeve
21. Gasket

Connecting rod . . . . 58-70 in.-lbs. (6.6-8 N·m)
Cylinder to crankcase . . . 105-115 in.-lbs. (12-13 N·m)
Engine mounting screws . . . . 142-170 in.-lbs. (16-19 N·m)
Tank to shroud . . . . 63-75 in.-lbs. (7-8 N·m)
Flywheel (C series) . . . . . 190-225 in.-lbs. (21-25 N·m)
Flywheel (D series) . . . . . 335-400 in.-lbs. (38-45 N·m)
Starter mounting . . . . 58-63 in.-lbs. (6-6.7 N·m)
Starter pulley . . . . 16-19 in.-lbs. (1.8-2 N·m)
Blade nut . . . . 450-550 in.-lbs. (51-62 N·m)
Engine to muffler plate . . . . 200-240 in.-lbs. (22.5-27 N·m)
Muffler to muffler plate . . . 58-63 in.-lbs. (6.6-7 N·m)

CONNECTING ROD. Rod and piston assembly can be removed after removing reed valve plate and cylinder from crankcase. Early C-series engine were equipped with an aluminum connecting rod with cast-in crankpin bearing surfaces in rod and cap. Late C-series engines and all D-series engines have a needle roller crankpin bearing; the aluminum connecting rod is fitted with renewable steel inserts.

Crankpin diameter is 0.7495-0.7500 inch (19.04-19.05 mm) for C-series engines with plain connecting rod bearing and recommended clearance between bearing and crankpin journal is 0.0025-0.0035 inch (0.064-0.089 mm).

Crankpin diameter is 0.7425-0.7430 inch (18.86-18.87 mm) for C-series engines with needle bearing connecting rod bearing and all D-series engines.

When reassembling plain bearing connecting rod to crankpin, oil bearing surfaces and be sure alignment marks on rod and cap match up as shown in Fig. LB28. After tightening cap retaining screws, be sure rod is free on crankpin. If screws have recesses in head in addition to screwdriver slot, stake cap metal into recesses with dull screwdriver as shown in Fig. LB29.

On needle bearing rods, press needle bearing liners into rod and cap so dovetail ends of liners (8L–Fig. LB30 or LB31) will fit together when rod and cap alignment marks match up. If installing new needle roller set, lay strip of 33 rollers on forefinger and carefully peel backing off rollers. Curl forefinger around crankpin to transfer rollers from finger to journal. Grease on rollers will hold them together and to crankpin. If reinstalling use needle rollers, inspect and renew any of the 33 rollers that are damaged. Coat surfaces with a layer of OMC Needle Bearing Grease (number 378642) or a suitable equivalent and stick roller to connecting rod end and cap. Fit 17 rollers to rod cap and 16 to rod end.

Carefully fit rod and cap to crankpin bearing and install cap retaining screws and screw lock tabs. After tightening screws, check to be sure rod assembly is free on crankpin and none of the rollers dropped out during assembly. Bend lock tabs up against flat of screw heads as shown in Fig. LB29.

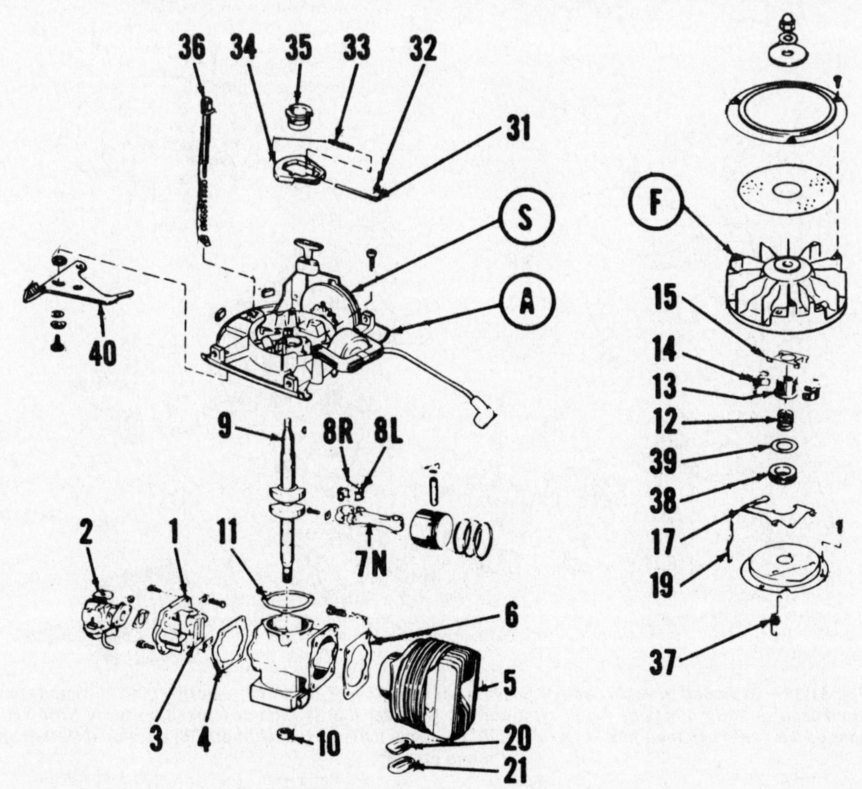

Fig. LB31—Exploded view of typical D-series engine.

1. Reed plate
2. Carburetor
3. Reed valve
4. Gasket
5. Cylinder
6. Gasket
7N. Connecting rod
8L. Bearing liners
8R. Bearing rollers (33)
9. Crankshaft
10. Lower seal
11. Shim
12. Governor spring
13. Governor collar
14. Governor weights
15. Governor yoke
17. Governor lever
19. Carburetor link rod
20. Sleeve
21. Gasket
31. Flyweight pin
32. Retainer
33. Ignition advance spring
34. Ignition advance flyweight
35. Ignition breaker cam
36. Primer plunger & shut-off bar
37. Variable speed spring
38. Thrust collar
39. Thrust washer
40. Governor control lever

PISTON, PIN AND RINGS. Piston, pin and rings are available in standard size only. Recommended piston to cylinder clearance is 0.004-0.0055 inch (0.10-0.14 mm) for models with 1.94 inch 49.2 mm) bore, 0.0045-0.006 inch (0.11-0.15 mm) for models with 2.13 inch (54.1 mm) bore or 0.005-0.0065 inch (0.13-0.17 mm) for models with 2.38 inch (60.4 mm) bore.

Piston pin is a push fit in piston. Pin is retained by snap rings at each end of pin bore in piston. Heat piston to facilitate removal and installation of pin.

Piston pin diameter is 0.4270-0.4272 inch (10.84-10.85 mm) for models with 1.94 inch (49.2 mm) bore or 0.4898-0.4900 inch (12.44-12.45 mm) for all models.

Piston pin clearance in connecting rod pin bore is 0.0003-0.0010 inch (0.008-0.025 mm) for all models. Piston pin clearance in piston pin bore is 0.0005 inch (0.013 mm) for models with 1.94 inch (49.2 mm) bore or 0.0007 inch (0.018 mm) for all other models.

Piston ring end gap is 0.015-0.025 inch (0.38-0.64 mm) for all models. Renew piston rings if end gap exceeds 0.025 inch (0.64 mm).

Prior to 1970, all engines were equipped with 3-ring pistons and a plain bronze bearing in wrist pin end of connecting rod. Since 1970, two rings are fitted to all pistons and connecting rod small end contains 27 loose needle roller bearings as shown in Fig. LB32. Great care must be taken to prevent loss of needles during removal as they are not serviced separately.

CYLINDER. Piston and rings are available in standard size only. Renew cylinder if bore is excessively worn or

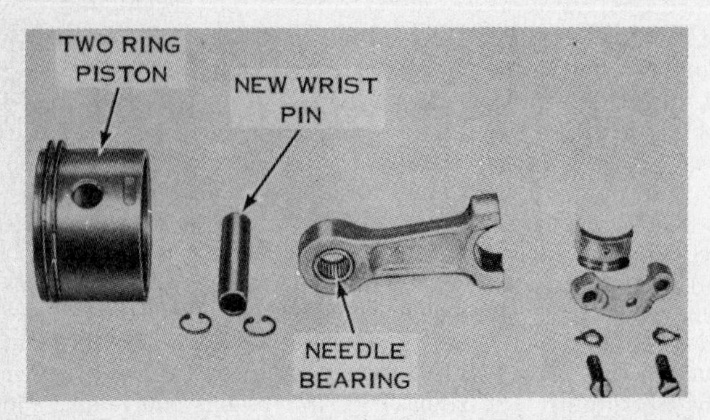

Fig. LB32—New style piston and connecting rod assembly with needle bearing set in small end of connecting rod. Assembly shown is interchangeable for all "C" and "D" engines, but components (wrist pin, bearings, etc.) DO NOT interchange. Service as a complete assembly.

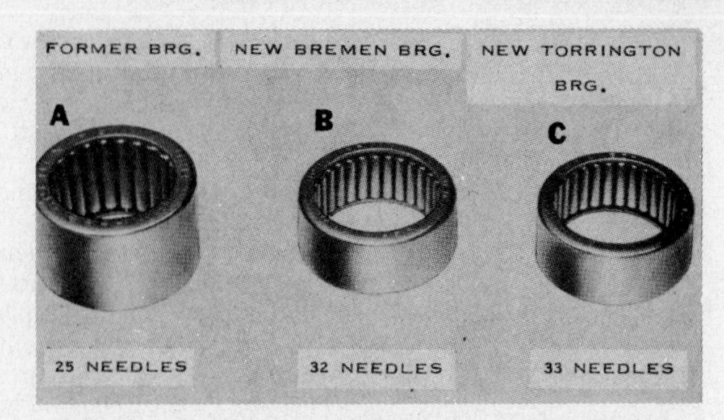

Fig. LB33—Needle roller bearing A (25 needles) was installed in armature plates through 1969 production. Bearings B and C have been used since 1970. Note that new Bremen bearing has 32 needles and that Torrington bearing has 33. Be sure to place installing tool against lettered side of cage.

scored. Standard and (new) cylinder bore diameters are 1.940-1.941 inch (49.28-49.30 mm), 2.125-2.126 inch (53.98-54.00 mm) or 2.377-2.378 inch (60.38-60.40 mm). In 1967 production only, Model C-78 and entire "D" series have 2.380-2.381 inch (60.45-60.48 mm) cylinder bore.

When reinstalling cylinder, be sure that the new cylinder to crankcase gasket is properly aligned so that it will not close the transfer ports. Loose cylinder retaining screws can cause considerable power loss. Lawn Boy recommends use of regular split type lock washers under screw heads even if screw is fitted with a serrated type washer.

Because of similarity of appearance, it is possible that cylinders for D-400 and D-600 engines might be confused. D-600 (high compression) cylinders have an "H"-shaped web cast between horizontal and vertical cooling fins. If this cylinder were installed on a D-400 engine, overheating and seizure would result.

CRANKSHAFT AND SEALS. All C-series engines have plain nonrenewable main bearings; magneto end bearing is plain bore in armature plate and pto end bearing is in plain bore of crankcase. Crankshaft main journal diameter is 0.8737-0.8742 inch (22.19-22.20 mm) for both journals. Recommended clearance at magneto end bearing is 0.002-0.0033 inch (0.05-0.08 mm) with maximum allowable clearance of 0.005 inch (0.13 mm). Recommended clearance at pto end bearing is 0.0038-0.0068 inch (0.10-0.17 mm). Crankshaft must be renewed if crankpin journal or main bearing journals are out-of-round 0.0015 inch (0.038 mm) or more.

The production top main bearing on 1964 and earlier D-series engines was a nonrenewable bronze bushing in the armature plate (A – Fig. LB31). Excessive clearance between the crankshaft and the top main bearing will affect both armature air gap and breaker point gap and result in faulty ignition. If the early

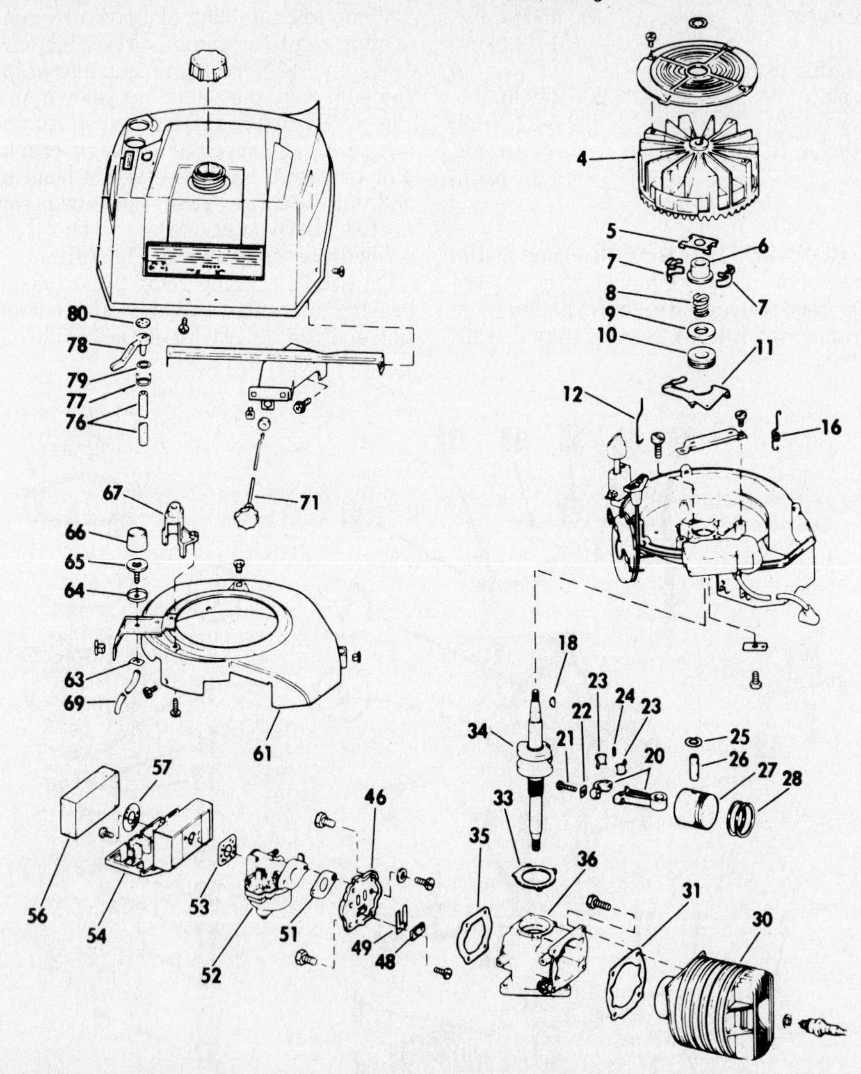

Fig. LB34—Exploded view of typical D-600 series engine. If installed, electric starter mounts on crankcase directly opposite recoil type starter. Cylinder (30) is high compression type. Note "H" shaped web cast between fins at top of cylinder. Primer (66) and fuel shut-off (78) are also different on these models.

4. Flywheel
5. Governor yoke
6. Governor collar
7. Governor weights
8. Governor spring
9. Washer
10. Thrust collar
11. Governor lever
12. Governor rod
16. Variable speed spring
18. Flywheel key
20. Connecting rod assy.
21. Screw
22. Lockplate
23. Bearing liners
24. Needle bearings (33)
25. Piston pin retainer (2)
26. Piston (wrist) pin
27. Piston
28. Piston rings (2)
30. Cylinder
31. Cylinder gasket
33. Armature plate gasket
34. Crankshaft
35. Reed plate gasket
36. Crankcase (includes bearings & seals)
46. Reed plate
48. Retainer
49. Reed assy.
51. Carburetor gasket
52. Carburetor
53. Gasket
54. Air filter case
56. Filter element
57. Washer
61. Air baffle
63. Fastener
64. Cap
65. Primer base
66. Primer bulb
67. Mounting bracket
69. Fuel line
71. Control rod & knob
76. Fuel line
77. Spring clip
78. Fuel shut-off
79. Washer
80. Gasket

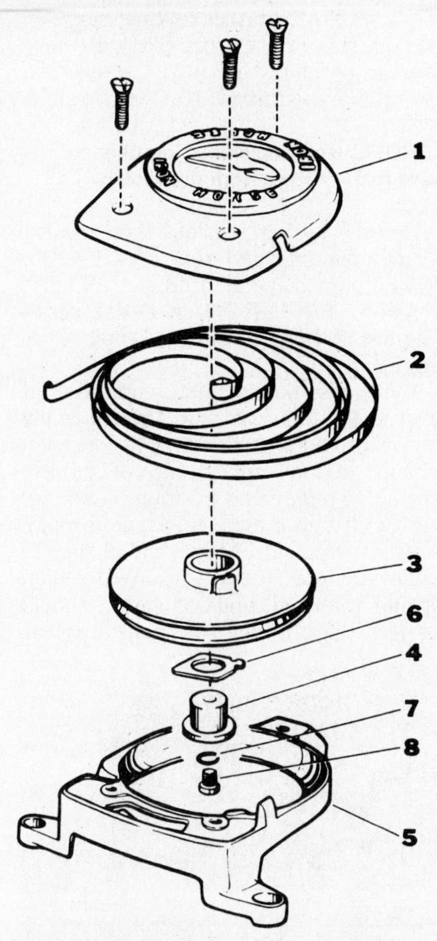

Fig. LB35—Exploded view of early type rewind starter used on some models. Very early models used bead on rope as retainer instead of plate (6). Refer to Fig. LB36 for method of adapting rope retaining plate to early production units.

1. Cover
2. Rewind spring
3. Rope pulley
4. Pulley bearing
5. Housing
6. Rope retainer
7. Lockwasher
8. Screw, pulley to cover

Fig. LB36—When renewing rope in early rewind starter, it is recommended that rope retainer plate (clamp) be installed. To install clamp, first remove 3/64 inch (1 mm) material from boss indicated by arrow; then assemble unit as shown.

bushing type top main bearing is worn, the later armature plate and needle bearing must be installed. Refer to Fig. LB33 for current change in armature plate bearings. Bearing removal tool 605082 is recommended for driving out old bearings. Hold armature plate in palm of hand for support, not on hard surface during removal to prevent breakage. Use tool 605081 to install new bearing assembly.

Crankpin diameter is 0.7495-0.7500 inch (19.04-19.05 mm) for C-series engines with plain connecting rod bearing and recommended clearance between bearing and crankpin journal is 0.0025-0.0035 inch (0.064-0.089 mm).

Crankpin diameter is 0.7425-0.7430 inch (18.86-18.87 mm) for C-series engines with needle bearing connecting rod bearing in all D-series engines. Crankshaft must be renewed if crankpin journal or main bearing journals show signs of wear, scoring or overheating.

On C and D-series engines, recommended crankshaft end play is 0.012 inch (0.30 mm). Gasket (11–Fig. LB30 or LB31) used between crankcase and armature plate is available in thicknesses of 0.005 and 0.10 inch for controlling end play.

As in all two-stroke engines, crankshaft seals must be maintained in good condition to prevent loss of crankcase compression and resulting loss of power. Install new seals with lips (grooved side of seal) to inside of crankcase. When installing crankshaft, use an installing sleeve or wrap tape over keyways and sharp shoulders on crankshaft.

RECOIL STARTERS (C-Series Engines). On early recoil starters, starter rope should be attached with retainer as shown at (6–Fig. LB36). Rope installation on late type starters is shown in Fig. LB39. Refer to Fig. LB37 when installing late type recoil spring in early type starter.

On QUIETFLITE mowers, starter is similar to that shown in Fig. LB38; however, top cover (1) is not used. The rewind spring is anchored on starter bracket as shown in Fig. LB40.

Be sure rewind spring is not overtightened. Never wind pulley over two turns after spring tension begins to be felt.

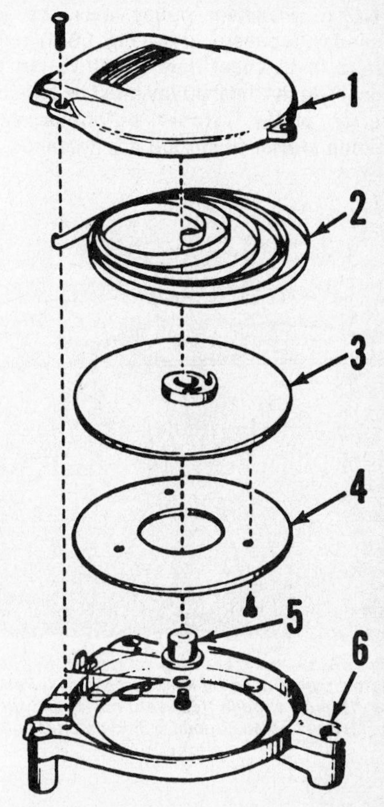

Fig. LB38—Exploded view of later type rewind starter. Refer to Fig. LB39 for rope installation.

1. Cover
2. Rewind spring
3. Rope pulley
4. Plate
5. Bushing
6. Housing

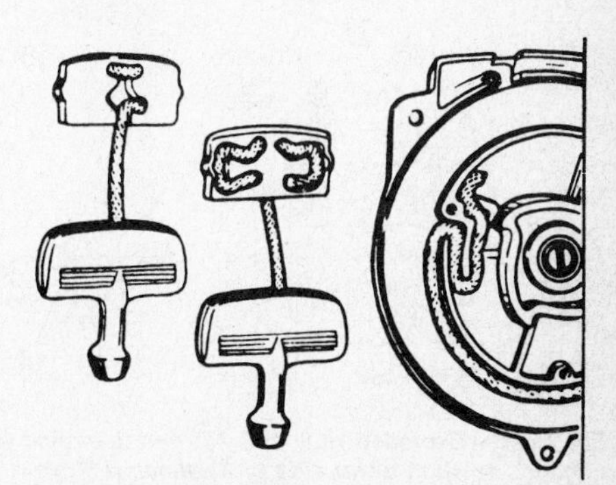

Fig. LB39—View showing method of retaining rope ends in rewind starter shown in Fig. LB36. Singe (heat) about ¾ inch (19 mm) of each end of nylon rope to be sure it will hold securely.

Fig. LB40—On QUIETFLITE models, be sure outer end of rewind spring is anchored at position indicated by arrow.

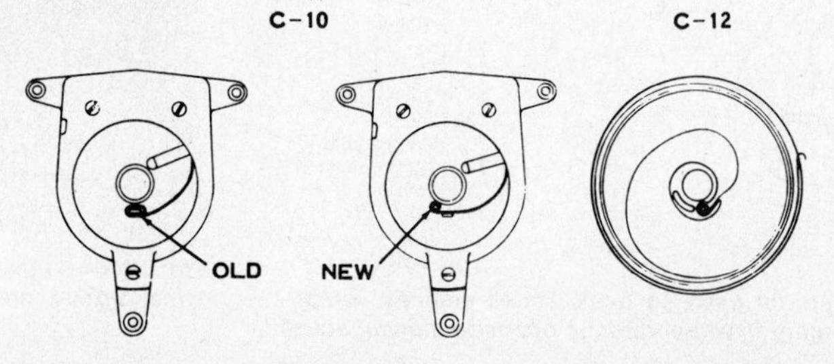

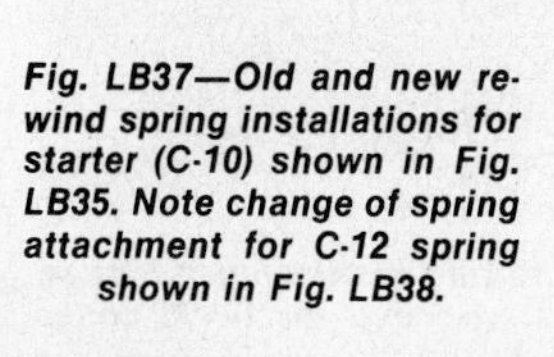

Fig. LB37—Old and new rewind spring installations for starter (C-10) shown in Fig. LB35. Note change of spring attachment for C-12 spring shown in Fig. LB38.

NOTE: Flywheel pulley must be installed with pulley lug (1—Fig. LB41) positioned in flywheel hole (2). Drive pin or pins (3) on flywheel pulley must be outside starter pulley ratchet before screws holding starter to shroud are tightened.

Fig. LB41—View showing a typical flywheel starter pulley used on C-series engines. Pulley lug (1) must engage flywheel hole (2) and drive pin (3) must be positioned outside starter pulley ratchet.

D-Series Engines. To remove recoil starter, remove cooling shroud and starter handle, then allow rope to slowly recoil. On D-600 series engines, carburetor and reed plate assembly must first be removed from crankcase to allow access to starter retainer screw in bottom side of armature plate. Remove starter assembly retaining screw and clamp, then withdraw starter assembly.

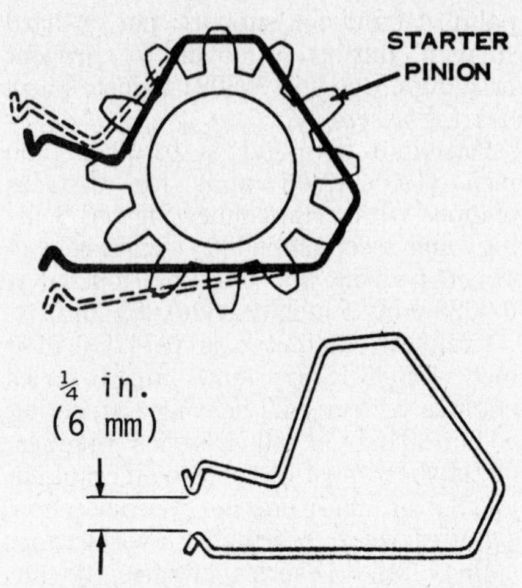

Fig. LB44—Ends of starter pinion spring should be less than 1/4 inch (6.4 mm) apart when spring is removed.

NOTE: Hold assembly together to prevent recoil spring from unwinding.

Release spring tension, then inspect all components and renew as needed. For reference and identification of parts refer to Fig. LB42 for D-400 series engines and Fig. LB43 for D-600 series engines.

When assembling, apply a light coat of grease to starter spring. **Do not** apply lubricant on starter worm gear or nylon pinion. Attach inner end of starter spring to pulley and position on starter cup with outer end of spring through slot in spring cup. Attach end of rope to pulley an install pulley plate. With pinion spring removed, ends of spring should be less than 1/4-inch (6.4 mm) apart as

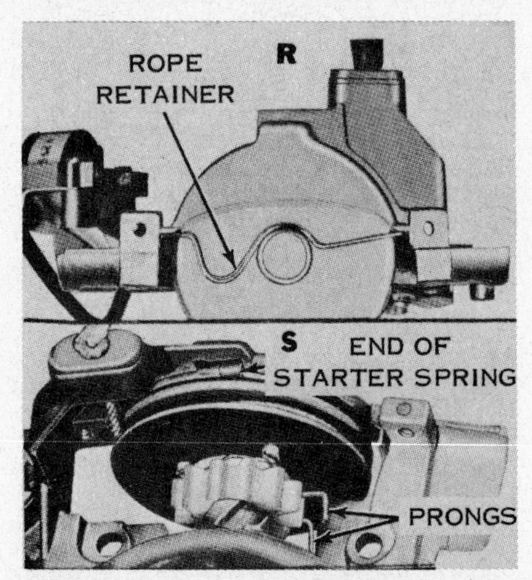

Fig. LB45—View R shows placement of D-400 series starter rope retainer locked in position with starter installed in armature plate. View S shows correct installation of starter pinion spring with one prong above and one below armature plate. Note location of end of recoil spring.

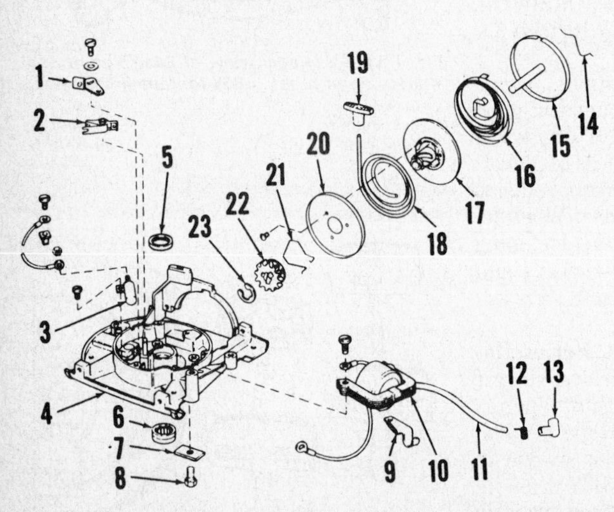

1. Breaker point base
2. Breaker point arm
3. Condenser
4. Armature plate
5. Crankshaft oil seal
6. Needle bearing
7. Starter pin clamp
8. Clamp screw
9. High tension wire clip
10. Coil & lamination assy.
11. High tension wire
12. Spring terminal
13. Spark plug cover
14. Rope retainer spring
15. Starter spring cup
16. Starter spring
17. Starter pulley
18. Starter rope
19. Starter handle
20. Starter pulley plate
21. Starter pinion spring
22. Starter pinion
23. Push-on retainer

Fig. LB42—Exploded view of D-400 series engine starter and magneto assembly. Starter pinion (22) engages teeth in lower side of flywheel; refer to view of flywheel in Fig. LB17. Needle bearing (6) is used in late engines; early production engines have plain bearing bore in armature plate (4) for upper crankshaft journal.

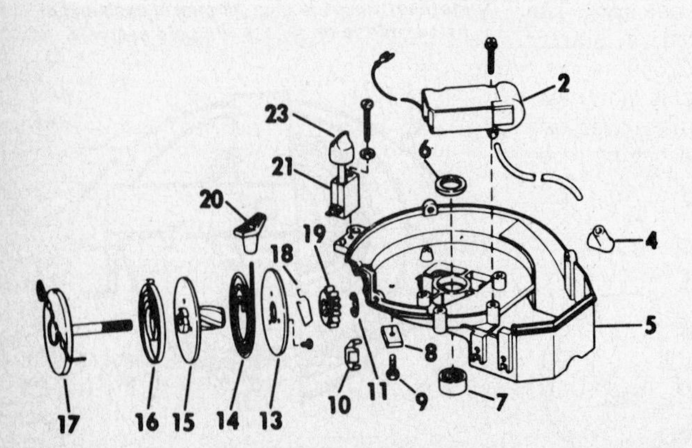

2. CD ignition spark
4. Spark plug boot
5. Armature plate
6. Crankshaft seal
7. Upper main bearing
8. Starter clamp
9. Clamp screw
10. Rope clip
11. Retainer
13. Plate
14. Rope
15. Pulley
16. Recoil spring
17. Cup & pin
18. Pinion spring
19. Pinion
20. Rope handle
21. Shorting switch
23. Knob

Fig. LB43—Exploded view of armature plate and starter used on D-600 series engines. When installed, electric starter is fitted to crankcase, engaging flywheel directly opposite manual starter.

Fig. LB46—View showing positioning of rope retainer spring on starter cup for D-400 series engines.

Fig. LB47—Position pinion spring prongs so one is above and one is below armature plate.

shown in Fig. LB44. Install spring in pinion groove with prongs toward pinion teeth as shown in Fig. LB45, being careful not to spread ends of spring excessively apart. Position pinion on pulley, then install retaining snap ring. Wind rope around pulley, then on D-400 series engines position rope retainer springs as shown in Figs. LB45 and LB46 and install assembly. On D-600 series engines rope retainer spring is permanently attached to starter cup. Turn pulley until starter spring "hook" is positioned at top. Position pinion spring prongs so one is above and one is below armature plate as shown in Fig. LB47. Preload starter spring ½-1½ turns, then route starter rope and install handle.

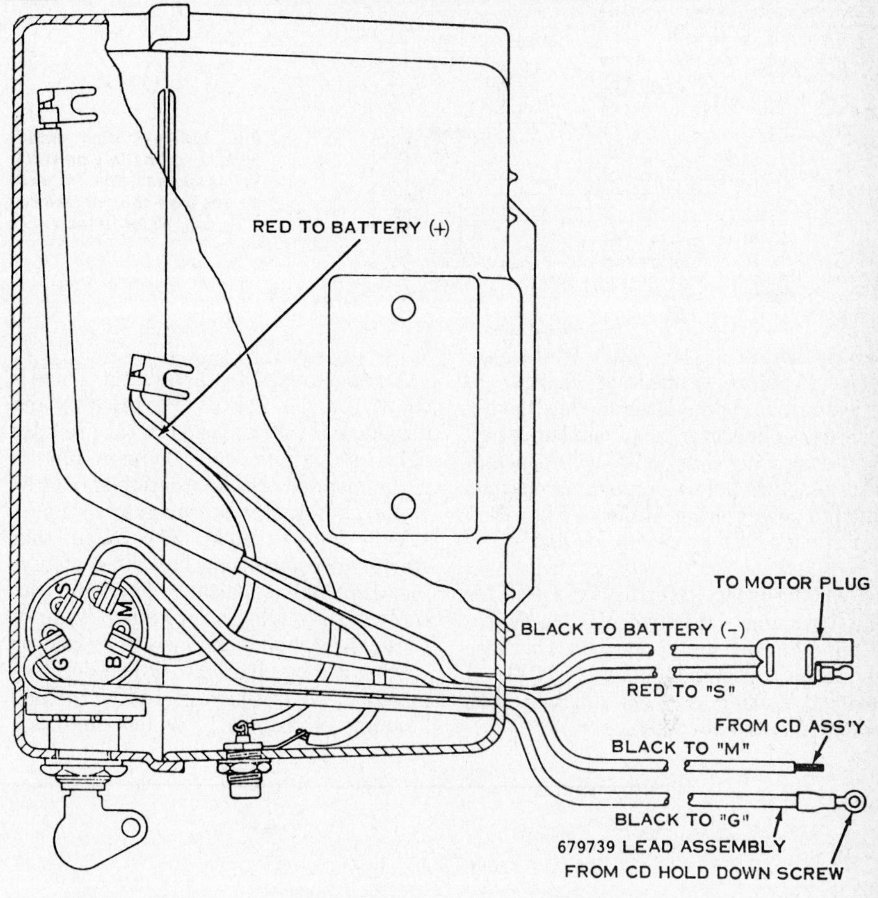

Fig. LB49—Wiring diagram for D-600 series engines with electric start. Note charging jack adjacent to igntion key switch. Refer to text.

12-VOLT ELECTRIC STARTER. Models equipped with electric start are identified by letter "E" following model number.

D-400 Series. The electric starter pinion (27–Fig. LB48) is located across from recoil starter pinion (22). The electric starter pinion assembly can be removed after removing drive belt, flywheel and set screw (24–Fig. LB50). When installing, clearance between snap ring and pulley should be 0.010 inch. Clearance is adjusted by moving pulley shaft before tightening set screw (24). The starter drive belt should have ⅛-inch (3.2 mm) deflection with 1½ lbs. (0.68 Kg.) pressure between pulley. Adjustment is accomplished by moving the starter and bracket after loosening the three screws.

To charge battery, connect trickle charger leads (red to positive post) to

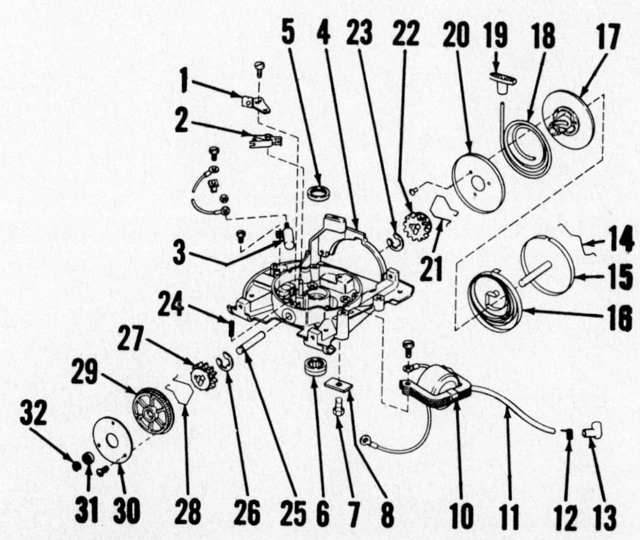

1. Breaker point base
2. Breaker point arm
3. Condenser
4. Armature plate
5. Crankshaft oil seal
6. Needle bearing
7. Starter pin clamp
8. Clamp screw
9. High tension wire clip
10. Coil & lamination assy.
11. High tension wire
12. Spring terminal
13. Spark plug cover
14. Rope retainer spring
15. Starter spring cup
16. Starter spring
17. Starter pulley
18. Starter rope
19. Starter handle
20. Starter pulley plate
21. Starter pinion spring
22. Starter pinion
23. Push-on retainer
24. Set screw
25. Starter shaft
26. Snap ring
27. Starter pinion
28. Pinion spring
29. Pulley
30. Pulley plate
31. Pulley bearing
32. Snap ring

Fig. LB48—Exploded view of armature plate and starter pinions used on D-400 series engines with 12-volt electric starter.

Fig. LB50—End play of starter assembly on starter shaft should be 0.010 inch (0.25 mm). End play is set by inserting feeler gage (F) between snap ring and pulley then pushing shaft in before tightening set screw (24).

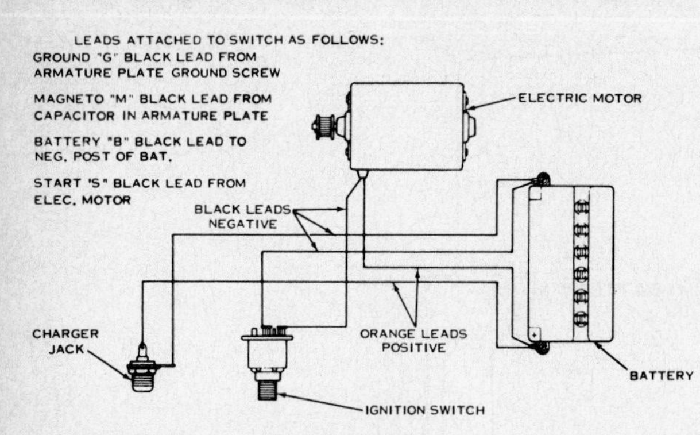

Fig. LB51—Starter circuit with 12-volt battery on D-400 series engines. Some models do not have charger jack circuit. Refer to text.

battery before connecting charger to 100V outlet. Some later models have a charger connecting jack installed in battery case. Plug charger cable into jack of these models before connecting charger cord to power outlet. Refer to Fig. LB51 for starter-battery circuit layout.

D-600 Series. Refer to Fig. LB49 for wiring diagram of starter-battery charger circuits with ignition switch. On these models, ignition key must be "ON" if manual starter is used and will then function as a kill switch.

Arrangement of starter and drive is shown in Fig. LB52. Pinion (82) and driven gear (83) enclosed in case require light lubrication; OMC Type A lube is recommended. No lubrication should be applied to starter worm gear or nylon starter pinion gear (87). Note that starter moter housing has three mounting legs which attach by long cap screws to engine crankcase. Two of these three bolt holes are oversized so casting can be pivoted to adjust depth of pinion mesh in flywheel. Allow a small amount of backlash when mounting starter assembly so pinion can engage teeth of flywheel without clashing or binding.

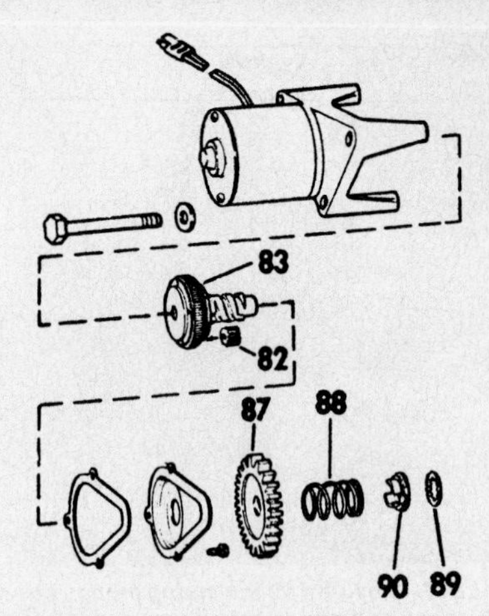

Fig. LB52—View of electric starting motor and drive used on D-600 series engines.

82. Motor pinion
83. Driven gear
87. Starter pinion
88. Spring
89. Retainer
90. Pinion stop washer

LAWN BOY

Model	Bore	Stroke	Displacement
F-100, F-101, F-140AE, F-140M, F-141, F-141AE, F-142AF, F-200, F-201, F-240, F-241, F-340	2.38 in. (60.4 mm)	1.75 in. (44.4 mm)	7.78 cu. in. (127 cc)

ENGINE INFORMATION

Engine model number is located on starter spring cap as shown in Fig. LB69. Always furnish engine model number when ordering parts or service information.

MAINTENANCE

SPARK PLUG. Recommended spark plug is a Champion CJ14, or equivalent.

Electrode gap should be set at 0.035 inch (0.9 mm) for all models. Tighten spark plug to 12-15 ft.-lbs. (16-20 N•m).

NOTE: Caution should be exercised if abrasive type spark plug cleaner is used. Inadequate cleaning procedure may allow the abrasive cleaner to be deposited in engine cylinder causing rapid wear and part failures.

CARBURETOR. Lawn Boy modular float type carburetor is used. Carburetors are fully automatic. The only adjusment offered is for proper operation at a given altitude using an adjustment needle that regulates the correct amount of fuel with the incoming air. Adjustment of needle is covered later in this section.

Fig. LB69—View of starter spring cap showing location of engine model number.

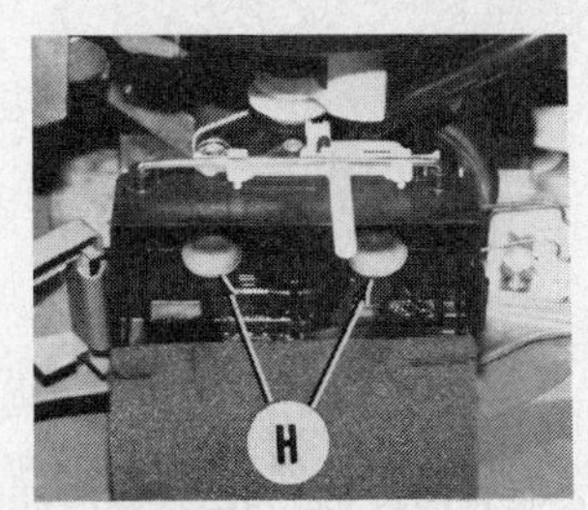

Fig. LB70—View showing carburetor air filter and access holes (H) in filter chamber.

Removal of carburetor requires removal of filter cover and element. Reach through carburetor filter chamber holes (H–Fig. LB70) with a suitable Phillips screwdriver and unscrew two mounting screws.

Carburetor is made of injection molded plastic as shown in Fig. LB71. There are two types of carburetors, "Old Style" and "New Style". "New Style" carburetor is identified by fuel flow adjustment screw as shown in Fig. LB72 while "Old Style" carburetor is equipped with an altitude adjustment needle located in side of carburetor. Both type carburetors have an air vane (pneumatic) governor that is an integral part of the carburetor throttle shaft and plate.

Initial adjustment for "New Style" carburetors is as follows: Turn fuel flow needle in (clockwise) until lightly seated, then turn needle out (counterclockwise) two turns. Start engine and allow to warm up. Set speed control lever to "High Speed" position, turn fuel flow needle slowly in (clockwise) in 1/8-turn increments until engine starts surging or slowing down, then turn needle slowly out (counterclockwise) until engine runs smoothly. Allow engine to run a few minutes at this setting to make certain fuel mixture is not too lean. Readjust "High Speed" setting as outlined in GOVERNOR section to 3100-3300 rpm. Shut off engine and immediately attempt to restart. Engine should restart easily in both "High Speed" and "Low Speed" settings. If difficulty in restarting is encountered, turn fuel flow needle out (counterclockwise) an additional 1/8-1/4 turn to richen fuel mixture.

Fig. LB71—View showing a typical modular type carburetor. Refer to text for differences in "Old Style" and "New Style" carburetor.

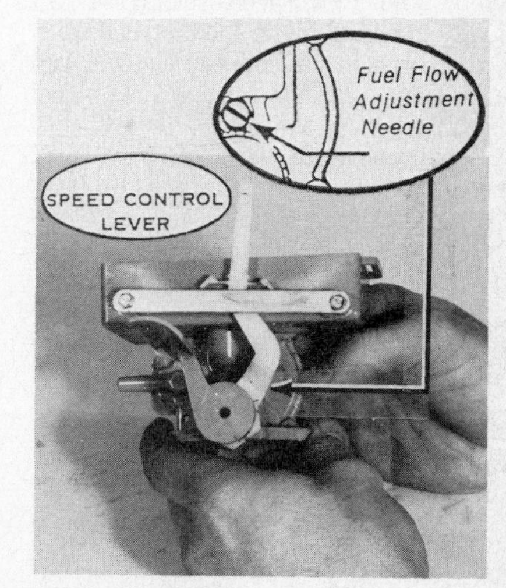

Fig. LB72—Top view of a F-series carburetor showing "New Style" fuel flow adjustment needle. Refer to text for adjustment procedure.

Initial adjustment of altitude needle on "Old Style" carburetor for elevations of 500-1500 feet above sea level should

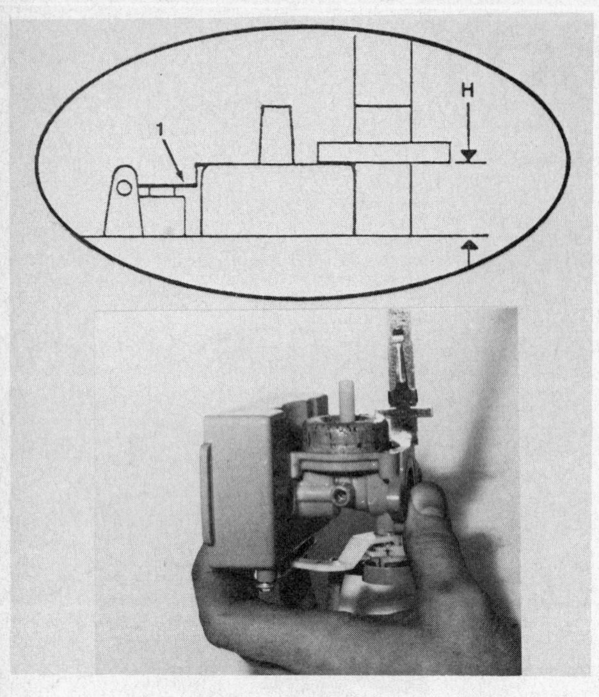

Fig. LB73—Correct float setting (H) is 15/32 inch (12 mm). Adjust float setting by bending float arm (1).

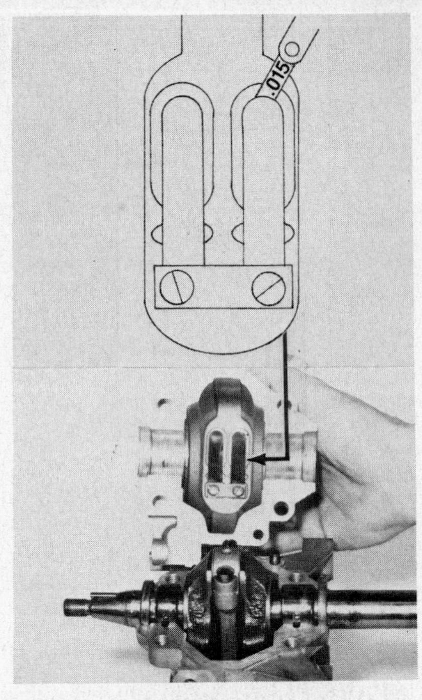

Fig. LB75—View showing positioning of reed valves in crankcase cover. Maximum clearance of 0.015 inch (0.38 mm) between reed tip and machined surface is allowed.

be ¾-turn out from fully seated position. For final adjustment of altitude needle proceed as follows: Start engine and allow to warm up. Turn needle in (clockwise) in ⅛-turn increments until engine starts surging or slowing down, then turn needle slowly out (counterclockwise) until engine runs smoothly. Allow engine to run a few minutes at this setting to make certain fuel mixture is not too lean. Readjust "High Speed" setting as outlined in GOVERNOR section to 3100-3300 rpm. Shut off engine and immediately attempt to restart. Engine should restart easily in both "High Speed" and "Low Speed" settings. If difficulty in restarting is encountered, turn altitude needle out (counterclockwise) an additional ⅛-¼ inch turn to richen fuel mixture.

Float setting (H – Fig. LB73) is 15/32 inch (11.9 mm) above carburetor body flange. Adjust float level only by bending float arm (1) using needle nose pliers. **DO NOT** apply strain or pressure to cork float. Make certain hinge on float arm is secured to pin. Pin should swivel in carburetor housing; float arm should not turn on pin.

Clean carburetor components with a mild solvent and blow dry with compressed air. **DO NOT** use a standard carburetor cleaner. **DO NOT** dry carburetor parts with a cloth, as lint from cloth could plug fuel passage and cause a performance problem.

PRIMER. A pneumatic bulb type primer as shown in Fig. LB74 is used. Depressing primer bulb forces air into float chamber which forces fuel through main jet into carburetor venturi. **DO NOT** overprime; one priming stroke is normally sufficient.

REED VALVE. Reed valves permit fuel-air mixture to enter crankcase on the compression stroke and seal fuel-air mixture in crankcase on the power stroke. Reed valves are mounted in crankcase cover as shown in Fig. LB75. Reeds can be cleaned with solvent or carburetor cleaner. **DO NOT** use compressed air on reed assemblies or distortion may occur resulting in hard starting or loss of power. Bent or distorted reeds cannot be repaired. Rough edge of reed must be positioned away from machined surface of mounting plate. Check reed-to-plate clearance as shown in Fig. LB75. Clearance of 0.015 inch (0.38 mm) between reed tip and machined surface is allowable. Renew reed if clearance is excessive. Coat reed mounting screws with Lawn Boy Nut & Screw Lock (number 682301) or a suitable equivalent and tighten to 10-13 in.-lbs. (1 N·m).

GOVERNOR. Governor is an air vane (pneumatic) type that is an integral part of the carburetor throttle shaft. Air vane (A – Fig. LB76) extends through

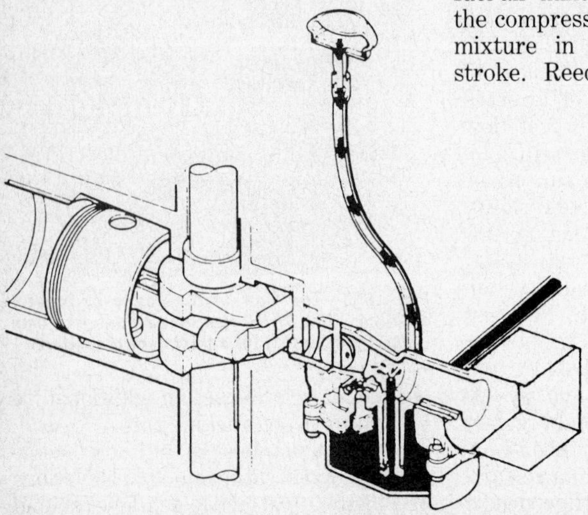

Fig. LB74—Cutaway view of a typical carburetor showing priming pump assembly.

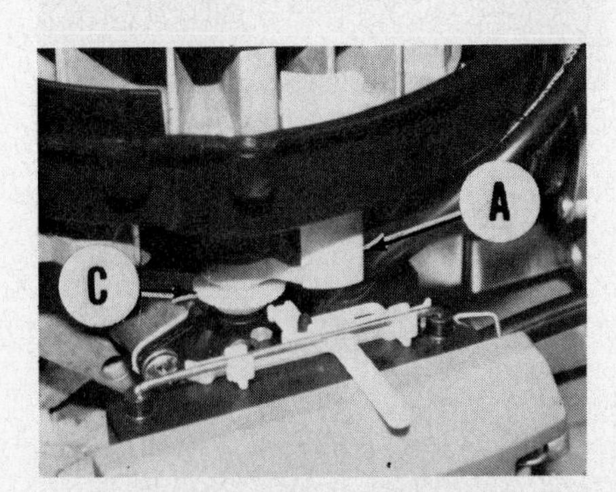

Fig. LB76—Governor air vane (A) extends through cooling shroud base and responds to air flow created by flywheel fan. View shows round-shaped adjusting collar (C). Refer to text.

Fig. LB77—View showing procedure for using Air Gap Gage to set air gap between flywheel magnets and coil heels. Air gap should be 0.010 inch (0.25 mm). Refer to text.

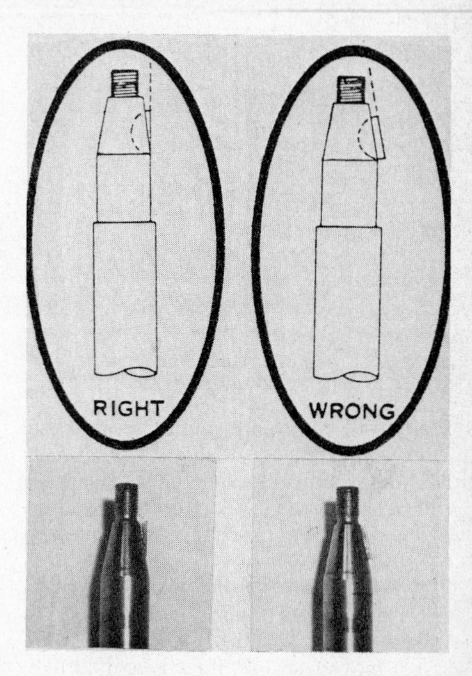

Fig. LB78—When installing flywheel, be sure Woodruff key is installed correctly in crankshaft groove as shown.

cooling shroud base and responds to air flow created by the flywheel fan. Fluctuation in engine speed, due to change in engine load, opens or closes throttle plate to maintain desired engine rpm. Recommended governor high speed is 3100-3300 rpm. Two types of governor adjustment mechanisms are used and are identified by either a hexagonal or round adjusting collar.

To adjust governor with hex-shaped collar, lift collar slightly (releasing it from splines on speed control lever) and turn either direction to obtain desired speed. From zero tension on collar spring, turn collar counterclockwise approximately ⅓-turn to obtain approximate high running speed. Each click on splines represents approximately 50-75 rpm. Clockwise rotation of collar **DECREASES** engine speed, while counterclockwise rotation **INCREASES** engine speed.

To adjust round-shaped collar (C – Fig. LB76), hold base of air vane and turn collar to desired engine speed. Each click represents approximately 50 rpm. Clockwise rotation of collar **INCREASES** engine speed, while counterclockwise rotation **DECREASES** engine speed.

Disassembly of governor with hex-shaped adjusting collar is as follows: With carburetor removed, remove throttle disc with needle nose pliers. Lift adjusting collar and turn clockwise until all tension is released from spring. Remove throttle shaft/governor vane, spring and adjusting collar. Pry up on hub of speed control lever until lever is free of lug.

Disassembly of governor with round-shaped collar is similar to hex-shaped except for the following: Rotate adjusting collar counterclockwise to release tension on spring. On models equipped with speed control lever, lower end of spring fits into lever. For models without speed control lever, spring end secures in carburetor body. Note position of spring end prior to disassembly.

Reassemble governor assemblies by reversing disassembly procedure. Adjust engine speed as previously outlined.

CAPACITIVE DISCHARGE IGNITION SYSTEM. Lawn Boy's solid-state (capacitive discharge) ignition design eliminates all moving parts which are normally associated with magneto ignition. On these models there is no spark advance assembly, breaker cam, breaker point set, condenser or coil. Complete system is electronic and contained in a single, sealed module which is dust and moisture proof and serviced only as an assembly.

Maintenance is confined to setting air gap between flywheel magnets and CD module, checking for spark output by use of a test spark plug grounded to cylinder and isolation test of ignition switch for continuity. No procedures for troubleshooting or testing ignition pack are offered by manufacturer. If unit will not produce a spark at electrodes of test spark plug and ignition switch and spark plug cable check out good, then renewal of module assembly is necessary. Substitution of a spare ignition pack known to be good is an ideal test procedure.

Ignition timing is fixed. System design is set up for 9° retarded spark timing for ease of starting. After engine starts, spark is advanced electronically to 29° BTDC as crankshaft speed reaches 800 rpm.

To service ignition module, remove five screws holding air baffle to shroud base and separate kill switch lead from ignition switch. Primer and starter need not be removed. With CD ignition pack exposed, 0.010 inch (0.25 mm) air gap may be checked. To do so, rotate flywheel to align flywheel magnets with laminated core heels of ignition module, then insert Lawn Boy Air Gap Gage 604659 or a suitable piece of 0.010 inch (0.25 mm) nonmetallic shim stock as shown in Fig. LB77. With gage in position, loosen module mounting screws so magnets will pull module snug against gage. Retighten screws and remove gage.

Should flywheel be removed from crankshaft, be sure during reassembly Woodruff key is installed correctly in crankshaft groove as shown in Fig. LB78.

Operators of this engine should be cautioned to allow 10-15 seconds after shutdown before removing spark plug lead. This is to allow high voltage (approximately 30,000 volts) from CD pack secondary to leak off.

LUBRICATION. Engine is lubricated by premixing oil and gasoline in a separate container before filling engine tank. Manufacturer recommends using Special Lawn Boy lubricant or a suitable good quality two-stroke oil. **DO NOT** use oil additives or standard automotive oils.

The use of regular gasoline (minimum octane rating of 89), unleaded or no-lead (minimum octane rating of 86) is recommended. **DO NOT** use gasoline additives except OMC 2 + 4 fuel conditioner. When using Special Lawn Boy lubricant mix fuel at a ratio of 32:1 or two gallons (7.6 L) of gasoline to eight ounces (236.6 mL) of lubricant. When using a different brand of two-stroke oil mix fuel at a ratio of 16:1.

CARBON. If ignition fuel supply to combustion chamber and compression check out as satisfactory, but engine will not run or runs poorly, it is likely that there is heavy carbon build-up in exhaust ports and muffler. A common

Fig. LB79—View showing muffler plate (1) and engine exhaust ports (2).

symptom is "four-cycling" or firing every other power stroke.

Manufacturer recommends cleaning out carbon every 35 operating hours. Carbon cleaning procedure is as follows:

NOTE: Drain fuel from tank and on electric start models remove battery.

Tip mower uint on its side and secure in position by blocking. Remove mower blade nut, blade, washer (models so equipped), blade stiffener and collar. Remove three mounting cap screws securing crankshaft support and muffler to muffler plate. Figure LB79 shows muffler plate (1) and engine exhaust ports (2). Rotate crankshaft so piston wall blocks exhaust ports, then use a 7/16-inch wooden dowel to break carbon away from port openings. Clean muffler openings using a ¾-inch wooden dowel.

When reassembling crankshaft support and muffler to muffler plate use Lawn Boy Crankshaft Support Gage 609968 or a 0.015-inch (0.38 mm) piece of shim stock to set clearance between crankshaft and crankshaft support. Use Lawn Boy Nut & Screw Lock on mounting cap screws, then tighten to 12-14 ft.-lbs. (16-19 N•m).

REPAIRS

TIGHTENING TORQUES. Recommended tightening torques are as follows:

Spark plug . . . . 144-180 in.-lbs. (16-20 N•m)
Carburetor to crankcase cover . . . . 63-75 in.-lbs. (7-8 N•m)
C.D. mounting screws . . . . 20-25 in.-lbs. (2-3 N•m)
Reed to crankcase cover . . . 10-13 in.-lbs. (1-1.5 N•m)
Crankcase cover to crankcase . . . . 100-120 in.-lbs. (11-13 N•m)
Connecting rod bolts . . . . . 55-65 in.-lbs. (6-7 N•m)
Flywheel . . . . 375-400 in.-lbs. (42-45 N•m)
Crankshaft support . . . . 140-170 in.-lbs. (16-19 N•m)
Engine to muffler plate . . . . 140-170 in.-lbs. (16-19 N•m)
Muffler to muffler plate . . . . 140-170 in.-lbs. (16-19 N•m)
Blade nut . . . . 450-550 in.-lbs. (51-62 N•m)

CONNECTING ROD. Rod and piston assembly can be removed after removing carburetor and crankcase cover half. Engines have a needle roller crankpin bearing, aluminum connecting rod is fitted with renewable steel inserts. Figure LB80 shows a view of connecting rod assembly. Crankpin journal diameter should be 0.7425-0.7430 inch (18.86-18.87 mm).

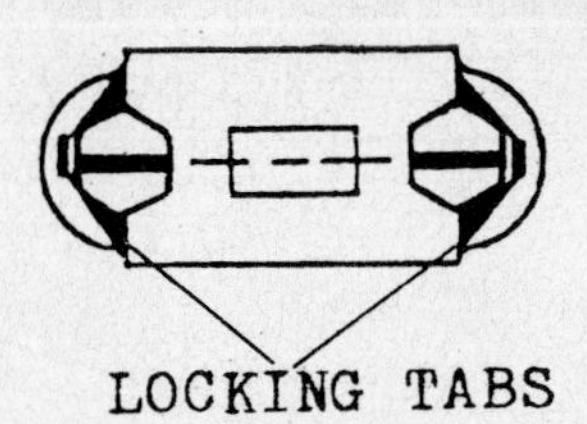

Fig. LB82—Bend locking tabs snugly up against connecting rod cap screws to secure.

When reassembling connecting rod to crankpin, install steel inserts in connecting rod end and cap making sure they are centered and dovetail guides will mate correctly when assembled. If bearing needles are being reused, be sure that none of the 33 rollers are damaged. Coat insert surfaces with a layer of OMC Needle Bearing Grease 378642 or a suitable equivalent and stick rollers to connecting rod end and cap. Fit 17 rollers to rod cap and 16 to rod end.

If installing new needle roller set, lay strip of 33 rollers on forefinger and carefully peel backing off rollers. Curl forefinger around crankpin to transfer rollers from finger to journal. Grease on roller will hold them together and to crankpin journal.

Install connecting rod cap; BE SURE rod end and cap mating marks align as shown in Fig. LB81. Tighten rod cap screws to initial torque of 15-20 in.-lbs. (1.7-2.3 N•m), then alternately tighten rod cap screws in 20 in.-lbs. (2-3 N•m) steps to a final torque of 55-65 in.-lbs. (6-7 N•m). Bend locking tabs (Fig. LB82) snugly against screwheads. Check to be sure rod assembly is free on crankpin and none of the rollers dropped out during assembly.

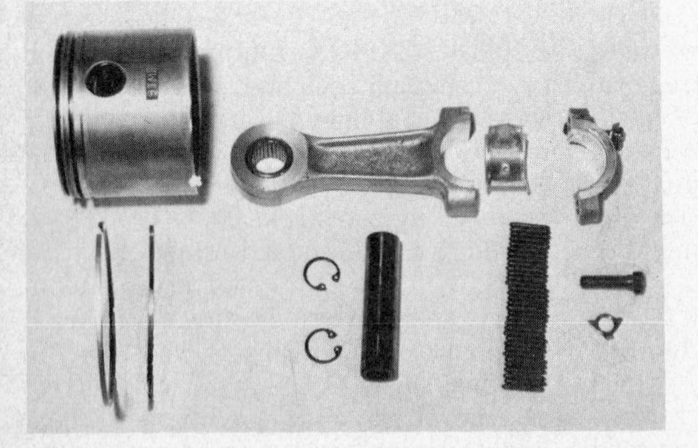

Fig. LB80—View showing connecting rod and piston assembly. Note semi-keystone ring used in top groove.

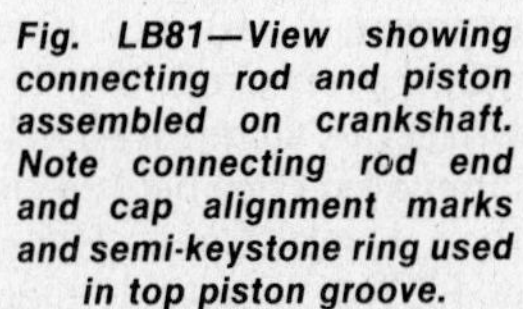

Fig. LB81—View showing connecting rod and piston assembled on crankshaft. Note connecting rod end and cap alignment marks and semi-keystone ring used in top piston groove.

PISTON, PIN AND RINGS. Piston, pin and rings are available in standard size only. Recommended piston to cylinder clearance is 0.003-0.004 inch (0.08-0.10 mm).

Piston pin is retained by snap rings at each end of pin bore in piston. Use Lawn Boy Special Tool 602884 or a suitable wooden dowel to drive piston pin out. Piston pin diameter is 0.4998-0.5000 inch (12.695-12.700 mm). Connecting rod piston pin bore is 0.5001-0.5005 inch (12.702-12.713 mm). Clearance between piston pin and bore should be 0.001-0.005 inch (0.03-0.13 mm). Top piston ring end gap should be 0.007-0.017 inch (0.18-0.43 mm) and bottom ring end gap should be 0.015-0.025 inch (0.38-0.64 mm). Renew rings if end gap is excessive. When installing rings, stagger ring gaps at least 30° from one another. Note semi-keystone ring used in top ring groove as shown in Fig. LB81. Reassemble connecting rod to piston and install piston pin retaining rings. Position retaining ring so beveled side is towards pin and opening faces upward toward piston dome to prevent ring from popping out during operation.

When installing piston and rod assembly, place Lawn Boy Stop 677389 or a suitable equivalent in spark plug hole.

NOTE: Without piston stop, piston could enter cylinder too far allowing top ring to become stuck in cylinder.

Oil piston thoroughly, then using Lawn Boy Ring Compressor 609967 or a suitable equivalent install piston and rod assembly in cylinder.

NOTE: Piston is cast with letters "BTM" on skirt. Install piston with letters "BTM" facing downward toward exhaust port.

CRANKCASE. Integral crankcase/cylinder head should be renewed if cylinder bore is worn so piston skirt to cylinder clearance exceeds 0.004 inch (0.10 mm), or ring end gap is excessive with new piston rings. Oversize pistons or rings are not available. Crankcase cover houses reed valve assembly.

Thoroughly clean and inspect all parts for excessive wear, cracks or other damage. Remove any gasket material on mating surfaces of crankcase and cover, then apply a thin coat of Lawn Boy Gasket Maker 682302 or a suitable equivalent to sealing surface of crankcase cover.

BE SURE crankshaft main bearing dowels align in crankcase locator slots as shown in Fig. LB83 before installing cover.

Install crankcase cover and securing screws, then using a criss-cross pattern, tighten screws in small increments until a final torque of 100-120 in.-lbs. (11-13 N·m) is obtained. Check crankshaft for free rotation.

CRANKSHAFT AND SEALS. Engines are equipped with caged roller bearings supporting crankshaft. Bearings contain 20 needle rollers, which should be renewed if excessively worn or damaged in any way. Top main bearing is shorter in height than lower bearing. Top and bottom crankshaft journal diameter is 0.8773-0.8778 inch (22.283-22.296 mm). Diametral clearance between crankshaft journal and roller bearing should be 0.002-0.003 inch (0.05-0.08 mm). Maximum allowable clearance is 0.005 inch (0.13 mm). Crankpin diameter is 0.742-0.743 inch (18.85-18.87 mm) with a diametral

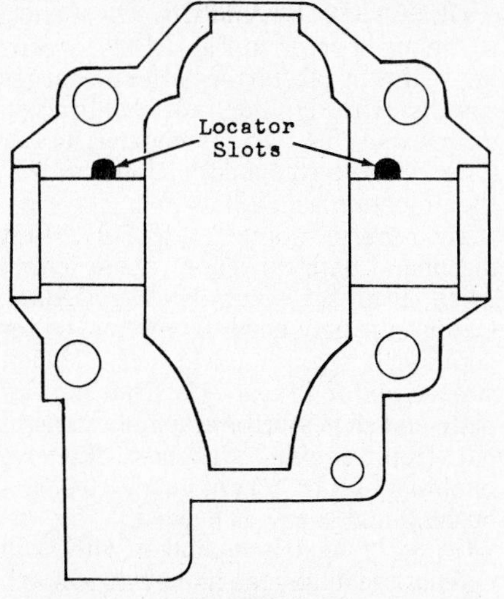

Fig. LB83—View showing crankcase locator slots used in aligning crankshaft main bearings. Be sure main bearing dowels align in slots.

clearance of 0.0025-0.0035 inch (0.064-0.089 mm) between pin and connecting rod assembly. Renew crankshaft if crankpin is out-of-round more than 0.0015 inch (0.38 mm). Crankshaft end play should be 0.006-0.016 inch (0.15-0.41 mm).

Crankshaft seals must be maintained in good condition to prevent loss of crankcase compression, resulting in loss of power. Install new seals with lips (grooved side of seal) to inside of crankcase. When installing crankshaft seals, use a protective sleeve or wrap tape over keyway and sharp shoulders on crankshaft to prevent damage to seals.

When installing crankshaft assembly make sure main bearing dowels are aligned in crankcase locator slots as shown in Fig. LB83. Remove any gasket material on mating surfaces of crankcase and cover, then apply a thin coat of Lawn Boy Gasket Maker 682302 or a suitable equivalent to sealing surface of crankcase cover.

Install crankcase cover and securing screws, then using a criss-cross pattern, tighten screws in small increments until a final torque of 100-120 in.-lbs. (11-13 N·m) is obtained. Check crankshaft for free rotation.

RECOIL STARTER. To remove recoil starter assembly, first remove starter rope handle, then allow rope to recoil slowly through hole in air baffle. Loosen crankcase Allen head screw securing starter, then withdraw starter assembly.

CAUTION: Starter spring is wound in cup housing and may cause personal injury if spring is released accidentally. Wearing eye protection is recommended during disassembly and reassembly.

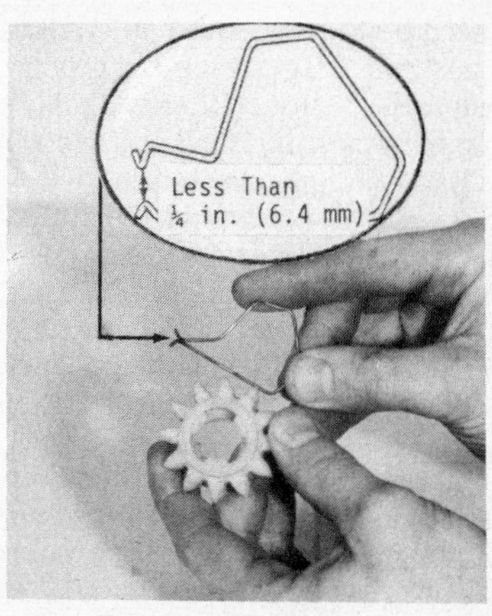

Fig. LB84—When pinion spring is removed from pinion gear, ends of spring should be less than ¼ inch (6 mm) apart.

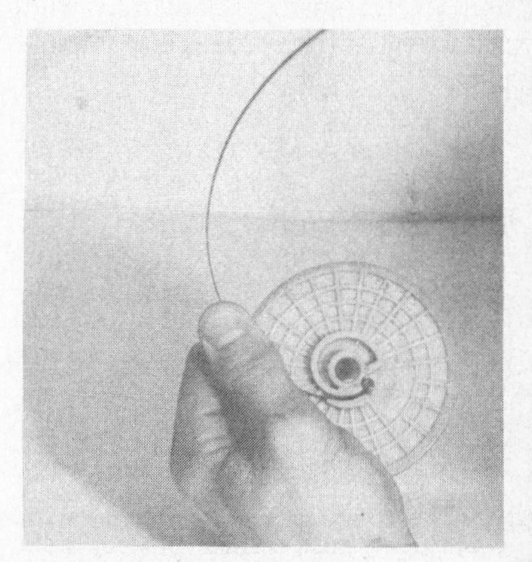

Fig. LB85—During reassembly, attach inner end of starter spring to pulley as shown.

When disassembling recoil starter, hold starter cup and pulley together to prevent starter halves from separating. Using needle nose pliers, withdraw retaining clip from cup shaft. Slide washer, pinion gear and drag spring off shaft end. To release spring tension, point shaft end downward and allow starter halves to fall onto a suitable solid surface. Halves will separate on impact and spring will unwind.

Inspect all parts for excessive wear or any other damage and renew as needed.

With pinion spring removed, ends of spring should be less than ¼-inch (6 mm) apart as shown in Fig. LB84.

When assembling, apply a thin coat of Lawn Boy "A" grease 610726 or a suitable equivalent over entire length of starter spring. Attach inner end of spring to pulley as shown in Fig. LB85. Install cup and shaft assembly with spring guided through slot in cup. Install pinion gear, thrust washer and retaining clip. Hold assembly securely and turn pinion gear clockwise to wind spring into housing. Wind spring until hook on spring is secured in cup slot as shown in Fig. LB86. Carefully allow spring to unwind against inside of cup.

Install starter assembly in crankcase, the position drag spring so prongs on spring are positioned one above one below flange of CD bracket as shown in Fig. LB87. Align rope retainer (1–Fig. LB88) with CD bracket (2), then tighten crankcase Allen head screw to secure starter.

Wind rope fully on pulley, then note if rope end is in the "A" or "B" portion of the pulley as shown in Fig. LB89. If rope end is in portion "A", pull rope out enough to allow one additional wrap around pulley.

If rope end is in portion "B", pull rope out enough to allow two additional wraps around pulley. Thread rope through guide and attach handle. Be sure starter handle has tension against it and starter spring is not fully wound tight when starter rope is completely drawn.

Fig. LB86—During reassembly wind starter spring until spring hook is secured in cup slot.

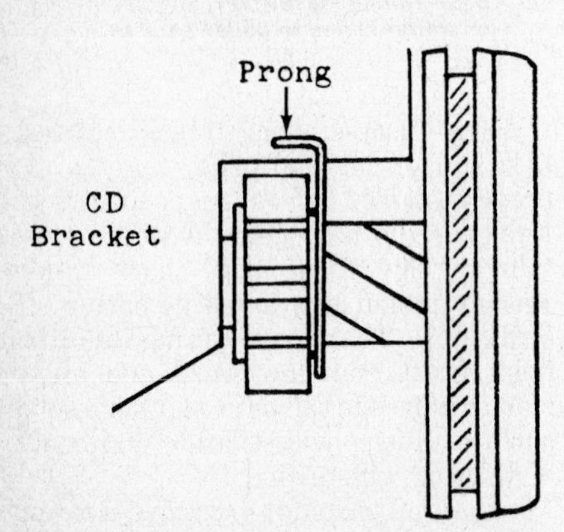

Fig. LB87—Position pinion spring prongs so one is above and one is below flange of CD bracket.

Fig. LB88—Align rope retainer (1) with CD bracket (2) during reassembly.

ELECTRIC STARTER. On models so equipped refer to Fig. LB90 for wiring diagram of starter-battery charger circuits with ignition switch. Ignition switch must be "ON" if manual starter (on models so equipped) is used and will then function as a kill switch.

To remove starter assembly, first diconnect battery leads, then loosen Allen head set screw (1–Fig. LB91). Carburetor may need to be removed for additional wrench clearance. Unbolt starter motor bracket (2) from muffler plate and slide starter assembly straight out from engine crankcase. Remove bushing from crankcase. Inspect bushing and renew if needed.

On early models equipped with both manual recoil starter and electric start, disassembly is as follows: Remove clip, washer, spring and pinion gear from motor shaft. Hold pulley securely to gear cover and remove clutch assembly. Carefully release spring tension, remove pulley, then remove screws securing gear cover to motor. Remove gear cover, pinion gear, starter gear and thrust washer. Inspect all parts for excessive wear or any other damage and renew as needed.

To reassemble, install recoil spring in pulley and assemble starter components, then wind spring into cup. Wind rope on pulley, then pull rope out enough to allow one additional wrap on pulley.

Install starter assembly on engine and leave 0.025-0.030 inch (0.64-0.76 mm) clearance between end of crankcase mounting boss and end of assembled starter components, then tighten all mounting screws.

Late model electric starters are not equipped with a manual recoil starter option as shown in Fig. LB91. Electric starter must be in good operating condition for engine to be started. To inspect and/or repair starter components, disassemble as follows: Remove retaining clip, washer, spring and pinion gear from motor shaft. Inspect all parts for excessive wear or any other damage and renew as needed.

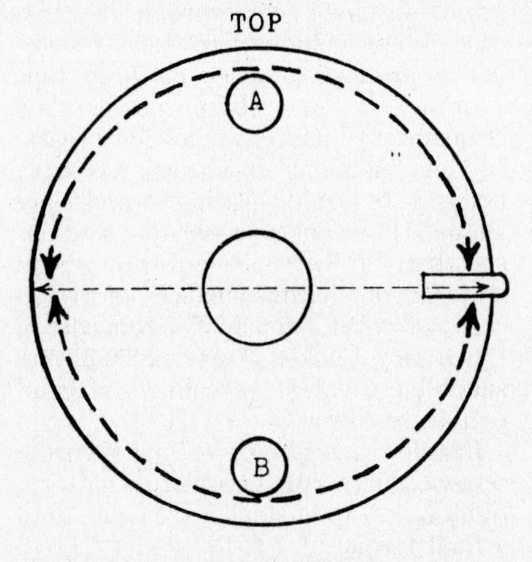

Fig. LB89—Note "A" and "B" halves of starter cup. Refer to text.

Fig. LB91—View of electric start assembly used on late model engines. Note starter motor is not equipped with manual start option as used on early models. To remove starter motor assembly, loosen Allen head set screw (1) and unbolt mounting bracket (2) from muffler plate.

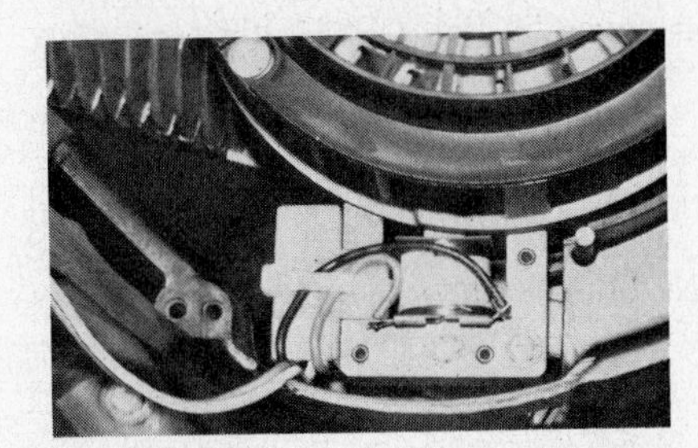

Fig. LB92—View showing "New Style" alternator used on electric start models.

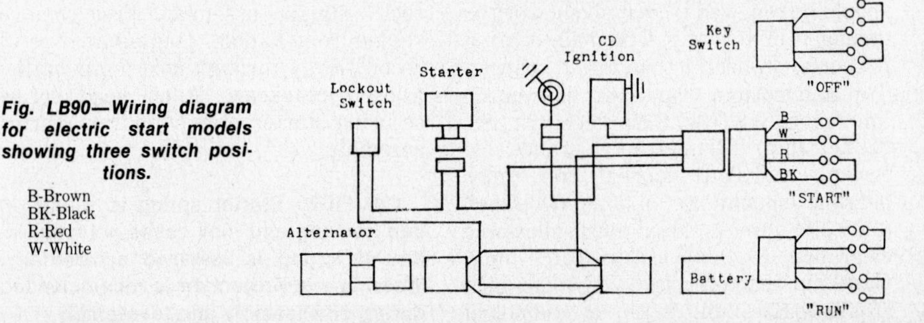

Fig. LB90—Wiring diagram for electric start models showing three switch positions.

B-Brown
BK-Black
R-Red
W-White

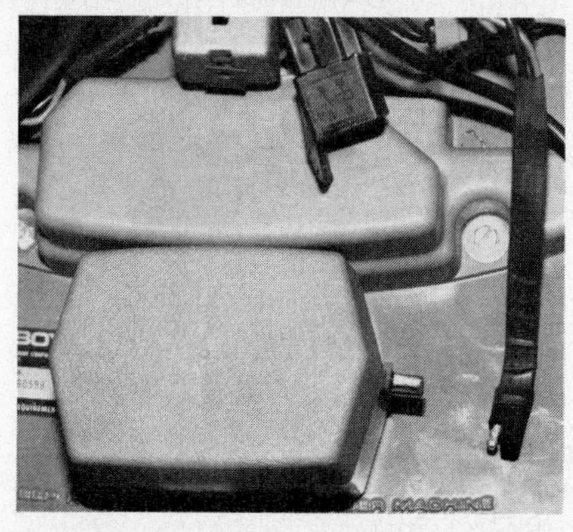

Fig. LB93—View showing battery location and electrical plug-in. Refer to text for alternator output testing procedure.

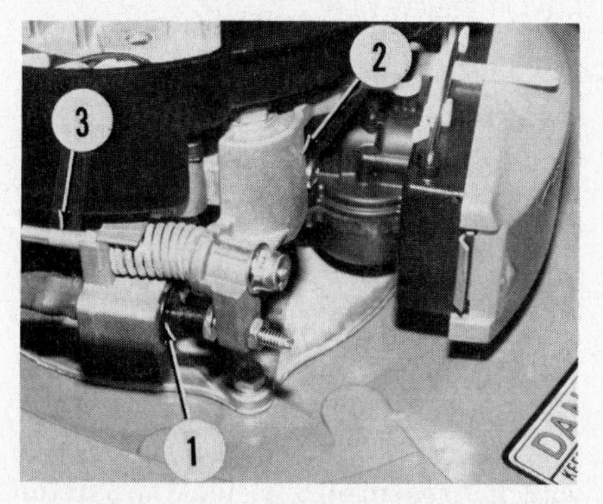

Fig. LB94—View showing safety interlock switch (1), flywheel brake (2) and adjustment cable (3) on models so equipped.

Reassemble in reverse order of disassembly, then install starter assembly in engine and leave 0.025-0.030 inch (0.64-0.76 mm) clearance between end of crankcase mounting boss and end of assembled starter components. Tighten all mounting screws.

A view of "New Style" charging alternator is shown in Fig. LB92. Air gap should be adjusted as outlined for CD module. Correct air gap setting is 0.010 inch (0.25 mm). To check alternator output, use a VOM meter with a 500MA capacity. Disconnect plug at battery as shown in Fig. LB93. Reconnect black wires, then connect test meter in series with red wire. Connect positive side of meter towards alternator and negative side of meter towards battery. Start engine and allow to run at high speed (3100-3300 rpm) while observing needle on meter. Correct reading should be between 200 and 450MA.

For "Old Style" alternator assembly (not shown), testing of alternator output is identical to procedure outlined for "New Style" alternator. Correct meter readings for "Old Style" alternator at high speed should be between 70 and 120MA. Correct air gap setting is 0.010 inch (0.25 mm).

SAFETY EQUIPMENT

INTERLOCK SWITCH. A view of a typical interlock switch is shown in 1–Fig. LB94. Switch is designed to open or close ignition circuit depending upon switch plunger. When safety handle is depressed, switch plunger will move inward causing contact points to come together allowing engine to be started and run. When safety handle is released, plunger is forced away from contact points preventing engine from being started or causing engine to die.

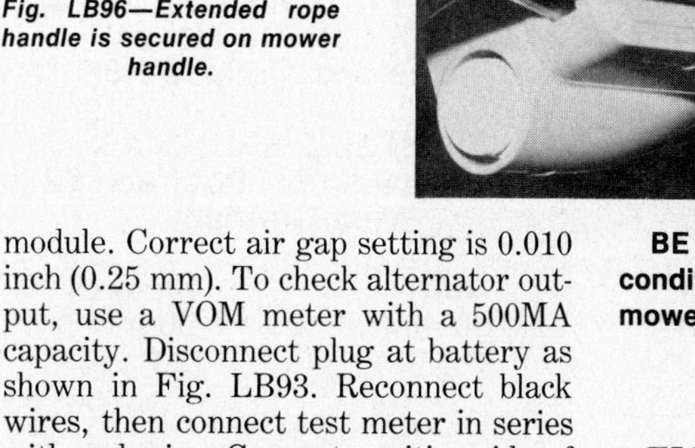

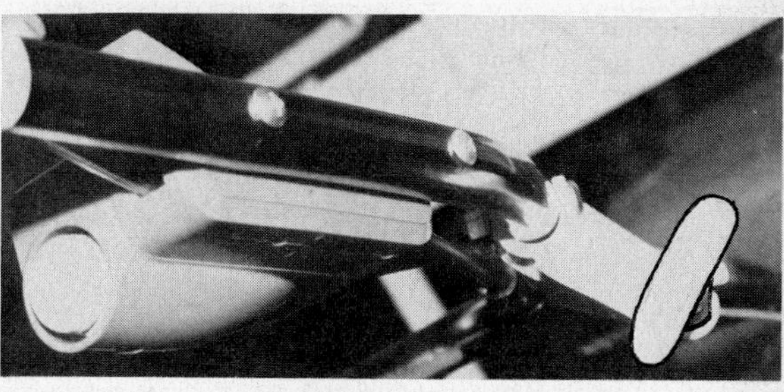

Fig. LB96—Extended rope handle is secured on mower handle.

BE SURE switch is in good operating condition to ensure safe operation of mower.

FLYWHEEL BRAKE. Flywheel brake is designed to stop the blade within three seconds after the operator releases the safety control handle. Shown in Fig. LB94 is a typical flywheel brake (2) located on the outside of the flywheel. When control handle is depressed, brake lever will move away from flywheel. When control handle is depressed, brake lever moves against flywheel to aid in blade stoppage.

Brake lever is adjusted by repositioning control cable (3) in mounting bracket. Inspect all components for excessive wear or any other damage and renew as needed. **BE SURE** all components are in good operating condition to ensure safe operation of mower.

EXTENDED ROPE START. An extended rope start (Zone Start) as shown in Fig. LB96 is normally combined with a flywheel type blade brake. The extended rope start serves two purposes, they are: It allows the operator to remain within the considered "Safe Zone" when starting the engine and it allows the operator to easily depress the safety control handle while pulling the starter rope handle.

SACHS

SACHS MOTOR CORP., LTD.
9615 Cote De Liesse
Dorval, Quebec, Canada H9P 1A3

FICHTEL & SACHS AG
852 Schweinfurt Postfach 52
West Germany

Model	Bore	Stroke	Displacement
SB140	60 mm (2.4 in.)	50 mm (2.0 in.)	142 cc (8.7 cu. in.)

ENGINE INFORMATION

Model SB140 engines are two-stroke, vertical crankshaft type rated at 2.8kW (3.8 hp) at 2850 rpm.

Engine model and serial number identification plate is attached to engine crankcase as shown in Fig. SA1. Always furnish engine model and serial number when ordering parts or service material.

Fig. SA1—Engine model and serial number plate location.

MAINTENANCE

SPARK PLUG. Spark plug should be removed, cleaned and inspected at 100-hour intervals.

Recommended spark plug is a Champion RJ17LM, or equivalent. Recommended spark plug electrode gap is 0.5 mm (0.020 in.).

CAUTION: Manufacturer does not recommend using abrasive blast method to clean spark plugs as this may introduce some abrasive material into the engine which could cause extensive damage. Use commercial cleaning solvent only.

CARBURETOR. All models are equipped with a float type carburetor with a fixed main fuel jet. Carburetor butterfly valve and upper body are an integral part of the fuel tank and cooling shroud assembly. Float bowl assembly can be removed by inserting screwdriver and pushing lock tabs inward while pulling downward on float bowl. Refer to Fig. SA2.

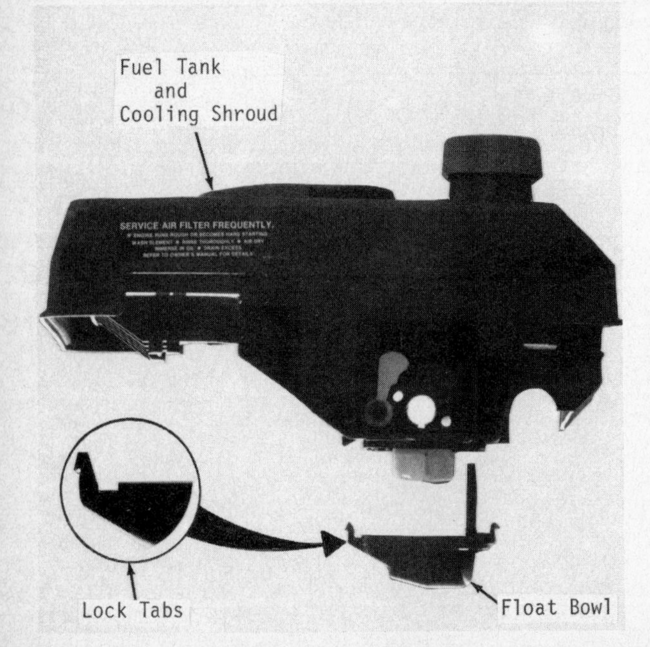

Fig. SA2—Fuel tank, carburetor float section and cooling shroud are one complete assembly. Inset shows lock tabs which retain carburetor float bowl. Refer to text.

Fig. SA3—View showing location of governor adjusting screw.

Engine speed should be 2850 rpm and is adjusted by pushing governor adjustment screw (Fig. SA3) in slightly and turning it clockwise to increase engine speed or counterclockwise to reduce engine speed. Refer to GOVERNOR paragraphs in REPAIRS section.

Main jet size is 0.93 and is an integral part of jet holder (Fig. SA4). Correct jet holder and jet are identified by "1.1" stamped in flange. Main jet and jet holder are serviced as an assembly only. When reinstalling jet holder, make certain "O" ring is in good condition.

To check float level, invert fuel tank, cooling shroud and upper carburetor assembly. Float edge should be parallel with carburetor to float bowl mating surface. Refer to Fig. SA5. Carefully bend float hinge to obtain correct float level.

Choke lever should be adjusted at cable nuts at control so choke just contacts the metal tang on main jet holder with choke control at "ON" position. Place choke control at "OFF" position. Choke should be fully open.

FUEL FILTER. A fuel filter screen is an integral part of the fuel tank and is not visible. Clean by removing fuel tank and cooling shroud assembly and flush with clean fuel. Use low air pressure to blow through fuel line with fuel cap removed.

AIR CLEANER. Engine is equipped with a coco fiber air cleaner element (Fig. SA7). Filter should be inspected at 8- to 10-hour intervals of normal operation.

To clean felt frame, carefully blow dust off with low air pressure. Clean coco fiber element in warm water and detergent solution. Rinse thoroughly, shake out excess water and allow to air dry. Immerse fiber element in SAE 30 oil for 5 minutes, then allow excess oil to drain before reinstalling element. Make certain cover is firmly seated.

GOVERNOR. The mechanical ball type governor is located inside engine crankcase and is attached to the crankshaft. Use an accurate tachometer to determine engine speed. Correct engine speed is 2850 rpm. To adjust, push governor adjustment screw (Fig. SA3) in slightly and turn screw clockwise to increase engine speed or counterclockwise to reduce engine speed. Also refer to GOVERNOR paragraphs in REPAIRS section.

IGNITION SYSTEM. The breakerless ignition system requires no regular maintenance. Ignition coil unit is mounted outside the flywheel and air gap between flywheel and coil should be 0.05-0.2 mm (0.002-0.008 in.).

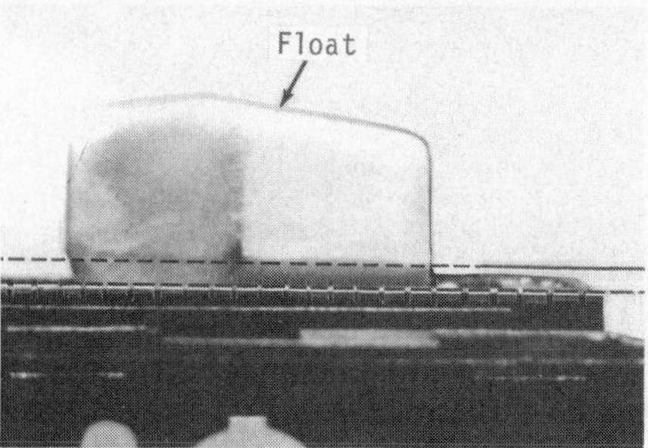

Fig. SA5—Lower edge of float must be parallel with carburetor to float bowl mating surface for correct float level.

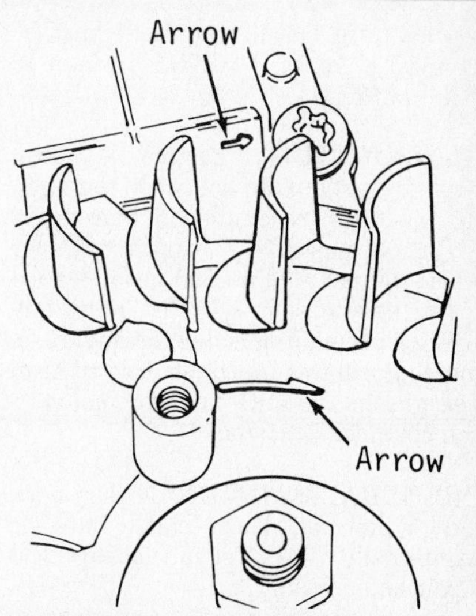

Fig. SA8—When installing ignition coil, arrow on the ignition coil and arrow on flywheel must point in the same direction.

If ignition coil has been removed, make certain arrow on ignition coil and arrow on flywheel are pointing the same direction after installation (Fig. SA8).

To check ignition coil, disconnect lead from stop switch and remove spark plug insulating cap. Position spark plug lead 4-6 mm (0.157-0.236 in.) from engine ground and spin engine flywheel. If strong spark does not occur, check insulation of stop switch lead, spark plug lead or ground contact of ignition coil at crankcase. If insulation and connections are good, and spark does not occur, renew ignition coil.

CARBON. Carbon should be removed when engine performance drops or when engine begins to four-stroke despite correct carburetor settings. To clean, remove muffler and carefully remove carbon deposits from combustion chamber and transfer ports. DO NOT attempt to remove all deposits from piston to obtain a bright surface. Remove only the large carbon deposits. Use a 3 mm (0.118 in.) drill bit to clean decompression hole (Fig. SA10). Muffler

Fig. SA4—View showing main jet holder and location of jet identification number "1.1" stamped on flange.

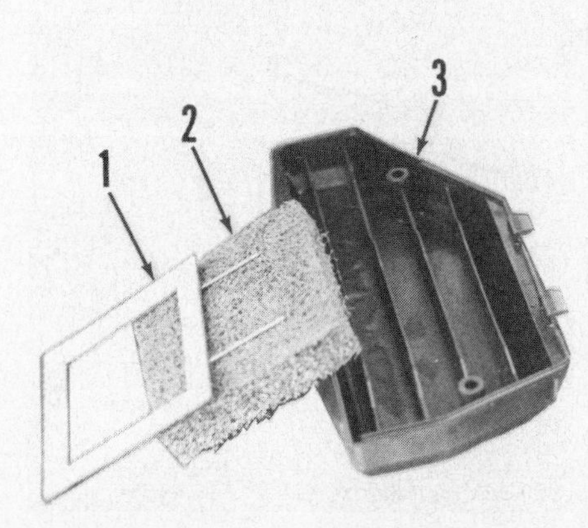

Fig. SA7—View of air filter component parts.
1. Felt frame 2. Element 3. Cover

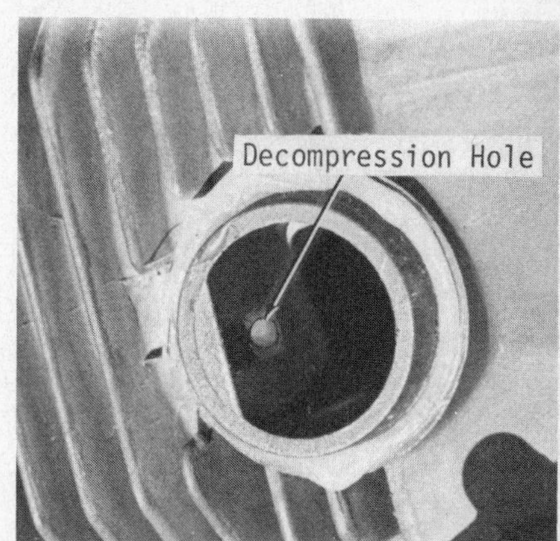

Fig. SA10—View showing location of decompression hole. Refer to text.

is filled with basalt wool and must be renewed if faulty. DO NOT attempt to clean muffler.

LUBRICATION. Engine is lubricated by premixing oil with the fuel. Manufacturer recommends thoroughly mixing one part BIA TC approved oil for every 100 parts of a good grade leaded or lead free gasoline. DO NOT use any form of premium gasoline or any gasoline that contains alcohol. Use of lead free gasoline results in fewer combustion chamber deposits.

GENERAL MAINTENANCE. Check and tighten all loose bolts, nuts or clamps daily. Check for fuel leakage and repair as necessary.

Clean dust, dirt, grease or any foreign material from cylinder head and cylinder block cooling fins at 100-hour intervals. Inspect fins for damage and repair as necessary.

REPAIRS

TIGHTENING TORQUES. Recommended tightening torque specifications are as follows:

Flywheel nut............30-35 N·m (20-25 ft.-lbs.)
Crankcase screws..........9-12 N·m (6-9 ft.-lbs.)

CYLINDER. Cylinder is an integral part of one of the crankcase halves (Fig. SA11). Cylinder should be renewed if scored, flaking or severely out-of-round.

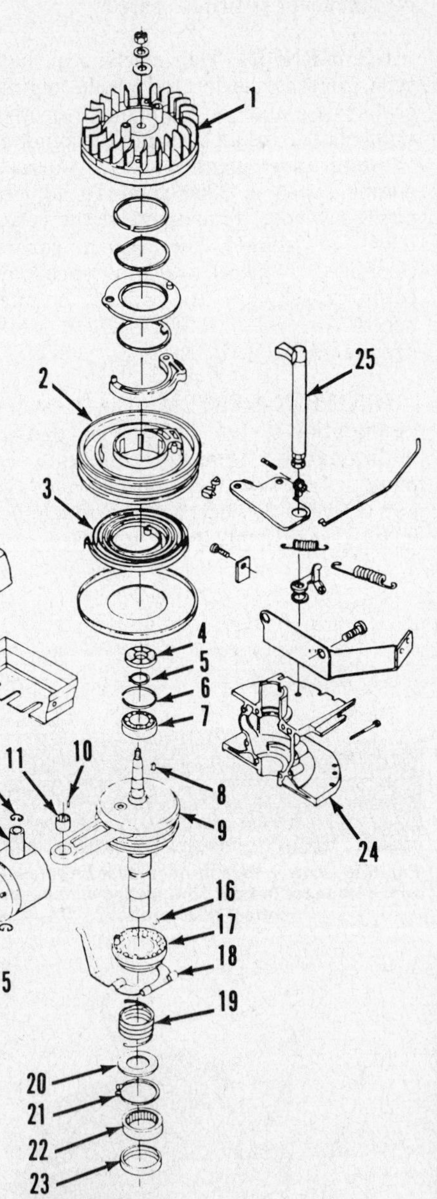

Fig. SA11—Exploded view of Sachs Model SB140 engine.

1. Flywheel
2. Pulley
3. Recoil spring
4. Seal
5. Snap ring
6. Retaining ring
7. Bearing
8. Key
9. Connecting rod & crankshaft assy.
10. Needle bearing (25)
11. Needle bearing race
12. Retaining ring
13. Piston pin
14. Piston
15. Ring
16. Governor balls
17. Governor cup
18. Governor fork
19. Spring
20. Washer
21. Snap ring
22. Needle bearing
23. Seal
24. Crankcase half
25. Brake lining & shaft
26. Fuel tank & cooling shroud assy.
27. Choke
28. Main jet holder
29. Float
30. Gasket
31. Float bowl
32. Ignition coil
33. Governor adjustment screw
34. Crankcase half (cylinder side)

PISTON, PIN AND RING. A single ring piston is used and ring is pinned into position to prevent ring rotation (Fig. SA12). It is necessary to split crankcase and remove piston, connecting rod and crankshaft as an assembly to renew piston. Refer to CRANKCASE section.

CRANKCASE. A split crankcase design with the cylinder as an integral part of one crankcase half is used. It is necessary to split crankcase to renew piston, connecting rod or crankshaft.

To split crankcase, remove fuel tank and cooling shroud assembly. Remove flywheel and recoil starter, ignition coil and flywheel brake assembly. Remove the eight socket head cap screws (Fig. SA13) and carefully separate crankcase halves leaving crankshaft assembly in the cylinder portion of crankcase. Very carefully lift crankshaft and piston assembly from cylinder portion of crankcase while guiding governor lever up off of its pin. Use care during this procedure so ring and piston are not damaged.

To remove piston, remove piston pin retaining rings (12–Fig. SA17) and use suitable pin extractor to remove piston

Fig. SA12—View showing location of pin on piston to prevent ring rotation. Make certain ring end gap aligns with pin.

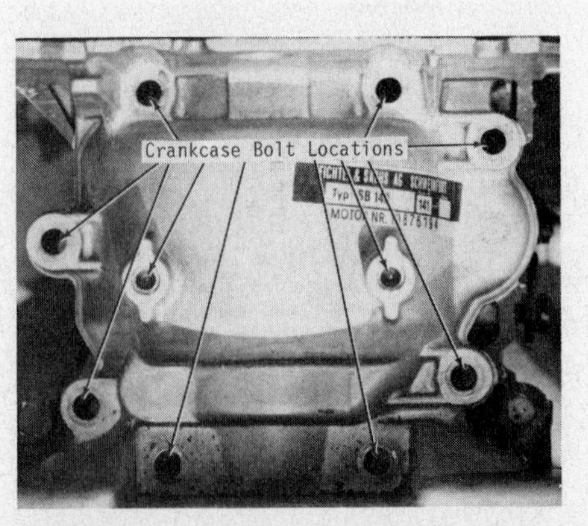

Fig. SA13—View showing location of crankcase retaining bolts. Loosen or tighten in a criss-cross pattern. Tighten to specified torque.

pin (13). Remove the 25 needle bearings (10) and press out the needle bearing race (11). New needle bearing is supplied with a plastic sleeve as an assembly aid and bearing should be pressed into connecting rod until it is centered in bore. When reassembling piston to connecting rod, arrow on piston top (Fig. SA14) should be toward exhaust side of engine. Piston pin should be a slight push fit in piston pin bore and needle bearing. Warm piston if necessary to aid pin installation.

CAUTION: Piston top will be stamped with either an "A", "B" or "C" (Fig. SA14) and cylinder top will be stamped with a corresponding "A", "B" or "C" (Fig. SA15). If piston or cylinder are renewed, piston and cylinder MUST HAVE the same letter identification. When ordering piston or cylinder, always specify letter identification needed.

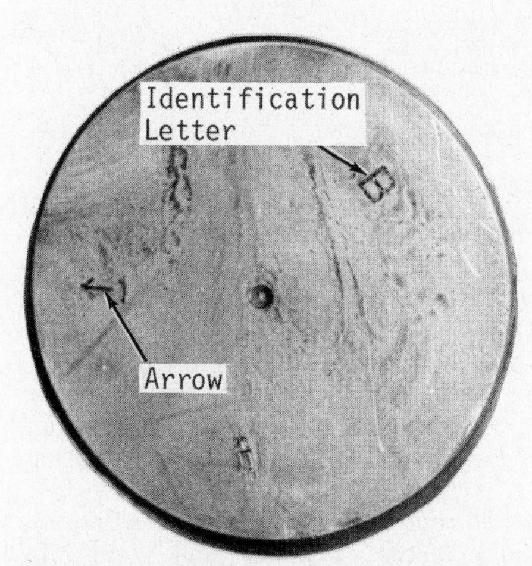

Fig. SA14—After piston installation, arrow must point toward exhaust side of engine. Piston identification letter will be either an "A", "B" (as shown) or "C". Correctly match piston with cylinder. Refer to Fig. SA15.

Fig. SA15—View showing location of cylinder identification letter required to correctly match piston to cylinder. Refer to text.

To reassemble crankcase, lightly oil cylinder wall. Make certain side of ring marked "TOP" is up, ring end gap is around positioning pin (Fig. SA12) and governor fork is on crankshaft. Carefully work piston into cylinder (cylinder is tapered to compress ring) while guiding governor fork onto pin. Governor fork lever is positioned against pin of throttle valve shaft as shown in Fig. SA16. Coat mating surfaces of crankcase halves with Loctite 572. Make certain seals and retaining rings are properly seated in grooves and place remaining crankcase half in position. Tighten crankcase retaining screws in a criss-cross pattern to specified torque. Continue to reverse disassembly procedure for reassembly.

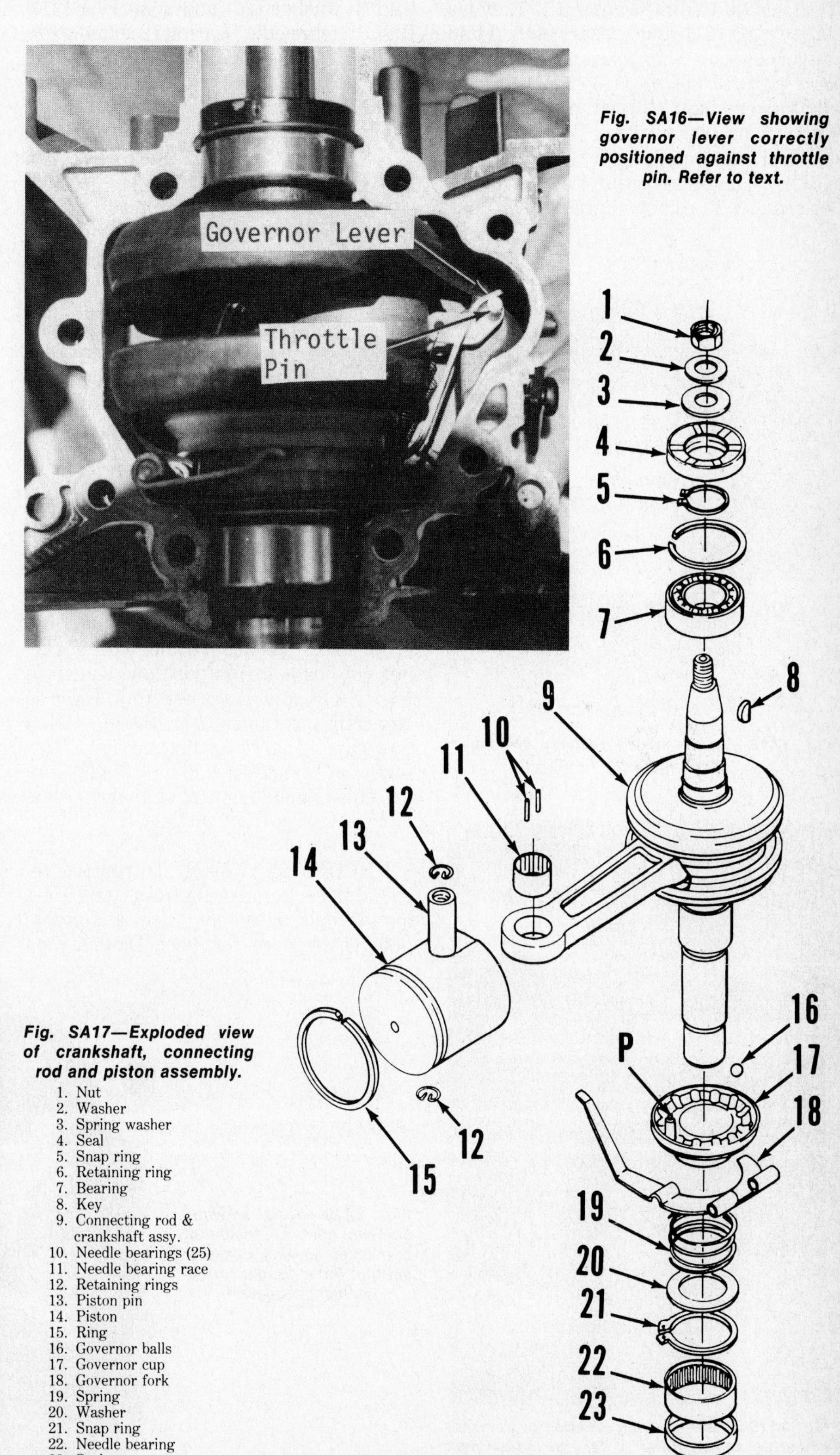

Fig. SA16—View showing governor lever correctly positioned against throttle pin. Refer to text.

Fig. SA17—Exploded view of crankshaft, connecting rod and piston assembly.

1. Nut
2. Washer
3. Spring washer
4. Seal
5. Snap ring
6. Retaining ring
7. Bearing
8. Key
9. Connecting rod & crankshaft assy.
10. Needle bearings (25)
11. Needle bearing race
12. Retaining rings
13. Piston pin
14. Piston
15. Ring
16. Governor balls
17. Governor cup
18. Governor fork
19. Spring
20. Washer
21. Snap ring
22. Needle bearing
23. Seal

CRANKSHAFT, MAIN BEARINGS AND SEALS. Crankshaft and connecting rod are serviced as an assembly only and crankshaft and connecting rod are not available separately.

To remove crankshaft and connecting rod assembly, refer to CRANKCASE section. Remove governor fork (18–Fig. SA17). Slide seal (23) and needle bearing (22) from crankshaft. Remove snap ring (21) and washer (20). Hold governor cup (17) against crankshaft and remove spring (19). Stand crankshaft up so balls (16) will not fall out of cup (17) and remove cup (17). Slide seal (4) and retaining ring (6) off of crankshaft. Remove snap ring (5). Use a suitable puller to remove bearing (7).

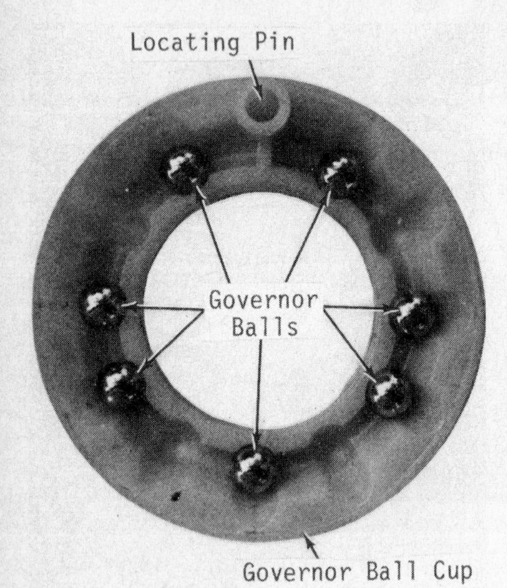

Fig. SA18—View showing governor ball placement in governor cup.

Fig. SA19—View showing location of governor adjusting screw. Refer to text for adjustment procedure.

To reassemble, slightly heat bearing (7) and press onto crankshaft. Install snap ring (5), retaining ring (6) and seal (4). Install balls (16) into cup (17) as shown in Fig. SA18. Hold crankshaft in vertical position and slide cup and ball assembly up against crankshaft making sure locating pin (Fig. SA18) enters hole in crankshaft. Install spring (19–Fig. SA17), washer (20) and snap ring (21). Lightly oil needle bearing (22) and slide bearing and seal (23) onto crankshaft. Place governor fork (18) in position and install crankshaft assembly as outlined in CRANKCASE section.

GOVERNOR. A ball type governor located on crankshaft and inside crankcase is used to maintain 2850 rpm engine speed. Speed adjustments are made by pushing governor adjustment screw (Fig. SA19) in slightly against spring pressure and turning screw clockwise to increase engine speed or counterclockwise to reduce engine speed. Governor adjustment screw setting is maintained by spring pressure holding lever seated in notches cast on the inside of crankcase (Fig. SA20).

To remove governor assembly refer to CRANKCASE section. Governor adjustment screw is removed by removing "E" clip and sliding component parts out of crankcase bore.

During reassembly, make certain governor fork ends are parallel with governor cup, balls are in positions shown in Fig. SA18 and governor fork lever is correctly positioned (Fig. SA16). When governor is correct, throttle valve butterfly will be open 0.5 mm (0.020 in.). Carefully bend long arm of governor lever to obtain correct setting.

RECOIL STARTER. To remove recoil starter, remove fuel tank and cooling shroud assembly. Use a suitable puller to remove flywheel. Detach rope

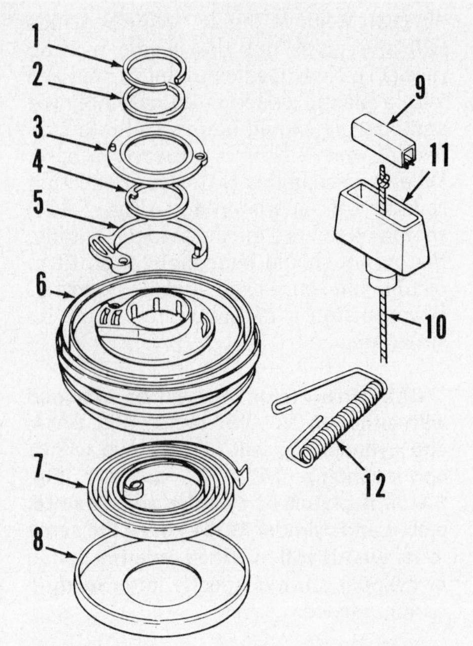

Fig. SA22—Exploded view of recoil starter assembly.

1. Retaining ring
2. Wave washer
3. Brake disc
4. Tab washer
5. Pawl
6. Pulley
7. Recoil spring
8. Cover
9. Knot retainer
10. Rope
11. Handle
12. Guide spring

handle, then remove snap ring (1–Fig. SA22), spring washer (2) and brake disc (3). Remove tab washer (4) and pawl (5). Remove rope pulley (6), rope, spring (7) and cover (8) as an assembly.

To remove spring (7), remove cover (8). Lift spring end while holding spring. Slowly release pressure to allow spring to expand to the outside.

To renew spring, insert end of spring into recess of pulley and wind into pulley. Inner end of spring must be correctly preloaded. Loop should be bent toward center of pulley. Fill lubricating grooves in bearing bore with light grease. Spring must be flat after in-

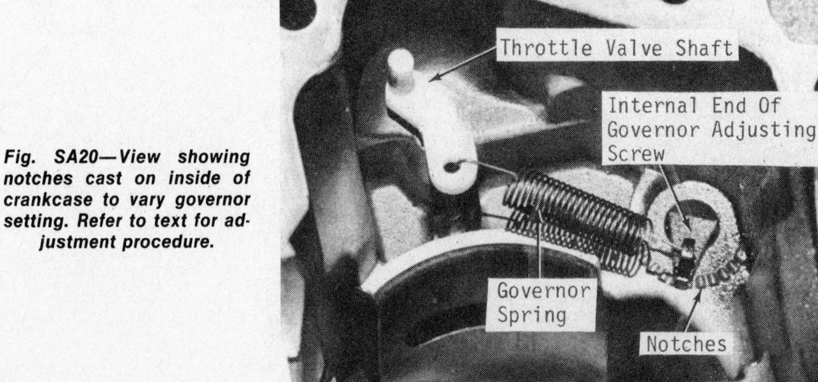

Fig. SA20—View showing notches cast on inside of crankcase to vary governor setting. Refer to text for adjustment procedure.

Fig. SA23—View showing location of flywheel brake lining path.

stallation. Lightly coat spring with oil and install cover plate. Knot rope end and pull rope tightly into eye of pulley. Wind rope around pulley 5¾ turns. Reverse removal procedure for remainder of reassembly.

FLYWHEEL BRAKE. To disassemble flywheel brake assembly (Fig. SA23), remove fuel tank and cooling shroud assembly. Disconnect springs (8 and 9–Fig. SA24). Remove special screws (12) and remove bracket (13), washer (11), brake lever (10) and lever (3). Pull shaft and brake lining assembly (1) upward out of shaft bore in crankcase half.

To reassemble, insert shaft and brake lining assembly (1) into shaft bore in crankcase half. Install lever (3), brake lever (10), washer (11) and bracket (13) onto shaft (1). Loosely install special screws (12) through bracket (13) and into crankcase. Position brake lining so it will run on braking path of flywheel (Fig. SA23). Secure bracket in this position by tightening the two special screws (12).

To adjust flywheel brake, release engine control. Turn brake adjustment screw shown in Fig. SA23 or SA25 so there is 2-6 mm (0.079-0.236 in.) between adjustment screw and brake lever tang. Refer to Fig. SA25. If brake lining is worn to a point that 2 mm (0.079 in.) cannot be obtained, hook spring into upper hole (U) and turn adjustment screw counterclockwise until measurement is 6 mm (0.236 in.).

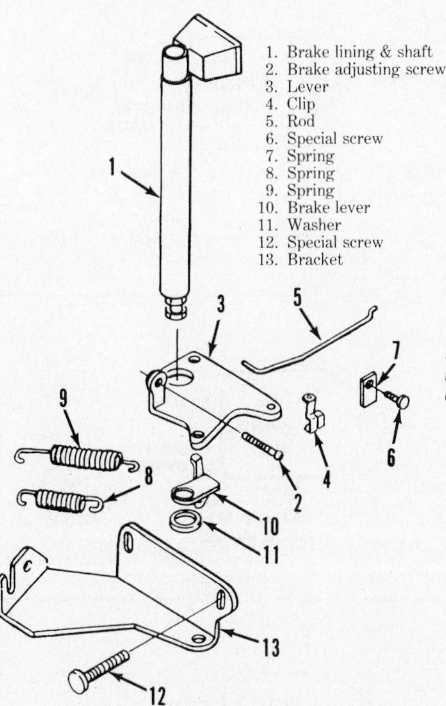

Fig. SA24—Exploded view of flywheel brake component parts.

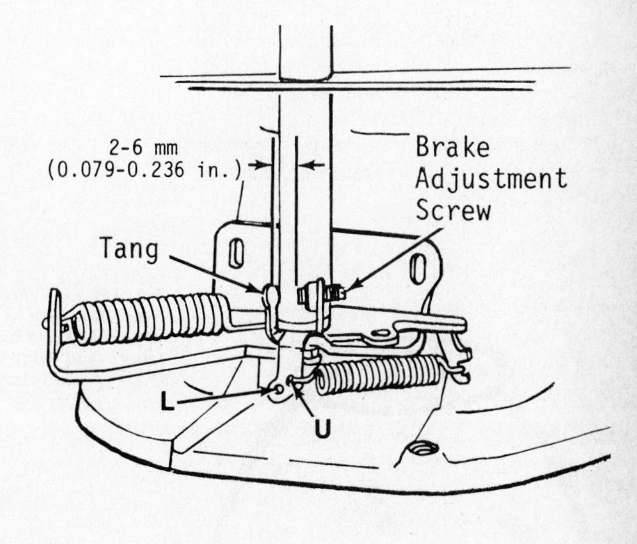

Fig. SA25—View of flywheel brake assembly showing correct dimensions for brake adjustment. Refer to text.

TECUMSEH 2-STROKE

UNIT BLOCK MODELS

Model	Bore	Stroke	Displacement
AH440	2.09 in. (50.8 mm)	1.25 in. (31.8 mm)	4.30 cu. in. (65 cc)
AH480	2.09 in. (50.8 mm)	1.41 in. (35.8 mm)	4.85 cu. in. (73 cc)
AH490	2.09 in. (50.8 mm)	1.41 in. (35.8 mm)	4.85 cu. in. (73 cc)
AH520, AV520	2.09 in. (50.8 mm)	1.50 in. (38.1 mm)	5.16 cu. in. (77 cc)
AV600, TVS600	2.09 in. (50.8 mm)	1.75 in. (44.5 mm)	6.00 cu. in. (90 cc)
AH750, AV750	2.38 in. (60.3 mm)	1.68 in. (42.7 mm)	7.50 cu. in. (122 cc)
AH817, AV817	2.44 in. (62.0 mm)	1.75 in. (44.5 mm)	8.17 cu. in. (134 cc)

ENGINE IDENTIFICATION

All models are two-stroke, single cylinder air-cooled engines with a unit block type construction.

Prefixes before type and serial number indicate the following information.

T-Tecumseh
A-Aluminum
V, VS-Vertical crankshaft
H-Horizontal crankshaft

Engine prefix, type number and serial numbers are located as shown in Fig. TP2-1. Figs. TP2-1A and TP2-1B show number interpretations.

Always furnish engine model, type and serial numbers when ordering parts or service material.

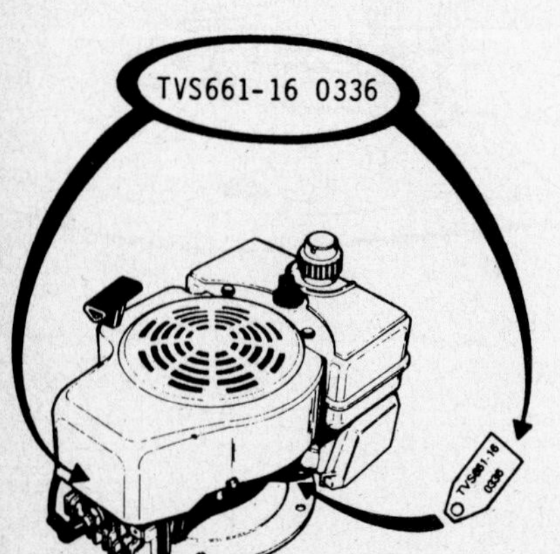

Fig. TP2-1—View showing location of engine model, type and serial number. Refer also to Figs. TP2-1A and TP2-1B.

MAINTENANCE

SPARK PLUG. Recommended spark plug for all engines installed on lawnmowers and snowblowers is a Champion J17LM or equivalent. Recommended spark plug for all other applications is a Champion CJ8 or J8J for Models AH440, AH480, AH290, AH520, AV520, AV600 and TVS600; UJ12Y for Model AH750; UJ10Y for Model AV750; J13Y for Models AH817 and AV817.

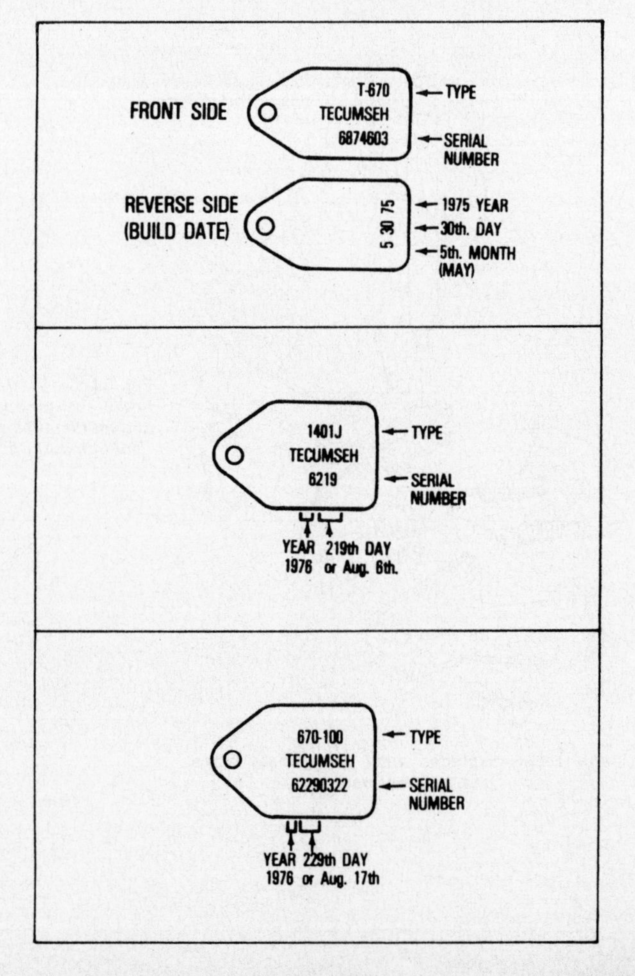

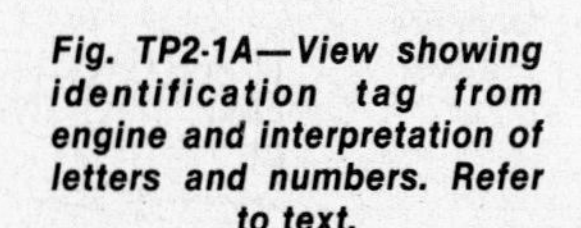
Fig. TP2-1A—View showing identification tag from engine and interpretation of letters and numbers. Refer to text.

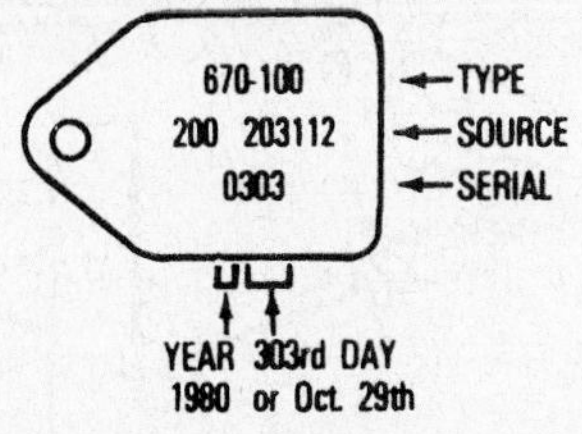

Fig. TP2-1B—View of identification tag from replacement short block and interpretation of letters and numbers.

Electrode gap should be 0.030 inch (0.76 mm) for Model TVS600 and Model AH520 with type numbers 1600 through 1617. Electrode gap for all other models is 0.035 inch (0.89 mm).

CARBURETOR. Tillotson HS, Tecumseh diaphragm and Tecumseh float type carburetors are used. Refer to the appropriate following paragraphs for information on specific carburetors.

Tillotson HS Type Carburetor. Refer to Fig. TP2-1C for identification and exploded view of Tillotson HS type carburetor.

Initial adjustment of idle and main fuel mixture screws from a lightly seated position is 1 turn open for idle mixture screw (6) and 1¼ turns open for main fuel mixture screw (5).

Final adjustments are made with engine at operating temperature and running. Operate engine at idle speed and adjust idle mixture screw to obtain smoothest engine idle. Operate engine at rated speed, under a load and adjust main fuel mixture screw for smooth engine operation. If engine fails to accelerate smoothly, it may be necessary to open idle mixture screw slightly past previous setting until engine accelerates properly.

Tecumseh Diaphragm Type Carburetor. Refer to Fig. TP2 for identification and exploded view of Tecumseh diaphragm type carburetor. Carburetor identification numbers are also stamped on carburetor as shown in Fig. TP2-2A.

Initial adjustment of idle mixture and main fuel mixture screws from a lightly seated position for models with solid adjustment (L and H) is 1 turn open for each mixture screw. There is no initial adjustment for idle and main fuel mixture screws which have a hole in inner end as hole size determines initial idle fuel mixture. When renewing adjustment screws, make certain hole diameter of replacement screw matches hole diameter of new screw.

Final adjustments are made with engine at operating temperature and running. Operate engine at idle speed and adjust idle mixture screw to obtain smoothest engine idle. Operate engine

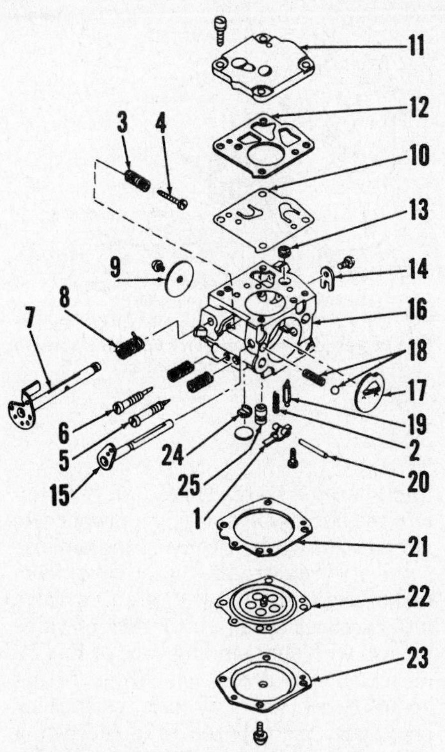

Fig. TP2-1C—Exploded view of Tillotson HS diaphragm type carburetor used on some models.

1. Inlet control lever
2. Spring
3. Spring
4. Idle speed stop screw
5. High speed mixture screw
6. Idle mixture needle
7. Throttle shaft
8. Spring
9. Throttle plate
10. Fuel pump diaphragm
11. Pump cover
12. Gasket
13. Fuel screen
14. Throttle shaft
15. Choke shaft
16. Carburetor body
17. Choke plate
18. Choke detent
19. Inlet needle
20. Lever pin
21. Gasket
22. Diaphragm
23. Cover
24. Channel reducer
25. Main nozzle & check ball

at rated speed, under load and adjust main fuel mixture screw for smooth engine operation. If engine cannot be operated under load, adjust main fuel mixture needle so engine runs smoothly at rated rpm and accelerates properly when throttle is opened quickly.

When overhauling, observe the following: The fuel inlet fitting is pressed into bore of body of some models. On these models, the fuel strainer behind inlet fitting can be cleaned by reverse flushing with compressed air after inlet needle and seat are removed. The inlet needle seat fitting is metal with a neoprene seat, so fitting (and enclosed seat) should be removed before carburetor is cleaned with a commercial solvent. The throttle plate (3) should be installed with short line stamped on plate to top of carburetor and facing out. The choke plate (9) should be installed with flat toward fuel inlet side of carburetor as shown.

When installing diaphragm (18), head of rivet should be against fuel inlet valve (16) regardless of size or placement of washers around rivet. On carburetor

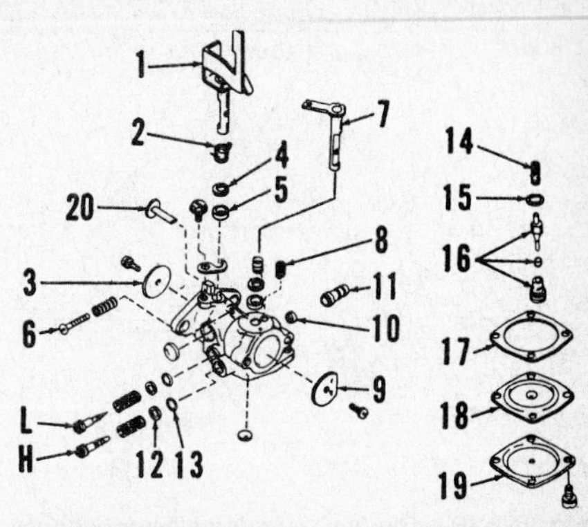

Fig. TP2-2—Exploded view of typical Tecumseh diaphragm type carburetor. Some models do not use fuel pump (20) and check valve (10).

H. High speed mixture screw
L. Idle mixture screw
1. Throttle shaft
2. Spring
3. Throttle plate
4. Felt washer
5. Flat washer
6. Idle stop screw
7. Choke shaft
8. Choke retainer
9. Choke plate
10. Outlet check valve
11. Inlet fitting
12. Washer
13. "O" ring
14. Spring
15. Gasket
16. Inlet valve
17. Gasket
18. Diaphragm
19. Cover
20. Pump element

Models 0234-252, 265, 266, 269, 270, 271, 282, 297, 303, 322, 327, 333, 334, 344, 345, 348, 349, 350, 351, 352, 356, 368, 371, 375, 378, 379, 380, 404, 405 and 441 and all models with "F" embossed on carburetor body, gasket (17–Fig. TP2-2) must be installed between diaphragm (18) and cover (19). All other models are assembled with gasket, diaphragm and cover positioned as shown in Fig. TP2-2.

On models equipped with a fuel pump, pumping element is a rubber boot (20) which expands and contracts due to changes in crankcase pressure. The pump inlet check valve is located in the fuel inlet fitting (11). The pump outlet check valve (10) is pressed into the car-

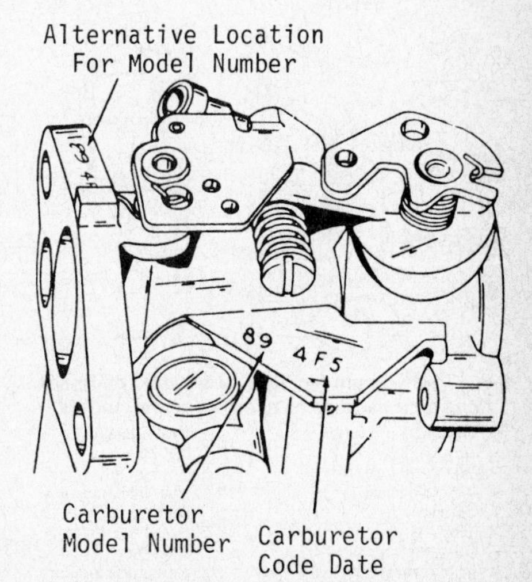

Fig. TP2-2A—View showing location of carburetor identification number on Tecumseh diaphragm type carburetor.

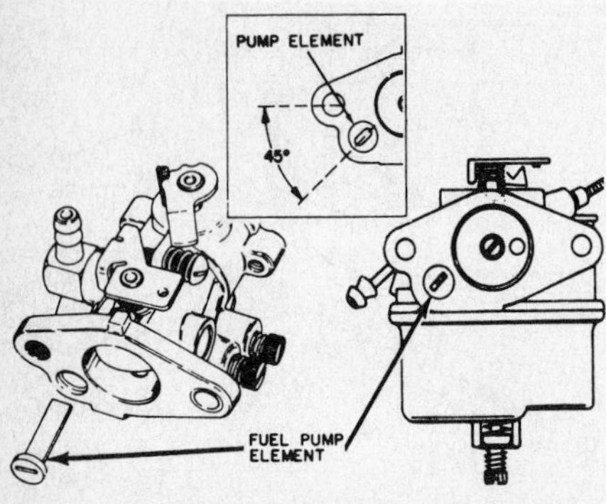

Fig. TP2-2B—The fuel pumping element should be installed at 45° angle as shown for Tecumseh diaphragm type carburetor (left) and float type carburetor (right).

buretor body behind the fuel inlet fitting. Engines equipped with this carburetor will operate in any position and the pump will deliver fuel to the carburetor when the fuel supply is below the carburetor.

NOTE: The fuel pumping element should be installed at 45° angle as shown in Fig. TP2-2B. Incorrect installation may interfere with pumping action.

Two types of fuel pump valves are used. Flap type of valve may be located behind plate attached to side of carburetor body. Renew flap type valves after detaching the plate from side of carburetor. Fuel pump valves are pushed into carburetor body of some models (Fig. TP2-2). Clamp the inlet fitting (11) in a vise and twist carburetor body from fitting. Using a 9/64 inch drill, carefully drill into outlet valve (10) to a depth of 1/8 inch (3.18 mm). Take care not to drill in-

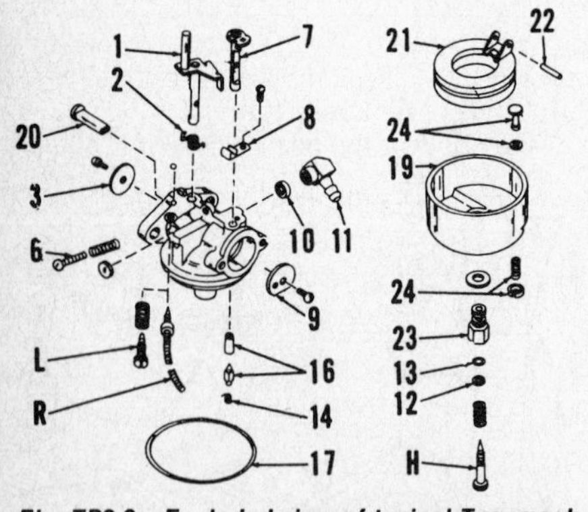

Fig. TP2-3—Exploded view of typical Tecumseh float type carburetor used on some models.

H. High speed mixture screw
L. Idle mixture screw
1. Throttle shaft
2. Spring
3. Throttle plate
6. Idle stop screw
7. Choke shaft
8. Choke retainer
9. Choke plate
10. Outlet check valve
11. Inlet fitting
12. Brass washer
13. "O" ring
14. Spring
16. Fuel inlet needle & seat
17. Seal
19. Fuel bowl
20. Pumping element
21. Float
22. Pivot pin
23. Bowl retaining nut
24. Bowl drain

Fig. TP2-4—View of late type fuel valves. Earlier type is pressed into carburetor body as shown in Figs. TP2-2 and TP2-3.

to carburetor body. Thread an 8-32 tap into the outlet valve. Using a proper size nut and flat washer, convert the tap into a puller to remove the outlet valve from carburetor body. Press new outlet valve into carburetor body until face of valve is flush with surrounding base of fuel inlet chamber. Press new inlet fitting about 1/3 of the way into carburetor body, coat the exposed 2/3 of the fitting shoulder with Loctite, then press fitting fully into carburetor body.

Tecumseh Float Type Carburetor. Refer to Fig. TP2-3 for identification and exploded view of Tecumseh float type carburetor.

Initial adjustment of idle and main fuel mixture screws from a lightly seated position is 1 turn open for each screw.

Final adjustments are made with engine at operating temperature and running. Operate engine at rated speed and adjust main fuel mixture screw (H) for smoothest engine operation. Operate engine at idle speed and adjust idle mixture screw for smoothest engine idle. Set idle speed at idle stop screw (6). If engine fails to accelerate smoothly, slight adjustment of main fuel mixture screw may be necessary.

When overhauling, check adjustment screws for excessive wear or other damage. The inlet valve fuel needle seats against a Viton rubber seat (16) which is pressed into carburetor body. Remove the rubber seat before cleaning carburetor in commercial cleaning solvent. The seat is removed using a 10-32 or 10-24 tap and must be renewed after removing. Install new seat using a punch that will fit into bore of seat and is large enough to catch the shoulder inside the seat. Drive the fuel inlet needle seat into bore until the Viton rubber seat is against bottom of bore. Install throttle plate (3) with the two stamped lines facing out and at 12 and 3 o'clock positions. Install choke plate (9) with flat side toward bottom of carburetor.

Float setting should be 7/32 inch (5.56 mm), measured with body and float assembly in inverted position, between

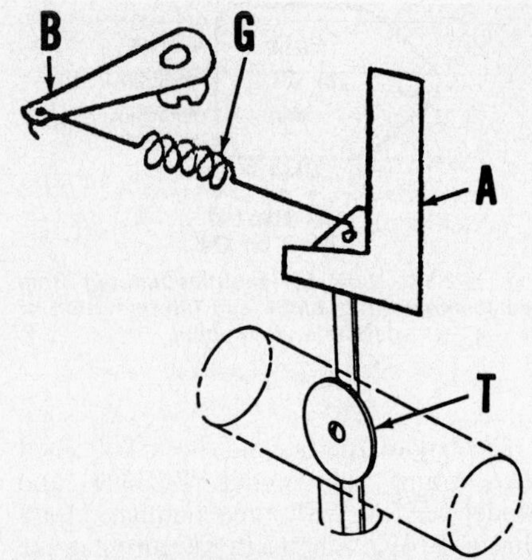

Fig. TP2-5—On some models, the governor air vane (A) is attached to the carburetor throttle shaft and tension of governor spring (G) holds throttle (T) open.

free end of float and rim on carburetor body.

On some models, the fuel inlet fitting (11) is pressed into body. When installing and fitting, start fitting into bore, then apply a light coat of Loctite to the shank and press the fitting into position.

When installing float bowl (19), make certain that correct "O" ring (17) is used. Some "O" rings are round section, others are square.

Fuel hole and the annular groove in retaining nut (23) must be clean. The flat stepped section of fuel bowl (19) should be below the fuel inlet fitting (11). Tighten retaining nut (23) to 50-60 in.-lbs. (6-7 N•m). The high speed mixture screw (H) must not be installed when tightening nut (23).

Some models are equipped with a fuel pump. The fuel pumping element (20) is a rubber boot which expands and contracts due to changes in crankcase pressure. The pumping element should be at 45° as shown in Fig. TP2-2B. Incorrect installation may interfere with pumping action.

On early models, fuel is drawn in through a check valve in the fuel inlet fitting (11). The outlet check valve (10) is pressed into bore behind the inlet fitting. The inlet fitting (11) is removed and installed in normal manner. To renew the outlet check valve (10) drill into the outlet check valve with a 9/64 inch drill to a depth of 1/8 inch (3.78 mm). Do not drill into carburetor body. Thread an 8-32 tap into the outlet valve and pull valve from the carburetor body. Press new outlet valve into carburetor body until face of valve is flush with base of the fuel chamber.

TVS Carburetor. Refer to Fig. TP2-6 for identification and exploded view of TVS carburetor. No idle or main fuel adjustment screws are provided on one

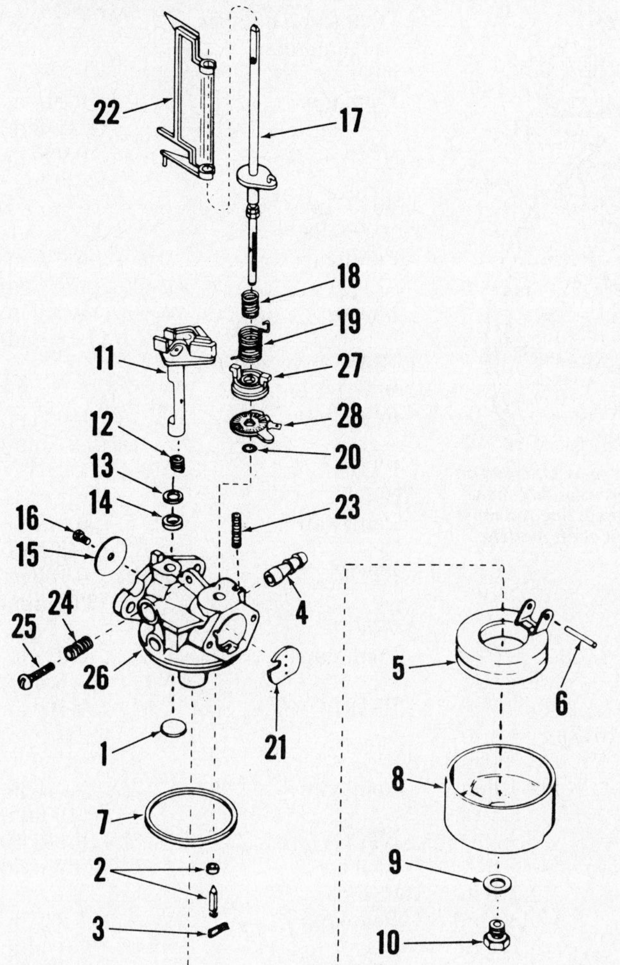

Fig. TP2-6—Exploded view of typical TVS type carburetor.

1. Welch plug
2. Fuel inlet valve & seat
3. Clip
4. Fuel inlet fitting
5. Float
6. Float shaft
7. Bowl gasket
8. Float bowl
9. Gasket
10. Bowl retainer
11. Throttle shaft
12. Return spring
13. Washer
14. Seal
15. Throttle plate
16. Screw
17. Choke/governor shaft
18. Spring
19. Governor spring
20. Washer
21. Choke plate
22. Air vane
23. Spring
24. Spring
25. Idle speed stop screw
26. Body
27. Sleeve
28. Serrated disc

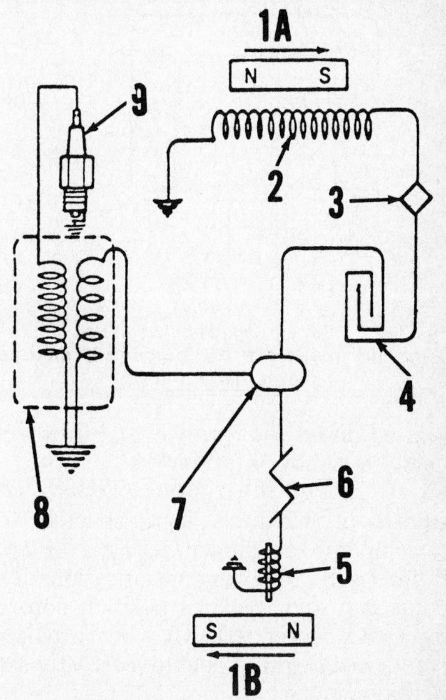

Fig. TP2-7—Diagram of Tecumseh solid state ignition system. Items (3, 4, 5, 6 and 7) are encased as shown at (10—Fig. TP2-8). Refer to text for description of operation.

model and only an idle fuel adjustment screw is provided on others.

Initial adjustment of idle mixture screw (Fig. TP2-6A as equipped) is 1 turn open from a lightly seated position. Idle fuel mixture on models without idle fuel mixture screw and main fuel mixture on all models is controlled by a fixed jet within carburetor. With engine at operating temperature and running, adjust idle mixture (as equipped) screw to obtain smooth engine idle.

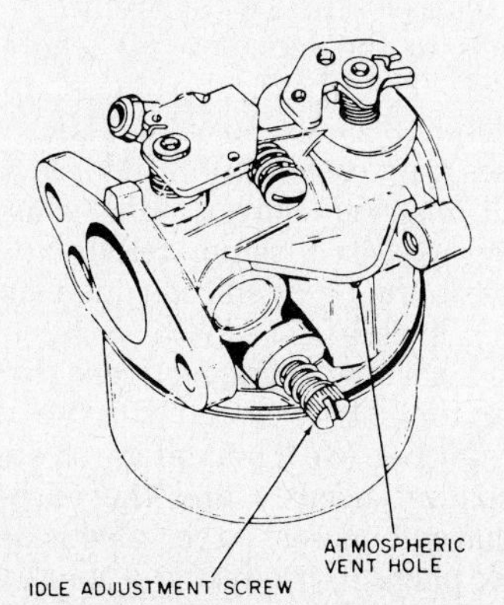

Fig. TP2-6A—View showing location of vent hole on TVS type fixed main jet carburetor.

To check float level, invert carburetor and measure from free end of float to float bowl mating surface of carburetor. Measurement should be 7/32 inch (5.56 mm). Carefully bend float lever tang which contacts fuel inlet needle as necessary to obtain correct float level.

GOVERNOR. An air vane type governor located on the carburetor throttle shaft is used on some models. On models with variable speed governor, the high speed (maximum) stop screw is located on the speed control lever. On models with fixed engine speed, rpm is adjusted by moving the governor spring bracket (B–Fig. TP2-5). To increase engine speed, bracket must be moved to increase governor spring (G) tension holding throttle (T) open.

On some TVS series engines a highly sensitive air vane (22–Fig. TP2-6) is attached to the choke shaft, independent of the choke shutter, but attached to the throttle arm. A spring on the choke shaft can be adjusted to vary the tension, causing the engine to run faster or slower as desired. One increment movement of sleeve (27) will vary engine speed approximately 100 rpm.

IGNITION SYSTEM. Engines may be equipped with either a magneto type (breaker point type) or a solid state ignition system. Refer to the appropriate paragraph for model being serviced.

Breaker Point Ignition System. Breaker point gap at maximum opening should be set before adjusting the ignition timing. On some models, ignition timing is not adjustable.

Some models may be equipped with external coil and magnet laminations. Air gap between external ignition coil laminations and flywheel magnet should be 0.015 inch (0.38 mm).

Ignition points and condenser are

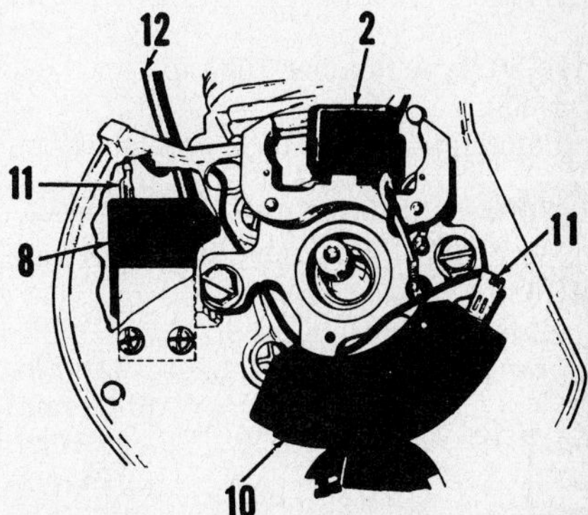

Fig. TP2-8—The solid state ignition charging coil, triggering system and mounting plate (2 and 10) are available only as an assembly and cannot be serviced.

2. Charging coil
8. Pulse transformer
10. Trigger system
11. Low tension lead
12. High tension lead

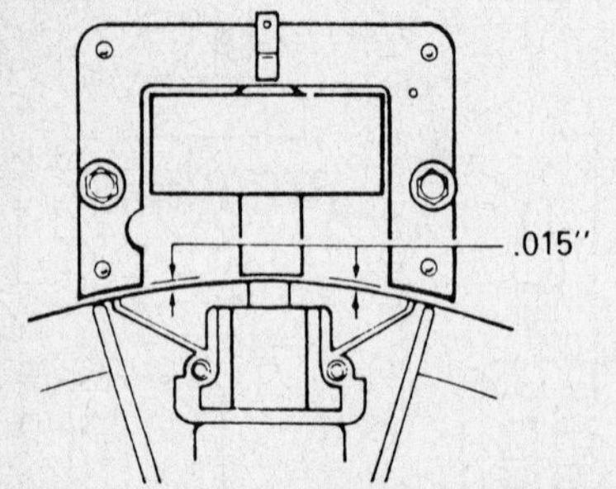

Fig. TP2-9 — Set air gap (G) between coil laminations and flywheel at 0.015 inch (0.38 mm) as shown. Refer to text.

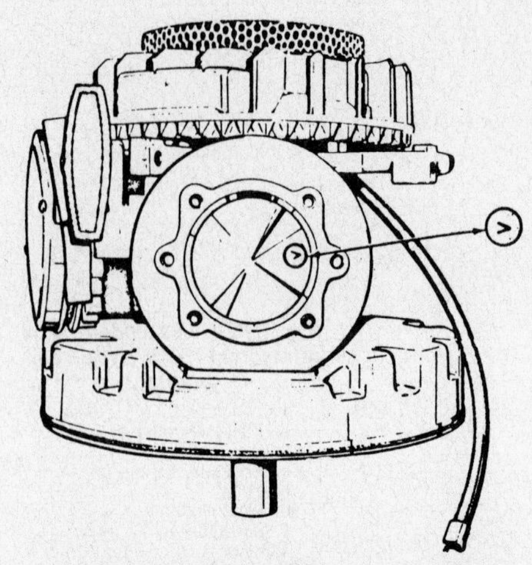

Fig. TP2-10 — The "V" or "1111" mark stamped on top of piston must be toward side shown. Lubrication hole in side of connecting rod must be toward top on all vertical shaft models.

located under the flywheel and flywheel must be removed for service.

Use Tecumseh gage 670259 or equivalent thickness plastic sheeting to set air gap as shown in Fig. TP2-9. Refer to the following specifications for point gap and flywheel position before top dead center (BTDC) when breaker points just begin to open to correctly set ignition timing.

AH440:
- Point gap . . . . . . 0.017 in. (0.43 mm)
- BTDC . . . . . . 0.122 in. (3.10 mm)

AH480, AH490:
- Point gap . . . . . . 0.017 in. (0.43 mm)
- BTDC (aluminum bushing rod) . . . . . . 0.100 in. (2.54 mm)
- BTDC (needle bearing rod) . . . . . . 0.135 in. (3.43 mm)

AH520 (type number 1500 to 1549):
- Point gap . . . . . . 0.020 in. (0.51 mm)
- BTDC . . . . . . 0.062 in. (1.575 mm)

AH520 (type number 1500 to 1577 & 1581 to 1582A):
- Point gap . . . . . . 0.017 in. (0.43 mm)
- BTDC . . . . . . 0.110 in. (2.79 mm)

AH520 (type number 1583 to 1589):
- Point gap . . . . . . 0.017 in. (0.43 mm)
- BTDC . . . . . . 0.062 in. (1.58 mm)

AH520 (aluminum bushing rod):
- Point gap . . . . . . 0.017 in. (0.43 mm)
- BTDC . . . . . . 0.110 in. (2.79 mm)

AH520 (needle bearing rod):
- Point gap . . . . . . 0.017 in. (0.43 mm)
- BTDC . . . . . . 0.185 in. (4.70 mm)

AH520 (type number 1601 to 1617):
- Point gap . . . . . . 0.020 in. (0.51 mm)
- BTDC . . . . . . *

AV520 (type number 638 to 650):
- Point gap . . . . . . 0.018 in. (0.48 mm)
- BTDC . . . . . . 0.100 in. (2.54 mm)

AV520 (type number 642-02E, F, 642-07C, 642-08, 642-13 to 14C & 642-15 to 23):
- Point gap . . . . . . 0.020 in. (0.51 mm)
- BTDC . . . . . . 0.085 in. (2.16 mm)

AV520 (type number 642 to 642-14C):
- Point gap . . . . . . 0.018 in. (0.46 mm)
- BTDC . . . . . . 0.110 in. (2.79 mm)

AV520 (type number 670 to 670-110):
- Point gap . . . . . . 0.020 in. (0.51 mm)
- BTDC . . . . . . 0.070 in. (1.78 mm)

AV600 (type number 641):
- Point gap . . . . . . 0.018 in. (0.46 mm)
- BTDC . . . . . . 0.100 in. (2.54 mm)

AV600 (type number 643-needle bearing rod):
- Point gap . . . . . . 0.020 in. (0.51 mm)
- BTDC . . . . . . 0.085 in. (2.16 mm)

AV600 (type number 643-aluminum bushing rod):
- Point gap . . . . . . 0.018 in. (0.46 mm)
- BTDC . . . . . . 0.090 in. (2.29 mm)

AV600 (type 660-2 to 660-38):
- Point gap . . . . . . 0.020 in. (0.51 mm)
- BTDC . . . . . . 0.070 in. (1.78 mm)

TVS600 (type 661-02 to 661-29):
- Point gap . . . . . . 0.020 in. (0.51 mm)
- BTDC . . . . . . *

AH750:
- Point gap . . . . . . 0.017 in. (0.43 mm)
- BTDC . . . . . . 0.100 in. (2.54 mm)

AV750:
- Point gap . . . . . . 0.018 in. (0.46 mm)
- BTDC . . . . . . 0.100 in. (2.54 mm)

AH817:
- Point gap . . . . . . 0.018 in. (0.46 mm)
- BTDC . . . . . . 0.113 in. (2.87 mm)

AV817:
- Point gap . . . . . . 0.020 in. (0.51 mm)
- BTDC . . . . . . 0.118 in. (2.80 mm)

*External coil, timing is nonadjustable. Set coil air gap to specified dimension.

Solid State Ignition. The Tecumseh solid state ignition system does not use ignition points. The only moving part of the system is the rotating flywheel with the charging magnets. As the flywheel magnet passes position (1A – Fig. TP2-7), a low voltage AC current is induced into input coil (2). The current passes through rectifier (3) converting this current to DC current. It then travels to the capacitor (4) where it is stored. The flywheel rotates approximately 180° to position (1B). As it passes trigger coil (5), it induces a very small electric charge into coil. This charge passes through resistor (6) and turns on the SCR (silicon controlled rectifier) switch (7). With SCR switch closed, the low voltage current stored in capacitor (4) travels to the pulse transformer (8). The voltage is stepped up instantaneously and the current is discharged across the electrodes of spark plug (9), producing a spark.

If system fails to produce a spark at spark plug, first check the high tension

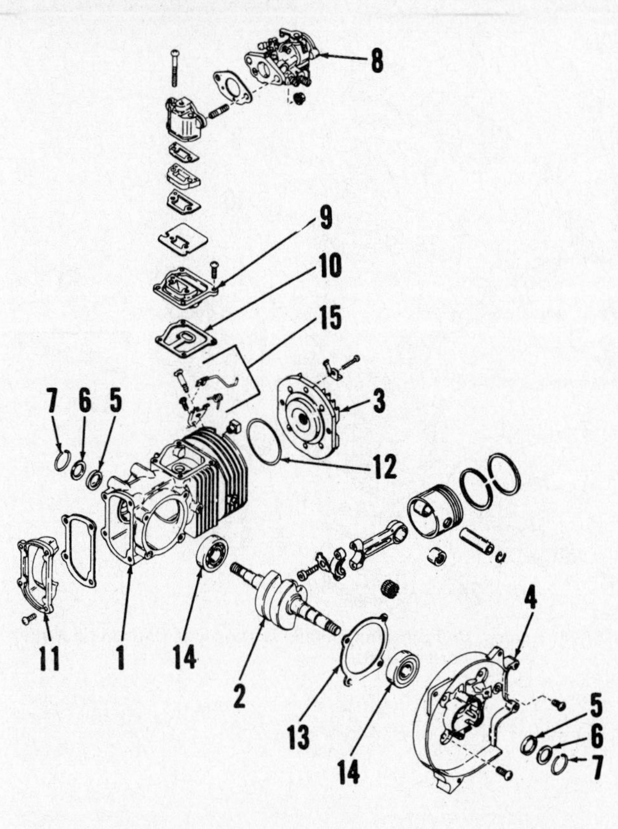

Fig. TP2-12—Exploded view typical of AH440 and AH490 models.

1. Cylinder & crankcase
2. Crankshaft
3. Cylinder head
4. Magneto end plate
5. Seal
6. Retainer
7. Snap ring
8. Carburetor
9. Third port cover
10. Gasket
11. End cover
12. Seal ring
13. Gasket
14. Main bearings (ball)
15. Compression release

lead (12–Fig. TP2-8). If condition of the high tension lead is questionable, renew the pulse transformer and high tension lead assembly. Check the low tension lead (11) and renew if insulation is faulty. The ignition charging coil, electronic triggering system and mounting plate are available only as an assembly. If necessary to renew this assembly, place the unit in position on the engine. Start the retaining screws, turn the mounting plate counterclockwise as far as possible, then tighten retaining screws to 5-7 ft.-lbs. (7-10 N•m).

LUBRICATION. Engines are lubricated by mixing a good quality two-stroke air-cooled engine oil with regular grade gasoline. For engines operating below 3600 rpm, mix 0.5 pint (0.24 L) of oil with each gallon of gasoline. For engines operating above 3600 rpm, mix 0.75 pint (0.36 L) of oil with each gallon of gasoline.

CARBON. Muffler and exhaust ports should be cleaned after every 50 hours of operation if engine is operated continuously at full load. If operated at light or medium load, the cleaning interval can be extended to 100 hours.

REPAIRS

DISASSEMBLY. Unbolt and remove the cylinder shroud, cylinder head and crankcase covers. Remove connecting rod cap and push the rod and piston unit out through top of cylinder. It may be necessary to remove the ridge from top of cylinder bore before removing piston and connecting rod assembly.

To remove the crankshaft, remove starter housing, flywheel and the bolts securing shroud base (bearing housing) to crankcase. Strike drive end of crankshaft with a leather mallet to dislodge crankshaft and bearing housing.

CONNECTING ROD. A steel connecting rod equipped with needle roller bearings at the crankpin and at the piston pin ends is used on some models.

An aluminum connecting rod which rides directly on crankpin journal is used on some models. Piston pin rides directly in connecting rod bore.

On other models with an aluminum connecting rod, a steel insert (liner) is used on inside of connecting rod and needle bearing rollers are used at the crankpin end.

On models with aluminum bushing type connecting rod, clearance on crankpin should be 0.0011-0.0020 inch (0.028-0.051 mm). Crankpin journal diameter should be 0.6857-0.6865 inch (17.417-17.437 mm). Piston pin diameter is 0.3750-0.3751 inch (9.525-9.528 mm). Only standard size parts are available.

On models with aluminum connecting rod with steel liners and bearing needles at crankpin, standard journal diameter is 0.8442-0.8450 inch (21.443-21.463 mm). There are 74 bearing rollers and ends of liners must correctly engage when match marks on connecting rod and cap are aligned. Piston pin diameter is 0.4997-0.4999 inch (12.692-12.698 mm) and piston pin rides in a cartridge type needle bearing that is pressed into connecting rod piston pin bore.

All models with steel connecting rod are provided with loose needle rollers at crankpin end of connecting rod and a

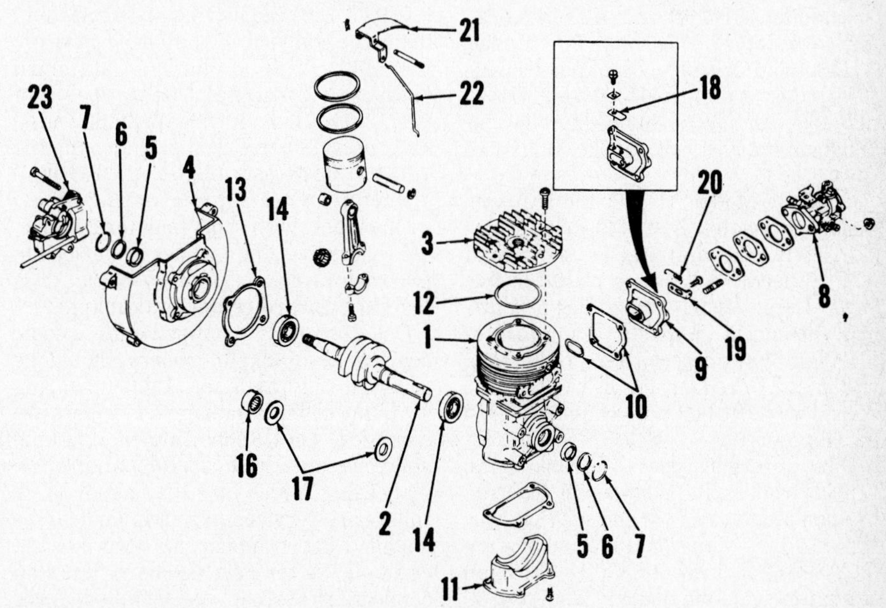

Fig. TP2-13—Exploded view of AH520 engine. Supplementary reed valve is shown at (18) and governor parts at (19 through 22).

1. Cylinder & crankcase
2. Crankshaft
3. Cylinder head
4. Magneto end plate
5. Seal
6. Retainer
7. Snap ring
8. Carburetor
9. Third port cover
10. Gaskets
11. End cover
12. Seal ring
13. Gasket
14. Main bearings (ball)
16. Main bearing (roller)
17. Thrust washers
18. Reed valve
19. Spring bracket
20. Governor spring
21. Air vane
22. Carburetor link
23. Magneto

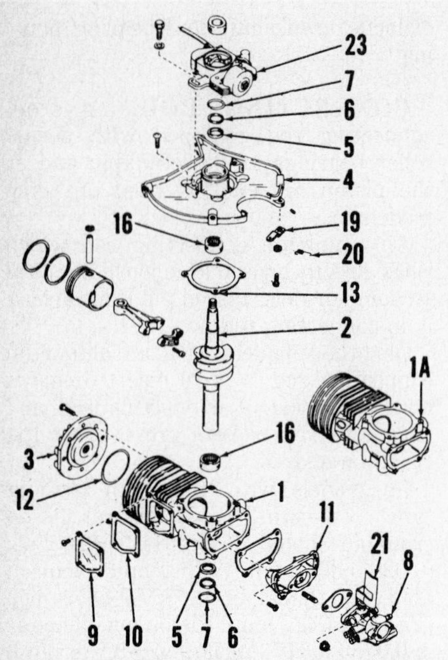

Fig. TP2-14—Exploded view typical of AV520 and AV600 engines. The cylinder and crankcase for some AV520 and all AV600 models is shown at (1A).

1. Cylinder & crankcase
2. Crankshaft
3. Cylinder head
4. Magneto end plate
5. Seal
6. Retainer
7. Snap ring
8. Carburetor
9. Cover
10. Gasket
11. Reed valve
12. Seal ring
13. Gasket
16. Main bearings (roller)
19. Spring bracket
20. Governor spring
21. Air vane (on carburetor)
23. Magneto

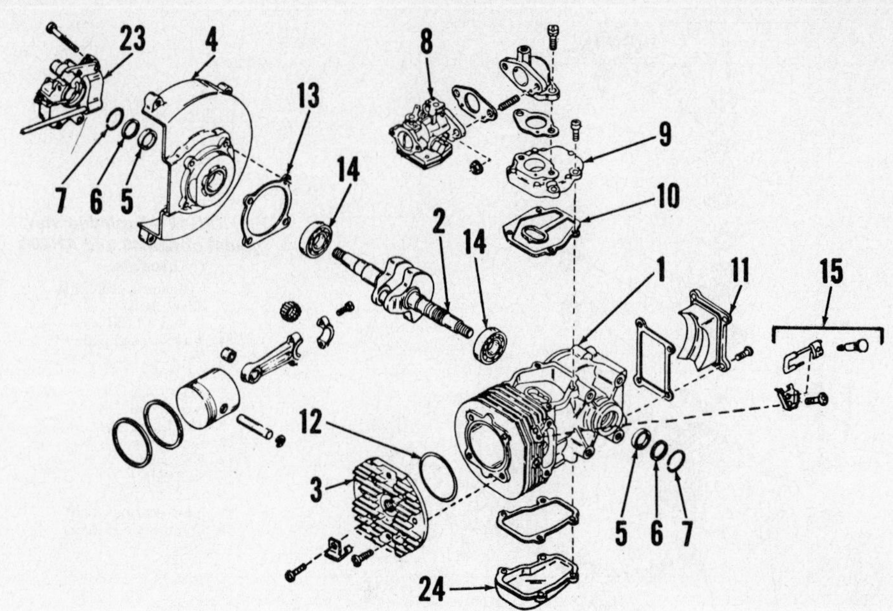

Fig. TP2-15—Exploded view of AH750 engine. Compression release is shown at (15). Model AH817 engines are similar.

1. Cylinder & crankcase
2. Crankshaft
3. Cylinder head
4. Magneto end plate
5. Seal
6. Retainer
7. Snap ring
8. Carburetor
9. Third port cover
10. Gasket
11. End cover
12. Seal ring
13. Gasket
14. Main bearings (ball)
15. Compression release
23. Magneto
24. Transfer port cover

cartridge needle bearing at piston pin bore.

Standard crankpin journal diameter is 0.5611-0.5618 inch (12.252-12.270 mm) for Model AH440 and 0.5614-0.5621 inch (12.260-12.277 mm) for Models AH480 and AH490. Crankpin bearing for Models AH440, AH480 and AH490 may use one row of 30 needle rollers or 60 short needle rollers placed in two rows.

On early Model AH520, the crankpin journal diameter is 0.5611-0.5628 inch (12.252-12.270 mm) and 56 short (half length) needle rollers are placed in two rows. Later Model AH520 has a standard crankpin journal diameter of 0.6922-0.6927 inch (17.582-17.595 mm) and uses 31 needle roller bearings in the 0.8524 inch (21.650 mm) connecting rod bearing bore.

Two different types of crankshafts, bearing rollers and connecting rods are used on Models AH750 and AV750. Early models are equipped with 0.6240-0.6243 inch (15.850-15.857 mm) diameter crankpin journal and use 32 needle rollers in the 0.7566-0.7569 inch (19.218-19.225 mm) diameter connecting rod bearing bore. Later Model AV750 is equipped with 0.6259-0.6266 inch (15.898-15.916 mm) diameter crankpin journal and uses 33 needle rollers in the 0.7588-0.7592 inch (19.274-19.284 mm) diameter connecting rod bearing bore.

Models AH817 and AV817 have standard crankpin journal diameters of 0.6259-0.6266 inch (15.898-15.916 mm) and use 66 short (half length) needle rollers placed in the two rows in the 0.7588-0.7592 inch (19.274-19.284 mm) diameter connecting rod bearing bore.

Model TVS600 has a standard crankpin journal diameter of 0.8113-0.8118 inch (20.607-20.620 mm) and uses 30 roller bearings in the 1.0443-1.0448 inch (26.525-26.538 mm) diameter connecting rod bearing bore.

On models with short (half length) needle rollers, bearing needles are placed in two rows around crankpin with flat ends together toward center of crankpin.

On all models equipped with needle bearing at crankpin, rollers should be renewed only as a set. Renew bearing set if any roller is damaged. If rollers are damaged, check condition of crankpin and connecting rod carefully and renew if bearing races are damaged. New rollers are serviced in a strip and can be installed by wrapping the strip around crankpin. After new needle rollers and connecting rod cap are installed, force lacquer thinner into needles to remove the beeswax, then lubricate bearing with SAE 30 oil.

On all models, make certain match marks on connecting rod and cap are aligned.

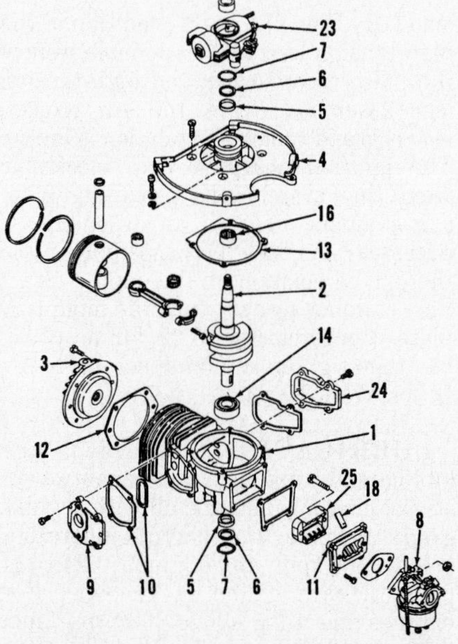

Fig. TP2-16—Exploded view of typical AV750 engine. Refer to text when assembling reed valve (11, 18 and 25). Model AV817 engines are similar.

1. Cylinder & crankcase
2. Crankshaft
3. Cylinder head
4. Magneto end plate
5. Seal
6. Retainer
7. Snap ring
8. Carburetor
9. Cover
10. Gasket
11. Reed plate
12. Head gasket
13. Gasket
14. Main bearing (ball)
16. Main bearing (roller)
18. Reed valve petal (8)
23. Magneto
24. Transfer port cover
25. Reed petal stop

On vertical shaft models, make certain lubrication hole in side of connecting rod is toward top. Some AV520 and AV600 models are stamped with "V" or "1111" mark on piston as shown in Fig. TP2-10. On models so equipped, make certain mark stamped on top of piston is toward the right side as shown.

On models with aluminum connecting rod, tighten connecting rod cap retaining screws to 40-50 in.-lbs. (5-6 N·m) and lock with the tab washer.

On all models with a steel connecting rod, tighten the connecting rod cap retaining screws (self locking) to 70-80 in.-lbs. (8-9 N·m).

PISTON, PINS AND RINGS. Refer to CONNECTING ROD section for piston removal procedure. Piston and cylinder bore specifications are as follows:

Models AH440, AH480, AH490, AH520, AV520, AV600 and TVS600

Cylinder bore diameter . . 2.093-2.094 in. (53.162-53.188 mm)

Piston-to-cylinder clearance:

AV520 (type number 642-02E & F, 642-07C, 642-08, 642-13 to 14C, 642-15 to 23) . . . . 0.005-0.012 in. (0.13-0.30 mm)

All other 440, 480, 490, 520, 600 models . . 0.005-0.007 in. (0.13-0.18 mm)

Ring end gap:

AV520 (type number 638 to 638-100 & 650) . . 0.006-0.014 in. (0.15-0.41 mm)

AV520 (type number 642-02E & F, 642-07C, 642-08, 642-13 to 14C, 642-15 to 23, 642-24 to 33, 661, 670-01 to 670-110) . . . . 0.006-0.016 in. (0.15-0.41 mm)

AV600 (type number 641 to 641-14) . . . . 0.006-0.014 in. (0.15-0.36 mm)

AV600 (type number 643-03B & C, 643-05B, 643-14A, B & C, 643-15A to 31) . . . 0.006-0.016 in. (0.15-0.41 mm)

TVS600 (type number 661-02 to 661-29) . . . . . 0.006-0.016 in. (0.15-0.41 mm)

AH480 (type number 1471 to 1471B, 1484 to 1484D, 1489 to 1490B, 1498, 1500, 1509, 1511, 1515 to 1516C, 1517, 1527, 1529A & B, 1531, 1535B, 1542, 1554, 1554A, 1573) . . . . 0.006-0.011 in. (0.15-0.28 mm)

AH520 (type number 1401 to 1401J, 1448 to 1450F, 1466 to 1466A, 1482 to 1482C, 1483, 1499, 1506 to 1507A, 1525A, 1534A, 1551, 1553, 1555, 1556, 1574 to 1577, 1581 to 1582A, 1583 to 1599, 1600 to 1620) . . . . 0.006-0.016 in. (0.15-0.41 mm)

All other 440, 480, 490, 520 & 600 models . . . . 0.007-0.017 in. (0.18-0.43 mm)

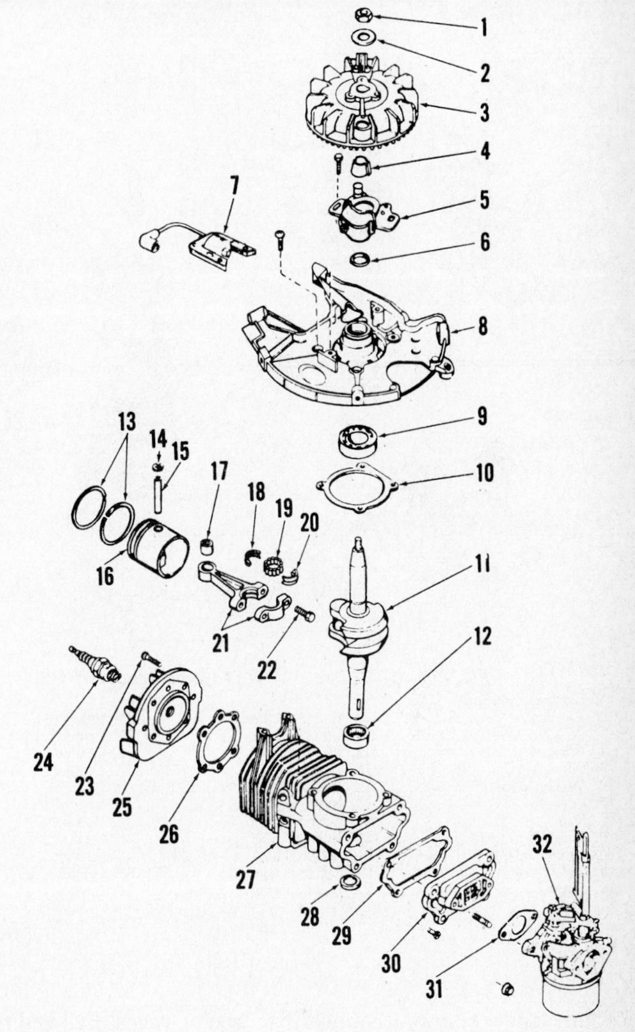

Fig. TP2-17—Exploded view of typical Model TVS600 type 661 engine.

1. Flywheel nut
2. Washer
3. Flywheel
4. Flywheel sleeve
5. Magneto
6. Seal
7. Coil
8. End plate
9. Bearing (ball)
10. Gasket
11. Crankshaft
12. Bearing (roller)
13. Piston rings
14. Clip
15. Piston pin
16. Piston
17. Bearing
18. Liner
19. Roller bearings
20. Liner
21. Connecting rod
22. Rod bolt
23. Head bolt
24. Spark plug
25. Cylinder head
26. Head gasket
27. Cylinder
28. Seal
29. Gasket
30. Reed valve
31. Gasket
32. Carburetor

Models AH750 and AV750

Cylinder bore diameter . . 2.375-2.376 in. (60.389-60.391 mm)

Piston-to-cylinder clearance:

AH750 . . . . 0.005-0.006 in. (0.13-0.15 mm)

AV750 . . . . 0.0055-0.0075 in. (0.140-0.191 mm)

Ring end gap . . . . 0.005-0.013 in. (0.13-0.33 mm)

Models AH817 and AV817

Cylinder bore diameter . . 2.437-2.438 in. (61.900-61.925 mm)

Piston-to-cylinder clearance . . . . 0.0063-0.0083 in. (0.160-0.211 mm)

Ring end gap . . . . 0.007-0.017 in. (0.18-0.43 mm)

Piston, rings and piston pin are available in standard size only. The piston pin should be a press fit in heated piston on models with needle bearing in connecting rod pin bore. On models without needle bearing, the piston pin should be a palm push fit in piston and thumb push fit in rod.

Refer to CONNECTING ROD section for standard piston pin diameters.

When assembling piston in connecting rod, on vertical crankshaft models, lubrication hole in side of connecting rod must be toward top of engine. On Models AV520 and AV600 equipped with offset piston, make certain that "V" mark or "1111" mark stamped on top of piston is toward the right side as shown in Fig. TP2-10.

Use the old cylinder head sealing ring and a ring compressor to compress piston rings, when sliding piston into cylinder.

NOTE: Make certain rings do not catch in recess at top of cylinder.

Always renew the cylinder head metal sealing ring. The cylinder head retaining screws should be tightened to 90-100 in.-lbs. (10-11 N·m). Refer to CONNECTING ROD section for installation of the connecting rod.

CRANKSHAFT AND CRANKCASE. The crankshaft can be removed

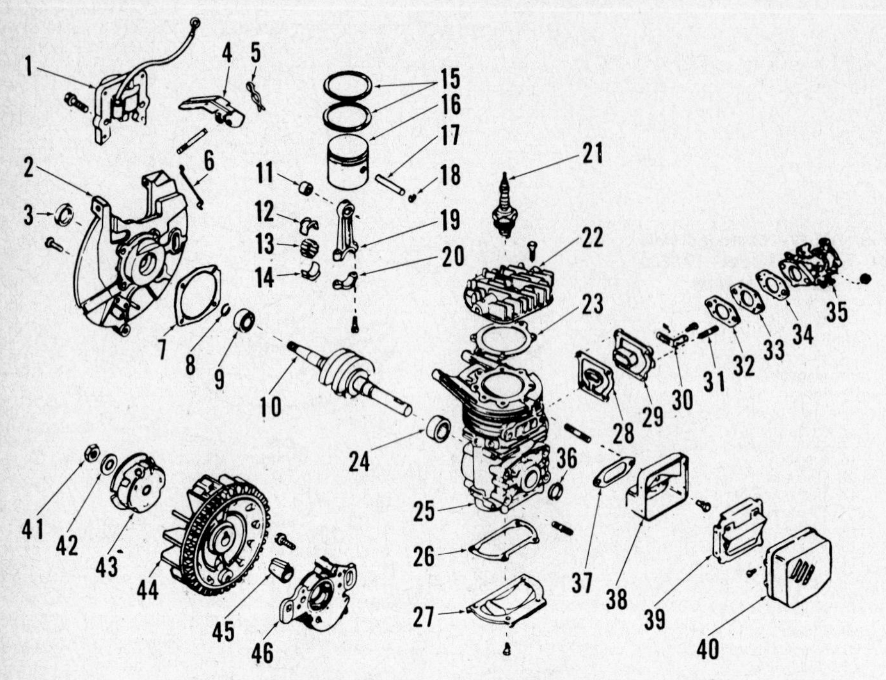

Fig. TP2-18—Exploded view of typical Model AH520 type 1600 engine.

1. External ignition
2. End plate
3. Seal
4. Air vane
5. Clip
6. Governor link
7. Gasket
8. Clip
9. Bearing
10. Crankshaft
11. Bearing
12. Liner
13. Roller needles
14. Liner
15. Piston rings
16. Piston
17. Piston pin
18. Retainer
19. Connectijng rod
20. Rod cap
21. Spark plug
22. Cylinder head
23. Head gasket
24. Bearing
25. Crankcase
26. Gasket
27. Crankcase cover
28. Gasket
29. Reed plate
30. Bracket
31. Stud
32. Gasket
33. Gasket
34. Gasket
35. Carburetor
36. Seal
37. Gasket
38. Muffler base
39. Baffle
40. Cover
41. Nut
42. Washer
43. Starter cup
44. Flywheel
45. Flywheel sleeve
46. Magneto

after the piston, connecting rod, flywheel and the magneto end bearing plate are removed.

Crankshaft main bearings may be either ball type or cartridge needle roller. If ball type main bearings are used, it should be necessary to bump the crankshaft out of the bearing inner races. Ball and roller bearing outer races should be a tight fit in bearing bores. If new ball bearings are to be installed, heat the crankcase when removing or installing bearings. Do not use open flame.

On all models, bearings should be installed with printed face on race toward center of engine.

If the crankshaft is equipped with thrust washers at ends, make certain they are installed when assembling.

Crankshaft end play should be ZERO for all AH440, AH480, AH490, AH520, AH750, AH817, AV750, AV817 and TVS600 models. Crankshaft end play should be 0.003-0.016 inch (0.08-0.41 mm) for Models AV520 and AV600 equipped with two needle roller main bearings.

It is important to exercise extreme care when renewing crankshaft seals to prevent their being damaged during installation. If a protector sleeve is not available, use tape to cover any splines, keyways, shoulders or threads over which the seal must pass during installation. Seals should be installed with channel groove of seal toward inside (center) of engine on all models except for use on outboard motors. If engine is used on an outboard motor, lip of lower seal should be toward outside (bottom) and the top (magneto end) seal should be toward inside (center) of engine.

REED VALVES. Some engines are equipped with a reed type inlet valve located in the lower crankcase cover. Reed petals should not stand out more than 0.010 inch (0.25 mm) from the reed plate and must not be bent, distorted or cracked. The reed plate must be smooth and flat. Renew petals (AV750 only) or complete valve assembly if valve does not seal completely.

Reed petals are available separately only for AV750 models. Petals must be installed with rounded edge against sealing surface of reed plate and the sharp edge must be toward reed stop. When installing the reed stop, apply Loctite to threads of retaining screws and tighten the two screws to 50-60 in.-lbs. (6-7 N•m).

TECUMSEH 4-STROKE

Model	Bore	Stroke	Displacement
LAV25, LAV30, TVS75, H25, H30	2.313 in. (58.74 mm)	1.844 in. (46.84 mm)	7.75 cu. in. (127 cc)
LV35, LAV35, TVS90, H35-prior 1983, ECH90	2.500 in. (63.50 mm)	1.844 in. (46.84 mm)	9.05 cu. in. (148 cc)
H35 after 1983	2.500 in. (63.50 mm)	1.938 in. (49.23 mm)	9.51 cu. in. (156 cc)
LAV40, TVS105, HS40, ECV105	2.625 in. (66.68 mm)	1.938 in. (49.23 mm)	10.49 cu. in. (172 cc)
TNT100, ECV100	2.625 in. (66.68 mm)	1.844 in. (46.84 mm)	9.98 cu. in. (164 mm)
V40, VH40, H40, HH40	2.500 in. (63.50 mm)	2.250 in. (57.15 mm)	11.04 cu. in. (181 cc)
ECV110	2.750 in. (69.85 mm)	1.938 in. (49.23 mm)	11.50 cu. in. (189 cc)
LAV50, ECV120, TNT120, TVS120, HS50	2.812 in. (71.43 mm)	1.938 in. (49.23 mm)	12.04 cu. in. (197 cc)
H50, HH50, VH50, TVM125	2.812 in. (71.43 mm)	2.250 in. (57.15 mm)	12.18 cu. in. (229 cc)
TVM140	2.625 in. (66.68 mm)	2.500 in. (63.50 mm)	13.53 cu. in. (222 cc)

ENGINE IDENTIFICATION

Engines must be identified by the complete model number, including the specification number in order to obtain correct repair parts. These numbers are located on the name plate and/or tags that are positioned as shown in Fig. T1 or T1A. It is important to transfer identification tags from the original engine to replacement short block assemblies so unit can be identified when servicing.

If selecting a replacement engine and the model or type number or the old engine is not known, refer to chart in Fig. T1B and proceed as follows:

1. List the corresponding number which indicates the crankshaft position.
2. Determine the horsepower needed. (Two-stroke engines are indicated by 00 in the second and third digit positions).
3. Determine the primary features needed. (Refer to the Tecumseh Engines Specification Book No. 692531 for specific engine variations.
4. Refer to Fig. T1C for Tecumseh engine model number interpretation.

The number following the letter code is the horsepower or cubic inch displacement. The number following the model number is the specification number. The last three digits of the specification number indicate a variation to the basic engine specification.

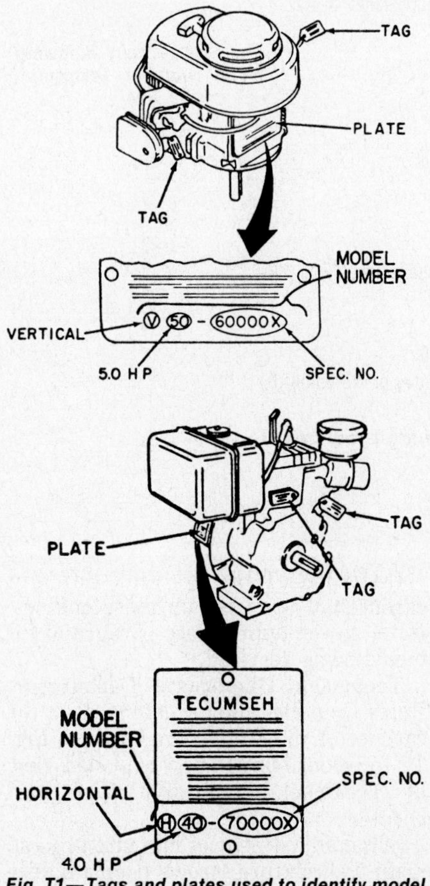

Fig. T1—Tags and plates used to identify model will most often be located in one of the positions shown.

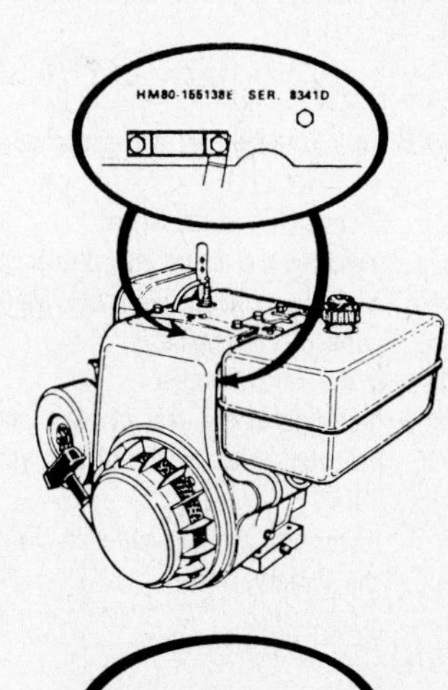

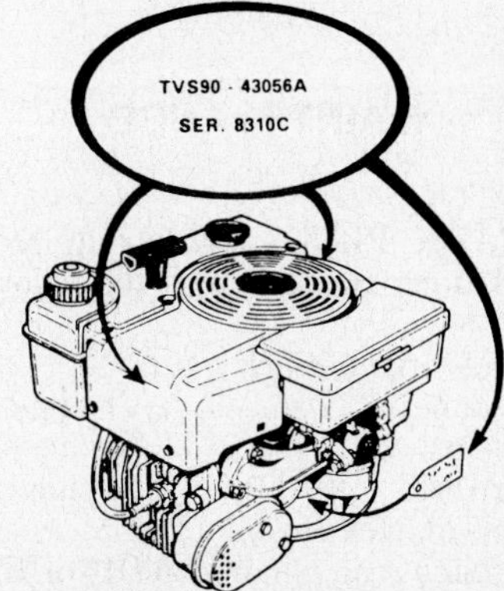

Fig. T1A—Locations of tags and plates used to identify later model engines.

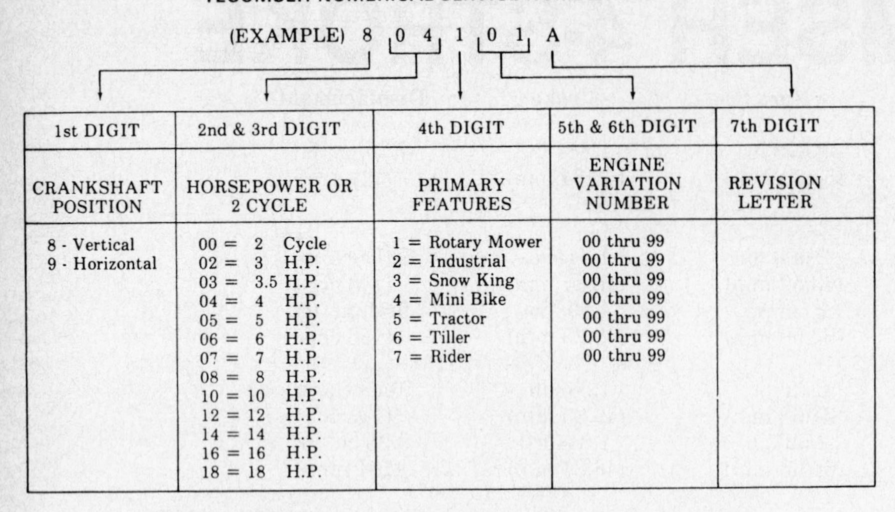

TECUMSEH NUMERICAL SERVICE NUMBER SYSTEM

(EXAMPLE) 8 0 4 1 0 1 A

1st DIGIT	2nd & 3rd DIGIT	4th DIGIT	5th & 6th DIGIT	7th DIGIT
CRANKSHAFT POSITION	HORSEPOWER OR 2 CYCLE	PRIMARY FEATURES	ENGINE VARIATION NUMBER	REVISION LETTER
8 - Vertical	00 = 2 Cycle	1 = Rotary Mower	00 thru 99	
9 - Horizontal	02 = 3 H.P.	2 = Industrial	00 thru 99	
	03 = 3.5 H.P.	3 = Snow King	00 thru 99	
	04 = 4 H.P.	4 = Mini Bike	00 thru 99	
	05 = 5 H.P.	5 = Tractor	00 thru 99	
	06 = 6 H.P.	6 = Tiller	00 thru 99	
	07 = 7 H.P.	7 = Rider	00 thru 99	
	08 = 8 H.P.			
	10 = 10 H.P.			
	12 = 12 H.P.			
	14 = 14 H.P.			
	16 = 16 H.P.			
	18 = 18 H.P.			

Fig. T1B—Reference chart used to select or identify replacement engines.

V - Vertical Shaft
LAV - Lightweight Aluminum Vertical
VM - Vertical Medium Frame
VH - Vertical Heavy Duty (Cast Iron)
TVS - Tecumseh Vertical Styled
TNT - Toro N' Tecumseh
ECV - Exclusive Craftsman Vertical
H - Horizontal Shaft
HS - Horizontal Small Frame
HM - Horizontal Medium Frame
HHM - Horizontal Heavy Duty (Cast Iron) Medium Frame
HH - Horizontal Heavy Duty (Cast Iron)
ECH - Exclusive Craftsman Horizontal

(EXAMPLE)

TVS90-43056A - is the model and specification number

TVS - Tecumseh Vertical Styled
90 - Indicates a 9 cubic inch displacement
43056A - is the specification number used for properly identifying the parts of the engine
8310C - is the serial number
8 - first digit is the year of manufacture (1978)
310 - indicates calendar day of that year (310th day or November 6, 1978)
C - represents the line and shift on which the engine was built at the factory.

Fig. T1C—Chart showing model number interpretation.

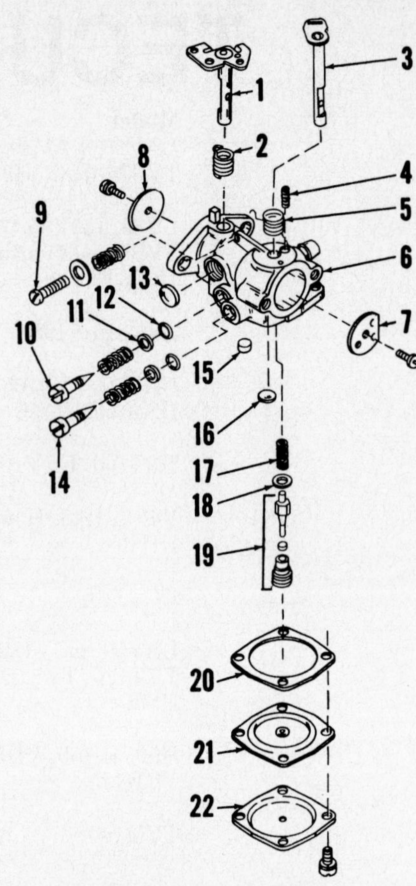

Fig. T2—Exploded view of typical Tecumseh diaphragm carburetor.

1. Throttle shaft
2. Return spring
3. Choke shaft
4. Choke stop spring
5. Return spring
6. Carburetor body
7. Choke plate
8. Throttle plate
9. Idle speed screw
10. Idle mixture screw
11. Washers
12. "O" rings
13. Welch plug
14. Main fuel mixture screw
15. Cup plug
16. Welch plug
17. Inlet needle spring
18. Gasket
19. Inlet needle & seat assy.
20. Gasket
21. Diaphragm
22. Cover

MAINTENANCE

SPARK PLUG. Spark plug recommendations are shown in the following chart.

14mm – 3/8 inch reach
Gasoline . . . . Champion J8
LP-Gas . . . . Champion J8
Kerosene . . . . Champion UJ12
18mm – 1/2 inch reach
Gasoline . . . . Champion D16 or MD16
Kerosene . . . . Champion D21
All 7/8 inch reach . . . . Champion W18

CARBURETOR. Several different carburetors are used on these engines. Refer to the appropriate paragraph for model being serviced.

Tecumseh Diaphragm Carburetor. Refer to model number stamped on the carburetor mounting flange and to Fig. T2 for identification and exploded view of Tecumseh diaphragm type carburetor.

Initial adjustment of idle mixture and main fuel mixture screws from a lightly seated position, is 1 turn open. Clockwise rotation leans mixture and counterclockwise rotation richens mixture.

Final adjustments are made with engine at operating temperature and running. Operate engine at rated speed and adjust main fuel mixture screw (14) for smoothest engine operation. Operate engine at idle speed and adjust idle mixture screw (10) for smoothest engine idle. If engine does not accelerate smoothly, slight adjustment of main fuel mixture screw may be required. Engine idle speed should be approximately 1800 rpm.

The fuel strainer in the fuel inlet fitting can be cleaned by reverse flushing with compressed air after the inlet needle and seat (19 – Fig. T2) are removed. The inlet needle seat fitting is metal with a neoprene seat, so the fitting (and enclosed seat) should be removed before carburetor is cleaned with a commercial solvent.The stamped line on carburetor throttle plate should be toward top of carburetor, parallel with throttle shaft

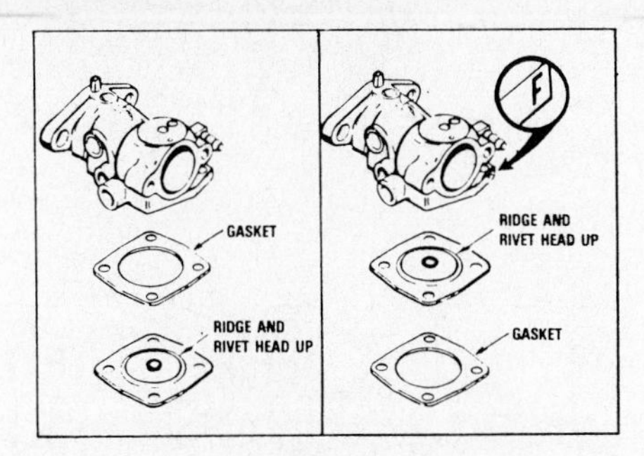

Fig. T2A—Illustration showing correct position of diaphragm on carburetors unmarked and marked with a "F".

and facing OUTWARD as shown in Fig. T3. Flat side of choke plate should be toward the fuel inlet fitting side of carburetor. Mark on choke plate should be parallel to shaft and should face INWARD when choke is closed. Diaphragm (21–Fig. T2) should be installed with rounded head of center rivet up toward the inlet needle (19) regardless of size or placement of washers around the rivet.

On carburetor Models 0234-252, 265, 266, 269, 270, 271, 282, 293, 303, 322, 327, 333, 334, 344, 345, 348, 349, 350, 351, 352, 356, 368, 371, 374, 378, 379, 380, 404 and 405, or carburetors marked with a "F" as shown in Fig. T2A, gasket (20–Fig. T2) must be installed between diaphragm (21) and cover (22). All other models are assemblies as shown, with gasket between diaphragm and carburetor body.

Tecumseh Standard Float Carburetor. Refer to Fig. T4 for identification and exploded view of Tecumseh standard float type carburetor.

Initial adjustment of idle mixture and main fuel mixture screws from a lightly seated position, is 1 turn open. Clockwise rotation leans mixture and counterclockwise rotation richens mixture.

Final adjustments are made with engine at operating temperature and running. Operate engine at rated speed and adjust main fuel mixture screw (34) for smoothest engine operation. Operate engine at idle speed and adjust idle mixture screw for smoothest engine idle. If engine does not accelerate smoothly, slight adjustment of main fuel mixture screw may be required.

Carburetor must be disassembled and all neoprene or Viton rubber parts removed before carburetor is immersed in cleaning solvent. Do not attempt to reuse any expansion plugs. Install new plugs if any are removed for cleaning.

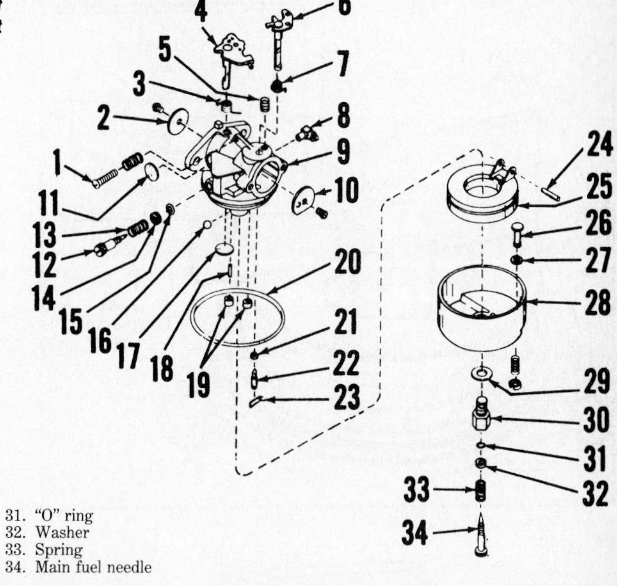

Fig. T4—Exploded view of standard Tecumseh float type carburetor.

1. Idle speed screw
2. Throttle plate
3. Return spring
4. Throttle shaft
5. Choke stop spring
6. Choke shaft
7. Return spring
8. Fuel inlet fitting
9. Carburetor body
10. Choke plate
11. Welch plug
12. Idle mixture needle
13. Spring
14. Washer
15. "O" ring
16. Ball plug
17. Welch plug
18. Pin
19. Cup plugs
20. Bowl gasket
21. Inlet needle seat
22. Inlet needle
23. Clip
24. Float shaft
25. Float
26. Drain stem
27. Gasket
28. Bowl
29. Gasket
30. Bowl retainer
31. "O" ring
32. Washer
33. Spring
34. Main fuel needle

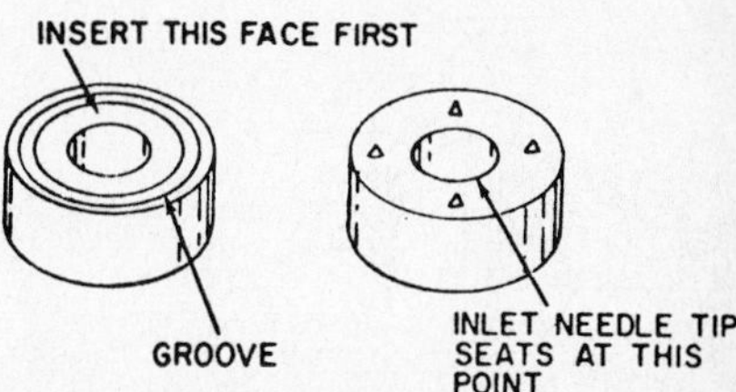

Fig. T7—The Viton seat used on some Tecumseh carburetors must be installed correctly to operate properly. All-metal needle is used with seat shown.

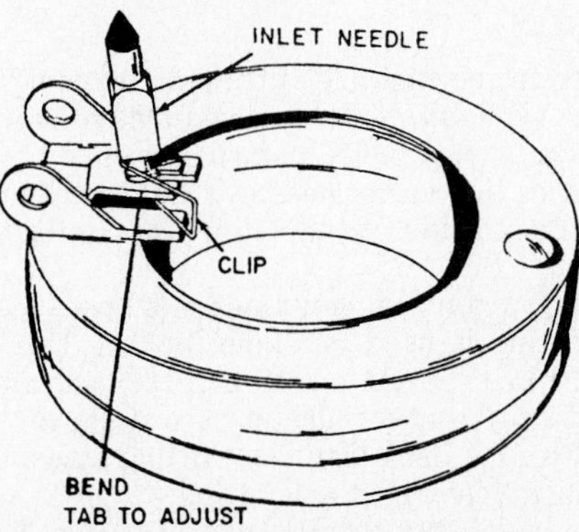

Fig. T5—View of float and fuel inlet valve needle. The valve needle shown is equipped with a resilient tip and a clip. Bend tab shown to adjust float height.

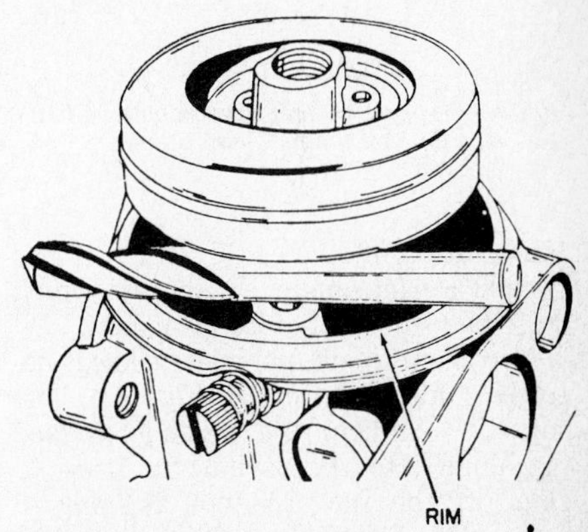

Fig. T8—Float height can be measured on some models by using a drill as shown. Refer to text for correct specifications.

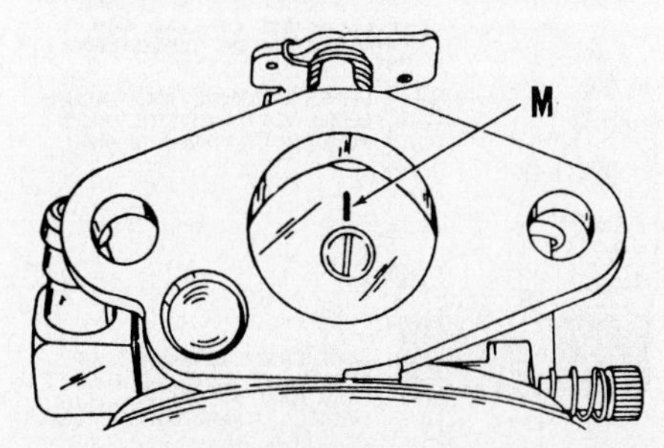

Fig. T3—The mark (M) on throttle plate should be parallel to the throttle shaft and outward as shown. Some models may also have mark at 3 o'clock position.

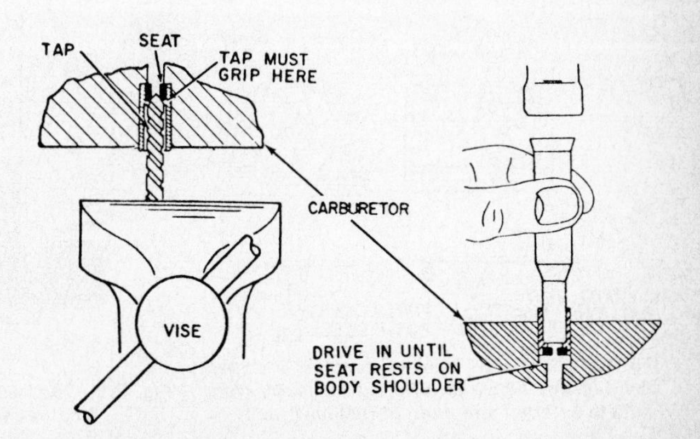

Fig. T6—A 10-24 or 10-32 tap is used to pull the brass seat fitting and fuel inlet valve seat from some carburetors. Use a close fitting flat punch to install new seat and fitting.

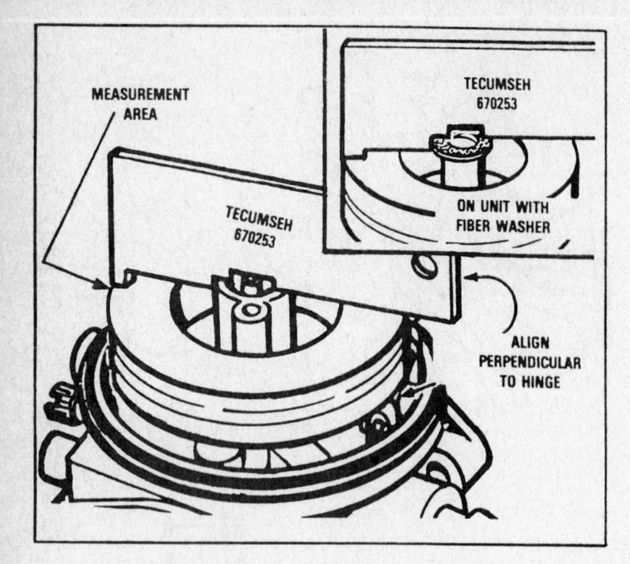

Fig. T8A—Float height can be set using Tecumseh float tool 670253 as shown.

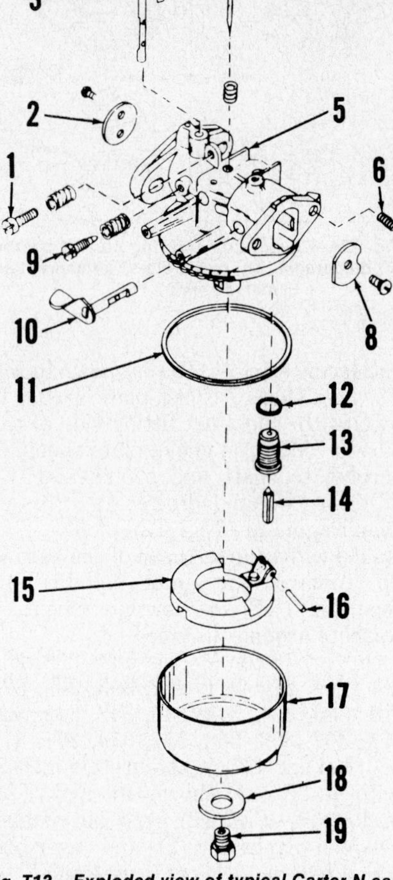

Fig. T13—Exploded view of typical Carter N carburetor.

1. Idle speed screw
2. Throttle plate
3. Throttle shaft
4. Main adjusting screw
5. Carburetor body
6. Choke shaft spring
7. Ball
8. Choke plate
9. Idle mixture screw
10. Choke shaft
11. Bowl gasket
12. Gasket
13. Inlet valve seat
14. Inlet valve
15. Float
16. Float shaft
17. Fuel bowl
18. Gasket
19. Bowl retainer

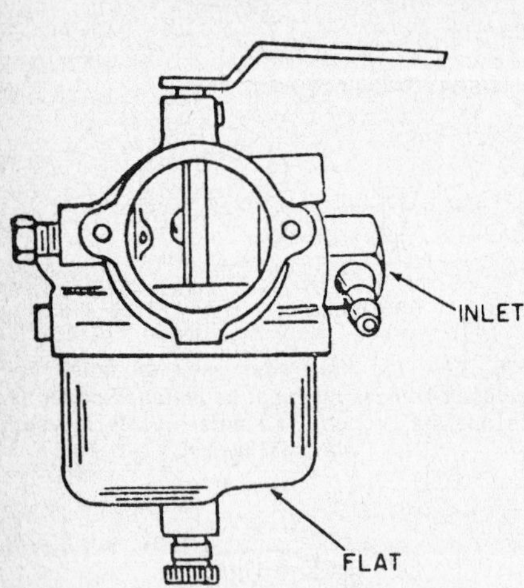

Fig. T9—Flat part of float bowl should be located under the fuel inlet fitting.

Fig. T11—Tecumseh "Automagic" carburetor is similar to standard float models, but can be identified by absence of choke and adjusting needles.

The fuel inlet needle valve closes against a neoprene or Viton seat which must be removed before cleaning.

A resilient tip on fuel inlet needle is used on some carburetors (Fig. T5). The soft tip contacts the seating surface machined into the carburetor body to shut off the fuel. Do not attempt to remove the inlet valve seat.

A Viton seat (21–Fig. T4) is used on some models. The rubber seat can be removed by blowing compressed air in from the fuel inlet fitting or by using a hooked wire. The grooved face of valve seat should be IN toward bottom of bore and the valve needle should seat on smooth side of the Viton seat. Refer to Fig. T7.

A Viton seat contained in a brass seat fitting is used on some models. Use a 10-24 or 10-32 tap to pull the seat and fitting from carburetor bore as shown in Fig. T6. Use a flat, close fitting punch to install new seat and fitting.

Install the throttle plate (2–Fig. T4) with the two stamped marks out and at 12 and 3 o'clock positions. The 12 o'clock line should be parallel with the throttle shaft and toward top of carburetor. Install choke plate (10) with flat side down toward bottom of carburetor. Float setting should be 0.200-0.220 inch (5.08-5.6

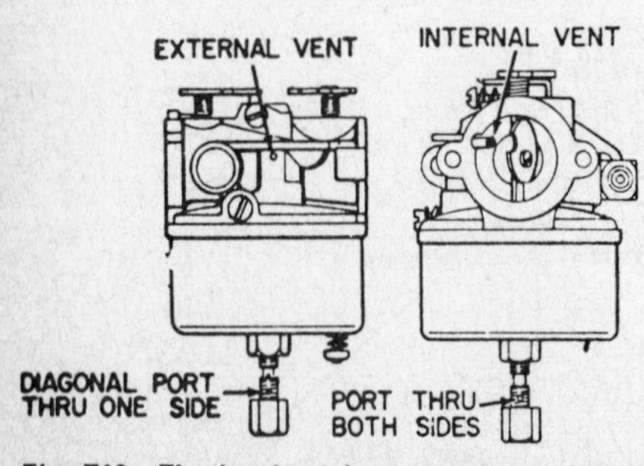

Fig. T10—The bowl retainer contains a drilled fuel passage which is different for carburetors with external and internal fuel bowl vent.

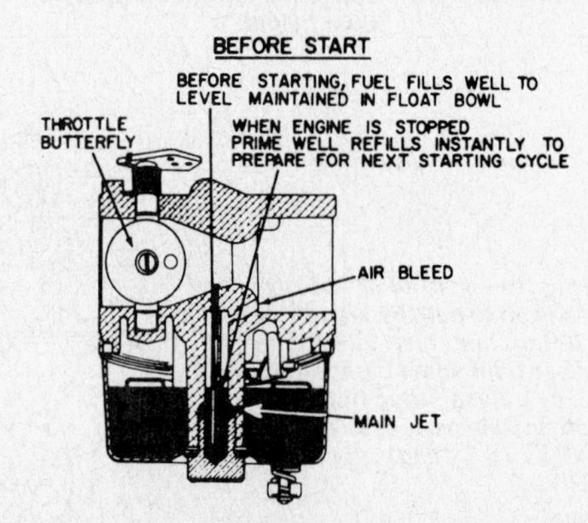

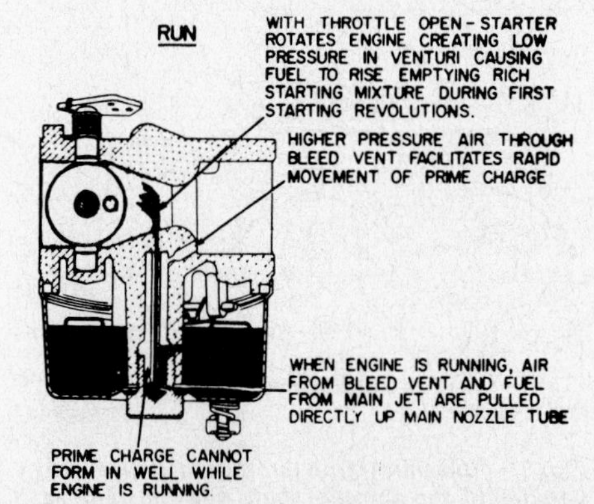

Fig. T12—The "Automagic" carburetor provides a rich starting mixture without using a choke plate. Mixture is changed by operating with a dirty air filter and by incorrect float setting.

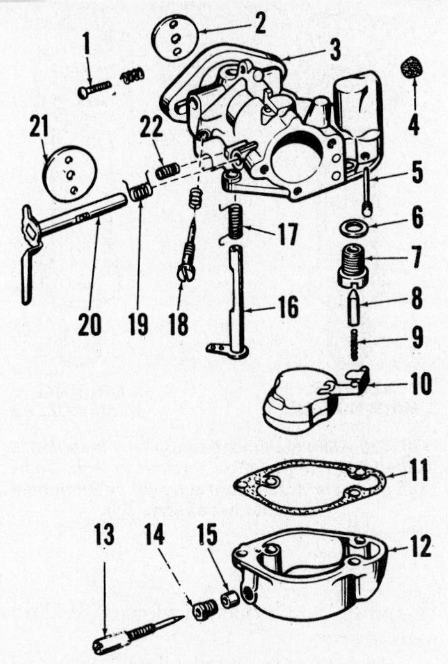

Fig. T14—Exploded view of Marvel-Schebler carburetor.

1. Idle speed screw
2. Throttle plate
3. Carburetor body
4. Fuel screen
5. Float
6. Gasket
7. Inlet valve seat
8. Inlet valve
9. Spring
10. Float
11. Gasket
12. Fuel bowl
13. Main adjusting screw
14. Retainer
15. Packing
16. Throttle shaft
17. Throttle spring
18. Idle mixture screw
19. Choke spring
20. Choke shaft
21. Choke ratchet spring

mm) and can be measured with a number 4 drill as shown in Fig. T8 or using Tecumseh float gage 670253 as shown in Fig. T8A. Remove float and bend tab at float hinge to change float setting. The fuel inlet fitting (8–Fig. T4) is pressed into body on some models. Start fitting, then apply a light coat of Loctite to shank and press fitting into

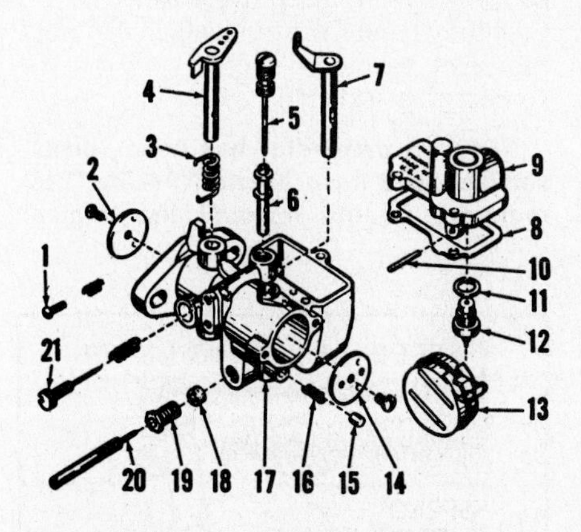

Fig. T15—Exploded view of typical Tillotson MT carburetor.

1. Idle speed screw
2. Throttle plate
3. Throttle spring
4. Throttle shaft
5. Idle tube
6. Main nozzle
7. Choke shaft
8. Gasket
9. Bowl cover
10. Float shaft
11. Gasket
12. Inlet needle & seat assy.
13. Float
14. Choke plate
15. Friction pin
16. Choke shaft spring
17. Carburetor body
18. Packing
19. Retainer
20. Main adjusting screw
21. Idle mixture screw

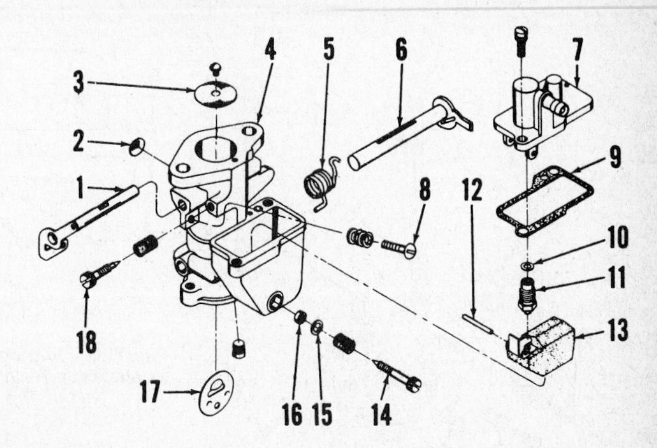

Fig. T16—Exploded view of typical Tillotson E carburetor.

1. Throttle shaft
2. Welch plug
3. Throttle plate
4. Carburetor body
5. Return spring
6. Choke shaft
7. Bowl cover
8. Idle speed stop screw
9. Bowl gasket
10. Gasket
11. Inlet needle & seat assy.
12. Float shaft
13. Float
14. Main fuel needle
15. Washer
16. Packing
17. Choke plate
18. Idle mixture screw

position. The flat on fuel bowl should be under the fuel inlet fitting. Refer to Fig. T9.

Be sure to use correct parts when servicing the carburetor. Some gaskets used as (20–Fig. T4) are square section, while others are round. The bowl retainer (30) contains a drilled passage for fuel to the high speed metering needle (34). A diagonal port through one side of the bowl retainer is used on carburetors with external vent. The port is through both sides on models with internal vent. Refer to Fig. T10.

Tecumseh "Automagic" Float Carburetor. Refer to Fig. T11 and note carburetor can be identified by the absence of the idle mixture screw, choke and main mixture screw.

Float setting is 0.210 inch (5.33 mm) or float may be set with a number 4 drill as shown in Fig. T8 or Tecumseh float gage 670253 as shown in Fig. T8A. Air filter maintenance is important in order to obtain the correct fuel mixture. Refer to notes for servicing standard Tecumseh float carburetors. Refer to Fig. T12 for operating principles.

Carter Carburetor. Refer to Fig. T13 for identification and exploded view of Carter carburetor.

Initial adjustment of idle mixture screw and main fuel mixture screw from a lightly seated position is 1½ turns open for idle mixture screw and 1¾ turns open for main fuel mixture screw. Clockwise rotation leans mixture and counterclockwise rotation richens mixture.

Final adjustments are made with engine at operating temperature and running. Operate engine at rated speed and adjust main fuel mixture screw (4) for smoothest engine operation. Operate engine at idle speed and adjust idle mixture screw (9) for smoothest engine idle. If engine acceleration is not smooth, main fuel mixture screw may have to be adjusted slightly.

To check float level, invert carburetor body and float assembly. There should be 11/64 inch (4.37 mm) clearance between free side of float and machined surface of body casting. Bend float lever tang to provide correct measurement.

Marvel-Schebler. Refer to Fig. T14 for identification and exploded view of Marvel-Schebler Series AH carburetor.

Initial adjustment of idle mixture screw and main fuel mixture screw from a lightly seated position, is 1 turn open for idle mixture screw and 1¼ turn open for main fuel mixture screw. Clockwise rotation leans mixture and counterclockwise rotation richens mixture.

Final adjustments are made with engine at operating temperature and running. Operate engine at rated speed and adjust main fuel mixture screw (13) for smoothest engine operation. Operate engine at idle speed and adjust idle fuel mixture needle (18) for smoothest engine idle. Adjust idle speed screw (1) to obtain 1800 rpm.

To adjust float level, invert carburetor throttle body and float assembly. Float clearance should be 3/32 inch (2.38 mm) between free end of float and machined surface of carburetor body. Bend tang on float which contacts fuel inlet needle as necessary to obtain correct float level.

Tillotson Type "MT" Carburetor. Refer to Fig. T15 for identification and exploded view of Tillotson type MT carburetor.

Initial adjustment of idle mixture screw and main fuel mixture screw from

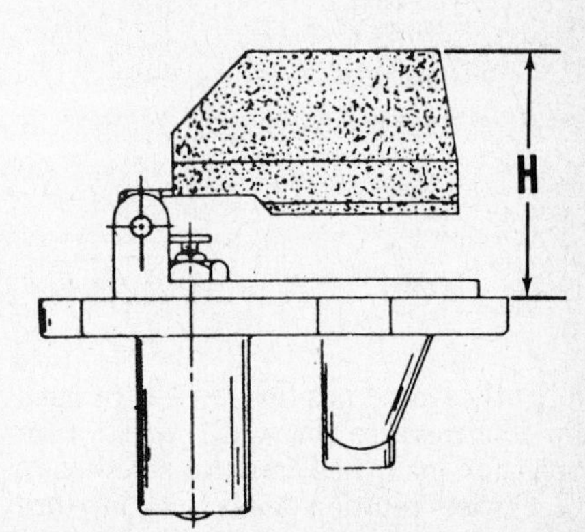

Fig. T17—Float height (H) should be measured as shown for Tillotson type "E" carburetors. Gasket should not be installed when measuring.

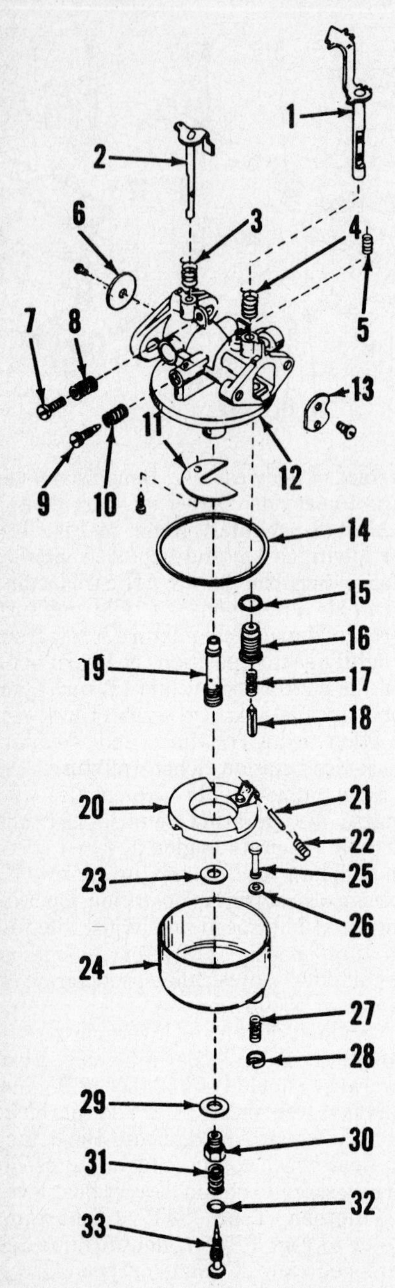

Fig. T18—Exploded view of Walbro LMG carburetor.

1. Choke shaft
2. Throttle shaft
3. Throttle return spring
4. Choke return spring
5. Choke stop spring
6. Throttle plate
7. Idle speed stop screw
8. Spring
9. Idle mixture screw
10. Spring
11. Baffle
12. Carburetor body
13. Choke plate
14. Bowl gasket
15. Gasket
16. Inlet valve seat
17. Spring
18. Inlet valve
19. Main nozzle
20. Float
21. Float shaft
22. Spring
23. Gasket
24. Bowl
25. Drain stem
26. Gasket
27. Spring
28. Retainer
29. Gasket
30. Bowl retainer
31. Spring
32. "O" ring
33. Main fuel screw

a lightly seated position, is ¾ turn open for idle mixture screw (21) and 1 turn open for main fuel mixture screw (20). Clockwise rotation leans fuel mixture and counterclockwise rotation richens fuel mixture. Final adjustments are made with engine at operating

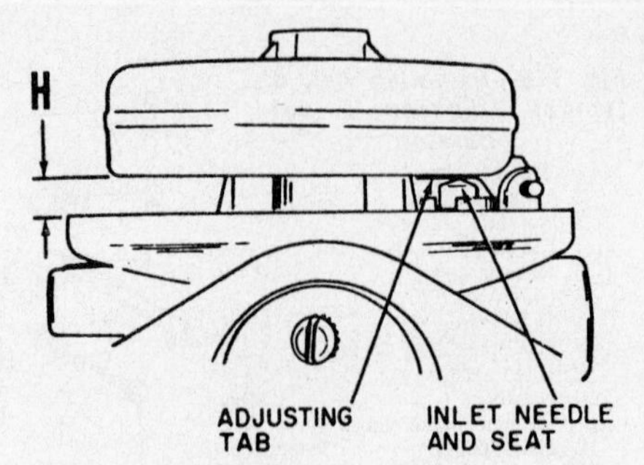

Fig. T19—Float height (H) should be measured as shown on Walbro float carburetors. Bend the adjusting tab to adjust height.

temperature and running. Operate engine at rated speed and adjust main fuel mixture screw for smoothest engine operation. Operate engine at idle speed and adjust idle mixture screw for smoothest engine idle. Adjust throttle stop screw (1) to obtain 1800 rpm. If engine acceleration is not smooth, main fuel mixture screw may have to be adjusted slightly.

To check float setting, invert carburetor throttle body and measure distance from top of float to carburetor body float bowl mating surface. Distance should be 1-13/32 inch (35.72 mm). Carefully bend float tang which contacts fuel inlet needle as necessary to obtain correct float level.

Tillotson Type "E" Carburetor. Refer to Fig. T16 for identification and exploded view of Tillotson type E carburetor.

Initial adjustment of idle mixture screw and main fuel mixture screw from a lightly seated position, is ¾ turn open for idle mixture screw (18) and 1 turn open for main fuel mixture screw (14). Clockwise rotation leans mixture and counterclockwise rotation richens mixture.

Final adjustments are made with engine at operating temperature and running. Operate engine at rated speed and adjust main fuel mixture screw for smoothest engine operation. Operate engine at idle speed and adjust idle mixture screw for smoothest engine idle. Adjust throttle stop screw (8) to obtain desired idle speed. If engine does not accelerate smoothly, it may be necessary to adjust main fuel mixture screw slightly.

To check float setting, invert carburetor throttle body and measure distance between top of float at free end to float bowl mating surface on carburetor. Distance should be 1-5/64 inch (27.38 mm). Refer to Fig. T17. Gasket should not be installed when measuring float height. Carefully bend float tang which contacts fuel inlet needle as necessary to obtain correct float level.

Walbro. Refer to Fig. T18 for identification and exploded view of Walbro carburetor.

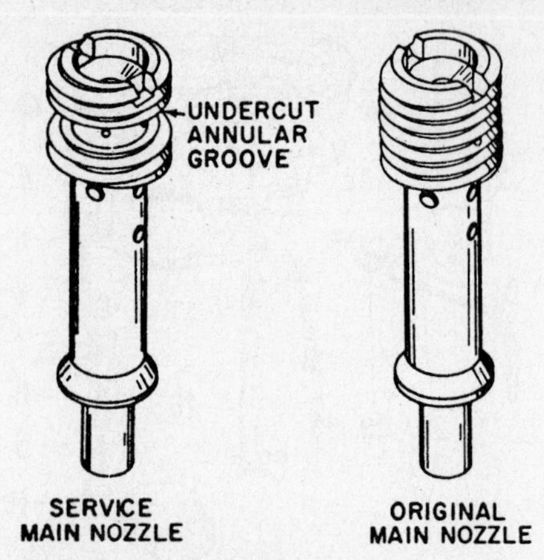

Fig. T20—The main nozzle originally installed is drilled after installation through hole in body. Service main nozzles are grooved so alignment is not necesasry.

Initial adjustment of idle mixture and main fuel mixture screws from a lightly seated position, is 1 turn open. Clockwise rotation leans fuel mixture and counterclockwise rotation richens fuel mixture.

Final adjustments are made with engine at operating temperature and running. Operate engine at rated speed and adjust main fuel mixture screw (33) for smoothest engine operation. Operate engine at idle speed and adjust idle mixture screw for smoothest engine idle. Adjust throttle stop screw (7) so engine idles at 1800 rpm.

To check float level, invert carburetor throttle body and measure clearance between free end of float and machined surface of carburetor. Clearance should be ⅛ inch (3.18 mm). Refer to Fig. T19. Carefully bend float tang which contacts fuel inlet needle as necessary to obtain correct float setting.

NOTE: If carburetor has been disassembled and main nozzle (19—Fig. T18) removed, do not reinstall the original

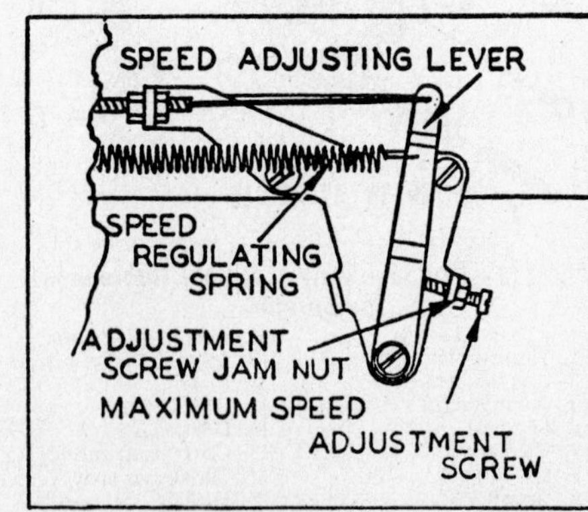

Fig. T21—Speed adjusting screw and lever. The no-load speed should not exceed 3600 rpm.

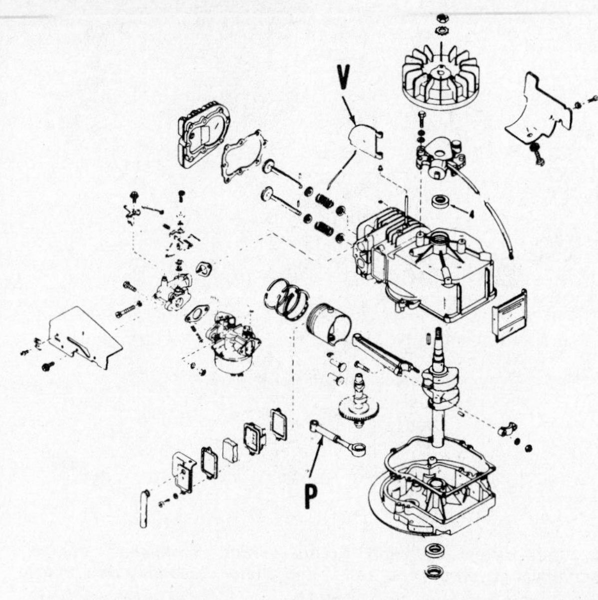

Fig. T21A—Exploded view of Light Frame vertical crankshaft engine. Model shown has plunger type oil pump (P) and air vane (V) governor.

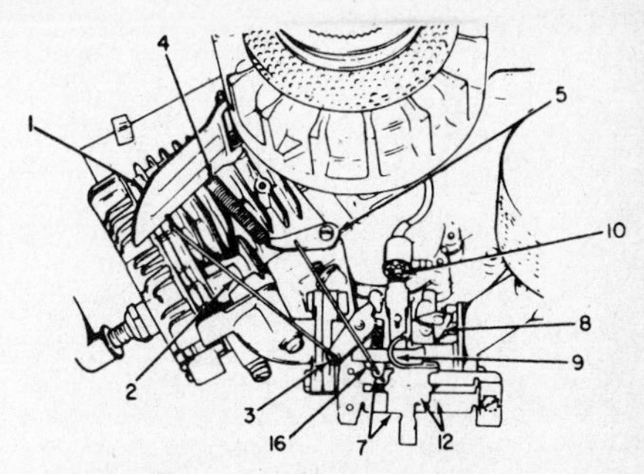

Fig. T24—View of one type of speed control with pneumatic governor. Refer also to Figs. T25, T26 and T27.

1. Air vane
2. Throttle control link
3. Carburetor throttle lever
4. Governor spring
5. Governor spring linkage
7. Speed control lever
8. Carburetor choke lever
9. Choke control
10. Stop switch
12. Alignment holes
16. Idle speed stop screw
18. High speed stop screw (Fig. T27)

equipment nozzle. Install a new service nozzle. Refer to Fig. T20 for differences between original and service nozzles.

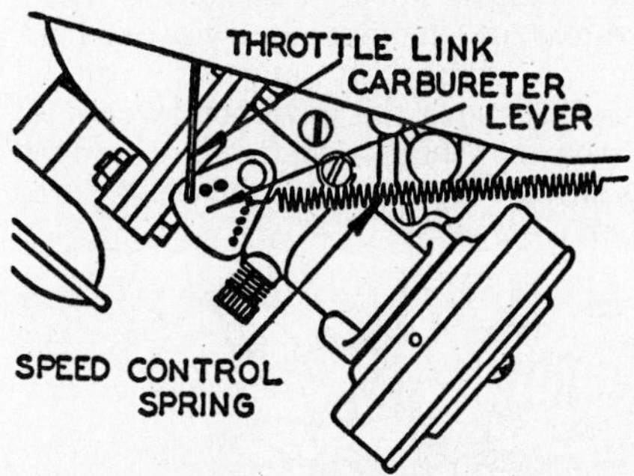

Fig. T22—Outer spring anchorage holes in carburetor throttle lever may be used to reduce speed fluctuations or surging.

PNEUMATIC GOVERNOR. Some engines are equipped with a pneumatic (air vane) type governor. On fixed linkage hookups, the recommended idle speed is 1800 rpm. Standard no-load speed is 3300 rpm. Operating speed range is 2600 to 3600 rpm. On engines not equipped with slide control (Fig. T23), obtain desired speed by varying the tension on governor speed regulating spring by moving speed adjusting lever (Fig. T21) in or out as required. If engine speed fluctuates due to governor hunting or surging, move carburetor end of governor spring into next outer hole on carburetor throttle arm shown in Fig. T22.

A too-lean mixture or friction in governor linkage will also cause hunting or unsteady operation.

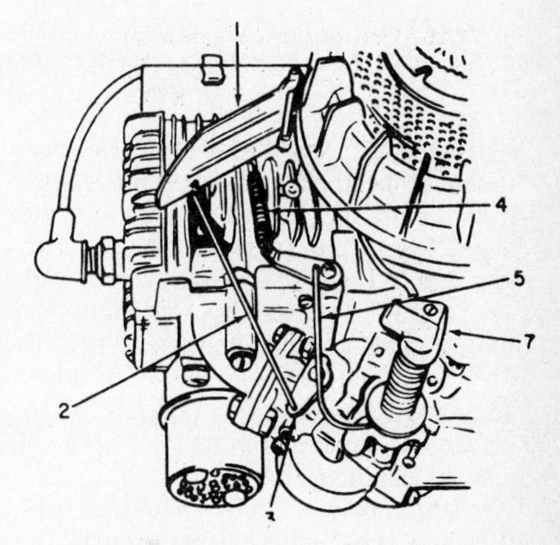

Fig. T25—Pneumatic governor linkage.

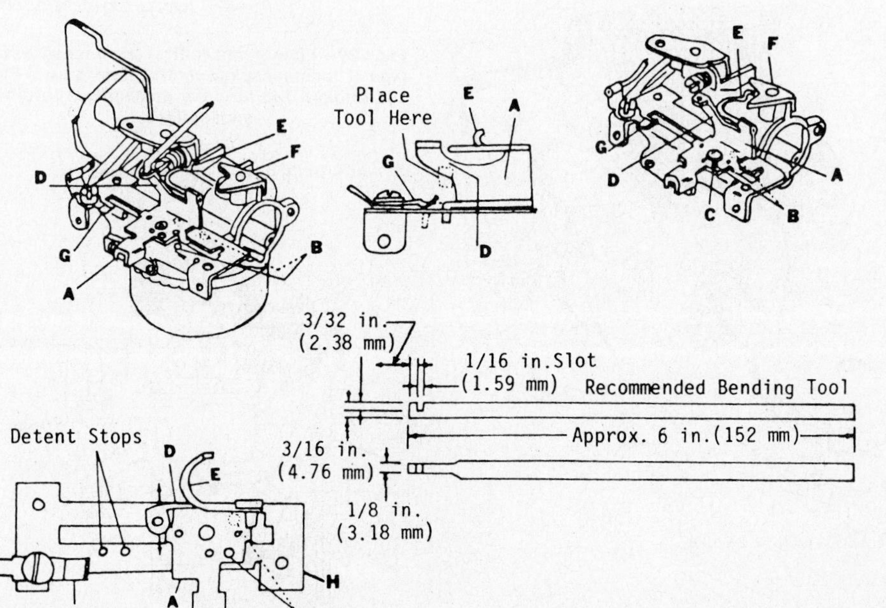

Fig. T23—When carburetor is equipped with slide control, governed speed is adjusted by bending slide arm at point "D". Bend arm outward from engine to increase speed.

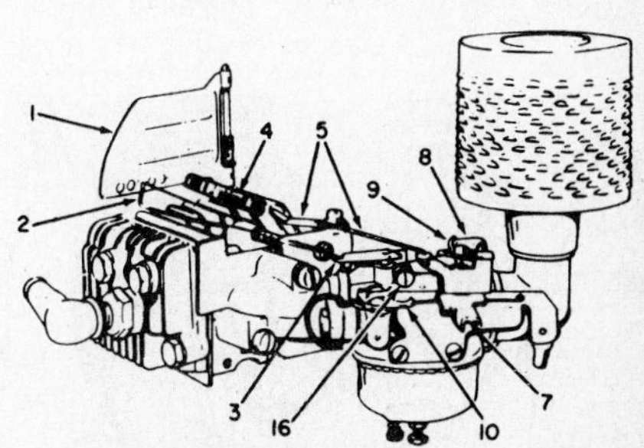

Fig. T26—Pneumatic governor linkage.

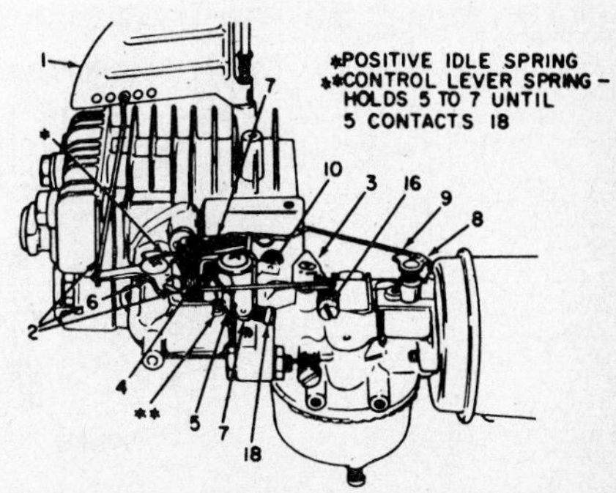

Fig. T27—Pneumatic governor linkage.

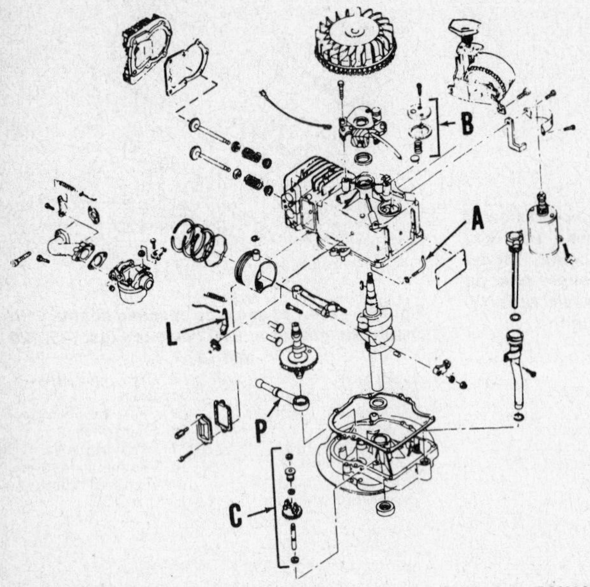

Fig. T27A—View of Light Frame engine typical of Models ECV100, ECV105, ECV110 and ECV120. Crankcase breather valve is shown at (B).

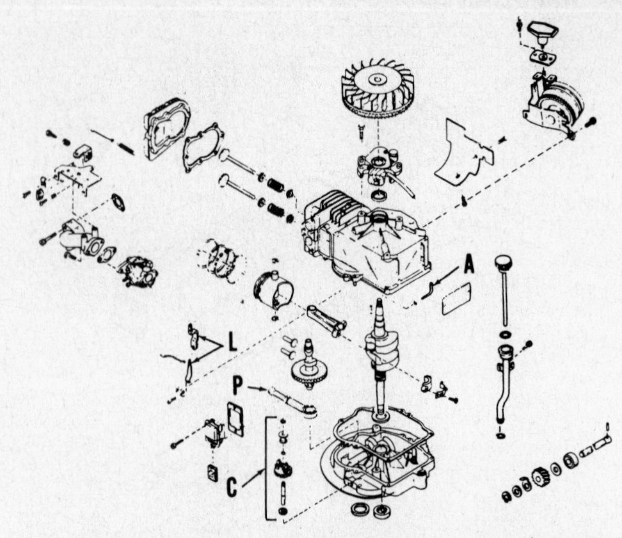

Fig. T28A—View of Light Frame vertical crankshaft engine. Mechanical governor (A, C and L) and plunger type oil pump (P) are shown.

On engines with slide control carburetors, speed adjustment is made with engine running. Remove slide control cover (Fig. T23) marked "choke, fast, slow, etc.". Lock the carburetor in high speed "run" position by matching the hole in slide control member (A) with hole (B) closest to choke end of bracket. Temporarily hold in this position by inserting a tapered punch or pin of suitable size into the hole.

At this time the choke should be wide open and the choke activating arm (E) should be clear of choke lever (F) by 1/64 inch (0.40 mm). Obtain clearance gap by bending arm (E). To increase engine speed, bend slide arm at point (D) outward from engine. To decrease speed, bend arm inward toward engine.

Make certain on models equipped with wiper grounding switch that wiper (G) touches slide (A) when control is moved to "STOP" position.

Refer to Figs. T24, T25, T26 and T27 for assembled views of pneumatic governor systems.

MECHANICAL GOVERNOR. Some engines are equipped with a mechanical (flyweight) type governor. To adjust the governor linkage, refer to Fig. T28 and loosen governor lever screw. Twist protruding end of governor shaft counterclockwise as far as possible on vertical crankshaft engines; clockwise on horizontal crankshaft engines. On all models, move the governor lever until carburetor throttle shaft is in wide open position, then tighten governor lever clamp screw.

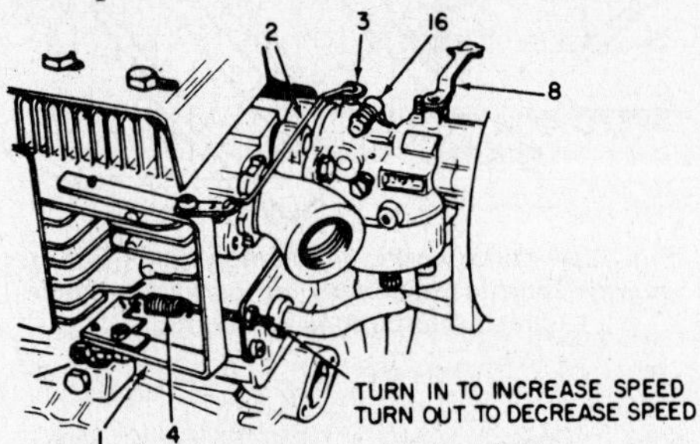

Fig. T29—View of mechanical governor with one type of constant speed control. Refer also to Fig. T30 through T43 for other mechanical governor installations.

1. Governor lever
2. Throttle control
3. Carburetor throttle lever
4. Governor spring
8. Choke lever
16. Idle speed stop screw

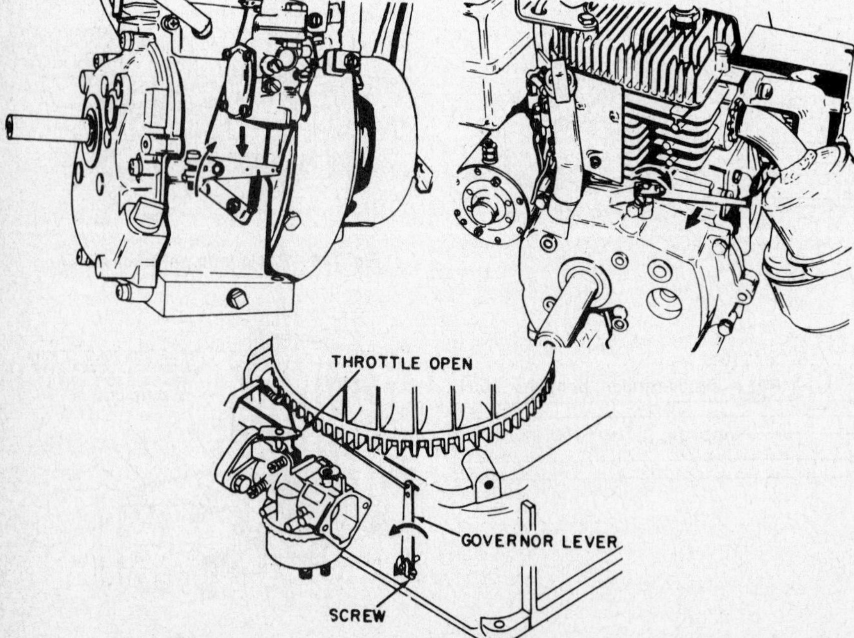

Fig. T28—Views showing location of mechanical governor lever and direction to turn when adjusting position on governor shaft.

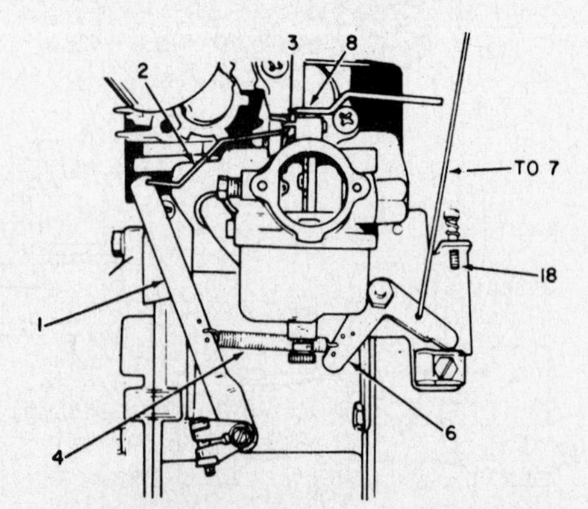

Fig. T30—Mechanical governor linkage. Refer to Fig. T29 for legend except for the following.

6. Bellcrank
7. Speed control lever
18. High speed stop screw

Binding or worn governor linkage will result in hunting or unsteady engine operation. An improperly adjusted carburetor will also cause a surging or hunting condition.

Refer to Figs. T29 through T42 for views of typical mechanical governor speed control linkage installations. The governor gear shaft must be pressed into bore in cover until the correct amount of the shaft protrudes. Refer to illustration and chart in Fig. T44.

IGNITION SYSTEM. A magneto ignition system with breaker points or

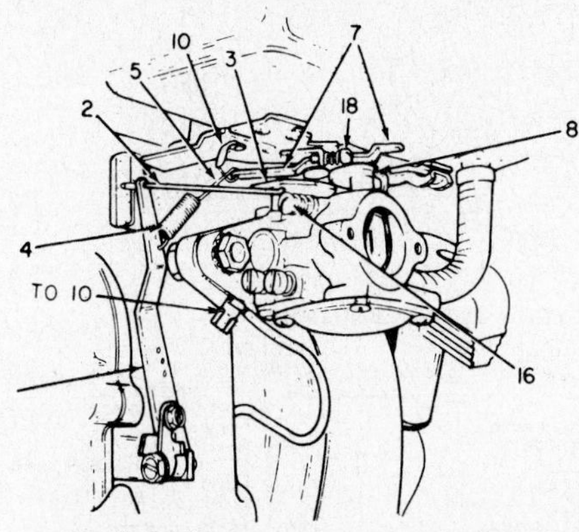

Fig. T31—Mechanical governor linkage.

1. Governor lever
2. Throttle control link
3. Carburetor throttle
4. Governor spring
5. Governor spring linkage
7. Speed control lever
8. Choke lever
10. Stop switch
16. Idle speed stop screw
18. High speed stop screw

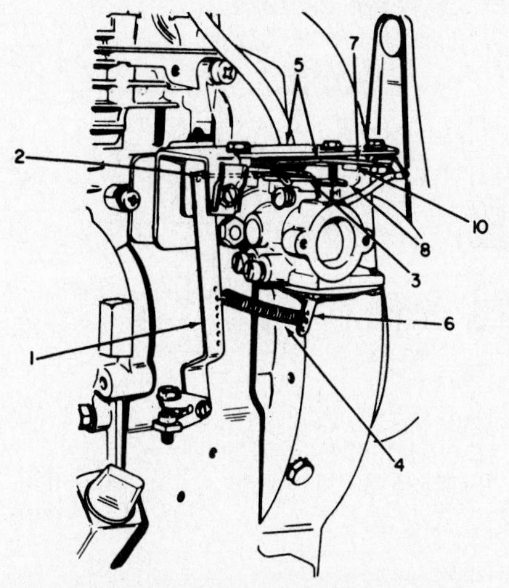

Fig. T32—Mechanical governor linkage. Refer to Fig. T31 for legend. Bellcrank is shown at (6).

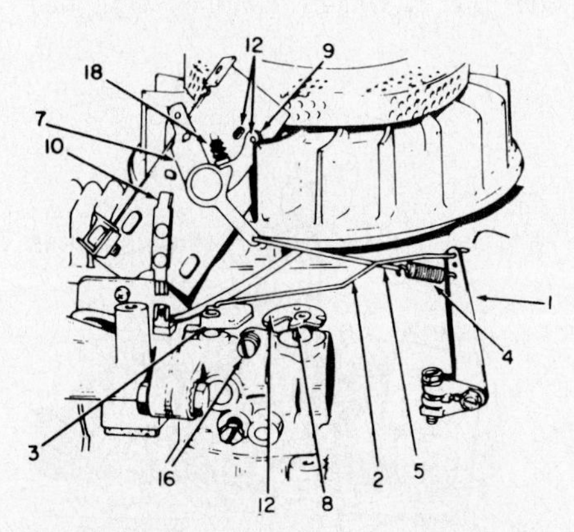

Fig. T33—Mechanical governor linkage. Control cover is raised to view underside.

1. Governor lever
2. Throttle control link
3. Carburetor throttle lever
4. Governor spring
5. Gover spring linkage
7. Speed control lever
8. Choke lever
9. Choke control
10. Stop switch
12. Alignment holes
16. Idle speed stop screw
18. High speed stop screw

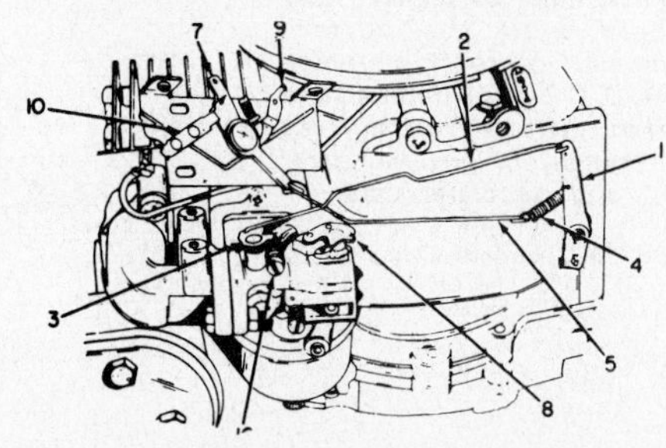

Fig. T34—Mechanical governor linkage with control cover raised. Refer to Fig. T33 for legend.

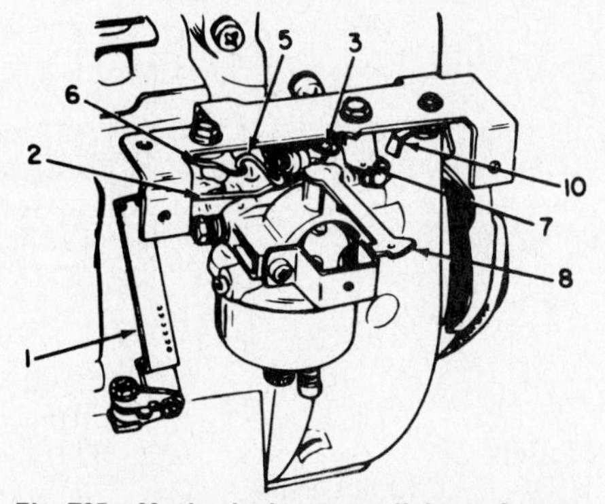

Fig. T35—Mechanical governor linkage. Governed speed of engine is increased by closing loop in linkage (5); decrease speed by spreading loop. Refer to Fig. T33 for legend.

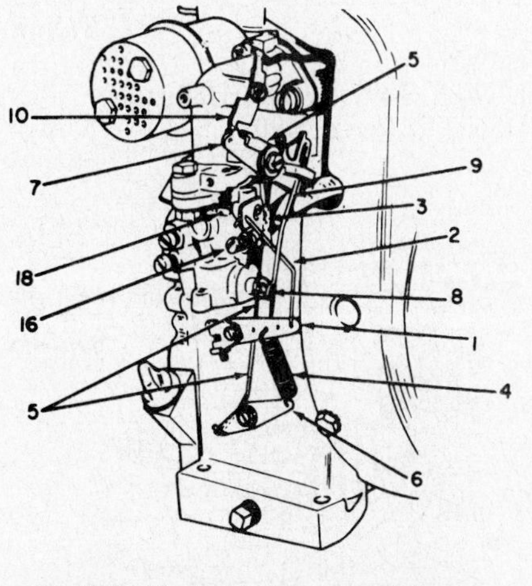

Fig. T36—Mechanical governor linkage.

1. Governor lever
2. Throttle control link
3. Throttle lever
4. Governor spring
5. Governor spring linkage
6. Bellcrank
7. Speed control lever
8. Choke lever
9. Choke control link
10. Stop switch
16. Idle speed stop screw
18. High speed stop screw

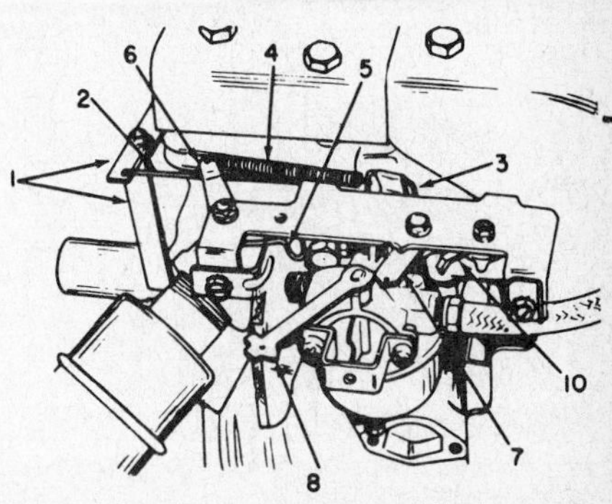

Fig. T37—View of control linkage used on some engines. To increase governed engine speed, close loop (5); to decrease speed, spread loop (5). Refer to Fig. T36 for legend.

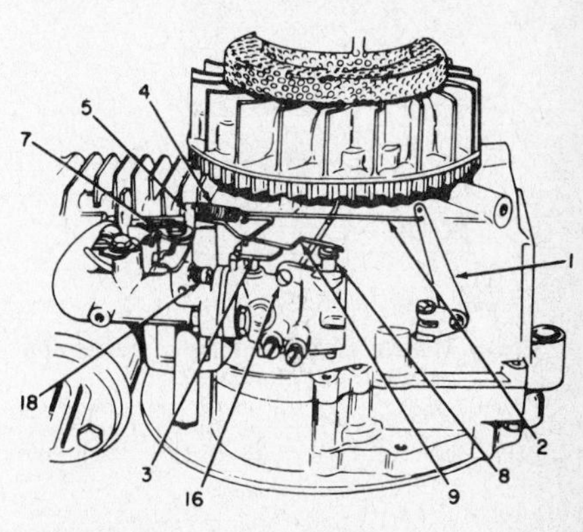

Fig. T38—View of mechanical governor control linkage. Governor spring (4) is hooked onto loop in link (2).

1. Governor lever
2. Throttle control link
3. Throttle lever
4. Governor spring
5. Governor spring linkage
7. Speed control lever
8. Choke lever
9. Choke control link
16. Idle speed stop screw
18. High speed stop screw

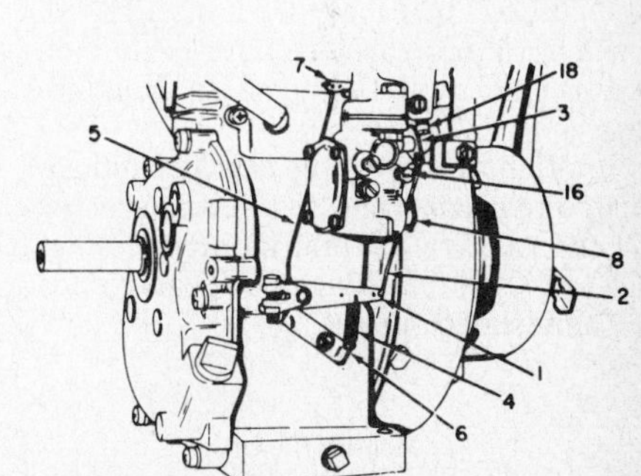

Fig. T39—Linkage for mechanical governor. Refer to have Fig. T38 for legend. Bellcrank is shown at (6).

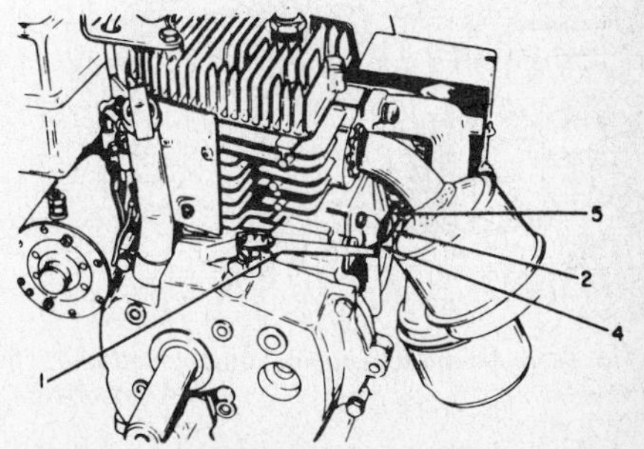

Fig. T40—Mechanical governor linkage. Refer to Fig. T38 for legend.

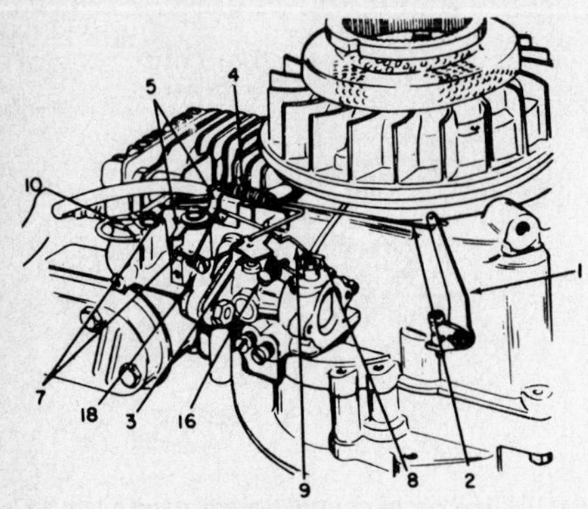

Fig. T41—Mechanical governor linkage. Refer to Fig. T38 for legend.

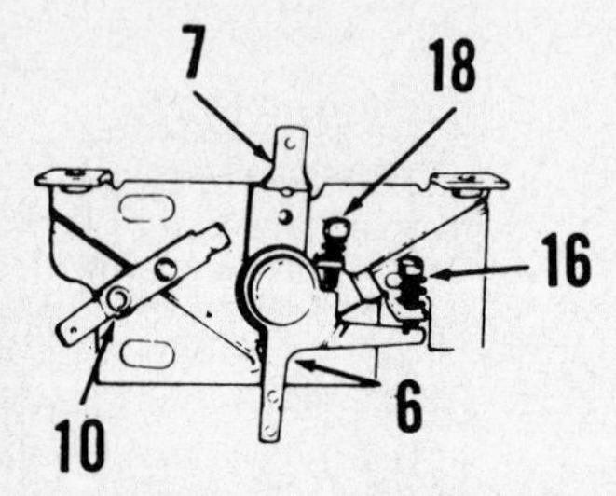

Fig. T42—Models with "Automagic" carburetor use control shown.

6. Governor spring bellcrank
7. Control lever
10. Idle speed stop screw
16. Idle speed stop
18. High speed stop screw

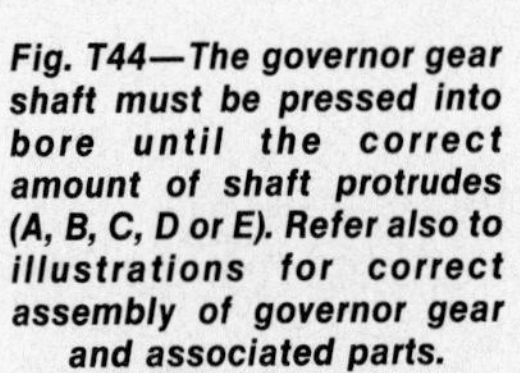
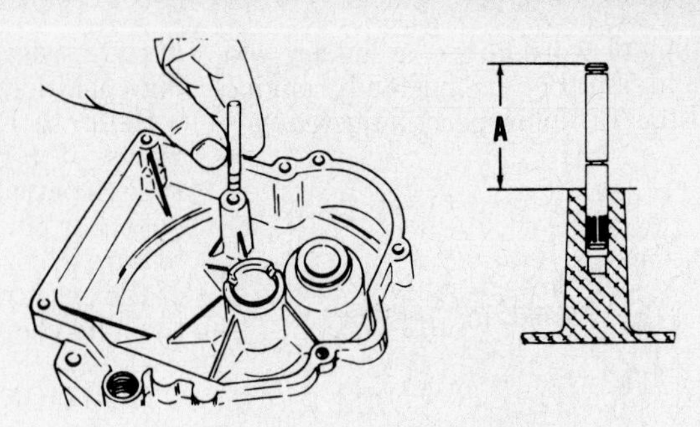

Fig. T44—The governor gear shaft must be pressed into bore until the correct amount of shaft protrudes (A, B, C, D or E). Refer also to illustrations for correct assembly of governor gear and associated parts.

B. 1-5/16 inches (33.338 mm)
C. 1-5/16 inches (33.338 mm)
D. 1-3/8 inches (34.925 mm)
E. 1-19/32 inches (40. 481 mm)

LIGHT FRAME ENGINES

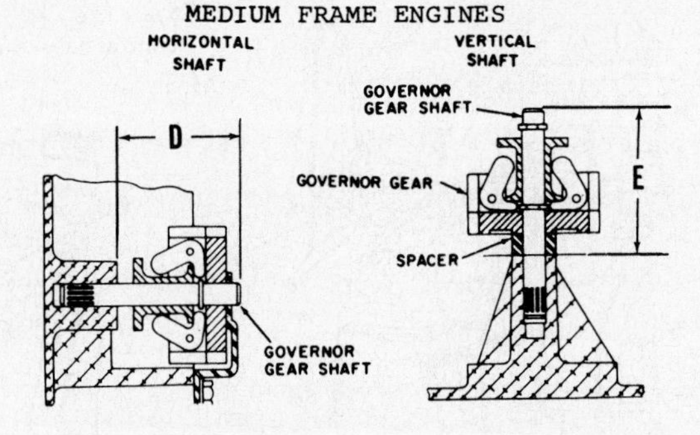

capacitor-discharge ignition (CDI) may have been used according to model and application. Refer to appropriate paragraph for model being serviced.

Breaker Point Ignition System. Breaker point gap at maximum opening should be 0.020 inch (0.51 mm) for all models. Marks are usually located on stator and mounting post to facilitate timing.

Ignition timing can be checked and adjusted to occur when piston is at specific location (BTDC) if marks are missing. Refer to the following specifications for recommended timing.

Models	Piston Position BTDC
HS40, LAV40, TVS105, ECV100, ECV105, ECV110, ECV120, TNT100, TNT120	0.035 in. (0.89 mm)
V40, VH40, LAV50, TVS120, H40, HH40, HS50	0.050 in. (1.27 mm)
LAV25, LAV30, LV35, LAV35, TVS90, H25, H30, H35, TVS75, ECH90	0.065 in. (1.65 mm)
V50, VH50, V60, VH60, H50, HH50, H60, HH60, TVM125, TVM140	0.080 in. (2.03 mm)

Some models may be equipped with the coil and laminations mounted outside the flywheel. Engines equipped

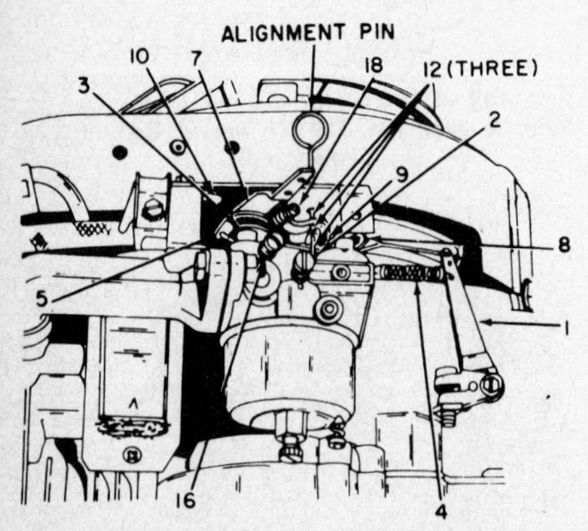

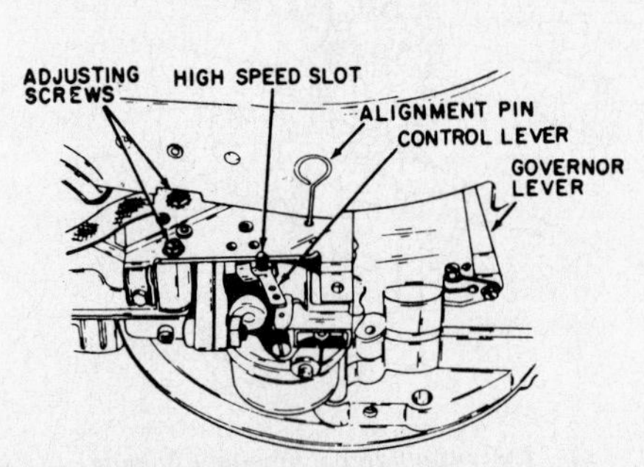

Fig. T43—On model shown, adjust location of cover so control lever is aligned with high speed slot and alignment holes are aligned.

1. Governor lever
2. Throttle control link
3. Throttle lever
4. Governor spring
5. Governor spring linkage
7. Control lever
8. Choke lever
9. Choke linkage
10. Stop switch
12. Alignment holes
16. Idle speed stop screw
18. High speed stop screw

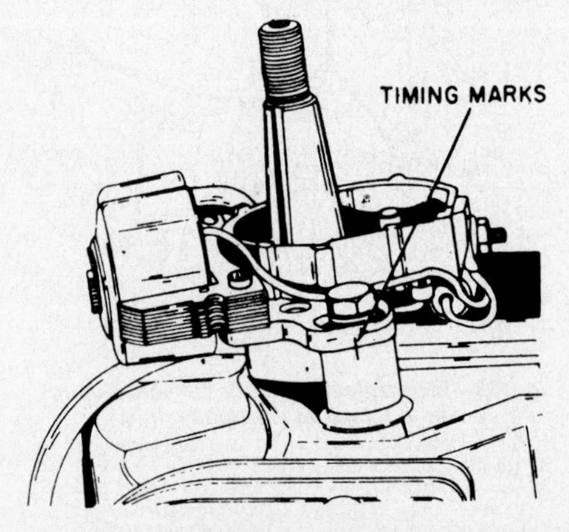

Fig. T45—Align timing marks as shown on magneto ignition system.

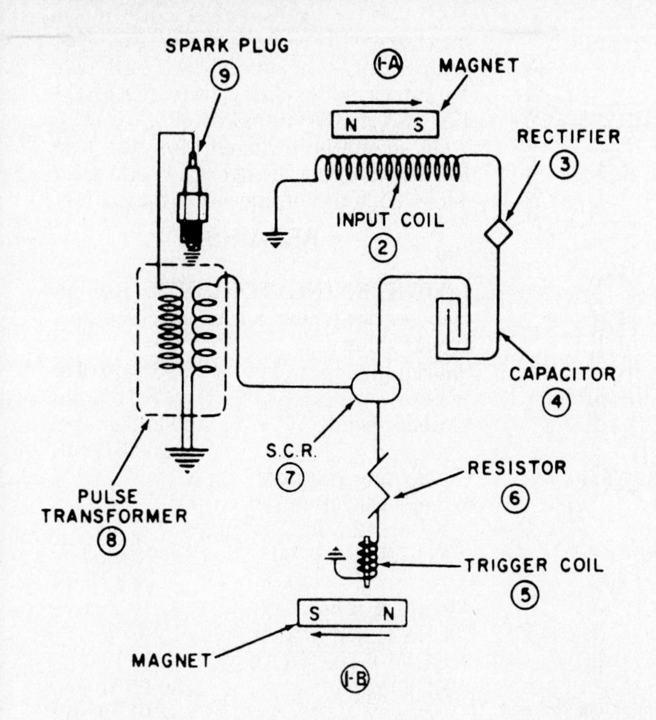

Fig. T46—Wiring diagram of solid state ignition system used on some models.

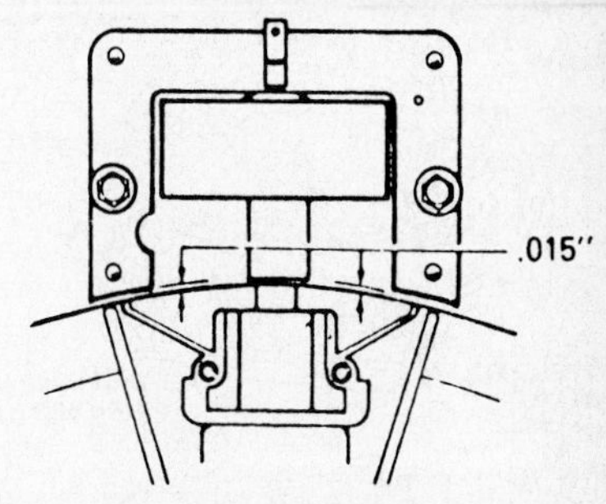

Fig. T47A—Set coil lamination air gap at 0.015 inch (0.38 mm) at points shown.

with breaker point ignition have the ignition points and condenser mounted under the flywheel with the coil and laminations mounted outside the flywheel. This system is identified by the round shape of the coil and a stamping "Grey Key" in the coil to identify the correct flywheel key.

The correct air gap setting between the flywheel magnets and the coil laminations is 0.015 inch (0.38 mm). Use Tecumseh gage 670259 or equivalent thickness sheet plastic to set gap as shown in Fig. T47A.

Solid State Ignition System. The Tecumseh solid state ignition system does not use ignition breaker points. The only moving part of the system is the rotating flywheel with the charging magnets. As the flywheel magnet passes position (1A–Fig. T46), a low voltage AC current is induced into input coil (2). Current passes through rectifier (3) converting this current to DC current. It then travels to capacitor (4) where it is stored. The flywheel rotates approximately 180° to position (1B). As it passes trigger coil (5), it induces a very small electric charge into the coil. This charge passes through resistor (6) and turns on the SCR (silicon controlled rectifier) switch (7). With the SCR switch closed, low voltage current stored in capacitor (4) travels to pulse transformer (8). Voltage is stepped up instantaneously and current is discharged across the electrodes of spark plug (9), producing a spark before top dead center.

Some units are equipped with a second trigger coil and resistor set to turn the SCR switch on at a lower rpm. This second trigger pin is closer to the flywheel and produces a spark at TDC for earlier starting. As engine rpm increases, the first (shorter) trigger pin picks up the small electric charge and turns the SCR switch on, firing the spark plug at TDC.

If system fails to produce a spark to the spark plug, first check high tension lead (Fig. T47). If condition of high tension lead is questionable, renew pulse transformer and high tension lead assembly. Check low tension lead and renew if insulation is faulty. The magneto charging coil, electronic triggering system and mounting plate are available only as an assembly. If necessary to renew this assembly, place unit in position on engine. Start retaining screws, turn mounting plate counterclockwise as far as possible, then tighten retaining screws to 5-7 ft.-lbs. (7-10 N·m).

Engines with solid state (CDI) ignition have all the ignition components sealed in a module and located outside the flywheel. There are no components

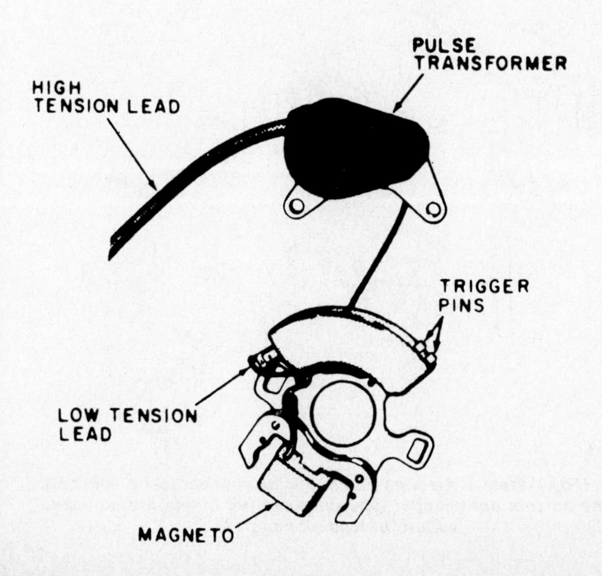

Fig. T47—Diagram of solid state ignition system used on some models.

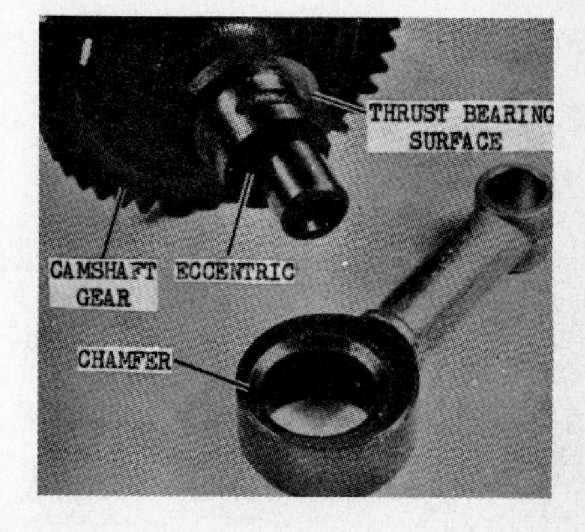

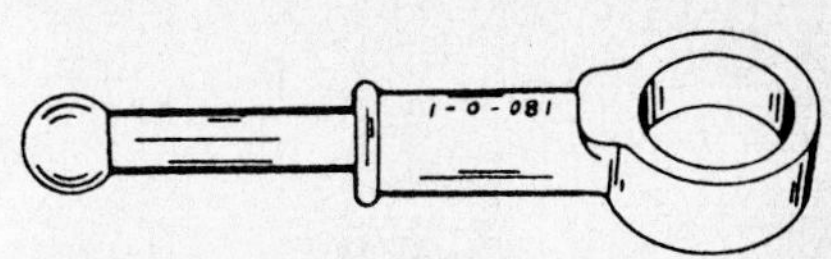

Fig. T48—Various types of barrel and plunger oil pumps have been used. Chamfered face of collar should be toward camshaft if drive collar has only one chamfered side. If drive collar has a float boss, the boss should be next to the engine lower cover, away from camshaft gear.

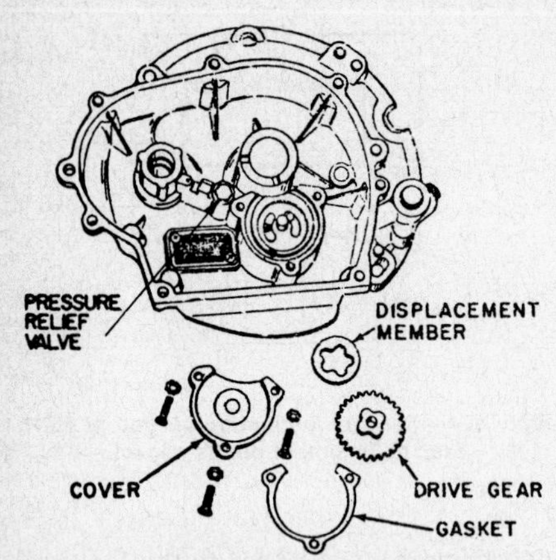

Fig. T49—Disassembled view of typical gear-driven rotor type oil pump.

under the flywheel except a spring clip to hold the flywheel key in position. This system is identified by the square shape module and a stamping "Gold Key" to identify the correct flywheel key.

The correct air gap setting between the flywheel magnets and the laminations on ignition module is 0.015 inch (0.38 mm). Use Tecumseh gage 670259 or equivalent thickness sheet plastic to set gap as shown in Fig. T47A.

LUBRICATION. Vertical crankshaft engines may be equipped with a barrel and plunger type oil pump or a gear driven rotor type oil pump. Horizontal crankshaft engines may be equipped with a gear-driven rotor type pump or with a dipper type oil slinger attached to the connecting rod.

Oil level should be checked after every five hours of operation: Maintain oil level at lower edge of filler plug or at "FULL" mark on dipstick.

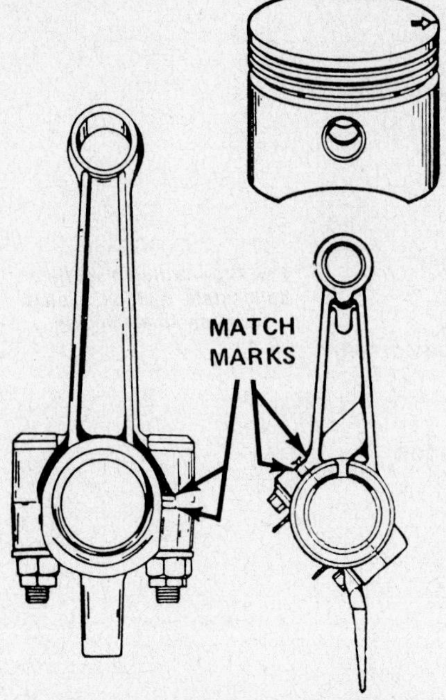

Fig. T50—Match marks on connecting rod and cap should be aligned and should be toward pto end of crankshaft.

Manufacturer recommends oil with an API service classification SC, SD, SE or SF. Use SAE 30 motor oil for temperatures above 32° F (0° C), SAE 5W-30 motor oil for temperatures between 32°F (0°C) and 0°F (–18°C) and a 90% SAE 10W oil-10% kerosene mixture or SAE 5W-30 for temperatures below 0° F (–18° C). Manufacturer explicity states: DO NOT USE SAE 10W-40 motor oil.

Oil should be changed after the first two hours of engine operation and after every 25 hours of operation thereafter.

REPAIRS

TIGHTENING TORQUES. Recommended tightening torque specifications are as follows:

Spark plug . . . . 250-360 in.-lbs. (28-41 N•m)

Cylinder head . . . . 160-200 in.-lbs. (18-23 N•m)

Connecting rod nuts (except Durlok nuts):
- 1.7 hp., 2.5 hp., 3.0 hp., ECH90, ECV100 . . . . . . 65-75 in.-lbs. (7-9 N•m)
- 4 hp. & 5 hp. light frame models, ECV105, ECV110, ECV120 . . . . 80-95 in.-lbs. (9-11 N•m)
- 5 hp. (3.7 kW) medium frame, 6 hp, (4.5 kW) . . . . 86-110 in.-lbs. (10-12 N•m)

Connecting rod bolts (Durlok):
- 5 hp. (3.7 kW) light frame . . . . 110-130 in.-lbs. (12-15 N•m)

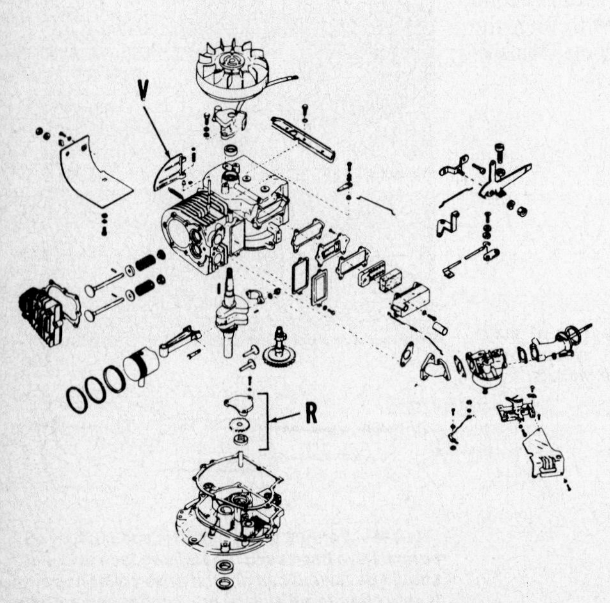

Fig. T49A—View of vertical crankshaft engine. Rotor type oil pump is shown at (R); governor air vane at (V).

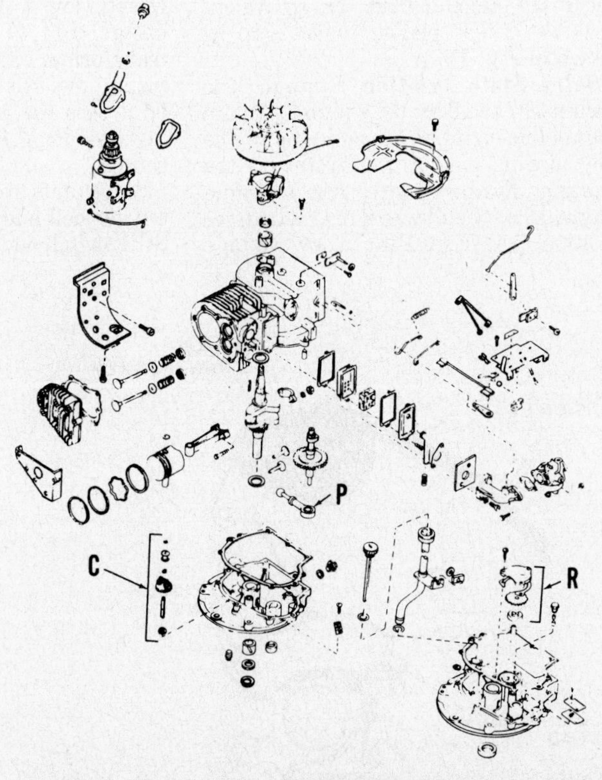

Fig. T50A—View of Medium Frame engine with vertical crankshaft. Some models use plunger type oil pump (P); others are equipped with rotor type oil pump (R).

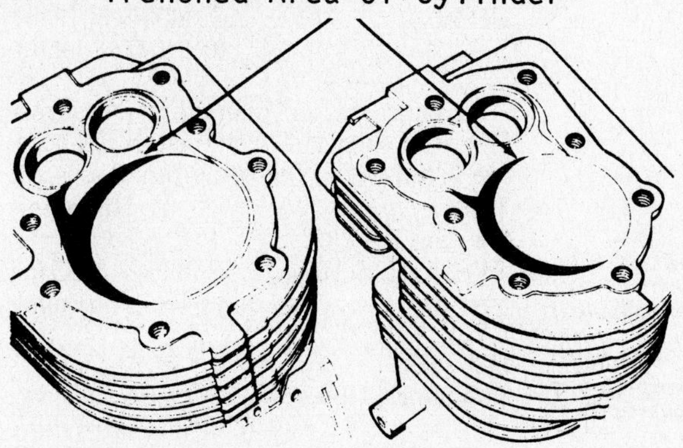

Fig. T51—Cylinder block on Models H50, HH50, VH50, TVM125 and TVM140 has been "trenched" to improve fuel flow and power.

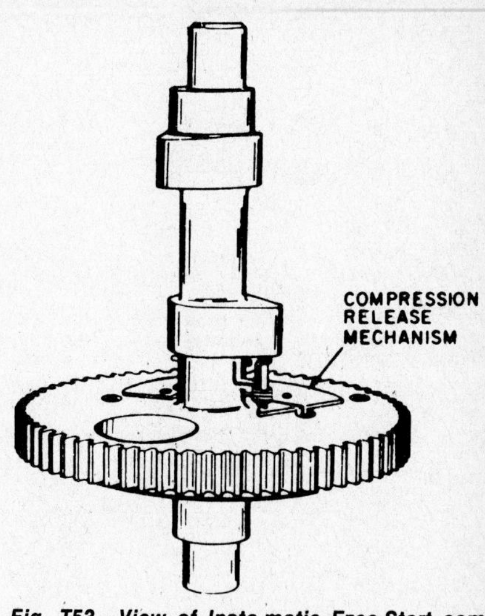

Fig. T53—View of Insta-matic Ezee-Start compression release camshaft.

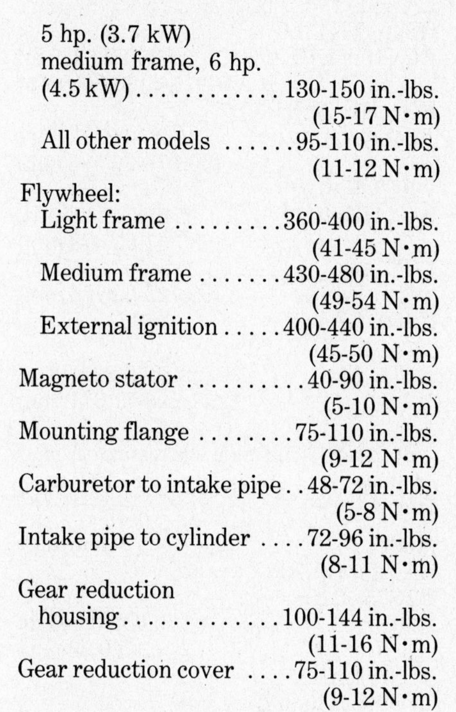

5 hp. (3.7 kW)
medium frame, 6 hp.
(4.5 kW) . . . . 130-150 in.-lbs. (15-17 N·m)
All other models . . . . . . 95-110 in.-lbs. (11-12 N·m)
Flywheel:
Light frame . . . . 360-400 in.-lbs. (41-45 N·m)
Medium frame . . . . 430-480 in.-lbs. (49-54 N·m)
External ignition . . . . . 400-440 in.-lbs. (45-50 N·m)
Magneto stator . . . . 40-90 in.-lbs. (5-10 N·m)
Mounting flange . . . . 75-110 in.-lbs. (9-12 N·m)
Carburetor to intake pipe . . 48-72 in.-lbs. (5-8 N·m)
Intake pipe to cylinder . . . . 72-96 in.-lbs. (8-11 N·m)
Gear reduction
housing . . . . 100-144 in.-lbs. (11-16 N·m)
Gear reduction cover . . . . 75-110 in.-lbs. (9-12 N·m)

CONNECTING ROD. Piston and connecting rod assembly is removed from cylinder head end of engine. The aluminum alloy connecting rod rides directly on crankshaft crankpin.

Refer to the following table for standard crankpin journal diameter.

Model	Diameter
LAV25, LAV30, TVS75, H25, H30, LV35, LAV35, TVS90, H35-prior 1983, ECH90, TNT100, ECV100	0.8610-0.8615 in. (21.869-21.882 mm)
H35-after 1983, LAV40, TVS105, HS40, ECV105, ECV110, ECV120, TNT120, LAV50, HS50, TVS120	0.9995-1.0000 in. (25.390-25.400 mm)
V40, VH40, H40, HH40, H50, HH50, VH50, TVM140	1.0615-1.0620 in. (26.962-26.975 mm)

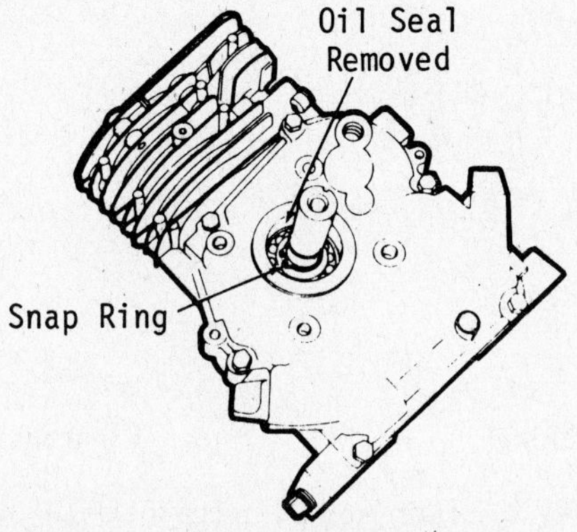

Fig. T52—On Models H30 through H50 it is necessary to remove snap ring under oil seal before removing crankcase cover.

Standard inside diameter for connecting rod crankpin bearing journal is shown in the following table.

Model	Diameter
LAV25, LAV30, TVS75, H25, H30, LV35, LAV35, TVS90, H35-prior 1983, ECH90, TNT100, ECV100	0.8620-0.8625 in. (21.895-21.908 mm)
H35-after 1983, LAV40, TVS105, HS40, ECV105, ECV110, ECV120, TNT120, LAV50, HS50, TVS120	1.0005-1.0010 in. (25.413-25.425 mm)
V40, VH40, H40, HH40, H50, HH50, VH50, TVM140	1.0630-1.0635 in. (27.000-27.013 mm)

Connecting rod bearing-to-crankpin journal clearance should be 0.0005-0.0015 inch (0.013-0.38 mm) for all models.

When installing connecting rod and piston assembly, align the match marks on connecting rod and cap as shown in

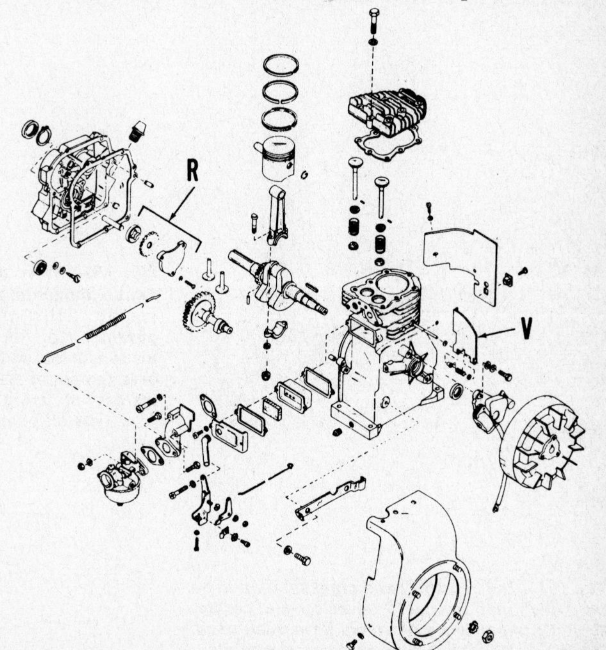

Fig. T52A—View of Light Frame engine with horizontal crankshaft. Air vane (V) type governor and rotor type oil pump (R) are used on model shown.

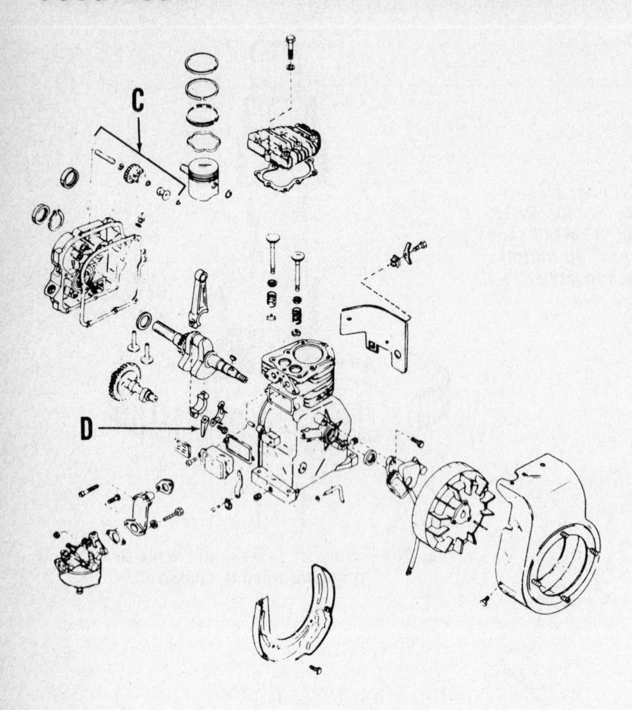

Fig. T53A—View of Light Frame horizontal crankshaft engine with mechanical governor and splash lubrication. Governor centrifugal weights are shown at (C) and lubrication dipper at (D).

Fig. T50. On some models, the piston pin hole is offset in piston and arrow on top of piston should be toward valves. On all engines, the match marks on connecting rod and cap must be toward power takeoff (pto) end of crankshaft. Lock plates, if so equipped, for connecting rod cap retaining screws should be renewed each time cap is removed.

PISTON, PIN AND RINGS. Aluminum alloy pistons are equipped with two compression rings and one oil control ring. Ring end gap for all models is 0.007-0.017 inch (0.18-0.43 mm).

Piston skirt-to-cylinder clearances are listed in the following table.

Model	Clearance
LAV25, LAV30, TVS75, H25, H30	0.0025-0.0040 in. (0.064-0.102 mm)
H50, HH50 VH50, TVM140	0.0035-0.005 in. (0.089-0.127 mm)
LV35, LAV35, TVS90, H35 (all), ECH90, LAV40, TVS105, HS40, TNT100, ECV100, ECV110, LAV50, TVS120, ECV120, TNT120, HS50	0.0045-0.0060 in. (0.114-0.152 mm)
ECV105	0.0050-0.0065 in. (0.127-0.165 mm)
V40, VH40, H40, HH40	0.0055-0.0070 in. (0.140-0.178 mm)

Standard piston diameters measured at piston skirt 90° from piston pin bore are listed in the following table.

Model	Diameter
LAV25, LAV30, TVS75, H25, H30	2.3090-2.3095 in. (58.649-58.661 mm)
LV35, LAV35, H35 (all), TVS90, ECH90	2.4950-2.4955 in. (63.373-63.386 mm)
LAV40, TVS105, HS40, TNT100, ECV100, EVC105	2.6200-2.6205 in. (66.548-66.560 mm)
H50, HH50, VH50, TVM140	2.6210-2.6215 in. (66.575-66.586 mm)
V40, VH40, H40, HH40	2.4945-2.4950 in. (63.363-63.373 mm)
ECV110	2.7450-2.7455 in. (69.723-69.736 mm)
HS50, LAV50, TVS120, ECV120, TNT120	2.8070-2.8075 in. (71.298-71.311 mm)

Standard ring side clearance in ring grooves are shown in the following table.

Model	Clearance
LAV25, LAV30, TVS75, H25, H30	0.002-0.005 in. (0.05-0.13 mm)
LV35, LAV35, TVS90, H35 (all), V40, VH40, H40, HH40	0.002-0.003 in. (0.05-0.08 mm)

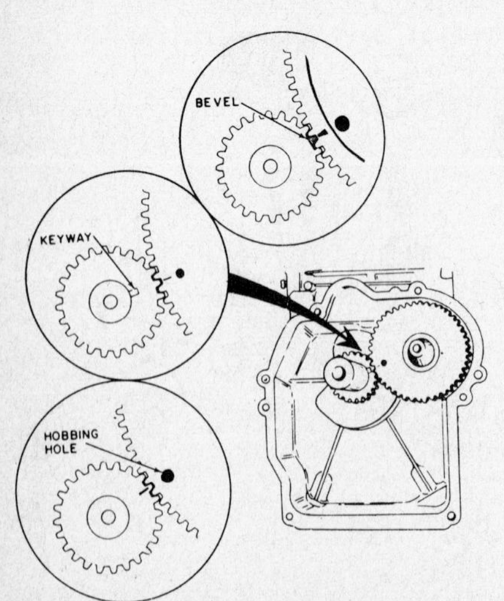

Fig. T54—The camshaft and crankshaft must be correctly timed to ensure valves open at correct time. Different types of marks have been used, but marks should be aligned when assembling.

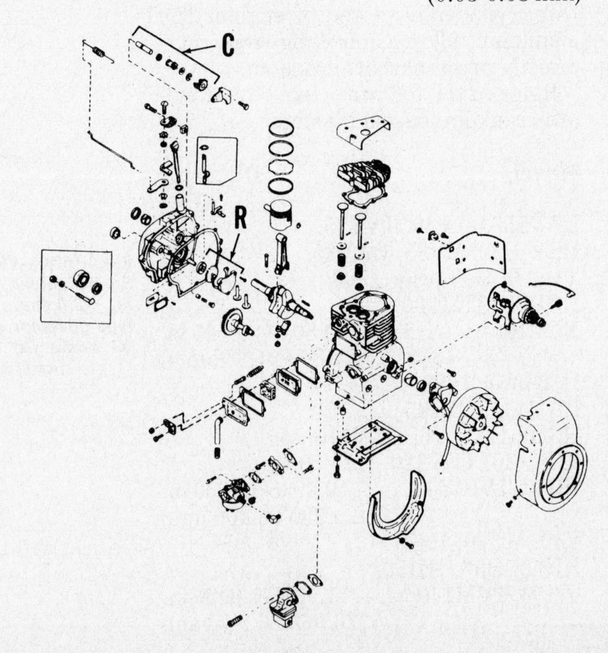

Fig. T55—View of Medium Frame horizontal crankshaft engine with mechanical governor (C). An oil dipper for splash lubrication is cast onto the connecting rod cap instead of using the rotor type oil pump (R).

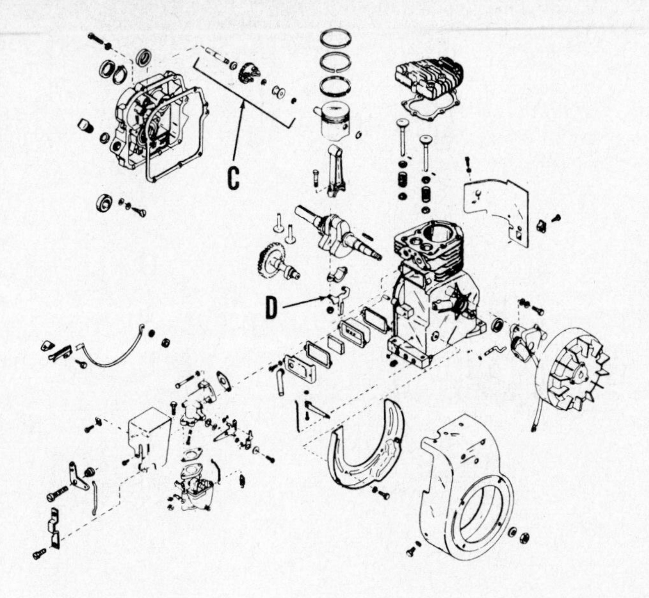

Fig. T56—View of Light Frame horizontal crankshaft engine with mechanical governor and splash lubrication. Notice that connecting rod cap on model shown is away from camshaft side of engine.

LAV40, TVS105,
HS40, ECV100:
Compression rings . . . . 0.002-0.004 in. (0.05-0.10 mm)
Oil control ring . . . . 0.001-0.004 in. (0.03-0.10 mm)

ECV110:
Compression rings . . . . 0.002-0.004 in. (0.05-0.10 mm)
Oil control ring . . . . 0.001-0.002 in. (0.03-0.05 mm)

LAV50, HS50, TVS120:
Compression rings . . . . 0.003-0.004 in. (0.08-0.10 mm)
Oil control ring . . . . 0.002-0.003 in. (0.05-0.08 mm)

ECV120:
Compression rings . . . . 0.003-0.004 in. (0.08-0.10 mm)
Oil control ring . . . . 0.001-0.002 in. (0.03-0.05 mm)

TNT120:
Compression rings . . . . 0.003-0.004 in. (0.08-0.10 mm)
Oil control ring . . . . 0.002-0.004 in. (0.05-0.10 mm)

H50, HH50,
VH50, TVM140 . . . . 0.002-0.004 in. (0.05-0.10 mm)

Refer to CONNECTING ROD section for correct piston to connecting rod assembly and correct piston installation procedure. Piston pin should be a tight push fit in piston pin bore and connecting rod pin bore and is retained by snap rings at each end of piston pin bore. Install marked side of piston rings up and stagger ring end gaps equally around circumference of piston during installation.

CYLINDER AND CRANKCASE. Cylinder and crankcase are an integral casting on all models. Cylinder should be honed and fitted to nearest oversize for which piston and ring set are available if cylinder is scored, tapered or out-of-round more than 0.005 inch (0.13 mm).

Standard cylinder bore diameters are shown in the following table.

Model	Diameter
LAV25, LAV30, TVS75, H25, H30	2.3125-2.3135 in. (58.738-58.763 mm)
LV35, LAV35, H35 (all), TVS90, ECH90, V40, VH40, H40, HH40	2.5000-2.5010 in. (63.500-63.525 mm)
LAV40, TVS105, HS40, ECV100, TNT100, ECV105, TVM140	2.6250-2.6260 in. (66.675-66.700 mm)
ECV110	2.7500-2.7510 in. (69.850-69.875 mm)
LAV50, TVS120, H50, HH50, VH50, HS50, ECV120, TNT120	2.8120-2.8130 in. (71.425-71.450 mm)

Refer to PISTON, PIN AND RINGS section for correct piston-to-cylinder block clearance.

Note also that cylinder block used on Models H50, HH50, VH50, TVM125 and TVM140 has been trenched to improve fuel flow and power (Fig. T51).

Standard main bearing bore diameters for both main bearings, or as indicated otherwise, are shown in the following chart.

Model	Diameter
LAV25, LAV30, TVS75, H25, H30 (prior 1983), LV35, LAV35, TVS90, H35 (prior 1983), ECH90, TNT100, ECV100	0.8755-0.8760 in. (22.238-22.250 mm)
H35 (after 1983), LAV40, TVS105, HS40, V40, VH40, H40, HH40, ECV110, LAV50, TVS120, HS50, H50	1.0005-1.0010 in. (25.413-25.425 mm)
H30 (after 1983): Crankcase side	1.0005-1.0010 in. (25.413-25.425 mm)
Cover (flange) side	0.8755-0.8760 in. (22.238-22.250 mm)
ECV105: Crankcase side	1.0005-1.0010 in. (25.413-25.425 mm)
Cover (flange) side)	1.2010-1.2020 in. (30.505-30.531 mm)

A special tool kit is available from Tecumseh to ream the cylinder crankcase and cover (mounting flange) main bearing bores to install service bushing for some models.

CRANKSHAFT, MAIN BEARINGS AND SEALS. Crankshaft main bearing journals on some models ride directly in the aluminum alloy bores in the cylinder block and the crankcase cover (mounting flange). Other engines are originally equipped with renewable steel backed bronze bushings and some are originally equipped with a ball type main bearing at the pto end of crankshaft.

Refer to CONNECTING ROD section for standard crankshaft crankpin journal diameters. Standard diameters for crankshaft main bearing journals are shown in the following chart.

Model	Diameter
LAV25, LAV30, TVS75, H25, H30 (prior 1983)*, LV35, TVS90, H35 (prior 1983), ECH90, ECV100, TNT100	0.8735-0.8740 in. (22.187-22.200 mm)
H35 (after 1983), H30 (after 1983)*, LAV40, TVS105, HS40, ECV105, TNT100, V40, VH40, H40, HH40, ECV110, LAV50, TVS120, HS50, ECV120, TNT120, H50, HH50, VH50, TVM140	0.9985-0.9990 in. (25.362-25.375 mm)

*H30 models produced after 1983 have a main bearing journal diameter at flywheel side of 0.9985-0.9990 inch (25.362-25.375 mm) and a main bearing journal diameter at pto side of 0.8735-0.8740 inch (22.187-22.200 mm).

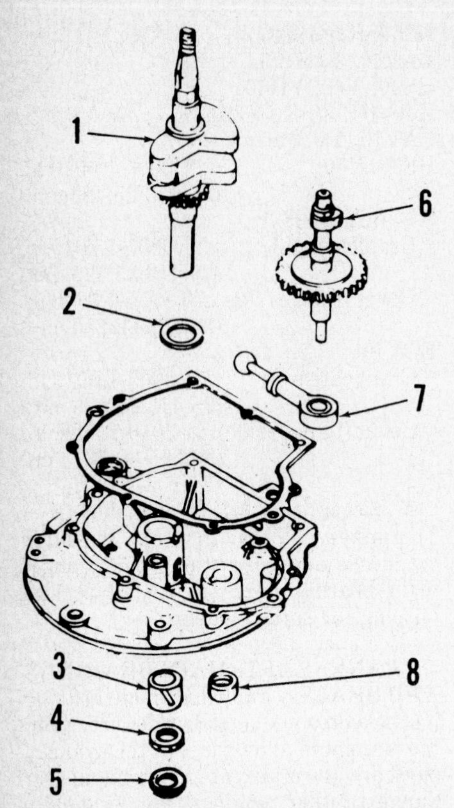

Fig. T61—An auxiliary pto shaft that turns at 1/2 the speed of the crankshaft is available by using a special extended camshaft. The hole in lower cover is sealed using lip type seal (8).

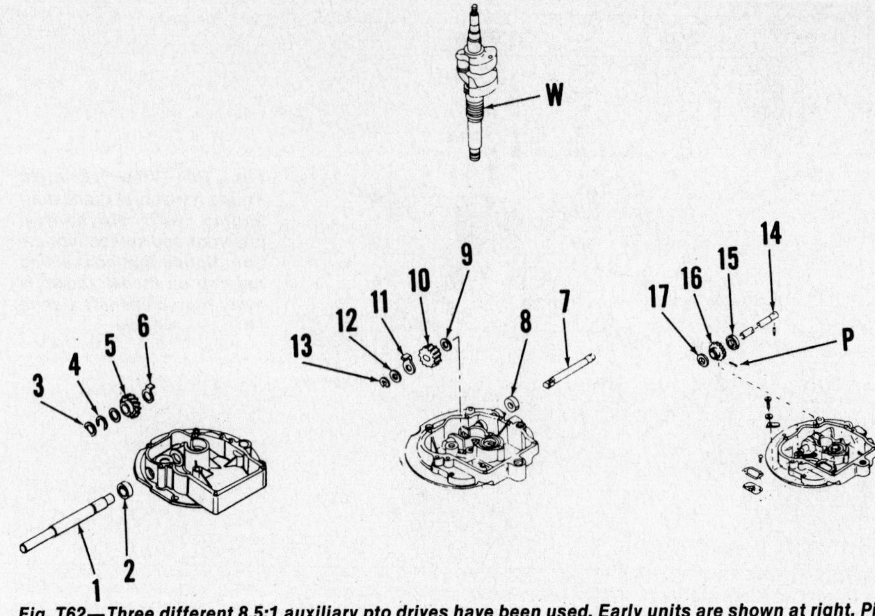

Fig. T62—Three different 8.5:1 auxiliary pto drives have been used. Early units are shown at right. Pin (P) attaches gear (16) to the pto shaft (14). On later units, the shaft is held in position by snap ring (4) or (13). Worm gear (W) on crankshaft drives the gear (5, 10 or 16) on all models.

1. Pto shaft
2. Seal
3. Washers (2)
4. Snap ring
5. Gear
6. Tang washer
7. Pto shaft
8. Seal
9. Thick washer
10. Gear
13. Snap ring
14. Pto shaft
15. Seal
16. Gear
17. Washer

Clearance between crankshaft main bearing journals and main bearing bushings should be 0.0010-0.0025 inch (0.025-0.064 mm) for 0.8735-0.8740 inch (22.187-22.200 mm) diameter journal of 0.0015-0.0025 inch (0.038-0.064 mm) for 0.9985-0.9990 inch (25.362-25.375 mm) journal.

Crankshaft end play for all models is 0.005-0.027 inch (0.13-0.69 mm).

On Models H30 through HS50, it is necessary to remove a snap ring (Fig. T52) which is located under oil seal. Note oil seal depth before removal of seal, remove seal and snap ring. Crankcase cover may not be removed. When installing cover, press new seal in to same depth as old seal before removal.

Ball bearing should be inspected and renewed if rough, loose or damaged. Bearing must be pressed on or off of crankshaft journal using a suitable press or puller.

Always note oil seal depth and direction before removing old seal from crankcase or cover. New seals must be pressed into seal bores to the same depth as old seal before removal on all models.

When installing crankshaft, align crankshaft and camshaft gear timing marks as shown in Fig. T54 except on certain engines manufactured for Craftsman. Refer to CRAFTSMAN engine section of this manual.

CAMSHAFT. The camshaft and camshaft gear are an integral part which rides on camshaft journals at each end of camshaft. Camshaft on some models also has a compression release mechanism mounted on camshaft gear which lifts exhaust valve at low cranking rpm to reduce compression and aid starting (Fig. T53).

Renew camshaft if lobes or journals are worn or scored. Spring on compression release mechanism should snap weight against camshaft. Compression release mechanism and camshaft are serviced as an assembly only.

Standard camshaft journal diameter is 0.6230-0.6235 inch (15.824-15.837 mm) for Models V40, VH40, H40, HH40, H50, HH50, VH50, TVM125 and TVM140 and 0.4975-0.4980 inch (12.637-12.649 mm) for all other models.

Note that exhaust cam follower (lifter) is longer on some models and that on models equipped with barrel and plunger type oil pump, the pump is operated by an eccentric on camshaft.

When installing camshaft, align crankshaft and camshaft timing marks as shown in Fig. T54 except on certain engines manufactured for Craftsman. Refer to CRAFTSMAN engine section of this manual.

OIL PUMP. Vertical crankshaft engines may be equipped with a barrel and plunger type oil pump or a gear-driven rotor type oil pump. Horizontal crankshaft engines may be equipped with a gear-driven rotor type pump or with a dipper type oil slinger attached to the connecting rod.

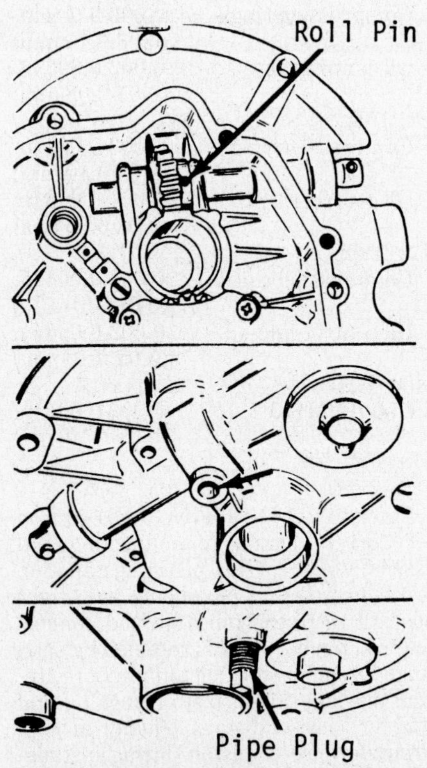

Fig. T63—The roll pin must be removed from early 8.5:1 auxiliary pto before the shaft and gear can be withdrawn. On some models (center) the boss is closed. The boss can be drilled (5/8 inch) and tapped to accept a 7-28 N.P.T. 1/8 inch pipe plug as shown in lower view.

The barrel and plunger type oil pump is driven by an eccentric on the camshaft. Chamfered side of drive collar (Fig. T48) should be toward engine lower cover. Oil pumps may be equipped with two chamfered sides, one chamfered side or with flat boss as shown. Be sure installation is correct.

On engines equipped with gear-driven rotor oil pump, check drive gear and rotor for excessive wear or other damage. End clearance of rotor in pump body should be within limits of 0.006-0.007 inch (0.15-0.18 mm) and is controlled by cover gasket. Gaskets are available in a variety of thicknesses.

On all models with oil pump, be sure to prime pump during assembly to ensure immediate lubrication of engine.

VALVE SYSTEM. Valve tappet clearance (cold) is 0.010 inch (0.25 mm) for intake and exhaust valves for Models V50, VH50, H50, HH50, TVM125 and TVM140. Valve tappet clearance for all other models is 0.008 inch (0.03 mm) for intake and exhaust valves.

Valve face angle is 45° and valve seat angle is 46° except on very early production 2.25 horsepower (7.6 kW) models which have a 30° valve face and seat angle.

Valve seat width should be 0.035-0.045 inch (0.89-1.14 mm) for all models.

Valve stem guides are cast into cylinder block and are nonrenewable. If excessive clearance exists between valve stem and valve guide, guide should be reamed and new valve with oversize stem installed.

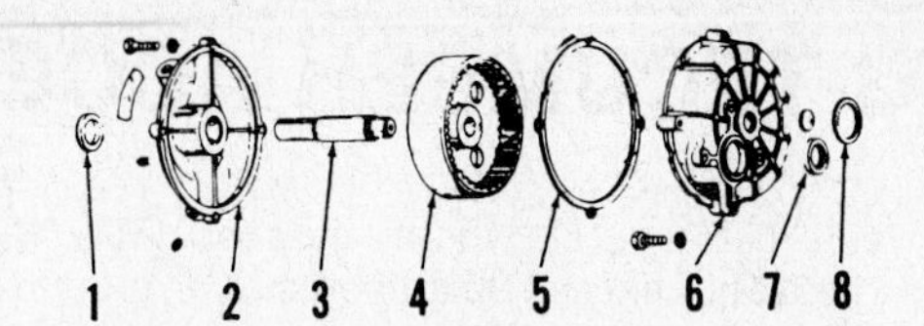

Fig. T64—Exploded view of the 6:1 gear reduction assembly. Housing (6) is bolted to cylinder block and the pinion gear is made onto end of crankshaft.

1. Seal
2. Cover
3. Output shaft
4. Gear
5. Gasket
6. Housing
7. Seal
8. Cork gasket

REDUCED SPEED PTO SHAFTS. A pto (power takeoff) shaft which rotates at ½ the speed of the crankshaft is available by extending the camshaft through the cover (lower mounting flange). Refer to Fig. T61. Except for the seal around the extended camshaft, service is similar to standard models.

A slow speed (8.5:1) auxiliary pto shaft is used on some vertical shaft engines (Fig. T62). A worm gear (W) on the crankshaft turns the pto gear and pto shaft. Several different versions of this unit have been used. A roll pin (Fig. T63) is used to hold the gear onto the shaft on early models. A pipe plug and a threaded boss are located as shown in center and lower views of some of these early models. The roll pin can be driven out of the gear and shaft through the threaded hole of models so equipped. Refer to Fig. T62 for order of assembly.

Disassembly and repair procedure for the 6:1 reduction will be evident after examination of the unit and reference to Fig. T64.

SERVICING TECUMSEH ACCESSORIES

12 VOLT STARTING AND CHARGING SYSTEMS

Engines with 3.5 horsepower (57 cc) or more may be equipped with a 12 volt direct current starting, generating and ignition unit system. The system includes the usual coil, condenser and breaker point for ignition, plus extra generating coils and flywheel magnets to generate alternating current. Also included is a silicon rectifier panel or regulator-rectifier for changing the generated alternating current to direct current, a series wound motor with a Bendix drive unit and a 12 volt wet cell storage battery.

Models LAV30, LAV35 and LAV40 may be equipped with a 12 volt starting motor with a right angle gear drive unit and a Nickel Cadmium battery pack. The "SAF-T-KEY" switch is located in the battery box cover. A battery charger which converts 110 volt ac house current to dc charging current is used to recharge the Nickel Cadmium battery pack.

Refer to the following paragraphs for service procedures on the units.

12 VOLT STARTER MOTOR (BENDIX DRIVE). Refer to Fig. TE1 for exploded view of 12 volt starter motor and Bendix drive unit used on some engines. This motor should not be operated continuously for more than 10 seconds. Allow starter motor to cool one minute between each 10 second cranking period.

To perform a no-load test, remove starter motor and use a fully charged 6 volt battery. Maximum current draw should not exceed 25 amperes at 6 volts. Minimum rpm is 6500.

When assembling a starter motor, use 0.003 and 0.010 inch spacer washers (14 and 15) as required to obtain an armature end play of 0.005-0.015 inch (0.13-0.38 mm). Tighten nut on end of armature shaft to 100 in.-lbs. (11 N•m). Tighten the through-bolts (30) to 30-34 in.-lbs. (3-4 N•m).

12 VOLT STARTER MOTOR (SAF-T-KEY TYPE). Refer to Fig. TE2 for exploded view of right angle drive gear unit and starter motor used on some Model LAV30, LAV35 and LAV40 models. Repair of this unit consists of renewing right angle gear drive parts or the starter motor assembly. Parts are not serviced separately for the motor (8). Bevel gears (9) are serviced only as a set.

When installing assembly on engine, adjust position of starter so there is a 1/16 inch (1.59 mm) clearance between crown of starter gear tooth and base of flywheel tooth.

NICKEL CADMIUM BATTERY. The Nickel Cadmium battery pack (Fig. TE3) is a compact power supply for the "SAF-T-KEY" starter motor. To test battery pack, measure open circuit voltage across black and red wires. If voltage is 14.0 or above, battery is good and may be recharged. Charge battery only with charger provided with the system. This charger (Fig. TE4) converts 110 volt AC house current to DC charging current. Battery should be fully charged after 14-16 hours charging time. A fully charged battery should test 15.5-18.0 volts.

A further test of battery pack can be made by connecting a 1.4 ohm resistor across black and red wires (battery installed) for two minutes. Battery voltage at the end of two minutes must be 9.0 volts minimum. If battery passes this test, recharge battery for service.

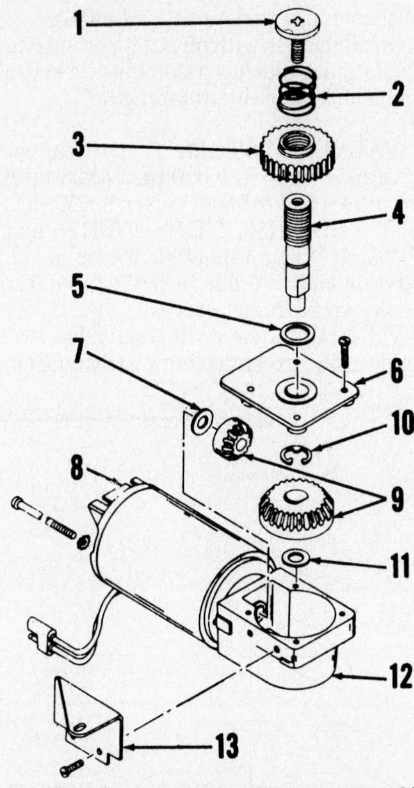

Fig. TE2—Twelve volt electric starter assembly used on some engines equipped with "SAF-T-KEY" Nickel Cadmium battery pack.

1. Screw L.H.
2. Spring
3. Starter gear
4. Shaft
5. Thrust washer
6. Cover
7. Shim washer
8. Starter motor
9. Gear set
10. Retaining ring
11. Shim washer
12. Gear housing
13. Mounting bracket

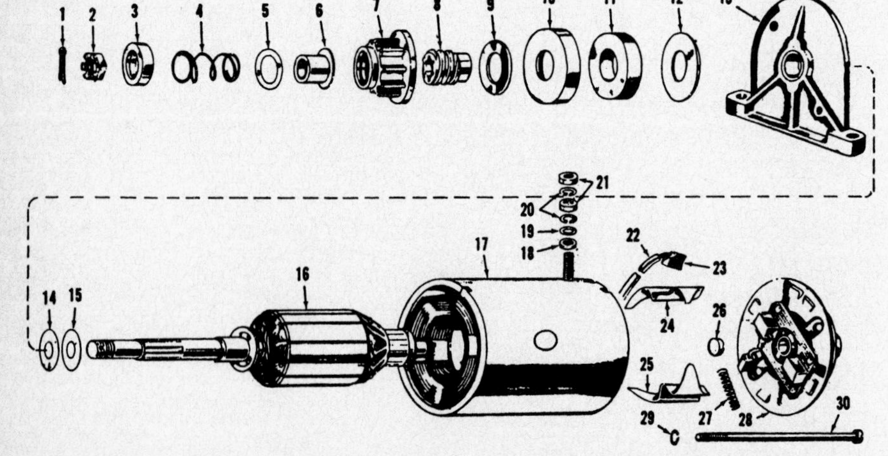

Fig. TE1—Twelve volt electric cranking motor used on some engines.

3. Pinion stop
4. Antidrift spring
5. Pinion washer
6. Antidrift sleeve
7. Pinion gear
8. Screw shaft
9. Thrust washer
10. Cushion
11. Rubber cushion
12. Thrust washer
13. Drive end cap
14. Spacer washer, 0.003 in.
15. Spacer washer, 0.010 in.
16. Armature
18. Insulation washer
22. Insulation tubing
23. Brush
24. Insulation
25. Insulation
26. Thrust spacer
28. Commutator end cap

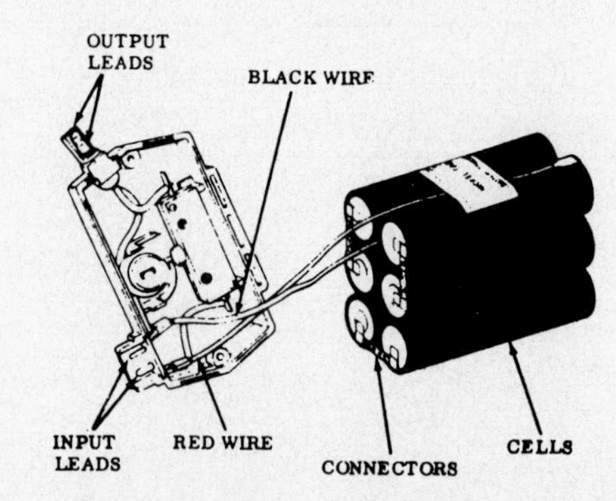

Fig. TE3—View of Nickel Cadmium battery pack connected to "SAF-T-KEY" starting switch.

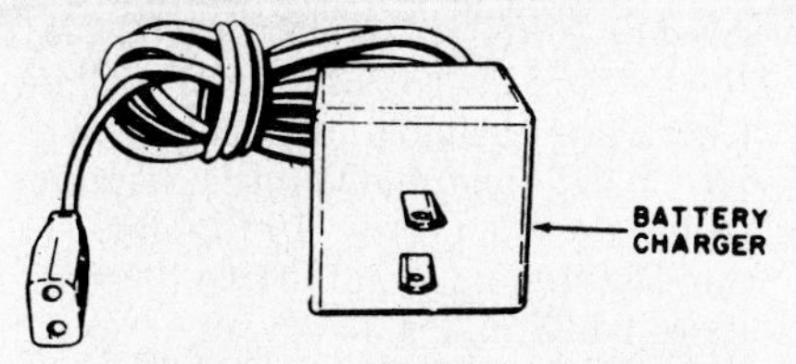

Fig. TE4—Recharge Nickel Cadmium battery pack only with the charger furnished with "SAF-T-KEY" starter system. Charger converts 110 volt ac to dc charging current.

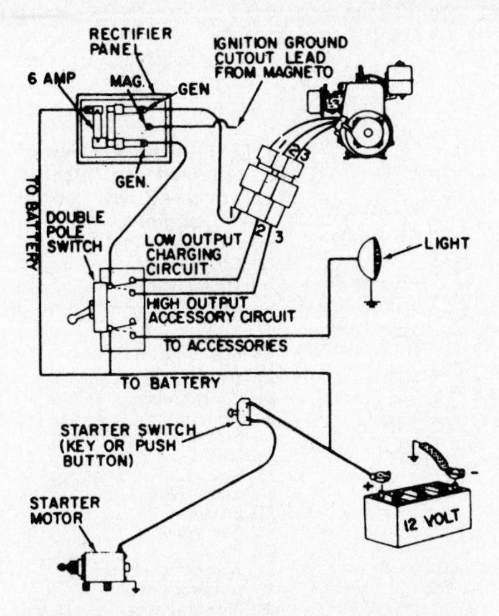

Fig. TE6—Wiring diagram of typical 7 ampere alternator and rectifier panel charging system. The double pole switch in one position reduces output to 3 amperes for charging or increases output to 7 amperes in other position to operate accessories.

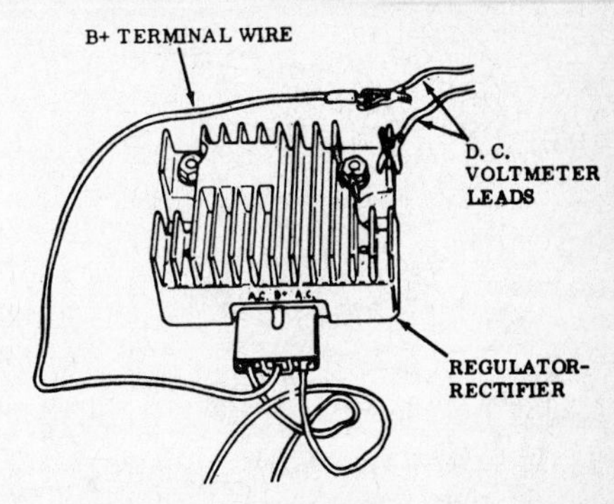

Fig. TE8—Connect dc voltmeter as shown when checking the regulator-rectifier.

ALTERNATOR CHARGING SYSTEMS. Flywheel alternators are used on some engines for the charging system. The generated alternating current is converted to direct current by two rectifiers on the rectifier panel (Figs. TE5 and TE6) or the regulator-rectifier (Fig. TE7).

The system shown in Fig. TE5 has a maximum charging output of about 3 amperes at 3600 rpm. No current regulator is used on this low output system. The rectifier panel includes two diodes (rectifiers) and a 6 ampere fuse for overload protection.

The system shown in Fig. TE6 has a maximum output of 7 amperes. To prevent overcharging the battery, a double pole switch is used in low output position to reduce the output to 3 amperes for charging the battery. Move switch to high output position (7 amperes) when using accessories.

The system shown in Fig. TE7 has a maximum output of 7 amperes and uses a solid state regulator-rectifier which converts the generated alternating current to direct current for charging the battery. The regulator-rectifier also allows only the required amount of current flow for existing battery conditions. When battery is fully charged, current output is decreased to prevent overcharging the battery.

Testing. On models equipped with rectifier panel (Figs. TE5 or TE6), remove rectifiers and test them with either a continuity light or an ohmmeter. Rectifiers should show current flow in one direction only. Alternator output can be checked using an induction ammeter over the positive lead wire to battery.

On models equipped with the regulator-rectifier (Fig. TE7), check the system as follows: Disconnect B+ lead and connect a dc voltmeter as shown in Fig. TE8. With engine running near full throttle, voltage should be 14.0-14.7 volts. If voltage is above 14.7 or below 14.0 but above 0, the regulator-rectifier is defective. If voltmeter reading is 0 the regulator-rectifier or alternator coils may be defective. To test the alternator coils, connect an ac voltmeter to the ac leads as shown in Fig. TE9. With engine running at near full throttle, check ac voltage. If voltage is less than 20.0 volts, alternator is defective.

110 VOLT ELECTRIC STARTER

110 VOLT AC-DC STARTER. Some vertical crankshaft engines are available with a 110 volt ac electric starting system as shown in Fig. TE11. When switch button (6) is depressed, switch (21) is turned on and the motor rotates driven gear (23). As more pressure is applied, housing (11) moves down compressing springs (19). At this time, clutch facing (24) is forced tight between driven gear (23) and cone (30) and engine is cranked.

Electric motor (3) is serviced only as an assembly. All other parts shown are available separately. Shims (18) are used to adjust mesh of drive gear (5) to driven gear (23).

110 VOLT AC-DC STARTER. Some engines may be equipped with a 110 volt ac-dc starting motor (Fig. TE12). The rectifier assembly used with this starter converts 110 volt ac house current to approximately 100 volts dc. A thyrector (8) is used as a surge protector for the rectifiers.

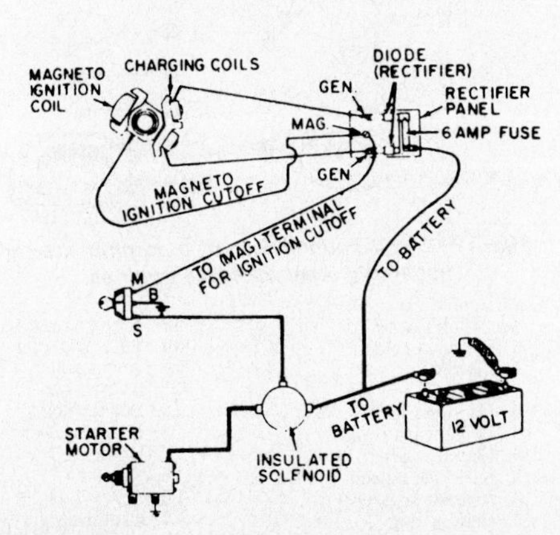

Fig. TE5—Wiring diagram of typical 3 ampere alternator and rectifier panel charging system.

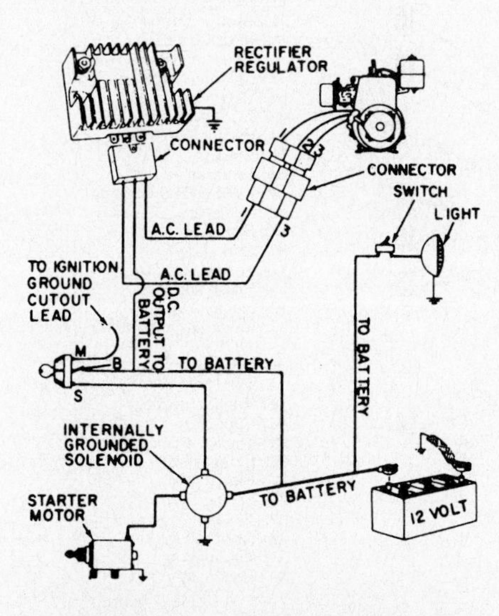

Fig. TE7—Wiring diagram of typical 7 or 10 ampere alternator and regulator-rectifier charging system.

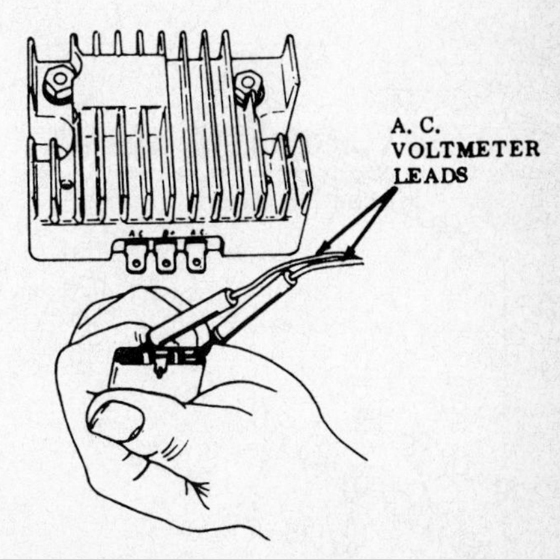

Fig. TE9—Connect ac voltmeter to ac leads as shown when checking alternator coils.

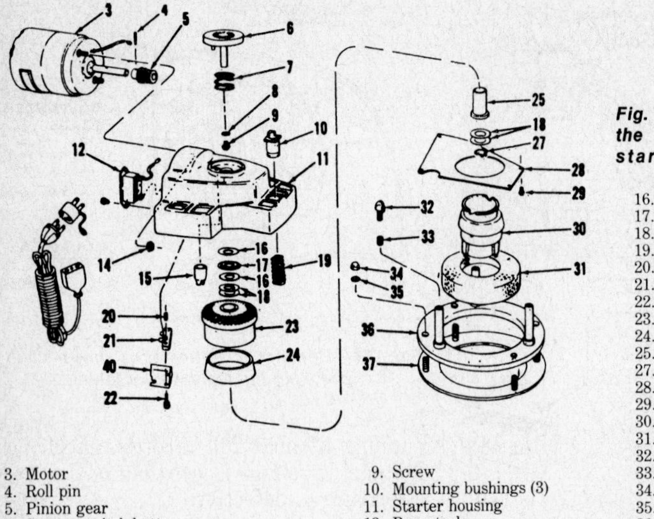

Fig. TE11—Exploded view of the 110 volt ac electric starter used on some engines.

3. Motor
4. Roll pin
5. Pinion gear
6. Starter switch button
7. Switch return spring
8. Washer
9. Screw
10. Mounting bushings (3)
11. Starter housing
12. Receptacle
14. Nuts (2)
15. Pinion bearing
16. Thrust races (2)
17. Thrust bearing
18. Thrust shims
19. Guide post springs (3)
20. Switch plunger spring
21. Switch
22. Screws (2)
23. Driven gear
24. Clutch facing
25. Bearing
27. Snap ring
28. Shield
29. Screws (4)
30. Lower cone
31. Screen
32. Cap studs (3)
33. Spring
34. Acorn nuts (4)
35. Lockwashers
36. Mounting ring
37. Reinforcing ring
40. Switch cover

To test starting motor, remove the dc output cable (9) from motor. Connect an ohmmeter between one of the flat receptacle terminals and motor housing. If a short is indicated, the motor is grounded and requires repair. Connect the ohmmeter between the two flat receptacle terminals. A resistance reading of 3-4 ohms must be obtained. If not, motor is faulty.

A no-load test can be performed with the starter removed. Do not operate the motor continuously for more than 15 seconds when testing. Maximum ampere draw should be 2 amperes and minimum rpm should be 8500.

Disassembly of starter motor is obvious after examination of the unit and reference to Fig. TE12. When reassembling, install shim washers (20) as necessary to obtain armature end play of 0.005-0.015 inch (0.13-0.38 mm). Shim washers are available in thicknesses of 0.005, 0.010 and 0.020 inch. Tighten elastic stop nut (32) to 100 in.-lbs. (11 N•m). Tighten nuts on the two through-bolts to 24-28 in.-lbs. (3 N•m).

To test the rectifier assembly, first use an ac voltmeter to check line voltage of the power supply, which should be approximately 115 volts. Connect the input cable (7) to the ac power supply. Connect a dc voltmeter to the two slotted terminals of the output cable (9). Move switch (3) to "ON" position and check the dc output voltage. Direct current output voltage should be a minimum of 100 volts. If a low voltage reading is obtained, rectifiers and/or thyrector could be faulty. Thyrector (8) can be checked after removal, by connecting the thyrector, a 7.5 watt ac light bulb and a 115 volt ac power supply in series. If light bulb glows, thyrector is faulty. Use an ohmmeter to check the rectifiers. Rectifiers must show continuity in one direction only.

WIND-UP STARTERS

RATCHET STARTER. On models equipped with the ratchet starter, refer to Fig. TE13 and move release lever to "RELEASE" position to remove tension

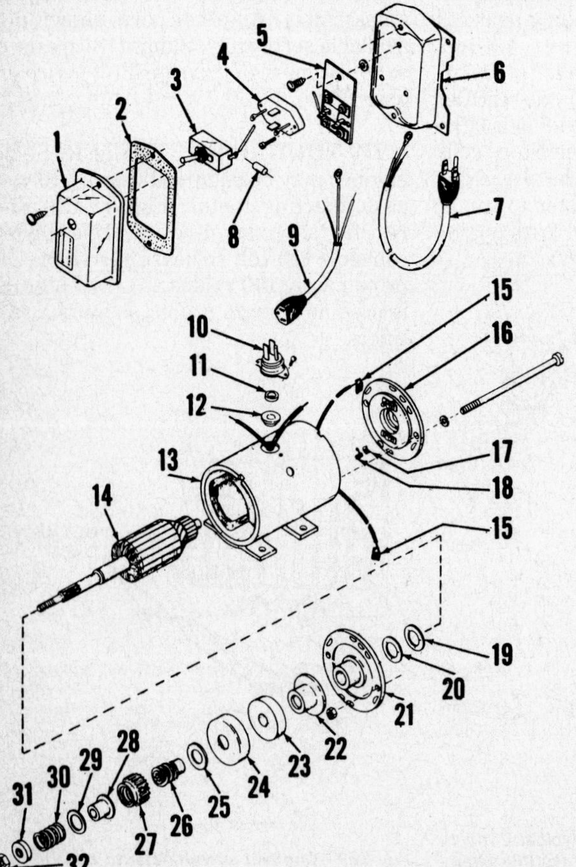

Fig. TE12—Exploded view of 110 volt ac-dc starter motor and rectifier assembly.

1. Cover
2. Gasket
3. Switch
4. Rectifier bridge
5. Rectifier base
6. Mounting base
7. AC input cable
8. Thyrector
9. DC output cable
10. Receptable
11. Washer
12. Insulator
13. Field coils & housing assy.
14. Armature
15. Brushes
16. End plate
17. Spring insulator
18. Brush spring
19. Nylon washer
20. Shim washers (0.005, 0.010 & 0.020 in.)
21. Drive end-plate
22. Thrust sleeve
23. Rubber cushion
24. Cup
25. Thrust washer
26. Screw shaft
27. Pinion gear
28. Spring sleeve
29. Washer
30. Antidrift spring
31. Pinion stop
32. Elastic stop nut

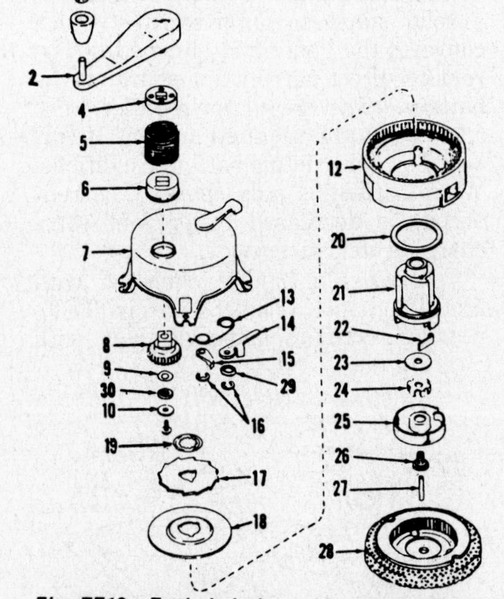

Fig. TE13—Exploded view of a ratchet starter assembly used on some engines.

2. Handle
4. Clutch
5. Clutch spring
6. Bearing
7. Housing
8. Wind gear
9. Wave washer
10. Clutch washer
12. Spring & housing
13. Release dog spring
14. Release dog
15. Lock dog
16. Dog pivot retainers
17. Release gear
18. Spring cover
19. Retaining ring
20. Hub washer
21. Starter hub
22. Starter dog
23. Brake washer
24. Brake
25. Retainer
26. Screw L.H.
27. Centering pin
28. Hub & screen
29. Spacer washers
30. Lockwasher

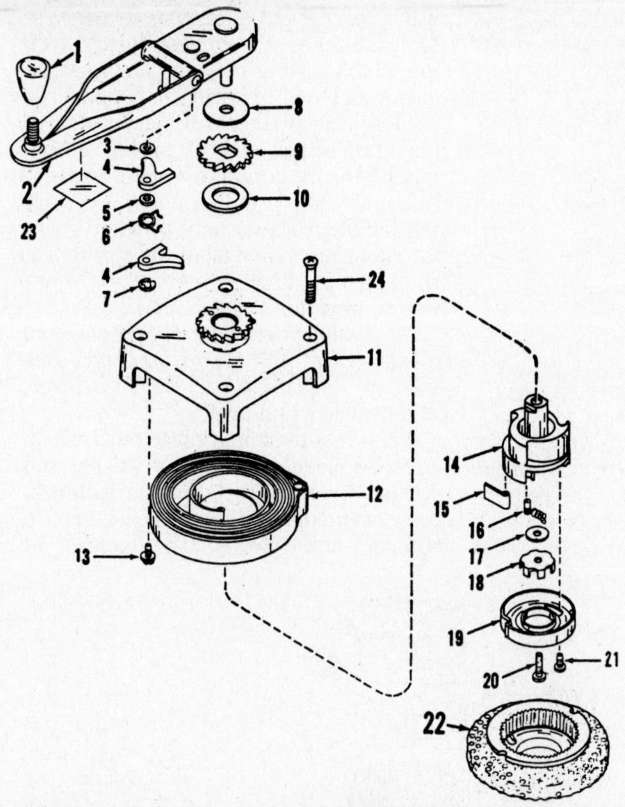

Fig. TE14 — Components of impulse mechanical starter. When handle (2) is used to wind spring (12) the energy put into the spring cranks the engine.

1. Handle knob
2. Starter handle
3. Dog washer
4. Dog release
5. Dog spacer
6. Dog release spring
7. Pivot retainer
8. Ratchet
9. Ratchet
10. Ratchet spacer
11. Housing assy.
12. Spring & keeper
13. Keeper screw
14. Spring hub assy.
15. Starter dog
16. Spring & eyelet
17. Brake washer
18. Brake
19. Retainer
20. Screw
21. Screw
22. Hub & screen
24. Screw

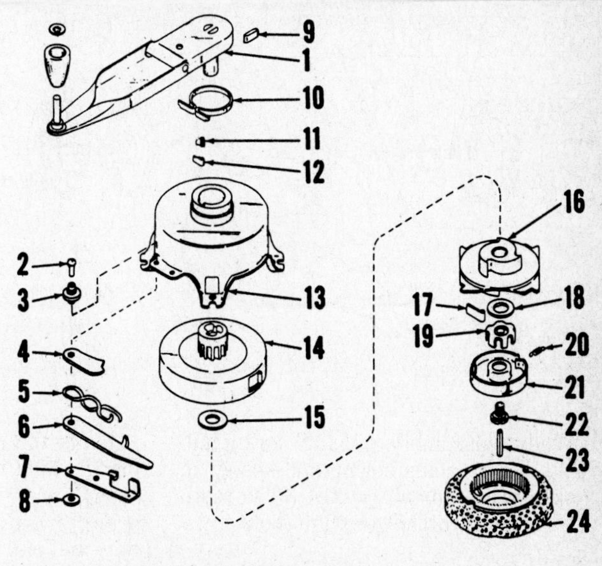

Fig. TE15 — Exploded view of remote release impulse starter used on some models. Refer also to Fig. TE16.

1. Cranking handle
2. Rivet
3. Pivot stud
4. Release lever
5. Release spring
6. Trip lever
7. Trip release
8. Washer
9. Wind dog
10. Brake band
11. Lock dog spring
12. Lock dog
13. Housing
14. Spring & keeper
15. Bearing washer
16. Hub
17. Starter dog
18. Thrust washer
19. Brake
20. Spring
21. Retainer
22. Screw L.H.
23. Centering pin
24. Hub & screen

from main spring. Remove starter assembly from engine. Remove left-hand thread screw (26), retainer hub (25), brake (24), washer (23) and six starter dogs (22). Note position of starter dogs in hub (21). Remove hub (21), washer (20), spring and housing (12), spring cover (18), release gear (17) and retaining ring (19) as an assembly. Remove retaining ring, then carefully separate these parts.

CAUTION: Do not remove main spring from housing (12). The spring and housing are serviced only as an assembly.

Remove snap rings (16), spacer washers (29), release dog (14), lock dog (15) and spring (13). Winding gear (8), clutch (4), clutch spring (5), bearing (6) and crank handle (2) can be removed after first removing the retaining screw and washers (10, 30 and 9).

Reassembly procedure is the reverse of disassembly. Centering pin (27) must align screw (26) with crankshaft center hole.

IMPULSE STARTER. To overhaul the impulse starter (Fig. TE14), first release tension from main spring. Unbolt and remove starter assembly from engine. Remove retaining screw (21), retainer (19), spring and eyelet (16) and starter dogs (15). Remove screw (20), brake (18), washer (17), cranking handle (2), bearing (8), ratchet (9) and spacer (10). Withdraw spring hub (14) and if spring and keeper assembly (12) are to be renewed, remove keeper screw (13). Carefully remove spring and keeper assembly.

CAUTION: Do not remove spring from keeper.

Remove pivot retainer (7), dog releases (4), spring (6), washer (3) and spacer (5) from handle (2).

Fig. TE16 — Exploded view of remote release impulse starter used on some models. Refer also to Fig. TE15.

1. Cranking handle
2. Wind dog
3. Brake band
4. Lock dog spring
5. Lock dog
6. Housing
7. Release lever
8. Wave washer
9. Release arm
10. Hub
11. Spacer washer
12. Power spring
13. Spring housing
14. Shim
15. Thrust washer
16. Brake
17. Starter dog
18. Spring
19. Retainer
20. Shoulder nut
21. Centering pin
22. Hub & screen

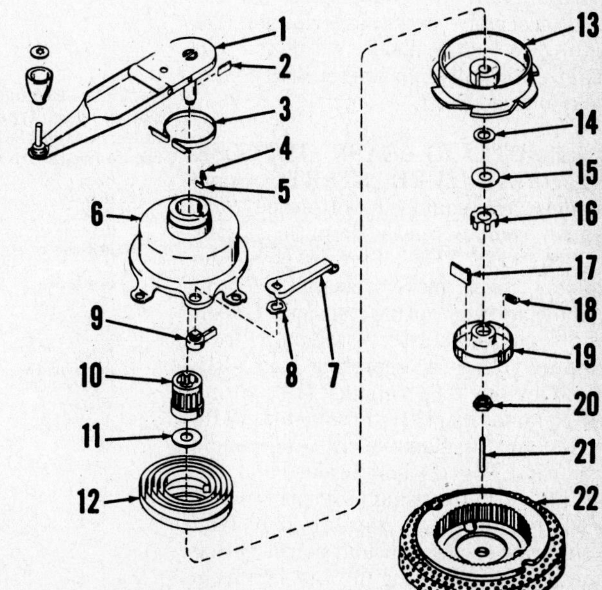

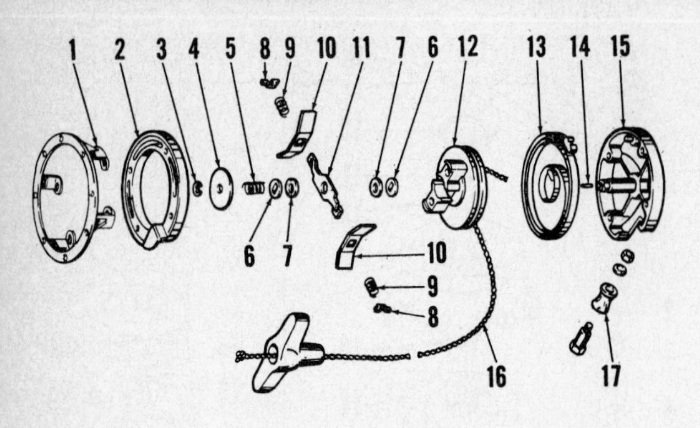

Fig. TE17—Exploded view of typical friction shoe rewind starter.
1. Mounting flange
2. Flange
3. Retaining ring
4. Washer
5. Spring
6. Slotted washer
7. Fiber washer
8. Spring retainer
9. Spring
10. Friction shoe
11. Actuating lever
12. Rotor
13. Rewind spring
14. Centering pin
15. Cover
16. Rope
17. Roller

When reassembling, install spring hub (14) before placing spring and keeper in housing (11). Install ratchet (9) so teeth are opposite ratchet teeth on housing.

REMOTE RELEASE IMPULSE STARTER (SURE LOCK). To disassemble the "Sure Lock" remote release impulse starter (Fig. TE15), first release tension from main spring. Unbolt and remove starter assembly from engine. Remove centering pin (23), then unscrew the left-hand threaded screw (22). Remove retainer (21), spring (20), brake (19), starter dog (17), thrust washer (18) and hub (16). Remove cranking handle (1), wind dog (9), brake band (10), lock dog spring (11) and lock dog (12). Carefully lift out spring and keeper assembly (14).

CAUTION: Do not attempt to remove spring from keeper. Spring and keeper are serviced only as an assembly.

Some models use pivot stud (3), release lever (4), release spring (5) and trip lever (6). Other models use rivet (2) and trip release (7). Removal of these parts is obvious after examination of the unit and reference to Fig. TE15.

Reassembly procedure is the reverse of disassembly. Press centering pin (23) into screw (22) about 1/3 of the pin length. Pin will align starter with center hole in crankshaft.

REMOTE RELEASE IMPULSE STARTER (SURE START). Some models are equipped with the "Sure Start" remote release impulse starter shown in Fig. TE16. To disassemble the starter, first move release lever to remove power spring tension. Unbolt and remove starter assembly from engine. Remove centering pin (21), shoulder nut (20), retainer (19), spring (18), starter dog (17), thrust washer (15) and shim (14). Remove cranking handle (1), wind dog (2) and brake band (3). Rotate spring housing (13) spring while pushing down on top of hub (10). Withdraw spring housing, power spring and hub from housing (6). Lift hub from spring.

This is the only wind-up type starter spring which may be removed from the spring housing (keeper). To remove spring, grasp inner end of spring with pliers and pull outward. Spring diameter will increase about 1/3 when removed. To install new power spring in spring housing, clamp spring housing in a vise. Anchor outer end of spring in housing. Install spacer washer (11) and hub (10), engaging inner end of spring in hub notch. Insert crankpin handle shaft in hub and while pressing down on cranking handle, rotate crank to wind spring into housing. When all of the spring is in the housing, allow cranking handle and hub to unwind.

To remove release lever (7) and release arm (9), file off peened-over sides of release arm. Remove lever, wave washer and arm.

When reassembling, place release arm on wood block, install wave washer and release lever. Using a flat face punch, peen over edges of new release arm to secure release lever. The balance of

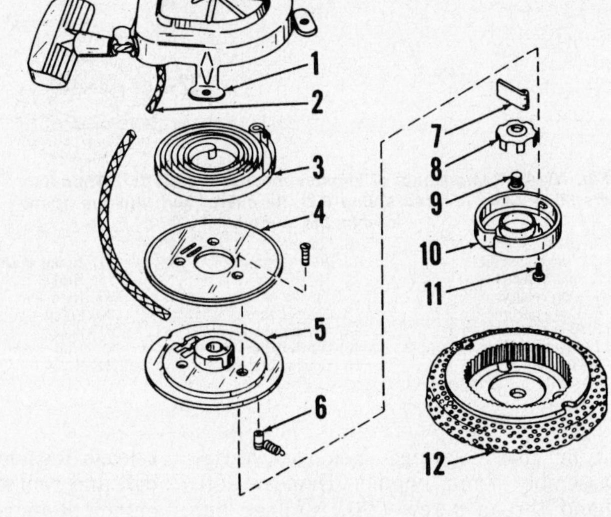

Fig. TE18—Exploded view of typical dog type rewind starter assembly. Some units of similar construction use three starter dogs (7).
1. Cover
2. Rope
3. Rewind spring
4. Pulley half
5. Pulley half & hub
6. Retainer spring
7. Starter dog
8. Brake
9. Brake screw
10. Retainer
11. Retainer screw
12. Hub & screen assy.

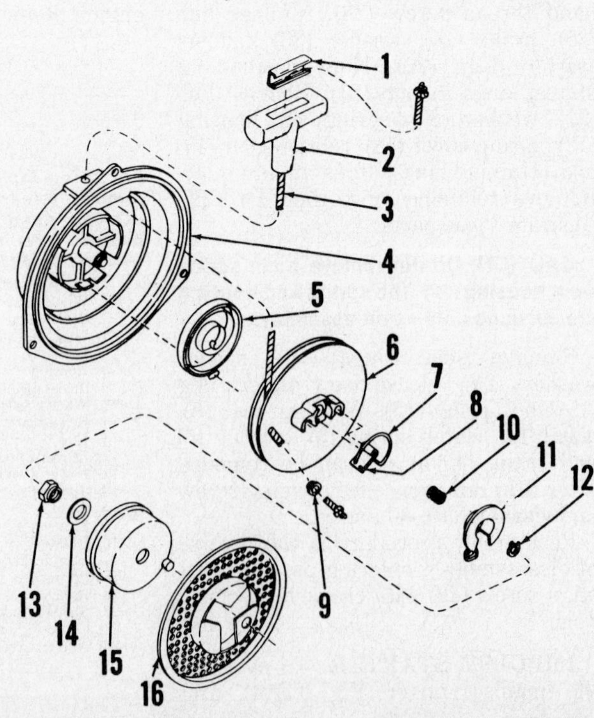

Fig. TE19—Exploded view of typical dog type rewind starter assembly used on some two-stroke engines.
1. Insert
2. Handle
3. Rope
4. Starter housing
5. Rewind spring
6. Pulley & hub assy.
7. Dog spring
8. Starter dog
9. Grommet
10. Brake spring
11. Brake
12. Retaining ring
13. Flywheel nut
14. Washer
15. Starter cup
16. Screen

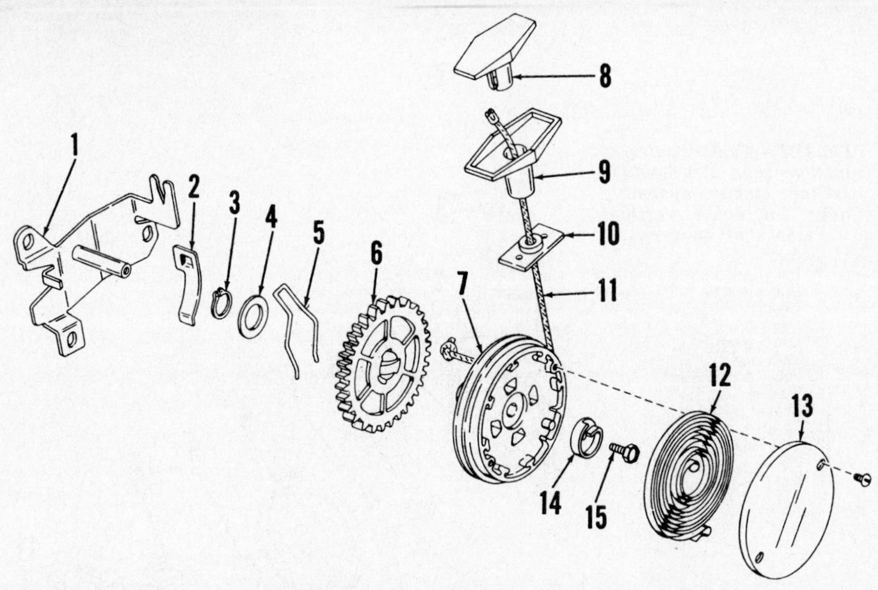

Fig. TE20—Exploded view of single-gear, side-mounted rewind starter used on some vertical crankshaft engines.

1. Mounting bracket
2. Rope clip
3. Snap ring
4. Washer
5. Brake
6. Gear
7. Pulley
8. Insert
9. Handle
10. Bracket
11. Rope
12. Rewind spring
13. Cover
14. Spring hub
15. Hub screw

reassembly is reverse of disassembly procedure. Install centering pin so end of pin extends below end of mounting legs. The pin will align the center of starter to center hole in end of crankshaft.

REWIND STARTERS

FRICTION SHOE TYPE. To disassemble the starter, refer to Fig. TE17 and hold starter rotor (12) securely with thumb and remove the four screws securing flanges (1 and 2) to cover (15). Remove flanges and release thumb pressure enough to allow spring to rotate pulley until spring (13) is unwound. Remove retaining ring (3), washer (4), spring (5), slotted washer (6) and fiber washer (7). Lift out friction shoe assembly (8, 9, 10 and 11), then remove second fiber washer and slotted washer. Withdraw rotor (12) with rope from cover and spring. Remove rewind spring from cover and unwind rope from rotor.

When reassembling, lubricate rewind spring, cover shaft and center bore in rotor with a light coat of Lubriplate or equivalent. Install rewind spring so windings are in same direction as removed spring. Install rope on rotor, then place rotor on cover shaft. Make certain inner and outer ends of spring are correctly hooked on cover and rotor. Preload the rewind spring by rotating the rotor two full turns. Hold rotor in preload position and install flanges (1 and 2). Check sharp end of friction shoes (10) and sharpen or renew as necessary. Install washers (6 and 7), friction shoe assembly, spring (5), washer (4) and retaining ring (3). Make certain friction shoe assembly is installed properly for correct starter rotation. If properly installed, sharp ends of friction shoes will extend when rope is pulled.

Remove brass centering pin (14) from cover shaft, straighten pin if necessary, then reinsert pin ⅓ of its length into cover shaft. When installing starter on engine, centering pin will align starter with center hole in end of crankshaft.

DOG TYPE (FOUR-STROKE) To disassemble the dog type starter, refer to Fig. TE18 and release preload tension of rewind spring by pulling starter rope until notch in pulley half (5) is aligned with rope hole in cover (1). Use thumb pressure to prevent pulley from rotating. Engage rope in notch of pulley and slowly release thumb pressure to allow spring to unwind. Remove retainer screw (11), retainer (10) and spring (6). Remove brake screw (9), brake (8) and starter dog (7). Carefully remove pulley assembly with rope from spring and cover. Note direction of spring winding and carefully remove spring from cover. Unbolt and separate pulley halves (4 and 5) and remove rope.

To reassemble, reverse the disassembly procedure. Then, preload rewind spring by aligning notch in pulley with rope hole in cover. Engage rope in notch and rotate pulley two full turns to properly preload the spring. Pull rope to fully extended position. Release handle and if spring is properly preloaded, the rope will fully rewind.

DOG TYPE (TWO-STROKE). Some engines are equipped with the dog type rewind starter shown in Fig. TE19. To disassemble the starter, first pull rope to fully extended position. Then, tie a slip knot in rope on outside of starter housing. Remove insert (1) and handle (2) from end of rope. Hold pulley with thumb pressure, untie slip knot and withdraw rope and grommet (9). Slowly release thumb pressure and allow pulley to rotate until spring is unwound. Remove retaining ring (12), brake (11), brake spring (10) and starter dog (8) with dog spring (7). Lift out pulley and hub assembly (6), then remove rewind spring (5).

Reassemble by reversing the disassembly procedure. When installing the rope, insert a suitable tool in the rope hole in pulley and rotate pulley six full turns until hole is aligned with rope hole in housing (4). Hold pulley in this position and install rope and grommet. Tie a slip knot in rope outside of housing and assemble handle, then insert on rope. Untie slip knot and allow rewind spring to wind the rope.

SIDE-MOUNTED TYPE (SINGLE GEAR). To disassemble the side-mounted starter (Fig. TE20), remove insert (8) and handle (9). Relieve spring tension by allowing spring cover (13) to rotate slowly. Rope will be drawn through bracket (10) and wrapped on pulley (7). Remove cover (13) and carefully remove rewind spring (12). Remove hub screw (15), spring hub (14), then withdraw pulley and gear assembly. Remove snap ring (3), washer (4), then withdraw pulley and gear assembly. Remove snap ring (3), washer (4), brake (5) and gear (6) from pulley.

Reassemble by reversing the disassembly procedure, keeping the following points in mind: Lubricate only the edges of rewind spring (12) and shaft on mounting bracket (1). Do not lubricate spiral in gear (6) or on pulley shaft. Rewind spring must be preloaded approximately 2½ turns.

When installing the starter assembly, adjust mounting bracket so side of teeth on gear (6) when in fully engaged position, has 1/16 inch (1.59 mm) clearance from base of flywheel gear teeth. Remove spark plug wire and test starter several times for gear engagement. A too-close adjustment could cause starter gear to hang up on flywheel gear when engine starts. This could damage the starter.

SIDE-MOUNTED TYPE (MULTIPLE GEAR). Some engines may be equipped with the side-mounted starter shown in Fig. TE21. To disassemble this type starter, unbolt cover (11) and remove items (7, 8, 9, 10, 11 and 14) as an assembly. Pull rope out about 8 inches (203 mm), hold pulley and gear assembly (8) with thumb pressure and engage rope in notch in pulley. Slowly release thumb pressure and allow spring to unwind. Remove snap ring (7), pulley and gear (8), washer (9) and rewind spring (10) from cover. Rope (14) and rewind spring can now be removed if necessary. Gears (3 and 6) can be removed after first removing snap ring (12), driving out pin (13) and withdrawing shaft (1). Remove brake (4).

To reassemble, reverse the disassembly procedure. Preload the rewind spring two full turns.

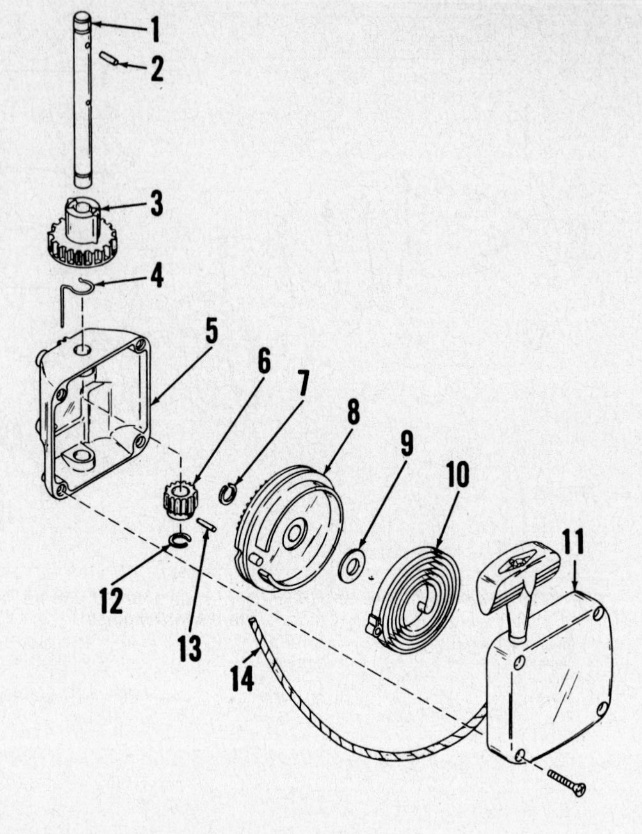

Fig. TE21—Exploded view of multiple-gear, side-mounted rewind starter assembly used on some vertical crankshaft engines.

1. Shaft
2. Pin
3. Upper gear
4. Brake
5. Housing
6. Lower gear
7. Snap ring
8. Pulley & gear assy.
9. Washer
10. Rewind spring
11. Cover
12. Snap ring
13. Pin
14. Rope

TORO TWO-STROKE

Model	Bore	Stroke	Displacement
47PZ2, 47PD3, 47PE4, 47PF5	58 mm (2.3 in.)	46 mm (1.8 in.)	121 cc (7.4 cu. in.)

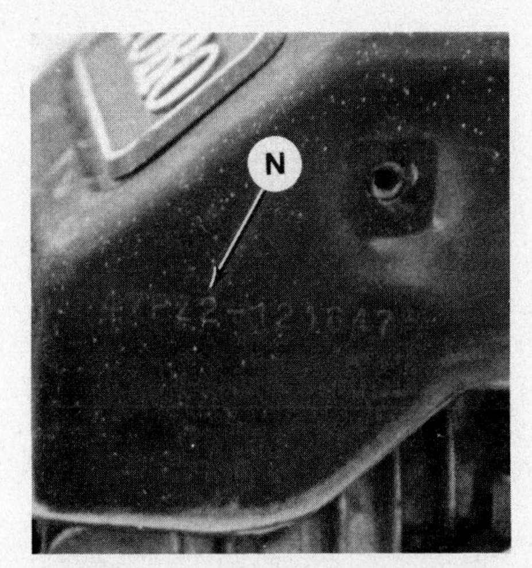

Fig. TO10—View showing location of engine model and serial numbers (N) on all models except Model 47PF5 built for 1986. Model 47PF5 model and serial numbers are located just above spark plug on cooling shroud.

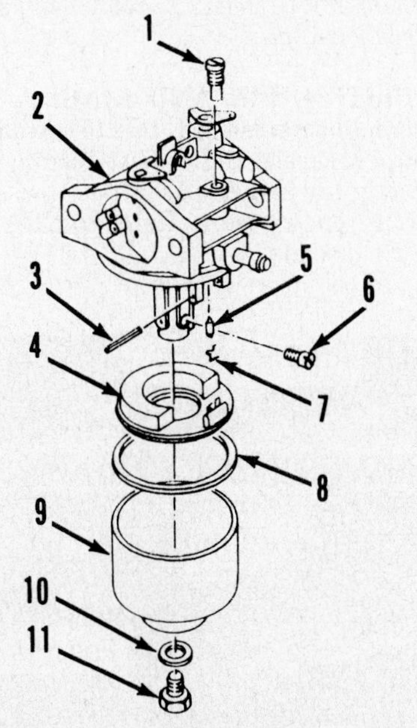

Fig. TO11—Exploded view of carburetor used on Model 47PZ2.

1. Pilot jet
2. Carburetor body
3. Float pin
4. Float
5. Needle valve
6. Main jet
7. Spring
8. Gasket
9. Float bowl
10. Gasket
11. Bolt

ENGINE INFORMATION

Toro (Suzuki) two-stroke vertical crankshaft engines are rated at 2.6 kW (3.5 hp) at 3600 rpm.

Engine model and serial numbers for all models except Model 47PF5 engines built for 1986 are stamped on blower housing behind the air cleaner (Fig. TO10). Model 47PF5 engines built for 1986 and after have the model and serial numbers stamped in the blower housing above the spark plug. Always furnish engine model and serial numbers when ordering parts or service material.

MAINTENANCE

SPARK PLUG. Spark plug should be removed, cleaned and inspected at 25-hour intervals.

Recommended spark plug is a Champion RCJ8Y, or equivalent. Recommended spark plug electrode gap is 0.8 mm (0.032 in.). Spark plug should be tightened to 9-13 N•m (80-120 in.-lbs.).

CARBURETOR. All models are equipped with a float type carburetor with a fixed main fuel jet (6 – Fig. TO11 or TO12). Model 47PZ2 engines are equipped with carburetor shown in Fig. TO11. Standard main jet size is a #80 and a high altitude #77.5 main jet is available (Toro part 81-1050). All other model engines are equipped with carburetor shown in Fig. TO12. Standard main jet size is a #76.3 and a high altitude #72.5 main jet is available (Toro part 81-2340).

To check float level, remove carburetor and separate float bowl (9 – Fig. TO11 or TO12) from carburetor throttle body (2). Invert carburetor throttle body and measure from carburetor body to edge of float as shown in Fig. TO13. Correct float height is 17.5 mm (11/16 in.).

Original equipment float for carburetors on 47PZ2 engines (Fig. TO11) should be replaced with a new black float with a metal hinge (Toro part 81-0970) during service. This float and original equipment float used on all

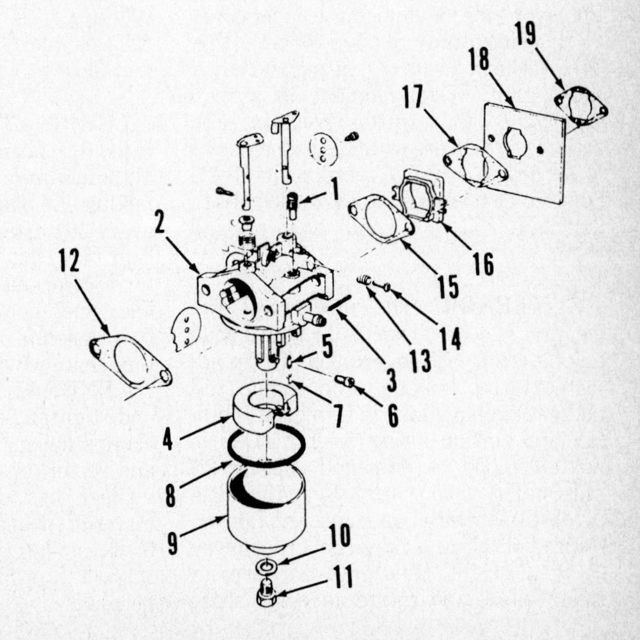

Fig. TO12—Exploded view of carburetor used on Models 47PD3, 47PE4 and 47PF5.

1. Pilot jet
2. Carburetor body
3. Float pin
4. Float
5. Needle valve
6. Main jet
7. Spring
8. Gasket
9. Float bowl
10. Gasket
11. Bolt
12. Gasket
13. Spring
14. Throttle stop screw
15. Gasket
16. Insulator
17. Gasket
18. Heat deflector
19. Gasket

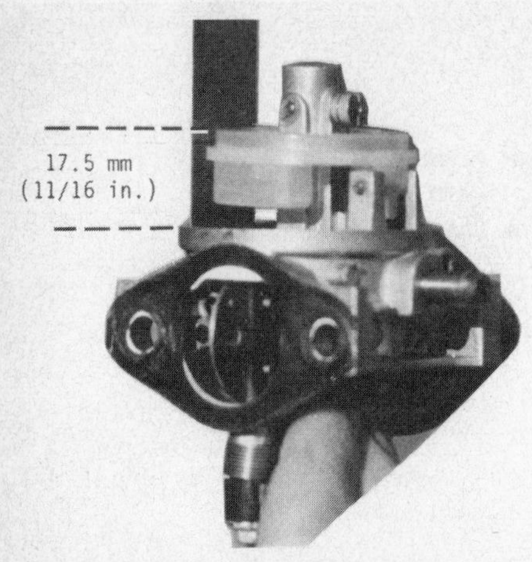

Fig. TO13—Float level should be 17.5 mm (11/16 in.).

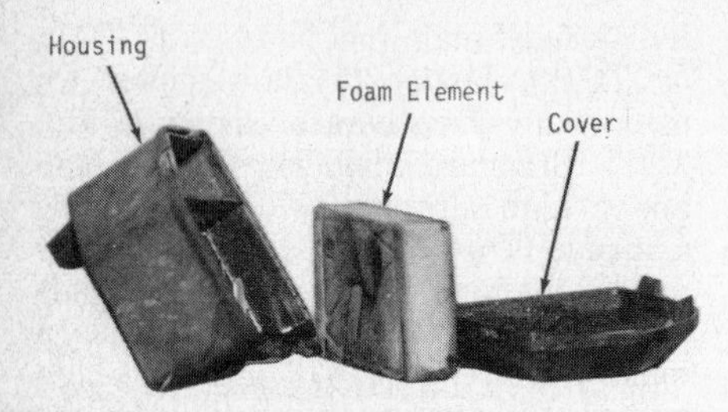

Fig. TO14—Air cleaner cover, element and housing assembly.

other models is alcohol resistant and adjustable. Adjust float height by carefully bending float hinge.

FUEL FILTER. A fuel filter is located in-line on some models. Toro part number is 56-6369. A sintered bronze fuel filter is also located in fuel tank outlet and is nonrenewable.

AIR CLEANER. Engine is equipped with a foam type air cleaner which should be removed and cleaned at 50-hour intervals. To remove element, lift cover tabs retaining air cleaner cover to housing and remove cover (Fig. TO14). Clean inside of housing and cover thoroughly. Wash element in a mild detergent and water solution and squeeze out excess water. Allow element to air dry. Apply five teaspoons of SAE 30 oil to element and squeeze element to distribute oil. Reinstall element and cover.

GOVERNOR. The mechanical flyweight type governor is located inside engine crankcase on crankshaft. To adjust external linkage, stop engine and make certain all linkage is in good condition and tension spring (5–Fig. TO15) is not stretched or damaged. Spring (2) must pull governor lever (3) and throttle pivot (4) toward each other. Loosen clamp bolt (7) and move governor lever (3) to the right. Hold governor lever in this position and rotate governor shaft (6) in clockwise direction until it stops. Tighten clamp screw.

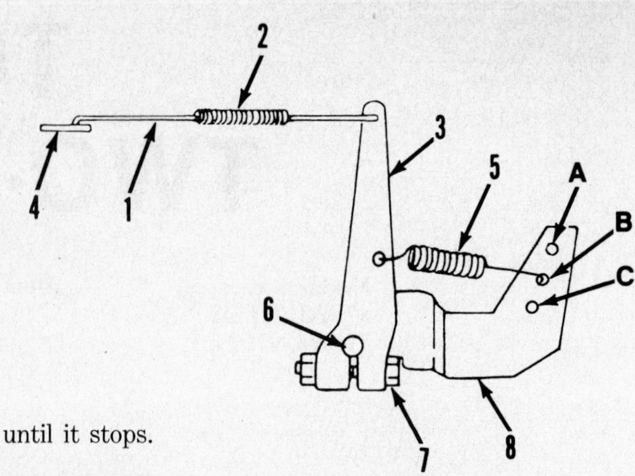

Fig. TO15—View showing location of external governor linkage. Refer to text for adjustment procedure.

1. Rod
2. Spring
3. Governor lever
4. Throttle pivot
5. Tension spring
6. Governor shaft
7. Clamp bolt
8. Tension spring bracket
A. Upper hole
B. Middle hole (3000 rpm)
C. Lower hole

Correct engine speed is 3000 rpm and is obtained with tension spring (5) connected in center hole (B) of bracket (8). Connecting spring in hole (A) increases engine speed by 150 rpm and connecting spring in hole (C) decreases engine speed by 150 rpm.

IGNITION SYSTEM. Model 47PZ2 engine is equipped with a breaker point type ignition. Breaker point gap should be 0.35 mm (0.014 in.). This should allow points to begin opening at 22° BTDC (Fig. TO16). Timing may be varied slightly by changing point gap.

All other models are equipped with a solid state ignition system which requires no regular maintenance.

Ignition coil air gap should be 0.38-0.50 mm (0.015-0.020 in.).

CARBON. Carbon should be removed when engine performance drops or when engine begins to four-stroke despite correct carburetor settings. To clean carbon from exhaust system, remove muffler and use a wooden dowel to remove carbon deposits from port. Clean the compression relief hole next to exhaust port. To clean piston and cylinder it is necessary to completely disassemble engine. Refer to REPAIR section.

LUBRICATION. Engine is lubricated by premixing oil with the fuel. Manufacturer recommends thoroughly mixing one part BIA TC approved oil for every 50 parts of a good grade leaded or lead free gasoline. DO NOT use any form of premium gasoline or any gasoline which contains alcohol. Use of lead free gasoline results in fewer combustion chamber deposits.

GENERAL MAINTENANCE. Check and tighten all loose bolts, nuts or clamps daily. Check for fuel or leakage and repair as necessary.

Clean dust, dirt, grease or any foreign material from cylinder head and cylinder block cooling fins at 100-hour intervals. Inspect fins for damage and repair as necessary.

REPAIRS

TIGHTENING TORQUES. Recommended tightening torque specifications are as follows:

Spark plug . . . . 9-14 N·m (7-10 ft.-lbs.)
Flywheel nut . . . . 39-49 N·m (29-36 ft.-lbs.)
Crankcase screws . . . . 9-12 N·m (6-9 ft.-lbs.)
Blade cap screw . . . . 61-81 N·m (45-60 ft.-lbs.)

CYLINDER. Cylinder is an integral part of one of the crankcase halves. Cylinder should be renewed if scored, flaking or severely out-of-round. Cylinder should be measured at top, just above intake port, and just below third port at bottom. Diameter should be 58.000-58.115 mm (2.284-2.288 in.) at all three locations.

PISTON, PIN AND RING. A cam ground piston is used. It is necessary to split crankcase and remove piston, connecting rod and crankshaft as an assembly to renew piston. Refer to CRANKCASE section.

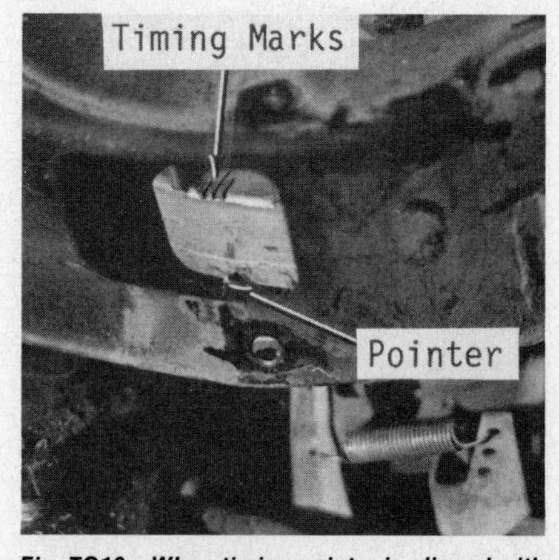

Fig. TO16—When timing pointer is aligned with middle mark as shown, timing is set at 22° BTDC. Refer to text.

Ring should be installed with the side marked "R" toward top of piston. Correct ring end gap is 0.7 mm (0.027 in.). Correct ring side clearance is 0.03-0.15 mm (0.001-0.006 in.) measured at bottom side of ring.

Piston to cylinder clearance should be measured 90° from piston pin bore, 2.5 mm (1 in.) from lower edge of skirt. Subtract piston diameter from largest cylinder diameter measurement (see CYLINDER section) to determine piston to cylinder clearance. Clearance should not exceed 0.1 mm (0.004 in.).

Piston pin diameter should be 11.955-12.005 mm (0.472-0.473 in.).

CRANKCASE. A split crankcase design with the cylinder as an integral part of one crankcase half is used. It is necessary to split crankcase to renew piston, connecting rod or crankshaft.

To disassemble crankcase, remove the recoil starter assembly, disconnect fuel line at carburetor and lift fuel tank from cooling shroud. Remove the muffler and heat shield. Lift air filter cover and remove air filter element. Remove the two locknuts from carburetor studs and the cap screw which retain air cleaner housing and remove housing. Remove the three remaining cap screws from cooling shroud assembly and remove cooling shroud.

Compress control bar against handle to compress flywheel brake spring, remove cap screw retaining ignition switch and brake assembly, disconnect switch and slowly release control bar to release flywheel brake spring. Disconnect brake control cable. Remove starter cup and flywheel (use Toro flywheel puller 41-7650). Remove point cover, points and condenser (if so equipped). Remove flywheel key and slide timing cam from engine. Remove the small timing cam locating pin. Remove governor lever, control rod and springs, carburetor, spacer and gaskets. Remove blade, blade adapter and self-propelled components (if so equipped). Remove engine from mower chassis. Remove engine adapter plate. Remove the six cap screws retaining crankcase halves and carefully separate crankcase. Remove crankshaft and piston assembly. Remove seals, bearing locating rings and bearings from crankshaft ends. Slide governor assembly from carnkshaft; use care not to lose locating pin for governor flyweight collar. Remove snap rings from piston pin bores and remove piston pin and piston.

Note that crankshaft and connecting rod are serviced as an assembly only. DO NOT DISASSEMBLE CRANKSHAFT AND CONNECTING ROD.

When reassembling engine, arrow stamped in top of piston must face toward exhaust port side of cylinder when piston and crankshaft assembly is installed. Apply a thin coat of Loctite 515 sealant to crankcase mating surfaces to seal crankcase during reassembly.

CRANKSHAFT, MAIN BEARINGS AND SEALS. Crankshaft and connecting rod are serviced as assembly only and crankshaft and connecting rod are not available separately.

To remove crankshaft and connecting rod assembly, refer to CRANKCASE section.

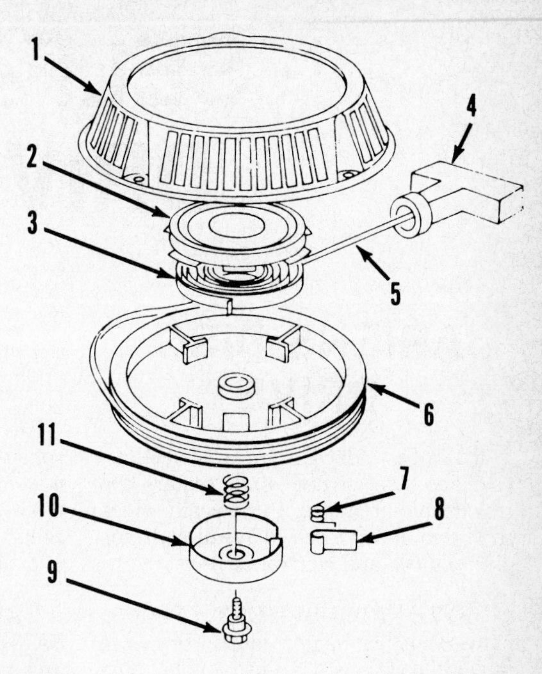

Fig. TO17—Exploded view of recoil starter.
1. Cover
2. Retainer
3. Recoil spring
4. Handle
5. Rope
6. Rope pulley
7. Spring
8. Pawl
9. Cap screw
10. Retainer
11. Spring

RECOIL STARTER. To remove recoil starter, remove the four retaining bolts and lift starter assembly from cooling shroud. Detach rope handle, then remove center cap screw (9–Fig. TO17) and remove pawl retainer (10). Remove spring (11), pawl (8) and pawl spring (7). Rope pulley (6), recoil spring retainer (2) and recoil spring (3) may now be removed. Exercise caution when removing spring; do not allow spring to uncoil uncontrolled. When reassembling recoil starter, use Loctite 242 on threads of center cap screw. Tighten center cap screw to 8-11 N•m (6-8 ft.-lbs.).

FLYWHEEL BRAKE. The flywheel brake and ignition switch are controlled by a common cable connected to the control bar at handle. When control bar is in "STOP" position, brake spring forces brake pad against flywheel. When control bar is in "START" position, brake pad must clear flywheel and ignition switch must be activated. Adjust cable as necessary.

SELF-PROPELLED DRIVE SYSTEMS

ARIENS DISC DRIVE

MAINTENANCE. Drive wheel assemblies are sealed and require no regular maintenance. Periodically check condition of drive belt, drive disc, driven disc, pulleys and control cable.

DRIVE ADJUSTMENT. Position mower cutting height control levers in middle positions. Adjust cable nuts (N–Fig. AR1) to provide a slight amount of slack. Place speed control lever in "SLOW" position. Loosen adjustment screws (S). Hold lever (L) against mower deck and adjust drive assembly to provide 1/32 to 1/16 inch (0.8-1.6 mm) clearance between drive wheel (D) and driven wheel (W). Tighten adjustment screws (S).

DRIVE BELT. To renew drive belt, drain fuel from fuel tank and oil from engine crankcase. Disconnect spark plug lead and properly ground lead. Remove battery, as equippped. Set right rear cutting height adjustment lever one notch forward of left rear cutting height adjustment lever to obtain clearance between drive disc and driven disc. Place speed control in "SLOW" position. Tip mower onto left side, disconnect idler spring and push idler pulley (I–Fig. AR1) to one side. Carefully work drive belt (DB) from drive wheel (D) and through mower deck opening under engine. Remove blade and work belt from engine pulley and shaft.

Reverse removal procedure for reinstallation. Make certain idler spring is connected and cutting height adjustment levers are set back in equal positions. Refill crankcase with specified oil and fill fuel tank. Connect spark plug.

DRIVE GEARS AND DRIVE ASSEMBLY. To disassemble drive gears and remove drive assembly, drain fuel tank and engine crankcase. Disconnect and properly ground spark plug lead. Block rear of mower up so drive wheels are off the ground. Remove snap ring and washer retaining left rear wheel to axle and remove wheel. Remove snap ring (SR–Fig. AR1) and gear cover (GC) retaining screws and remove gear cover. Slide differential assembly from axle and remove idler gear by sliding it from idler shaft. Use care not to lose needles from needle bearing pressed into center of idler gear. Support pinion gear and drive roll pin retaining pinion gear on hex shaft out and remove pinion gear.

Remove compression spring at front of drive mount. Remove the two screws securing clutch collar to drive mount and arm and remove clutch collar. Remove cotter pin, washer and shift link.

To remove drive mount for bearing renewal, remove snap ring from behind axle spline on left side of mower and remove axle bearing. Remove cotter pin retaining left side of trailing shield on bracket. Remove washer and pivot shield end to one side. Spread snap ring at right side of axle and pull right wheel and axle out right side of mower. Drive assembly may now be disassembled and bearings or driven disc renewed. Use Ariens bearing driver tool 000026, bearing puller and a press to remove bearing. Use Ariens bearing driver 000058 to reinstall bearing.

DRIVE DISC. To renew drive disc, remove drive belt as previously outlined. Remove nut, lock washer and flat washer at top center of drive disc and allow carriage bolt to drop out of disc. Remove the four bolts retaining bearing retainer to drive disc and remove retainer and bearing spindle. Press out the two ball bearings and shim washer from drive disc using Ariens bearing driver 000026. Use Ariens bearing driver 000046 to reinstall bearing.

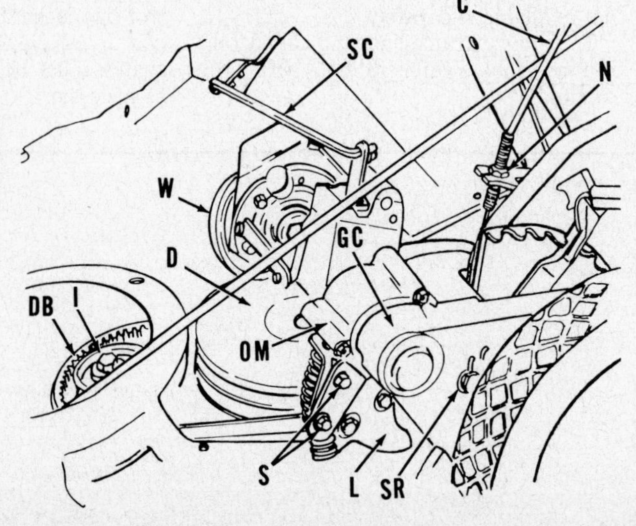

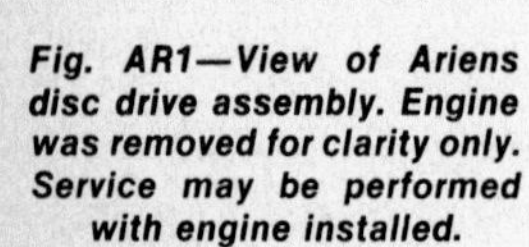

Fig. AR1—View of Ariens disc drive assembly. Engine was removed for clarity only. Service may be performed with engine installed.

C. Cable
D. Drive disc
DB. Drive belt
DM. Drive mount
GC. Gear cover
I. Idler pulley
L. Lever
N. Cable adjustment nuts
S. Adjustment screws
SC. Speed control rod
SR. Snap ring
W. Driven disc

CHAIN DRIVE SYSTEM

MAINTENANCE. Periodically check all nuts and bolts, bushings and pivots for tightness and smooth operation. Use light oil at pivots and bushings.

To adjust chain tension, mark position of left and right drive brackets. Loosen nuts retaining bracket at chain side and move bracket until all slack has just been removed from chain. Tighten nuts. Measure distance bracket was moved (distance from mark just made to brackets new position). Loosen opposite drive bracket retaining nuts and move bracket exact distance measured at chain side drive bracket. Tighten nuts. Unequally adjusted brackets may cause mower to drift to right or left from a straight line.

CAUTION: With drive rollers fully engaged, chain must be able to be easily moved side-to-side on drive sprockets. Excessive chain tension will cause rapid chain and sprocket wear as well as damage to auxiliary pto shaft and bushings in engine crankcase.

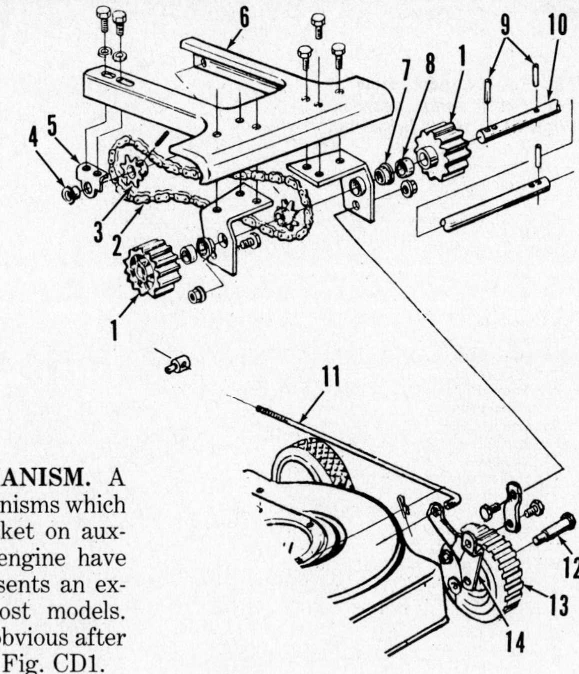

Fig. CD1—Exploded view of typical chain drive mechanism used by a variety of mower manufacturers.
1. Drive roller
2. Chain
3. Drive sprocket
4. Bushing
5. Bracket
6. Shield
7. Bushing
8. Spacer
9. Roll pins
10. Drive shaft
11. Control rod
12. Shoulder bolt
13. Wheel & tire
14. Pivot bracket

CHAIN DRIVE MECHANISM. A variety of chain drive mechanisms which derive power from a sprocket on auxiliary drive shafts on the engine have been used. Fig. CD1 represents an exploded view typical of most models. Service and disassembly is obvious after inspection and reference to Fig. CD1.

DRIVE ADJUSTMENT. Drive control cable should be adjusted so that gearbox fully engages and disengages. Loosen cable clamp (C) and adjust cable housing to obtain correct shift lever (S) movement. Tighten cable clamp.

DRIVE BELT. To renew drive belt (D), remove shield to expose gearbox. Disconnect and properly ground spark plug lead. Remove mower blade. Remove left and right wheel hub caps, retaining pins, washers, snap rings, drive gears, dust covers and wheels. Remove the two screws on each side which retain wheel height adjuster plates. Remove cable clamp and disconnect drive cable at shift lever. Drive belt may now be removed. Gearbox may be removed if necessary.

GEARBOX. Refer to DRIVE BELT paragraph for gearbox removal procedure. To disassemble gearbox, remove snap ring (S – Fig. CR2) from input shaft and remove pulley (P) and snap ring located just below pulley. Remove the four screws retaining gearbox halves and remove cable bracket (C). Separate gearbox halves.

Remove worm drive gear (6 – Fig. CR3) and shift fork retaining pin (5). Remove shift fork (4) and lift drive shaft (1) and gear assembly from gearbox half. Remove steel pin from input shaft (7) and push input shaft out of gearbox

CRAFTSMAN GEARBOX

MAINTENANCE. Bolt (B – Fig. CR1) should be removed yearly and a good quality lithium base grease pumped into gearbox to assure adequate gear and bearing lubrication. Light oil should be applied to wheel axles at 25-hour intervals.

Fig. CR1—View of gearbox used on Craftsman self-propelled walk behind lawn mowers.
B. Bolt
C. Cable clamp
D. Drive belt
G. Gearbox
H. Cable housing
S. Shift lever

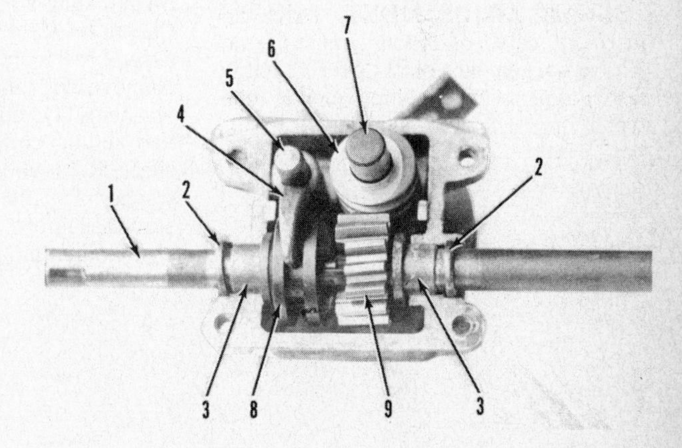

Fig. CR3—View of Craftsman gearbox showing components assembled correctly.
1. Drive shaft
2. Seal
3. Bushing
4. Shift fork
5. Shift fork pin
6. Worm gear
7. Input shaft
8. Sliding clutch ring
9. Driven gear

Fig. CR2—View of Craftsman gearbox (G) showing location of drive shaft (D), pulley (P), snap ring (S), cable bracket (C) and bolt (B).

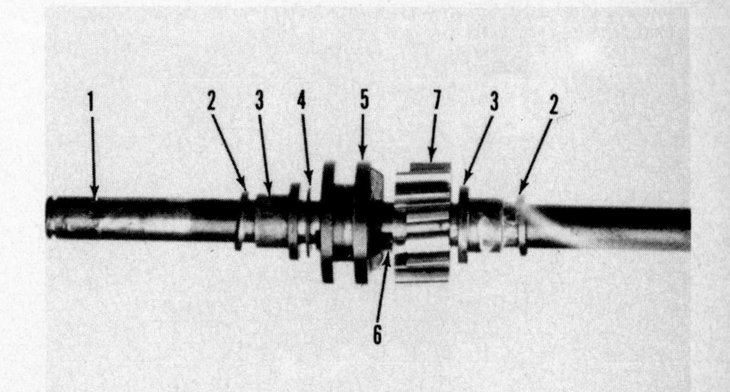

Fig. CR4—View of drive shaft and components removed from gearbox.
1. Drive shaft
2. Seal
3. Bushing
4. Snap ring
5. Sliding clutch ring
6. Key
7. Driven gear

half. Use care not to lose thrust washer between steel pin and gearbox.

To disassemble drive shaft and gear assembly, remove seal (2–Fig. CR4), bushing (3) and snap ring (4). Remove sliding clutch ring (5) and key (6). Remove driven gear (7), bushing (3) and seal (2).

Inspect all parts for wear and damage. Gear surfaces should be flat and dogs on sliding clutch ring and driven gear should be in good condition to prevent jumping out of gear. Reassemble by reversing disassembly procedure. Pack housing with a good quality lithium base grease prior to joining gearbox halves.

FOOTE GEARBOXES

Foote gearboxes are used on a variety of self-propelled walk-behind lawn mowers from several different mower manufacturers. Fig. F1 shows an external view of a Foote gearbox and shows serial number location (S). Fig. F2 shows an exploded view of a Foote 2820 series gearbox. Foote 2840 series gearboxes are similar.

MAINTENANCE. Foote gearboxes are a sealed unit and are packed with grease at the factory. Periodically check drive belt condition and make certain all bolts are tight.

DRIVE ADJUSTMENT. Drive adjustment cable of linkage must be adjusted so gearbox shift lever (L) fully disengages gearbox when mower control is in disengaged position.

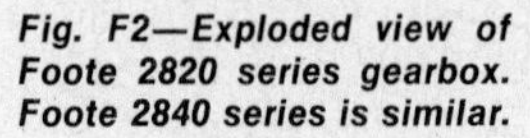

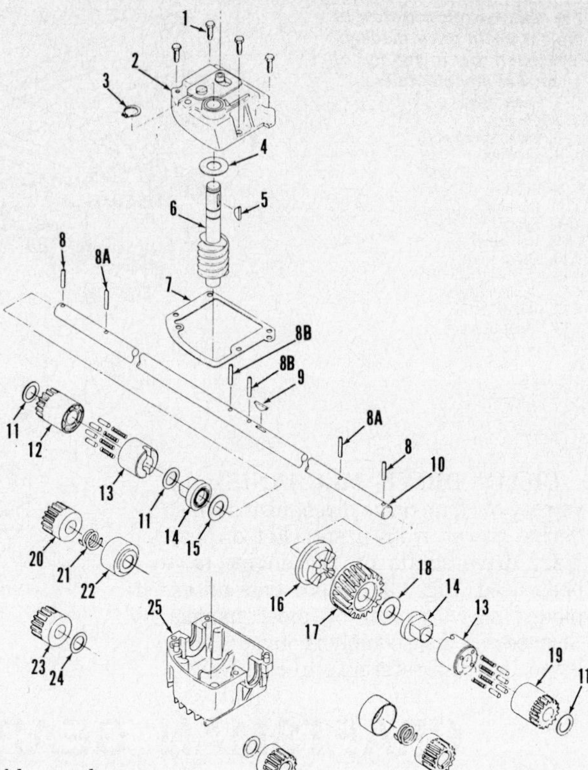

Fig. F2—Exploded view of Foote 2820 series gearbox. Foote 2840 series is similar.

1. Bolts
2. Upper housing
3. Snap ring
4. Washer (0.015, 0.020, 0.030 in.)
5. Key
6. Input shaft
7. Gasket
8. Roll pin

8A. Roll pin
8B. Roll pin

9. Key
10. Drive shaft
11. Washer
12. Ratchet gear (some models)
13. Ratchet assy.
14. Bushing
15. Washer
16. Sliding clutch gear
17. Driven gear
18. Washer
19. Ratchet gear (some models)
20. Ratchet gear (some models)
21. Spring (some models)
22. Dust cover
23. Ratchet gear (some models)
24. Washer

GEARBOX. To disassemble gearbox, remove snap ring, pulley and lower snap ring from input shaft (6–Fig. F2). Remove cap screw (C–Fig. F1), flat washer and shift lever (L). Remove the four cap screws from gearbox and carefully lift off upper housing (2–Fig. F2). Input shaft (6) may be removed from upper housing after removing key (5) and snap ring (3). Remove gasket (7). Carefully lift drive shaft (10) and gear assembly from lower housing (25). Remove roll pins (8B) and slide thrust washer (11), bushing (14), washer (15) and sliding clutch ring (16) from drive shaft. Remove key (9), driven gear (17), washer (18) and bushing (14). The plastic shift assembly is riveted to the upper case and cannot be disassembled.

Fig. F1—Foote gearbox serial numbers (S) are stamped into gearbox upper housing. Parts indicated are: L—Shift lever, I—Input shaft and C—Cap screw.

Reassemble by reversing disassembly procedure. Pack lower housing with a good quality multi-purpose grease before installing upper housing.

Reassemble by reversing disassembly procedure. Pack lower housing with a good quality multipurpose grease before installing upper housing.

HONDA SELF-DRIVE

SINGLE AND TWO-SPEED GEARBOX

MAINTENANCE. Both the single-speed and the two-speed gearboxes contain 130 cc (0.14 qt.) SAE 90 gear oil. Gearboxes are sealed. If leakage is apparent, correct leak, clean and refill gearbox.

CLUTCH. Forward mower motion is controlled by the engagement of clutch dogs on bevel gear and clutch dogs on driven clutch gear inside gearbox. Clutch control is located at mower handle and is connected to gearbox by cable.

Clutch cable should be adjusted yearly or at 300-hour intervals. To adjust, loosen jam nuts on control cable and adjust cable length to provide 5-10 mm (3/16-3/8 in.) drive clutch lever free play before clutch begins to engage. Tighten jam nuts.

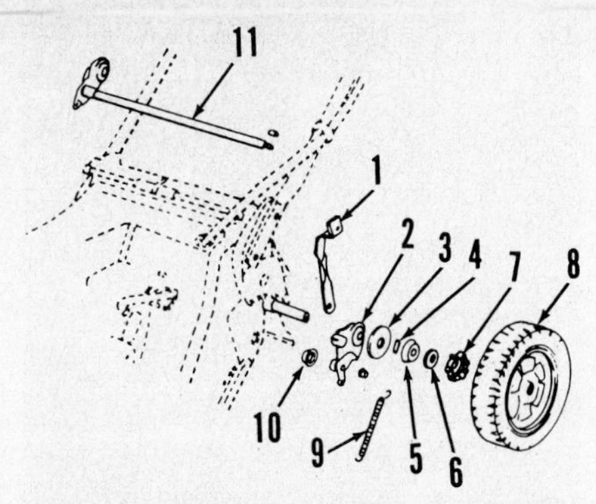

Fig. HN99—Exploded view of rear wheel and axle assembly on Honda self-propelled mowers.

1. Height adjustment lever
2. Adjustment arm
3. Cupped washer
4. Pin
5. Ratchet
6. Washer
7. Ratchet outer ring
8. Tire
9. Spring
10. Bushing
11. Cross-shaft

GEARBOX. To remove either the single-speed or the two-speed gearbox, remove wheels (8–Fig. HN99), washer (6), ratchet (5), pin (4) and washer (3). Note left and right side ratchets must not be interchanged. Remove height control spring (9) and cross-shaft (11). Remove gear selector knob (if so equipped). Disconnect clutch cable at gearbox. Slide rubber dust cover from pto coupler, remove spring and pin from coupler. Remove gearbox.

Refer to Fig. HN100 for exploded view of single-speed gearbox and to Fig. HN101 for exploded view of two-speed gearbox. Note that shift spring (9–Fig. HN101), spring retainer (10) and shifter (11) are installed inside bore of countershaft. Install shifter (11) and rotate 90° to lock in position.

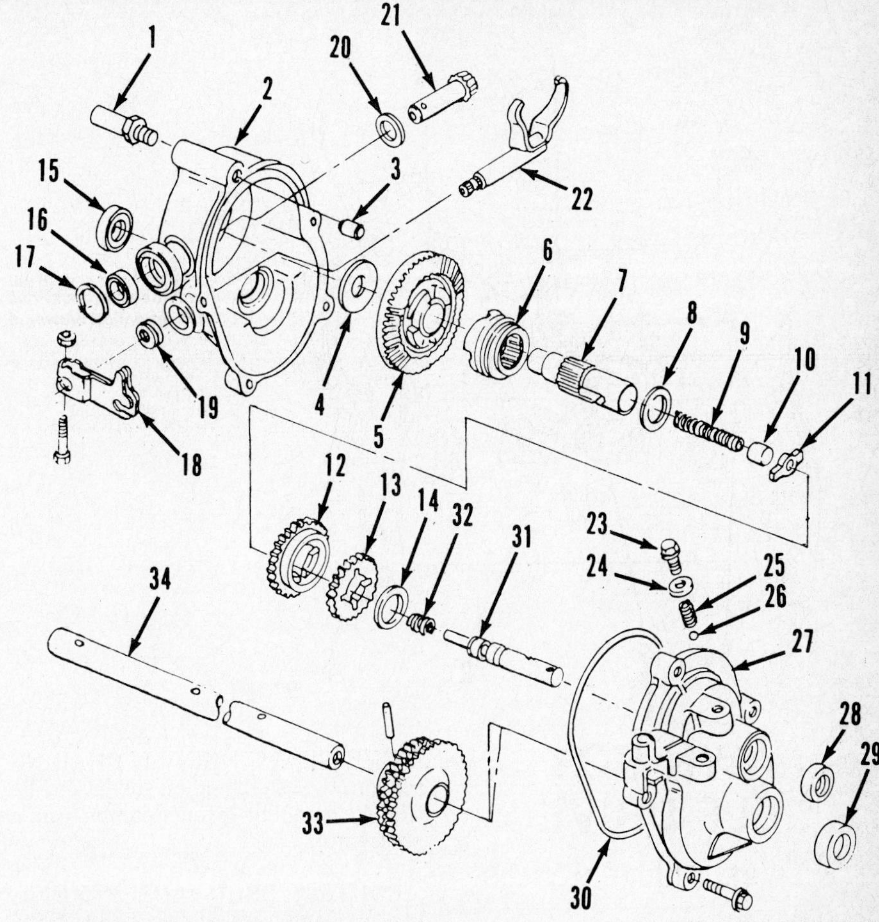

Fig. HN101—Exploded view of two-speed Honda gearbox used on some self-propelled models.

1. Guide stud
2. Case half
3. Dowel pin
4. Thrust washer
5. Bevel gear
6. Driven clutch gear
7. Countershaft
8. Thrust washer
9. Spring
10. Spring retainer
11. Shifter
12. High gear
13. Low gear
14. Thrust washer
15. Oil seal
16. Oil seal
17. Pin retaining spring
18. Clutch shift lever
19. Oil seal
20. Thrust washer
21. Pinion gear
22. Clutch shift fork
23. Bolt
24. Washer
25. Detent spring
26. Detent ball
27. Case half
28. Oil seal
29. Oil seal
30. Sealing ring
31. Shift sliding shaft
32. Shift spring
33. Axle gear
34. Axle

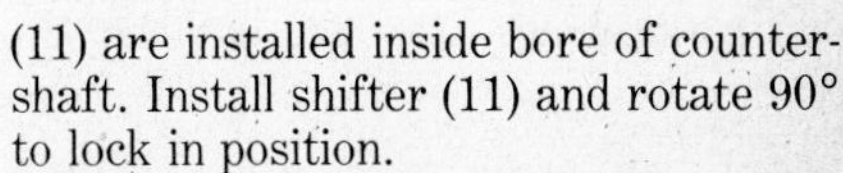

Reverse removal procedure for reinstallation. When installing clutch arm, punch mark (P–Fig. HN102) must be aligned with chisel mark (C) on end of clutch fork shaft. Washer (3–Fig. HN99) is installed with cupped side toward wheel. Adjust clutch control.

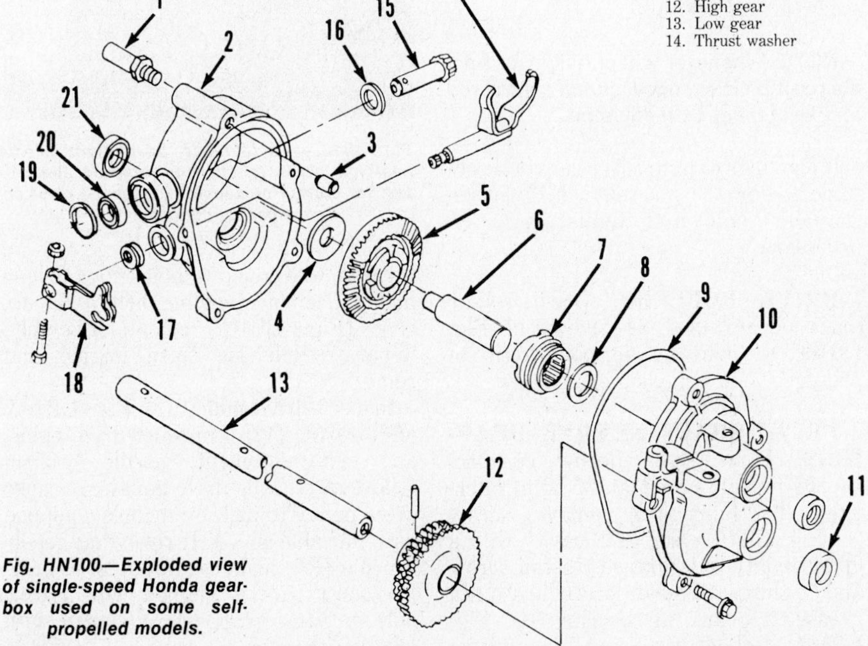

Fig. HN100—Exploded view of single-speed Honda gearbox used on some self-propelled models.

1. Guide stud
2. Case half
3. Dowel pin
4. Thrust washer
5. Bevel gear
6. Countershaft
7. Driven clutch gear
8. Thrust washer
9. Sealing ring
10. Case half
11. Oil seal
12. Axle gear
13. Axle
14. Clutch shift fork
15. Pinion gear
16. Thrust washer
17. Oil seal
18. Clutch shift lever
19. Pin retaining spring
20. Oil seal
21. Oil seal

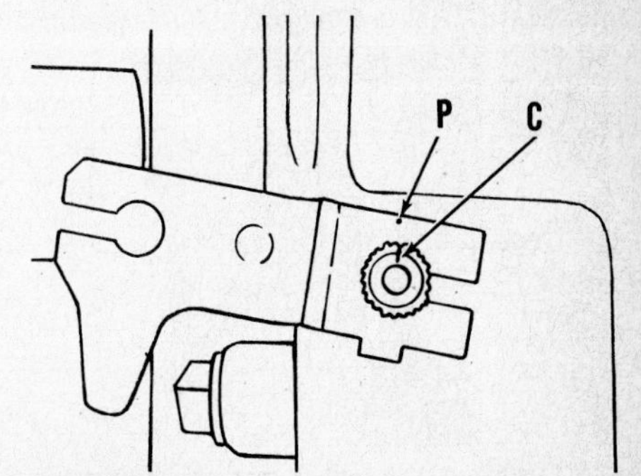

Fig. HN102—When installing clutch fork shift lever, punch mark (P) on shift lever must be aligned with chisel mark (C) on clutch fork shaft end.

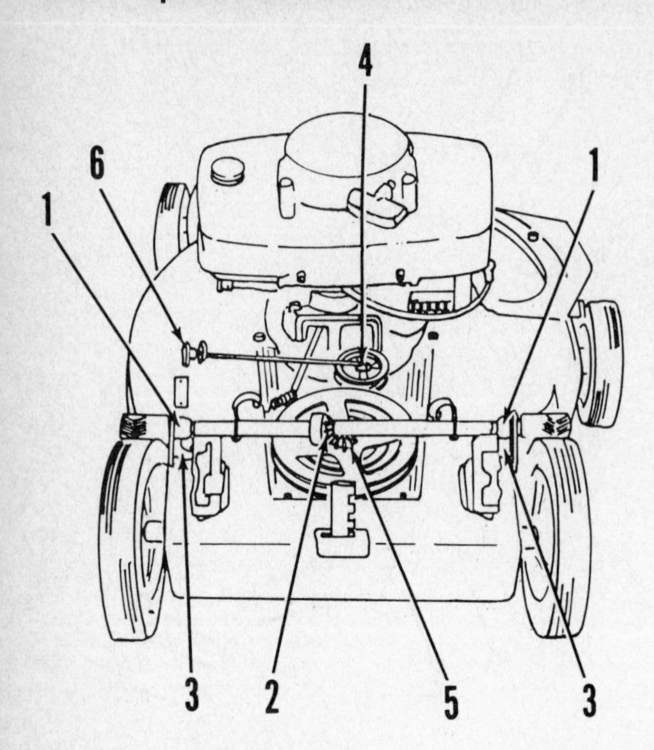

Fig. LB99—View showing maintenance point locations on self-propelled mower.
1. Drive shaft needle bearings
2. Gears
3. Clutch linkage
4. Idler pulley
5. Driven pulley bearing
6. Variable speed control

Fig. LB101—To lubricate drive shaft needle bearings, screw (1) must be removed. Refer to text for correct procedure. Grease (G) must appear at bearings.

LAWN BOY SELF-DRIVE

LOWER BELT DRIVE MODELS

MAINTENANCE. At 25-hour intervals, release spring clips and remove drive assembly cover. Light oil should be applied to idler belt pulley bearing (4-Fig. LB99), driven pulley bearing (5) and clutch linkage (3). A good quality axle bearing grease should be applied to drive gears (2). Lubricate drive shaft needle bearings (1) as outlined in DRIVE SHAFT NEEDLE BEARING section. Grease fitting on sub-base (Fig. LB103) should be given 4 pumps from a grease gun.

"V"
3/16 in.
(5 mm)

Fig. LB100—A 3/16 inch (5 mm) space must be maintained between tire and drive rollers. "V" cuts on drive roller point to tire when correctly installed.

CONTROL ROD ADJUSTMENT. Loosen clamp screw on control rod and position handle in "out-of-drive" position. With handle in this position, adjust control rod length to obtain 3/16 inch (5 mm) space between drive collar roller and tire (Fig. LB100). Tighten clamp screw in this position.

NOTE: Whenever wheel height or handle position is changed, clutch control rod and lever must be readjusted.

If 3/16 inch (5 mm) space cannot be obtained, remove right rear wheel, loosen shoulder bolt and adjust axle as necessary.

DRIVE ROLLERS. Drive rollers must be installed as shown in Fig. LB100. "V" cuts on roller point to tire to preserve their self-cleaning feature.

DRIVE SHAFT NEEDLE BEARINGS. Drive shaft needle bearings should be lubricated at 25-hour intervals. To lubricate, remove screw (1-Fig. LB101) and pack cavity with a good quality axle grease. Install screw and tighten. Repeat procedure until grease appears at bearing (G-Fig. LB101). Repeat procedure for opposite side.

DRIVE BELT. To renew drive belt, remove blade and drive assembly cover and right drive roller. Pull drive shaft to the left and remove belt from driven pulley. Remove engine, belt and sub-base (Fig. LB102) as an assembly. Separate sub-base from engine and remove belt.

Inspect drive pulley (3-Fig. LB102) and bearing (12). Improper drive pulley and belt alignment, needle bearing failure or sticking drive pulley can cause drive pulley to melt on models equipped with variable speed. If renewing needle bearing (12), make certain hole in bearing outer race is aligned with grease hole in base on models equipped with grease fitting on sub-base.

To reassemble models with variable speed, clean drive pulley and coat fixed pulley journal with Lubriplate. Install sliding pulley portion onto fixed pulley.

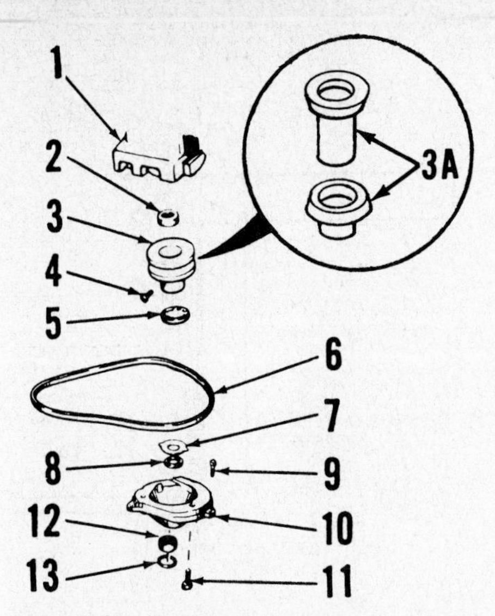

Fig. LB102—Exploded view of lower engine pulley and belt assembly on lower belt drive models.

1. Crankcase
2. Seal
3. Drive pulley
3A. Drive pulley (variable speed)
4. Set screw
5. Retainer
6. Belt
7. Washer
8. Felt seal
9. Sub-base to muffler plate screws
10. Sub-base
11. Sub-base to crankcase screws
12. Needle bearing
13. Seal

On all models, install pulley on crankshaft and clean sub-base and crankcase mating surfaces. Install drive belt on drive pulley, align sub-base with crankcase and install sub-base. Tighten retaining screws evenly. Lubricate sub-base needle bearing at grease fitting (Fig. LB103) if so equipped. Reinstall engine, drive belt and sub-base assembly. Install belt on driven pulley and idler pulley. Correct alignment of idler pulley may be obtained by placing pipe over end of idler pulley stud and slightly bending pulley bracket (Fig. LB104). Belt must ride in center on all pulleys. Pull drive shaft to the right and install drive roller. Install drive assembly cover and blade.

DRIVEN PULLEY. Two adjustments are required to ensure proper mesh of drive gears. A clearance of 0.018-0.019 inch (0.46-0.48 mm) must be maintained between bottom side of horizontal drive shaft and face of bevel gear above horizontal driven pulley. To

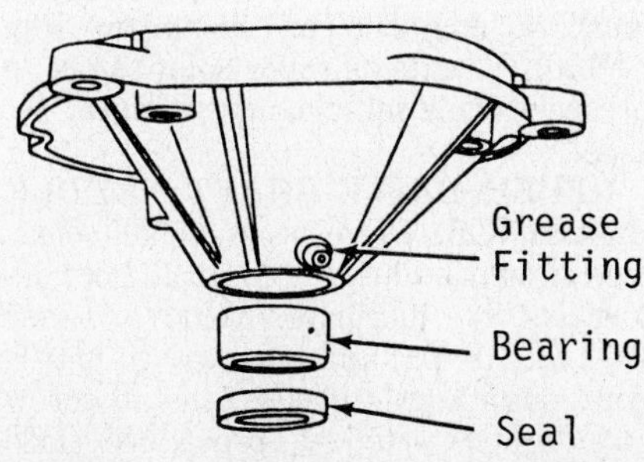

Fig. LB103—Later sub-base has a grease fitting installed to lubricate bearing. Refer to text.

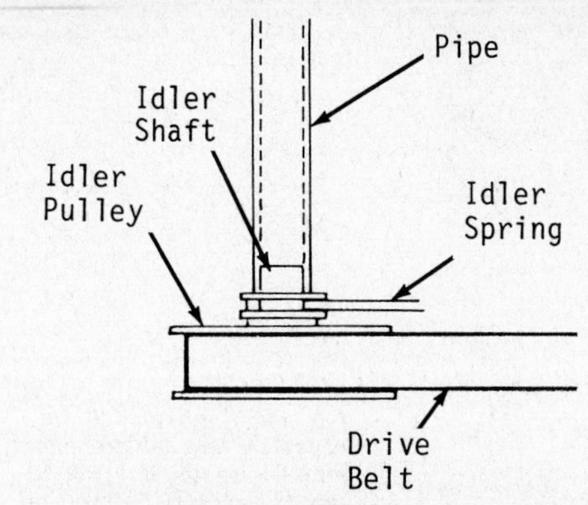

Fig. LB104—Belt must be centered in all pulleys. Idler shaft may be bent slightly using a pipe over the idler shaft if necessary.

obtain clearance, install or remove shims under pulley to raise or lower gear face in relation to shaft. Shims are available in a variety of thicknesses for this adjustment.

Adjust mesh of bevel gears as follows: Loosen set screw of left (from operator's position) drive roller so roller is loose on horizontal drive shaft, then move shaft as far right as possible and hold in this position. Insert a 0.010 inch (0.25 mm) feeler gage between nylon bushing-bearing assembly and inner side of right-hand roller. Adjust roller snug against gage blade and be sure set screw is tight and centered in place over flat sector of shaft. Bring left side drive roller snug against bearing, tighten set screw and gear clearance is set to provide required backlash.

UPPER BELT DRIVE MODELS

MAINTENANCE. At 25-hour intervals, light oil should be applied to clutch linkage and drive shaft needle bearings should be lubricated as outlined in DRIVE SHAFT NEEDLE BEARINGS section.

CONTROL ROD ADJUSTMENT. Place handle in "out-of-drive" position. Loosen jam nuts on control rod and adjust turnbuckle until there is a 3/16 inch (5 mm) space between drive rollers and tires (Fig. LB100). Tighten jam nuts against turnbuckle.

NOTE: Whenever wheel height or handle position is changed, clutch control rod and lever must be readjusted.

DRIVE ROLLERS. Proper drive roller engagement is obtained by loosening top nut and bolt on right side of link (B-Fig. LB101) and moving drive shaft in or out until rollers are equal distance from tires. Tighten nut and bolt.

Drive rollers must be installed as shown in Fig. LB100. "V" cuts on roller point to tire to preserve their self-cleaning feature.

DRIVE SHAFT NEEDLE BEARINGS. Drive shaft needle bearings should be lubricated at 25-hour intervals. To lubricate, remove screw (1–Fig. LB101) and pack cavity with a good quality axle grease. Install screw and tighten. Repeat procedure until grease appears at bearing (G). Repeat procedure for opposite side.

Fig. LB106—Exploded view of D series engines typical of self-propelled models. Refer to Fig. LB107 for view of mower base and remainder of drive.

1. Drive pulley
2. Idler bracket
3. Spring retainer bracket
4. Lever
5. Spring
6. Shim
7. Grommet
8. Lockwasher
9. Idler pulleys
10. Belt cover

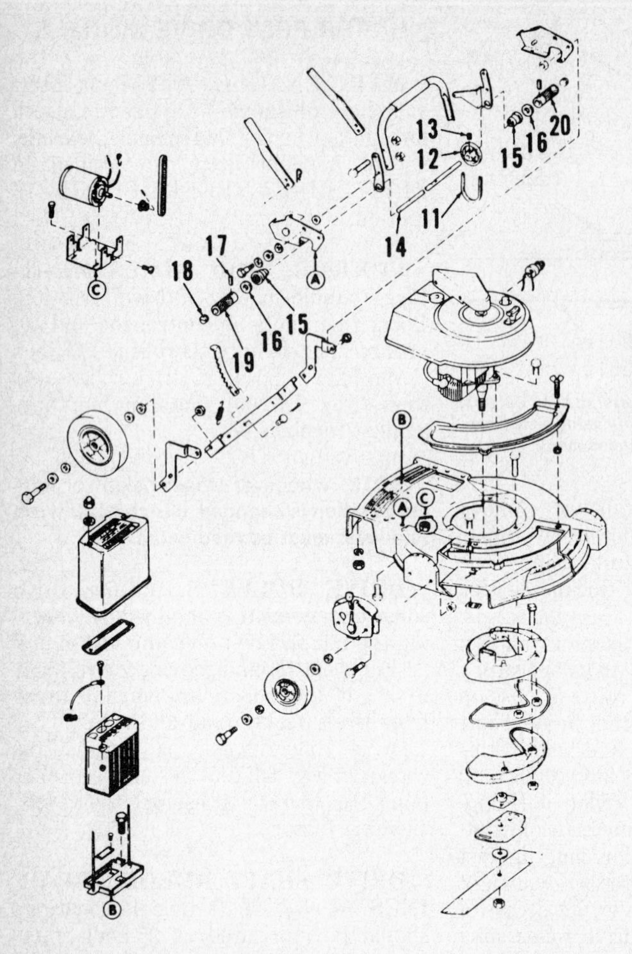

Fig. LB107—Exploded view of typical late self-propelled mower. Handle bracket attaches at (A), battery attaches at (B) and electric starter attaches at (C).

11. Drive belt
12. Driven pulley
13. Set screw
14. Drive shaft
15. Sleeve & bearing
16. Washer
17. Roll pin
18. Grease plug
19. Drive roller
20. Drive roller

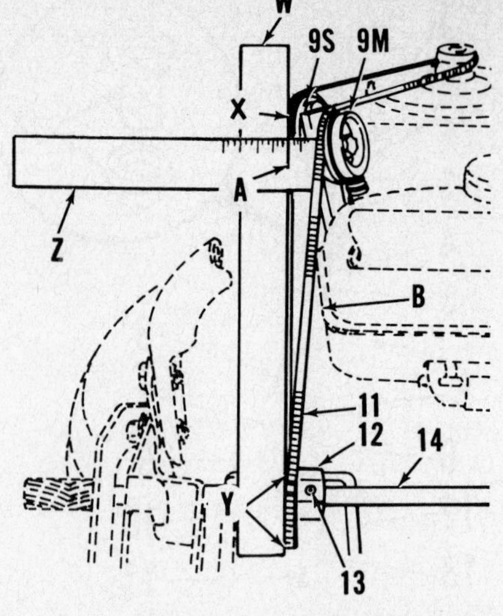

Fig. LB109—Self-propelled drive belt should be aligned as described in text.

9M. Movable idler pulley
9S. Stationary idler pulley
11. Drive belt
12. Driven pulley
13. Set screw
14. Drive shaft
W. Straightedge
X. Inside edge of pulley
Z. Ruler

DRIVE BELT. To renew drive belt, remove cover (10–Fig. LB106), drive roller (20–Fig. LB107), washer (16) and sleeve (15) from left side. Withdraw belt from opening in handle bracket where sleeve (15) was installed.

Due to changes in design of idler bracket (2–Fig. LB106) there are two belt alignment procedures. Refer to Fig. LB108 to determine if unit being serviced has old or new style bracket.

To correctly align the drive belt with old style bracket installed, position a straightedge against edge of driven pulley as shown at (Y–Fig. LB109). Locate driven pulley (12) on drive shaft (14) so straightedge is aligned with inside of pulley (9S) as shown at (X). Distance between straightedge and belt at point where belt just enters movable pulley (9M) should be 15/32 inch (12 mm) as shown at (A). If clearance (A) is incorrect, move pulley (12) either way as necessary. Minimum clearance between belt and fuel tank (B) should be 3/32 inch (2 mm) when mower is in drive position. If belt is too close to tank, loosen mounting bracket and move tank.

If new style bracket is used, procedure steps are as before, except dimension (A–Fig. LB109) becomes ⅛-5/32 inch (3-4 mm).

It may not be possible to adjust drive belt correctly due to excessive end play in drive shaft (14–Fig. LB109). If this condition appears, particularly in 1968-1969 models, remove wave washers fitted to shoulder bolt which holds bearing carrier bracket to handle bracket and substitute plain washers for use as spacers until 0.012-0.015 inch (0.30-0.38 mm) end play is obtained. If this does not correct looseness, install new type shoulder bolt, part number 606672.

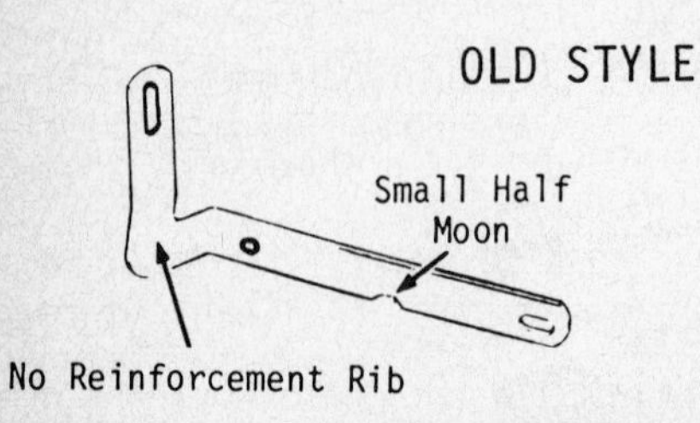

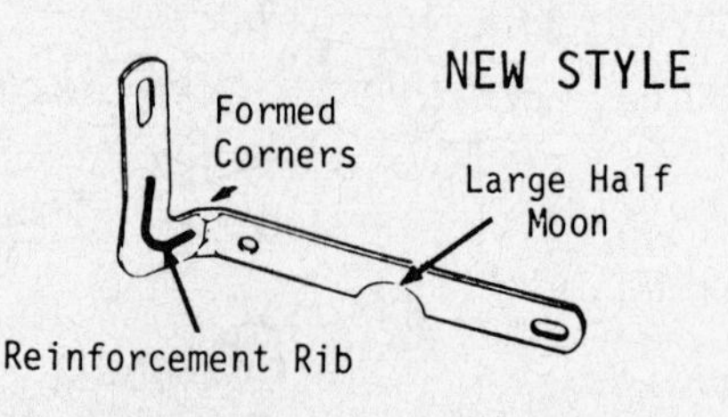

Fig. LB108—Comparison views of old and new style idler brackets used on self-propelled models with upper belt drive. Refer to text for details.

GEARBOX DRIVE MODELS

MAINTENANCE. Lubricate drive wheels and clutch linkage at 25-hour intervals. Make certain all bushings are tight and linkage is in good condition.

CLUTCH ADJUSTMENT. Loosen clamp screw on control rod and position handle in "out-of-drive" position. With handle in this position, adjust control rod length to obtain 3/16 inch (5 mm) space between drive roller and tire (Fig. LB100). Tighten clamp screw in this position.

NOTE: Whenever wheel height or handle position is changed, clutch control rod and lever must be readjusted.

If 3/16 inch (5 mm) space cannot be obtained, remove right rear wheel, loosen shoulder bolt and adjust axle as necessary.

DRIVE ROLLERS. Drive rollers must be installed as shown in Fig. LB100. "V" cuts on roller point to tire to preserve their self-cleaning feature.

UPPER DRIVE SHAFT NEEDLE BEARINGS. Drive shaft needle bearings should be lubricated at 25-hour intervals. To lubricate, remove screw (1–Fig. LB101) and pack cavity with a good quality axle grease. Install screw and tighten. Repeat procedure until grease appears at bearing (G). Repeat procedure for opposite side.

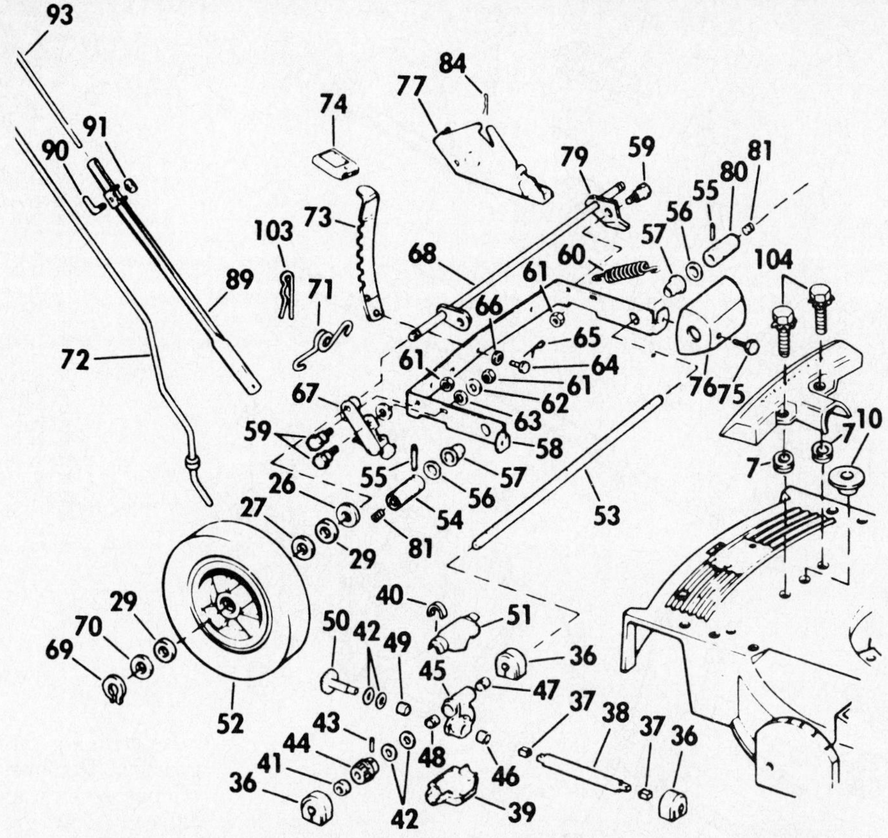

Fig. LB110—Exploded view of self-propelled drive used on later model Lawn Boy mowers.

7. Grommet
26. Washer
27. Felt washer
29. Spacer
36. Seal
37. Bushing
38. Drive tube
39. Bottom gear cover
40. Clamp
41. Felt washer
42. Thrust washer
43. Roll pin
44. Driven gear
45. Bracket & bearing
46. Sealed bearing
47. Sealed bearing
48. Needle bearing
49. Needle bearing
50. Drive gear
51. Gear top cover
52. Rear wheel
53. Drive shaft
54. Drive roller
55. Spirol pin
56. Washer
57. Sleeve & bearing
58. Bracket
59. Shoulder bolt
60. Spring
61. Nut
62. Washer
63. Washer
64. Pin
65. Cotter pin
66. Washer
67. Clutch
68. Axle
69. Retainer
70. Washer
71. Spring
72. Grass bag rod
73. Lever
74. Knob
75. Screw
76. Roller guard
77. Handle bracket
79. Bracket
80. Drive roller
81. Plug
84. Clip
89. Control lever
90. Screw
91. Nut
93. Control rod
103. Hairpin
104. Screw

GEARBOX AND DRIVE SHAFTS. Refer to Fig. LB110 for an exploded view of gearbox assembly.

To adjust lower drive shaft end play, measure between drive roller (54) and washer (56) for 1/16 inch (1.6 mm) end play at each end. If play exceeds 1/16 inch (1.6 mm), add an additional washer (56) on right side.

End play of drive shaft tube (38) should measure 0.035-0.135 inch (0.89-3.43 mm). If end play of tube measures 0.136-0.235 inch (3.45-5.97 mm), add one "O" ring, part number 303067, between nylon bushing (37) and square-end of driven shaft (50). If measurement exceeds 0.236 inch (5.99 mm), add an "O" ring at each end of drive tube. If nylon bushings (37) are worn to such extent that tube is loose in rotation, renew bushings.

If bronze driven gear is to be separated from worm gear of engine crankshaft, it must be marked for reinstallation exactly as removed. If not, a new gear must be installed. If bevel gear set in gearbox is worn or damaged, renew both gears. Needle bearings (46, 47, 48 and 49) are individually renewable. These bearings must not be driven in or out. Use a press.

SNAPPER DISC DRIVE

MAINTENANCE. Remove gearbox check plug (6 – Fig. S10) at 25-hour intervals to make certain grease is visible on input gear. If grease is not visible, add a new small amount of Snapper 00 grease. Amount added must not exceed 2 oz. (59 mL). Reinstall plug (6). Lightly oil height adjusting levers, wheel axles and shift levers.

CLUTCH ADJUSTMENT. With clutch handle released, there should be 1/16-1/8 inch (2-3 mm) clearance between clutch spring hook and inside eye of pull rod. To reduce clearance, turn outer spring clockwise; to increase clearance, turn outer spring counterclockwise.

DRIVEN DISC ADJUSTMENT. Disconnect and properly ground spark plug cable. Place ground speed control lever into "HIGH SPEED" position. If driven disc is properly adjusted, it will be 1/8-1/4 inch (3.6 mm) from outer edge of drive disc. Nut on connector may be loosened and disc correctly positioned as necessary. Connect spark plug lead.

ENGINE BELT. Disconnect and properly ground spark plug lead and remove mower blade. Cut and remove old belt. Disconnect wheel drive spring from driven disc and move disc assembly to one side. Install belt below drive disc pulley but do not place it in pulley groove. Insert belt through deck opening and place belt in groove of engine pulley. Work belt onto drive disc pulley.

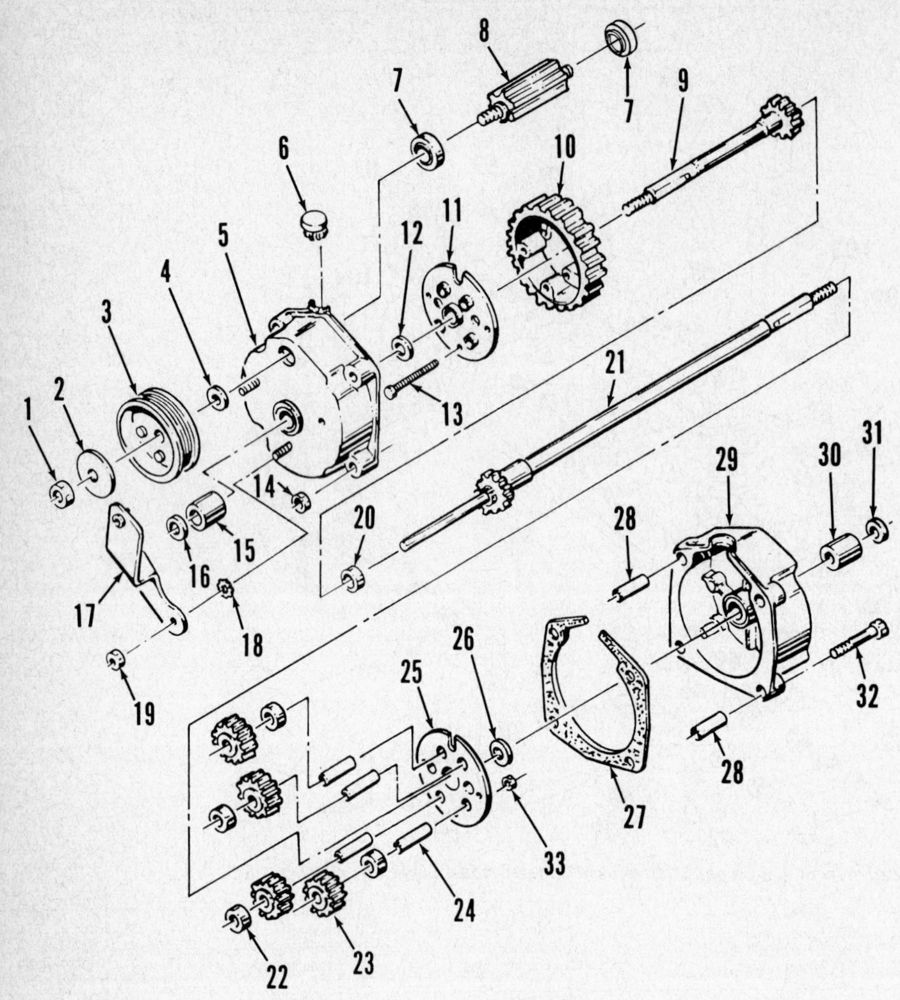

Fig. S10—Exploded view of gearbox used on some Snapper self-propelled mowers with disc drive.

1. Nut
2. Washer
3. Pulley
4. Seal
5. Housing
6. Plug
7. Bearings
8. Input shaft
9. Short axle
10. Bull gear
11. Plate
12. Thrust washer
13. Cap screws
14. Nut
15. Bearing
16. Seal
17. Bracket
18. Washer
19. Locknut
20. Spacer
21. Long axle
22. Pinion spacers
23. Pinion gears
24. Pinion shafts
25. Plate
26. Thrust washer
27. Gasket
28. Roll pins
29. Housing
30. Bearing
31. Seal
32. Cap screws

Install blade and tighten cap screw to 15-30 ft.-lbs. (20-40 N•m). Reconnect wheel drive spring and spark plug lead.

Bottom edge of engine pulley should be 1½ inch (38 mm) from end of crankshaft to ensure correct belt alignment. Loosen set screw to reposition pulley as necessary.

CLUTCH BELT. To renew clutch belt, remove clip from transfer rod and disconnect rod from speed control lever. Disconnect drive spring from driven disc and remove driven disc from hex shaft. Note location of old belt before removing. Install new belt over hex shaft and around hex shaft pulley. Twist belt sideways and pull it upward between the differential bracket and input pulley on gearbox. Make certain belt is above hex shaft pulley belt guide and install driven disc assembly and wheel drive spring. Connect transfer rod to speed control lever.

GEARBOX. To remove gearbox, remove cap screw retaining differential link bracket and remove belt from input pulley. Remove rear wheels and the wheel arm assembly from left side. Remove rear guard and slide gearbox/axle assembly left until axle clears differential link. Remove gearbox from mower.

Fig. TO18—External view of Toro gearbox used on some self-propelled Toro lawn mowers.

C. Clutch lever
D. Drive shaft
G. Gearbox housings
I. Input shaft
L. Pulley
P. Plug
S. Clutch fork spring

Refer to Fig. S10 for exploded view of gearbox. Remove gearbox housing retaining screws and separate gearbox halves. Allow grease to drain into a suitable container. Remove input shaft and bearing assembly from housing. On models with two-piece cast bull gear, place in vise and drive roll pins out of gear halves. Separate gear halves. On late gearbox, remove the four cap screws and disassemble bull gear. Continue disassembly as necessary by referring to Fig. S10.

Reassemble by reversing disassembly procedure. Tighten gearbox housing cap screws to 17 ft.-lbs. (23 N•m). Lubricate gearbox by installing 4 oz. (118 mL) of 00 Snapper grease through plug (6).

TORO GEARBOX

MAINTENANCE. Gearbox plug (P-Fig. TO18) should be removed at 25-hour intervals to check gear lubricant level. Gear lubricant should be to lower edge of plug. Add SAE 90 gear grease as necessary. Lightly oil all bushings and pivot points.

CLUTCH. Clutch cable should be adjusted so gears completely disengage. Clutch lever (C) movement between beginning engagement and disengagement position is very slight.

DRIVE BELT. To renew drive belt, disconnect and properly ground spark plug. Remove mower blade. Remove left and right wheels, height adjustment

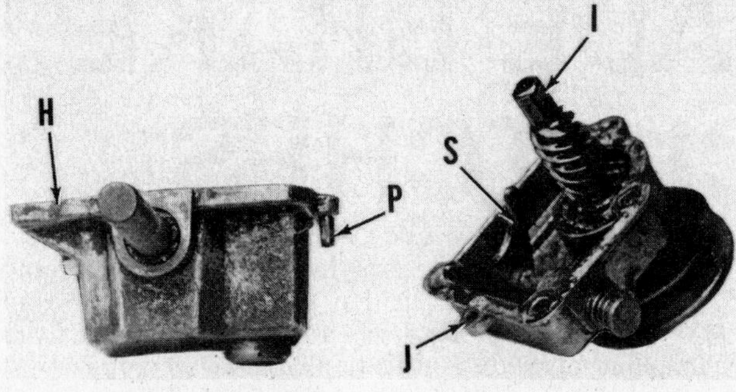

Fig. TO19—View of upper (H) and lower (J) gearbox housings after separation. Locating pins (P) must be driven out of upper housing before housings are separated.

Fig. TO20-View of component parts lined up in their respective positions.

1. Sliding clutch gear hub
2. Drive shaft
3. Roll pin
4. Sliding clutch gear
5. Driven gear
6. Thrust washer

mechanism and disconnect clutch engagement cable. Remove belt from loosened gearbox assembly and crankshaft pulley. Install new belt and reverse removal procedure.

GEARBOX. To disassemble gearbox, remove pulley (L–Fig. TO18) from input shaft (I). Drive the two locating pins (P–Fig. TO19) down out of upper gearbox housing. Remove the four cap screws and separate housing halves. Drain gear lubricant into a suitable container. Separate sliding clutch gear (4–Fig. TO20) from driven gear (5) and drive roll pin (3) from sliding gear hub (1) and shaft (2). Pull shaft (2) out of housing, hub, sliding clutch gear, driven gear and thrust water. Remove snap rings from input shaft and remove worm gear/input shaft assembly.

When reassembling, make certain shift fork (S–Fig. TO19) engages groove in sliding clutch gear.

BLADE BRAKE CLUTCHES

OPERATION

Blade clutches, blade brake clutches and flywheel engine brakes are safety features designed to stop blade rotation.

Blade clutches allow the operator to stop blade rotation without stopping mower engine. Clutches are usually some form of centrifugal clutch which rely on slow engine speed to stop blade rotation. While blade clutch stops engine power from turning blade, they do not have a positive stopping feature and blade may freewheel.

Blade brake clutches are designed to allow the operator to stop blade rotation quickly without stopping the engine. A control handle or bail in conjunction with a control cable is used to control the engagement and disengagement of mower blade. Blade should stop within three seconds after control handle is placed in disengaged position to meet government safety standards and offer maximum operator protection. Blade brake clutch should be checked periodically for correct engagement and disengagement of mower blade.

CAUTION: When checking blade brake clutch operation use extreme caution. Mower blade may be turning when you assume it has stopped.

Flywheel engine brakes are incorporated with an engine safety switch which sorts out ignition to stop engine power and flywheel brake stops engine (blade) rotation within three seconds. Refer to appropriate engine service section for flywheel brake service.

Fig. AR10—Exploded view of blade brake clutch used on some Ariens walking lawn mowers.

1. Engine crankshaft
2. Pulley
3. Idler pulley
4. Hex stud
5. Spring
6. Idler bracket
7. Ramp
8. Control rod
9. Brake actuator
10. Key
11. Blade brake flywheel
12. Leaf spring & bearing

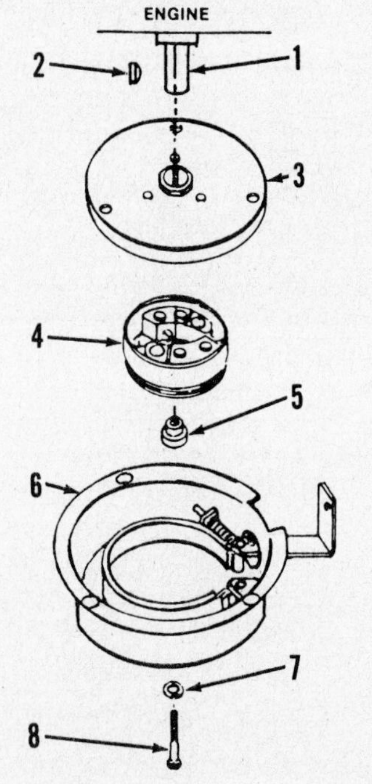

Fig. C1—Exploded view of a Comet blade brake clutch.

1. Crankshaft
2. Woodruff key
3. Flywheel
4. Clutch assy.
5. Clutch washer
6. Bowl & brake band assy.
7. Flat washer
8. Cap screw

ARIENS

ARIENS COMPANY
655 West Ryan St.
Brillion, WI 54110

Shown in Fig. AR10 is an exploded view of blade brake clutch used on some Ariens lawn mowers. Parts are renewable as separate components except the sealed bearing located in blade brake clutch leaf spring (12). No special tools are required for servicing, but good mechanical skills should be exercised during servicing to ensure safe and correct operation.

Control cable bracket should be adjusted so engine is stopped and blade brake clutch actuated when control ball is moved 3/8-5/8 inch (10-16 mm) from handle bar.

COMET

COMET INDUSTRIES
A DIVISION OF HOFFCO INC.
358 Northwest F
Richmond, Indiana 47374

Shown in Figure C1 is an exploded view of a blade brake clutch. Parts are renewable as three separate subassemblies; flywheel, rotor or clutch and bowl and brake band. Sealed bearing in clutch assembly is renewable only with assembly. No special tools are required for servicing, but good mechanical skills should be exercised during servicing to ensure safe and correct operation.

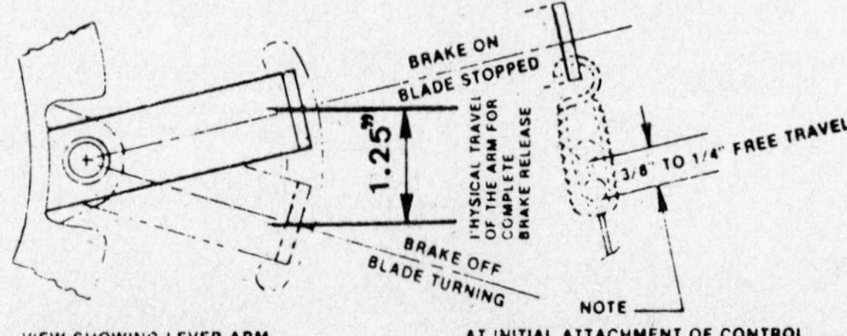

Fig. C2—View showing travel length of actuating lever and correct initial adjustment of control cable in spring.

NOTE: Cap screws with thread length longer than one inch (25.4 mm) should not be used to attach mower blade to clutch assembly.

Check actuating lever for correct travel length and control cable for correct initial adjustment as shown in Fig. C2. There should be 1¼ inch (32 mm) travel in actuating lever when control handle is moved from brake on to brake off position. If lever does not travel the full length, then check lever for freedom of movement, correct control cable adjustment and tension of control spring. Correct initial adjustment of control cable in spring is ¼-3/8 inch (6-10 mm) free movement with control handle placed in brake applied position.

HONDA

AMERICAN HONDA MOTOR CO., INC.
100 W. Alondra Blvd.
Gardena, California 90247

Shown in Fig. HN200 is an exploded view of Honda blade brake clutch. Parts are renewable separately. No special tools are required for servicing, but good mechanical skills should be exercised during servicing to ensure safe and correct operation.

Control lever cable should be adjusted to provide 5-10 mm (3/16-1/8 in.) free play between control lever and handle bar. Adjust cable length by loosening cable locknuts and turning adjuster in or out as necessary. Tighten locknuts securely.

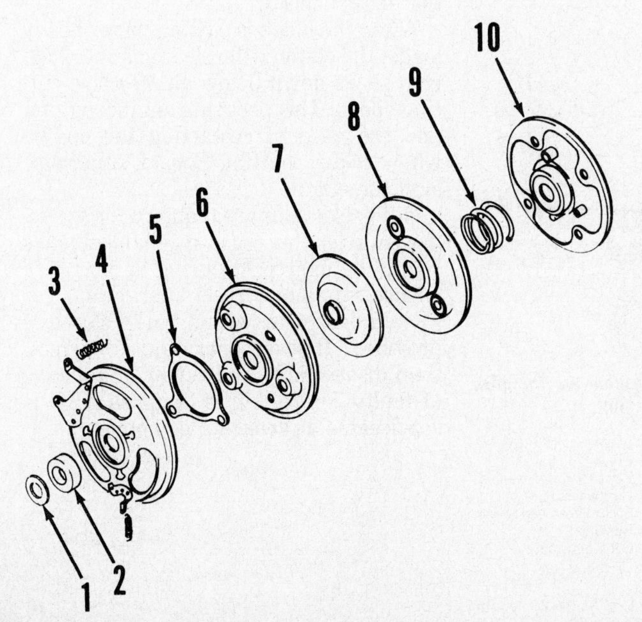

Fig. HN200—Exploded view of Honda blade brake clutch used on some Honda walk-behind lawn mowers.

1. Thrust washer
2. Bearing
3. Spring
4. Ramp plate
5. Ball carrier
6. Brake disc
7. Drive plate
8. Clutch plate
9. Spring
10. Blade plate

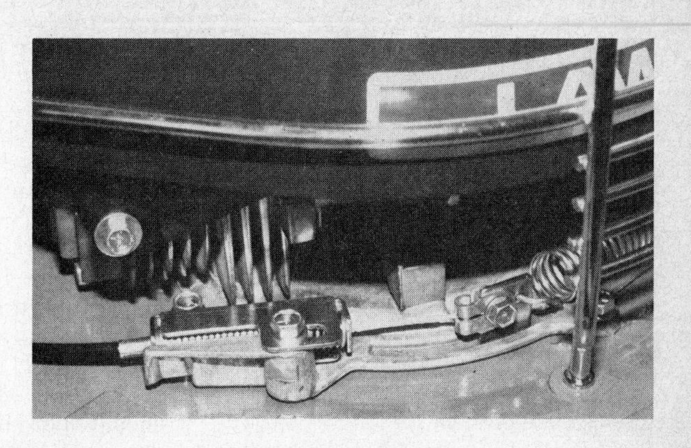

Fig. LB200—Typical view of the blade clutch control components used on models so equipped.

LAWN BOY

OUTBOARD MARINE CORPORATION
Post Office Box 82409
Lincoln, Nebraska 68501

Some Lawn Boy mowers are equipped with blade brake clutches with a single, spring-operated disc used to provide blade clutching and braking. Clutch is operated by a cable attached to a control handle. For a typical view of the blade clutch control components refer to Fig. LB200. Adjust control cable by loosening mounting bracket cap screws and sliding bracket in slots until proper engagement and disengagement is obtained.

Blade clutch should not be disassembled. Unit must be renewed as a complete assembly as no service parts are available.

MTD

MTD PRODUCTS, INC.
P.O. Box 36900
Cleveland, Ohio 44136

Figure M1 shows an exploded view of a typical "Action Guard" blade brake clutch. Unit is serviceable as separate components, with the exception of the brake pads (10). Brake pads are renewable only with brake/clutch housing assembly (14). Operation of clutch assembly is controlled by cable (3).

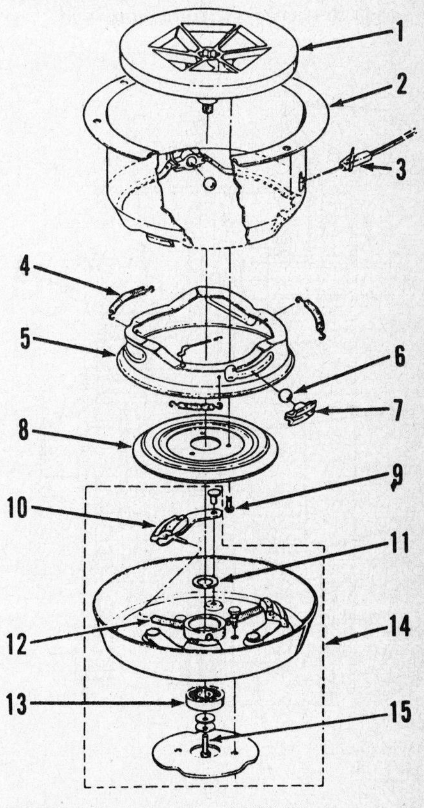

Fig. M1—Exploded view of MTD "Action Guard" blade brake clutch.

1. Fan adapter
2. Clutch housing
3. Clutch control cable
4. Extension spring
5. Brake cup cone
6. Steel ball
7. Ball block
8. Clutching cone
9. Self-tapping screw
10. Brake pad assy.
11. Flat washer
12. Compression spring
13. Ball bearing
14. Brake/clutch housing assy.
15. Cap screw

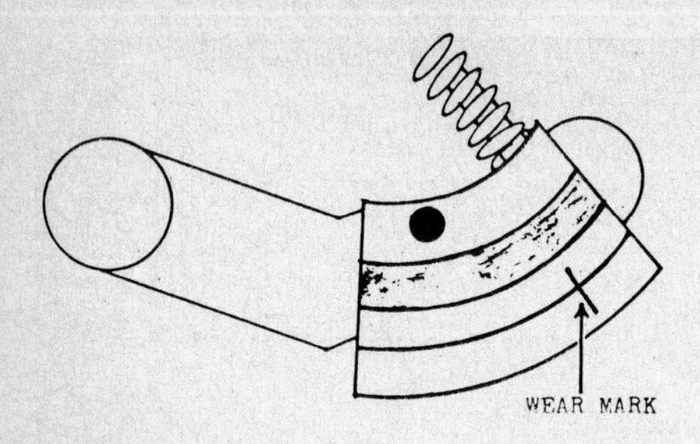

Fig. M2—Brake pads must be renewed if wear marks are not visible. Pads are only serviced with brake/clutch housing assembly.

Periodically inspect cable to be sure it is not frayed or damaged in any way.

No special tools are required for servicing. When servicing blade clutch, always disconnect spark plug wire and ground it against engine block. Drain fuel tank and engine crankcase if mower is tipped on its side.

An access slot located in clutch housing (2) where control cable enters is provided for assistance in removal of cap screw (15) or three self-tapping screws (9) securing clutch cone (8). Insert a straight blade screwdriver through slot, then using the proper size wrench turn cap screw until screwdriver blade catches in groove of clutch. Screwdriver blade will prevent clutch from turning and allow cap screw to be removed.

To withdraw brake/clutch housing assembly, first reinstall center bolt (15) finger tight. Place Belleville washers on head of center bolt, then position blade on top of washers and secure with nuts. Using a suitable tool tighten nuts down evenly until brake/clutch housing assembly is pulled loose.

Brake/clutch housing assembly must be renewed if brake pad wear marks are not visible as shown in Fig. M2. Bearing (13–Fig. M1) in housing should turn freely and compression springs (12) should not be broken, stretched or damaged in any way. Renew all parts as needed.

If excessive vibration of unit is noticed, inspect for a bent crankshaft or a damaged fan adapter.

OGURA

KANEMATSU-GOSHO INC.
P.O. Box 392
S. Plainfield, New Jersey 07080

Figure O1 shows an Ogura blade brake clutch typical of all models. To engage blade, lever (1) is moved by control cable to engaged position. Lever causes actuator cam (6) to move brake plate (5) axially in relation to engine crankshaft. Clutch spring (3) pushes friction disc (2) against clutch disc (7) which is pinned to engine crankshaft. Clutch disc (7), friction disc (2), blade carrier (4) and blade now rotate at engine speed.

To stop blade, handle control is released permitting return spring to pull lever (1) to released position. The actuator cam (6) reverses rotation pushing brake plate (5) against friction disc (2). This separates friction disc and clutch disc, thus stopping power flow to blade.

There is no adjustment for wear; clutch is completely self-adjusting. If lever operation is correct, then disassembly and repair or renewal of faulty component is required.

Unit is serviceable as separate components or as sub-assemblies. No special tools are required for servicing, but good mechanical skills should be exercised during servicing to ensure safe and correct operation.

Fig. O1—View showing a typical Ogura blade brake clutch.

1. Lever
2. Friction disc
3. Clutch spring
4. Blade carrier
5. Brake plate
6. Actuator cam
7. Clutch disc & flywheel

WORTHINGTON

WORTHINGTON INDUSTRIES
MOWSAFE PRODUCTS
6500 Huntley Road
Worthington, Ohio 43085

Shown in Fig. W1 is an exploded view of a typical blade clutch. Blade clutch uses two-piece articulated (jointed) flyweights. The jointed design allows clutch to be engaged when engine is running at low rpm. The engagement of clutch is dependent upon blade speed, not engine speed.

When the blade is cutting under heavy loads, the clutch will release as the blade rpm slows down below an effective cutting speed. This prevents engine lugging and the need of restarting the engine when heavy cutting would otherwise stall the engine.

No special tools are required for servicing clutches, but good mechanical skills should be exercised during servicing to ensure safe and correct operation.

No adjustment within blade clutch is provided. If lever operation is correct, then disassembly and repair or renewal of faulty component is required. Unit is serviceable as separate components.

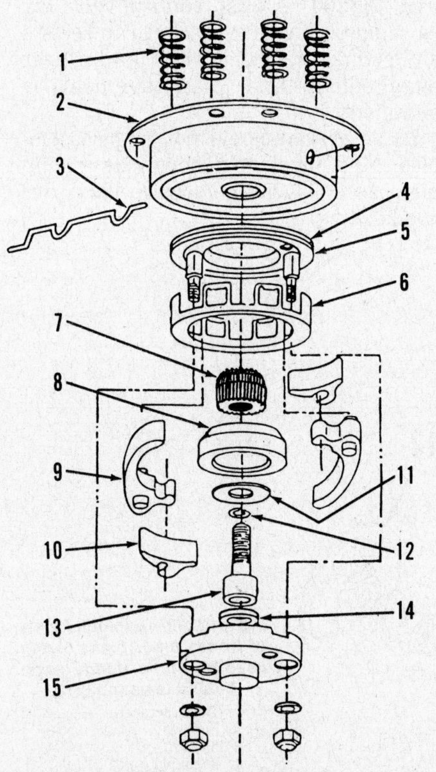

Fig. W1—Exploded view of typical Worthington blade brake clutch.

1. Spring
2. Housing
3. Brake lever
4. Brake disc assy.
5. Brake shoe
6. Spring pressure ring
7. Clutch cone insert
8. Clutch cone
9. Articulated flyweight
10. Flyweight toe
11. Friction clutch washer
12. Lockwasher
13. Socket head cap screw
14. Bushing
15. Cone housing

NOTES

SAFETY

The operator is directly responsible for the safe and practical use of the equipment; however, it is important that the operator know how to use and control the equipment. Manufacturers provide detailed instructions for operating their equipment. New operators should be taught correct procedures, then tested to determine their level of understanding. Observe equipment use to assure that operators are continuing to follow safe practices.

High speed movement, sharp cutting surfaces and very strong forces characterize the normal actions of many types of lawn and garden power equipment. Cautious operation by personnel who know and exercise approved, safe procedures can minimize the risk of injury. Safe operation of damaged equipment may not be possible. It is important that the equipment be maintained correctly in the interest of safety as well as ease of operation.

Always observe the following check list of STANDARD SAFETY PRACTICES.

1. Do not operate equipment if under the influence of alcohol or drugs.
2. Do not allow irresponsible people or people not trained in proper use (regardless of age) to operate powered equipment.
3. Know how to immediately disengage the power and stop the engine before starting engine.
4. Be especially cautious when operating equipment in area where children usually play. A child may suddenly enter the area causing a hazard.
5. Keep people and animals away from the working area.
6. Use only approved grip locations and always control unit as intended by the manufacturer.
7. Make sure that all guards are installed and all safety devices operate properly. NEVER by-pass any safety device provided by the manufacturer.
8. Know how to correctly operate all controls.
9. Wear appropriate close fitting clothing, shoes and eye protection. Consider appropriate breathing filter for operator especially in dry or dusty conditions.
10. Do not smoke while servicing gasoline or diesel powered equipment in any way. Smoking should also be discouraged while operating equipment.
11. Be careful of fire hazards.
12. Don't run any gasoline or diesel engine in an enclosed area.
13. NEVER permit any part of your body or clothing near a cutting blade or any other moving part. If clothing wraps around a rotating part, your body can be pulled in and injured. Always keep all protective shields in place.
14. Operate equipment ONLY with good daylight or artificial light. Be sure of what you are doing.
15. NEVER compromise safety by operating equipment in a way that you or the equipment could accidentally move and cause injury. This includes pulling instead of pushing, mowing extreme grades and hanging from a tree by one hand while operating a chain saw with the other.
16. Stop engine and properly ground the spark plug wire before cleaning, inspecting, adjusting or repairing anything driven by the engine.
17. Always stop the engine when equipment is left unattended. Allow engine to cool completely bevore covering or storing in any enclosure.
18. NEVER store gasoline or diesel powered equipment or extra fuel in an area where fuel or fuel vapor may accumulate or may travel to an open flame or spark.
19. Check parts regularly for looseness, wear, cracks or other damage and make sure equipment is properly adjusted before starting engine.
20. Practice routine maintenance.

Operators should be alert, constantly monitoring equipment performance. Any change could indicate equipment problems, but prompt action may prevent injury or damage.

All mowers are designed to cut and they operate with absolutely no regard for what is being cut. The cutting blade will try just as hard to cut fingers, hands and feet, as it does roots, grass, weeds and children's toys. Observing the following warnings will reduce the possibility of injury.

Observe the following check list in addition to STANDARD SAFETY PRACTICES.

1. NEVER insert any part of your body into or even near the cutting blade. This includes reaching into the discharge chute to dislodge whatever is clogging it, sliding your feet under the mower deck and lifting mower while it is running. All similar actions are far too dangerous to even consider.
2. NEVER compromise safety by operating the mower in a way that you or the mower could accidentally move and cause injury. This includes pulling instead of pushing and mowing extreme grades.
3. Mow ONLY with good daylight or artificial light. Be sure of what you are mowing.
4. NEVER run the operating mower over any loose objects.
5. Stop engine and ground spark plug whenever inspecting or adjusting anything near the cutting blade.
6. Stop mower and inspect carefully after striking any object or if it begins to vibrate. Repair damage before restarting.
7. Do not operate in wet grass or anywhere that traction for mower or operator is unsure.
8. Mow terrace or slope only in lengthwise direction. NEVER up or down the incline. Some inclines are too steep for safe mowing.
9. Be especially cautious when mowing in area where children usually play. A child may suddenly enter the area causing a hazard and toys are probably scattered in the grass.
10. Wear appropriate close fitting clothing, shoes and eye protection. Consider appropriate breathing filter for operator, especially in dusty conditions.
11. Check parts regularly for looseness, wear, cracks or other damage and make sure mower is properly adjusted before starting engine.
12. Keep cutter properly adjusted and sharp.
13. Use only approved grip locations and always control unit as intended by the manufacturer.
14. Operate mower only over grass or firm clean surface. Never operate mower over loose rock or area covered with sand.
15. Keep people and animals away from working area. Never aim discharge toward people that could be injured or property which could be damaged by an object being thrown.
16. Keep feet and hands in safe locations as indicated by manufacturer while starting engine. NEVER by-pass any safety device provided by the manufacturer.
17. Stop engine and properly ground the spark plug wire before cleaning, removing an object or performing any maintenance to the mower.

STARTING

Make sure that all people other than the operator are a safe distance from the unit and that mower discharge is directed away from any people or objects that could be injured or damaged. Mower should be located over a clean flat surface, but not in tall grass, rocks or sand. Before starting, make sure that all lubricants have been properly serviced, fuel tank filled and that all parts are tight.

Move throttle to "START" or "CHOKE" position and ignition switch should be "ON". Many models have a single control; however, some have a separate ignition switch. Engage starter. If equipped with rewind starter, pull handle until mechanism engages, then pull starter smartly. Do not let starter cord snap back because starter may be damaged.

It should be noticeable that starter is turning engine and attempting to start. Exact procedure differs from one mower to another, but most manual start mowers should start within the first 8 tries. If engine does not start or at least give encouraging sounds and puffs of smoke from exhaust fairly quickly, investigate problem before tiring yourself or damaging the equipment. Engine should warm enough to operate smoothly very soon after starting.

To stop engine, move controls to "STOP" position, turn ignition switch "OFF" or engage the engine kill switch as applicable.

When Beginning to Mow—

Check for sufficient amount of gasoline, gasoline and oil mix or other required fuel to finish mowing planned for that day. Fill tank with correct mixture.

Check for sufficient amount of lubricating oil.

Check air cleaner for cleanliness and condition.

Walk through the area to be mowed and remove all debris which could damage mower, be damaged by mower or thrown by the mower. In short try to remove everything except grass and weeds.

Start mower and check for proper operation.

⚠ DANGER

Always disconnect spark plug and ground the detached wire to the engine before doing anything under the mower, even inspecting.

Fuel may drain from tank when mower is tipped for examination of blade. Be careful of fire hazard. Fuel spilled on ground can also damage or kill grass.

SAFETY HINT

NEVER fill the fuel tank inside any enclosed building. Gasoline near an open flame water heater or an electrical switch, light or an appliance which is NOT Explosion Proof is a major cause of fires.

SAFETY HINT

Always stop the engine when making height adjustment on push or self-propelled mowers. Stop engine of riding mower if operator must leave seat to make the adjustment or if two people are required to make the adjustment.

SAFETY HINT

If the lawn mower vibrates noticeably, something is probably very wrong. A bent blade or shaft on rotary mowers can cause so much violent vibration that it's hard to control where it's going to jump next. Stop the mower and repair the cause of vibration. The vibration is dangerous, but it will also shake parts loose adding to the danger.

HELPFUL HINT

Keep engine speed in "FAST" position when using grass catcher. The increased engine speed will help increase suction that will move grass clippings further into the catcher. Slow engine speed will cause mower discharge to clog easily and often. Grass will be cut more ragged and grass will be damaged more with slow engine speed too.

Each Time Mower is Stopped to Fill with Fuel—

Clean around fuel filler cap and oil filler cap before removing either cap.

Fill oil reservoir (on 4 stroke engine) with correct type of oil, if low.

Fill fuel tank, but leave some room for expansion. Some manufacturers recommend about ¾ full.

Check blade condition. Sharpen if dull, install a new blade if bent, cracked, broken or otherwise unserviceable.

HELPFUL HINT

Fuel or oil spilled on ground can damage or kill grass. Never fill fuel tank inside a building, but locate mower over a clean flat surface. Wipe up any spills.

HELPFUL HINT

If you hit something with a rotary mower and the engine stops, you may have used a safety device. Briggs & Stratton and some other manufacturers of vertical shaft engines designed for rotary mowers use a soft metal drive key between the engine crankshaft and the flywheel. The soft key is used to prevent damage (and injury) if the blade is stopped suddenly. If this drive key is sheared, the engine will not start or will be very hard to start. It will be necessary to have a servicing dealer remove the flywheel and install a new key, but the blade or other parts may have also been damaged. Be sure to inspect for damage and repair if necessary.

At the End of Each Working Day—

Disconnect and ground engine spark plug to prevent acdental starting, that could result in injury.

Clean grass, dirt and other debris from mower blade.

Remove and clean air filter.

Check cutting blade for sharpness and condition.

Be sure to clean grass catcher if used. Any grassy remaining in bag will probably mildew. Hang clean bag up so that any remaining water will drain and catcher can dry quickly.

At the End of the Mowing Season—

Refer to the OFF SEASON STORAGE Section in the General Maintenance section, beginning on page

MOWING

Successful lawn care includes careful selection of grass type, control of desired type and amount of nutrients and grooming the blades of grass to desired height. Mower performance and method of using the mower also affect the appearance of the lawn, health of the grass and the safety of the operator.

HELPFUL HINT

Don't have the mower setting in one place with the engine running, if the exhaust is directed under the mower deck. The high temperature of the exhaust can cook the grass under the deck. Circles of dead grass the size and shape of the mower deck may have been caused by letting the mower run while emptying the grass catcher or other operation. The mower should be stopped for stopped for safety as well as to prevent damage to the lawn.

For best results, cut the grass frequently, so that only a small amount is cut from the ends of grass blade. Grass should be dry when cut, but should never be allowed to become too dry and the cutting blade should always be sharp so that grass is cut, not bent or torn.

Cut grass only when mower is moving forward. All mowers, including rotary type, are designed to cut in one (forward) direction only. The fact that a mower will cut at all when pulled backwards is only an accident. Safety of the operator is the primary reason that a mower should only be moving forward while cutting. **It is not safe to pull a lawn mower backwards.**

⚠DANGER

DO NOT operate a mower either up or down a slope and NEVER pull a mower up a hill. If the operator should slip, the mower could be pulled over his feet causing severe injury.

Whenever possible, mow long straight (or nearly straight) strips. Strips should overlap slightly to assure that all grass is cut evenly. Large, open, level areas can be used to train inexperienced people to control the mower safely.

When mowing around posts, trees and other objects, follow the contour of obstruction with the side of the mower. Many mowers have wheels offset to permit closer mowing on one side than the other. Side mounted bags and discharge chutes prevent close mowing on one side of some models.

Terraces and slopes should only be cut with mower tilted from side to side. If slope is too steep to be safely cut in this way, it is too steep to mow.

⚠DANGER

NEVER operate a riding mower on slope so steep that operator feels that it is necessary to lean over the side of the mower to keep the mower upright. The hands, feet and seat of the operator should stay in correct locations on mower.

The best pattern for the mower to travel depends upon whether clippings are to be raked or caught in a grass catcher and the shape/contour of the area to be cut. Patterns which are mostly long straight strips will probably be the fastest and easiest. Changing direction radically will require more time, may decrease quality of cut, increases the chances of lawn damage and is sometimes more dangerous, especially if mower is backed up.

Each time the lawn is cut, the mower should move in a different pattern for best development of the grass. The grass and soil may become corrugated and the grass will develop specific growing patterns if the mower travels the same path every time. Variation in pattern will also benefit growing and mowing in other ways.

Grass clippings can be discharged toward the uncut grass permitting them to be further pulverized. Raking can also be easier by using this method; however, with tall grass or large areas, it is usually necessary to make windrows of manageable size. Discharging cuttings into uncut grass will load the engine more, cause additional wear on the cutting blade and decrease the quality of cut. Grass cut by a rotary mower is much more uniform if the discharge is toward an area where grass has already been cut.

Most rotary and reel mowers can be equipped with grass catchers which will eliminate the need for most raking. If grass catcher is mounted on the side of mower, the grass catcher should extend over the strip of grass which has been cut whenever possible. The mower will be easier to move and the grass will not be bent down by the bag if the grass catcher is over an area which has already been cut.

Wet grass doesn't cut easily or well and sticks to everything under the mower. If grass must be cut while wet, reduce the round speed of the mower, but do not slow engine speed.

New grass should be cut in the afternoon, when dry. Mower should not be set too low and be sure that ground condition will not cause damage to the new grass root system if grass is cut.

Very thick, tall and wet grass is difficult to cut and may load the engine enough to slow engine speed. Dull cutting blades will load engine slowing speed even more.

SAFETY HINT

Mowers manufactured after June 30, 1982, must be able to stop the blade within 3 seconds by killing the engine or by stopping the blade with a clutch system. Earlier mowers must be considered more dangerous to use. NEVER remove or modify this or any other safety device.

CLEANING AND INSPECTING

Clean the mower (both above and below the mower deck) at least once a day after operation. Cut grass and dirt can collect, impairing safe operations and making it impossible to inspect for damage.
Stop the engine, switch all electrical controls to OFF, lock all brake levers, and block any parts that could accidently lower (or fall), causing injury.
Clean the underside of mower using a scraper, mild detergent, brush and water.
Inspect for cracks, missing bolts, missing nuts or bent parts.
Check to be sure that blades do not touch the housing or deflectors and that they track evenly when turned.
Repair or renew all bent or broken parts.
Check all bolts and nuts for tightness and make sure to install new fasteners of correct size and type if any are missing or damaged.

Excessive vibration can be caused by broken, missing, bent or improperly balanced blades, or damaged drive belts. Some multiple blade systems must have blades timed to be at certain positions to prevent hitting and to ensure correct balance.

Inspect the blade to determine if it is wearing normally, if it needs to be sharpened, or if a new blade should be installed.

Operating a mower in dry, sandy conditions can cause the blade to erode quickly. The cutting edge will dull faster and the trailing lip will have an unusual amount and type of wear. Erosion of the trailing lip may cause an uneven cut and the grass will not discharge properly.

If blades are damaged by hitting a hard object, inspect them carefully before sharpening and returning them to service. Replace twisted, bent, cracked or broken blades. Nicks caused by hitting small rocks, etc., can usually be repaired by careful sharpening and balancing. Be careful when inspecting and reusing a damaged blade because even slight twists or small cracks may cause greater damage or injury.

Inspect drive belts, pulleys and shafts for any unusual wear or damage. Refer to the mower operator's manual for checking and adjusting drive belt tension and for lubrication instructions.

BLADE SHARPENING AND BALANCING

Regardless of mower cutting type, more power will be required, more fuel will be used, the grass wil be damaged and turf growth recovery after mowing will be slower if the cutting blade is dull. Research indicates that bluegrass cut by a sharp mower blade will start to grow again within about 3 hours; however, nearly identical grass cut with a similar, but dull blade, required more than 24 hours to resume growing. Other varieties of grasses are similarly effected and all are more susceptible to disease and other damage after being injured by the dull blade. Reduced growth may be noticed by reduced amount of clippings and failure to respond to fertilizing or other treatments. Injury caused by a dull cutting blade can be identified by close inspection of cut end of leaf.